STATISTICAL ANALYSIS FOR GEOGRAPHERS

Daniel A. Griffith, B.S., M.A., M.S., Ph.D.
Syracuse University

and

Carl G. Amrhein, B.S., Ph.D.
University of Toronto

with contributions by

Joseph R. Desloges, B.E.S., M.S., Ph.D.
University of Toronto

Prentice Hall, Englewood Cliffs, New Jersey 07632

Library of Congress Cataloging-in-Publication Data

Griffith, Daniel A.
Statistical analysis for geographers/Daniel A. Griffith and Carl G. Amrhein; with contributions by Joseph R. Desloges
p. cm.
Includes bibliographical references and index.
ISBN 0-13-844184-7
1. Geography—Statistical methods. 2. Spatial analysis (Statistics) I. Amrhein, Carl G. II. Desloges, Joseph R. III. Title.
G70.3.G75 1991 90-45850
910'.21—dc20 CIP

Editorial/production supervision: *Kathleen M. Lafferty*
Cover design: *Lundgren Graphics, Inc.*
Pre-press buyer: *Paula Massenaro*
Manufacturing buyer: *Lori Bulwin*

Printed in the United States of America

10 9 8 7 6 5 4 3 2 1

ISBN 0-13-844184-7

Prentice-Hall International (UK) Limited, *London*
Prentice-Hall of Australia Pty. Limited, *Sydney*
Prentice-Hall Canada Inc., *Toronto*
Prentice-Hall Hispanoamericana, S.A., *Mexico*
Prentice-Hall of India Private Limited, *New Delhi*
Prentice-Hall of Japan, Inc., *Tokyo*
Simon & Schuster Asia Pte. Ltd., *Singapore*
Editora Prentice-Hall do Brasil, Ltda., *Rio de Janeiro*

This volume is dedicated to our wives, Diane and Ellen. Their unselfish support throughout the development of our respective careers, and their patience and willingness to suffer the long "vacation" days spent on this manuscript, are deeply appreciated.

CONTENTS

PREFACE

Geography was given its first serious introductory textbook on statistical applications by Gregory (1968). This publication was followed by a string of more sophisticated books, written by, among others, King (1969), Yeates (1973), Hammond and McCullagh (1974), Norcliffe (1977), Taylor (1977), Ebdon (1977 and 1985), Silk (1979), Matthews (1981), and Shaw and Wheeler (1985). During the evolution of this pedagogic literature geographers have become increasingly aware of unique statistical problems faced in analyzing spatially referenced data. Classical statistical techniques have often been found to be woefully inadequate for solving geographic problems; perhaps the discontent spurred by this drawback has been a major reason for the quantitative counterrevolution in the discipline. In addition, beneath any classical statistical analysis of geo-referenced data lurks the pitfall of spatial autocorrelation, which awakens many problems that lie dormant in classical statistical situations. This realization that geo-referenced data are special has given rise to an increased integration of spatial statistical concepts into classical statistical frameworks. No longer does the training of geographers strictly in techniques of aspatial statistical analysis constitute an adequate education, nor do brief excursions into the application of classical statistics to spatial problems, which characterize many of the earlier texts. More recent textbooks for geographers, such as Barber (1988) or Clark and Hosking (1986), provide a concerted effort to integrate classical and spatial statistics, to a certain extent, but still do not go far enough along these lines. New themes need to be incorporated and integrated into these sorts of texts, especially in light of the explosion of interest in geographical information systems (GIS) and analysis. This book is an effort to exploit new developments and statistical applications in spatial analysis. It is designed, in part, to serve as a precursor to GIS courses (especially Chapter 5).

The material in *Statistical Analysis for Geographers* is extracted from courses we have taught; much of it has been drafted from our revised class notes. Our backgrounds include the completion of numerous undergraduate and graduate courses in both theoretical and applied statistics, and research involving the use of various statistical techniques. In addition, together we have 24 years of experience teaching statistics to undergraduate students in geography. This experience represents a variety of settings, universities, and locations and involves students from a wide range of high-school preparations. It also includes several collaborative teaching efforts on our part during the early 1980s.

Our book is on applications of statistical techniques. There are no proofs, although selected algebraic manipulations are occasionally presented. We assume that students have had some high-school mathematical training (at least intermediate algebra) but no prior statistical training. Our concern is threefold. First, we endeavor to furnish students with an understanding of why some statistics are useful, whereas others are not. Second, we seek to earmark potential sources of error, together with their consequences, in a spatial statistical analysis. And, third, we aspire to foster a comprehension of the information content and redundancies latent in geo-referenced data. Consistent with a focus on applications, the book relies heavily on a battery of examples from both human and physical geography. Four data sets are introduced in Chapter 1. With a few exceptions that are based upon simulated data, all the examples in the book use one of these four data sets. The majority of the examples are based upon Griffith's Puerto Rico data, which have been compiled from his research efforts during the past decade. This data set provides a variety of economic and physical variables at several scales of measurement. Amrhein's Ontario automotive industry data set provides a variety of industrial variables, many of them at nominal or ordinal scales of measurement. With this data set, the peculiar problems encountered with secondary data derived from government surveys are explored; particularly emphasized are the problems encountered with grouped ratio data. Both Griffith and Amrhein are human geographers. Desloges provides balance to the book by making contributions that address the concerns of physical geographers. His data sets represent a range of physical variables and problems: one set contains drainage basin information from Ontario, the other geophysical and climatic data from a glacier in British Columbia. In every aspect of the book, we have spent considerable time addressing a balance between human and physical content.

The undergraduate statistics course at the University of Toronto is a year-long course; at Syracuse University the approximately equivalent course lasts one semester. The material in this book is presented so that both courses are serviced. For the one-semester course, Part III on multivariate statistics (Chapters 13 through 16) might be skipped, as well as selected topics in Chapters 11 (Complex Hypothesis Testing) and 12 (Simple Linear Regression). In the year-long course, every chapter might be presented, along with some of the exercises that reinforce and extend slightly the material in the text.

There is always a difficult trade-off in the creation of examples and problems involving the amount of computation required to be done by students. If all the examples and problems are to be solved with the aid of a hand calculator, then the data sets must be kept fairly small (and potentially trivial). At the same time, using large data sets that require a computer increases the possibility that students come to view the techniques as *black boxes*. We have tried to achieve a balance between these two approaches. In Part III, "A Transition to Multivariate Statistics," most of the examples and problems require a computer. The data sets and computer code for widely available statistical packages (SAS and MINITAB) are provided. In the first Part I, "Descriptive Statistics," most of the calculations can be done by hand. Each data set provides an opportunity for manual calculations on regionally defined subsets of the larger data sets, rather than on some arbitrarily chosen sample of observations. These possibilities often are supplemented by the presentation of results for an entire data set. Part II, "The Foundations of Inferential Statistics," mixes the use of manual calculations with computer applications. Throughout the book, computer-generated results are clearly labelled so students do not try to reproduce such results manually.

A list of additional readings is provided at the end of each chapter. Complete references are listed at the end of the book in topical groupings. These supplemental citations are not meant to be comprehensive. Some of the material cited is more advanced than the text, whereas some is redundant. Of particular interest are two series, CATMOG and SAGE, of small pamphlets. These materials are written at an introductory level. Reports from students over the years suggest that these pamphlets are very useful in explaining well-defined topics, especially to the novice. They may be particularly useful for the more advanced topics in Chapters 10 through 16.

Available software

A 5.25" diskette containing SAS computer code is available from the publisher. To order, submit the tear-out card found in the back of this book, together with the requested amount.

This diskette houses three directories, namely EXAMPLES.SAS, DATA, and HOMEWORK.SAS, and a READ.ME file that is located in the master directory. The READ.ME file contains the text of this section. Directory EXAMPLES.SAS includes ASCII files that can be transferred to PC or mainframe environments. The Names of files indicate which examples they pertain to. For instance, all of the Puerto Rico examples begin with PR, and all of the generic examples begin with PHEX. The first numeric characters indicate which chapter is involved (e. g., 1, 2, . . ., 16). Following the hyphen is a number denoting the example. Finally, the extension is SAS, since these files contain SAS code. Files relating to multiple examples have the example numbers separated by a dollar sign. For example, file PR7-3$4.SAS has SAS code for Puerto Rico examples PR-7.3 and PR-7.4, appearing in Chapter 7. In other words, the same numbering system is used to name these files that is used in the text to number examples. There are no files for Chapters 2, 6, and 8.

Non-interactive SAS code has been developed employing CMS SAS 5.18 for an IBM 3090 mainframe. As such, an uploaded file may require its RECFM to be set to F. Users wishing to adapt this code for other operating systems (e. g., AOS/VS, OS, PRIMOS, VM/PC, VMS, VSE) should consult the SAS User's Guide: Basics, Version 5 Edition (Cary, NC: SAS, 1985, ISBN 0-917382-65-X) for directions. The only change that should be necessary has to do with the INFILE statement, which accesses required data sets.

The data sets presented in Appendices Tables 1.1-A, 1.2-A, 1.3-A, 1.4-A, and 7.1-A are housed in the directory entitled DATA. For the IBM mainframe environment, these data sets will interface with file executions using the following system commands:

```
FILEDEF PUERTOR DISK GRIFFITH DAT **
FILEDEF PUERTOR1 DISK SANJUAN DAT **
FILEDEF PUERTOR2 DISK ARECIBO DAT **
FILEDEF PUERTOR3 DISK MAYAGUEZ DAT **
FILEDEF PUERTOR4 DISK PONCE DAT **
FILEDEF PUERTOR5 DISK CAGUAS DAT **
FILEDEF CONNSJ DISK CONN SJ **
FILEDEF CONNARE DISK CONN AR **
FILEDEF CONNMAY DISK CONN MR **
FILEDEF COORDMAY DISK COORD MR **
FILEDEF RUNIFORM DISK TABLE 7-1 **
```

The double asterisk must be replaced by the disk location where these data files are stored.

SAS code for selected homework problems is housed in directory HOMEWORK.SAS. The protocol used here is the same as that used for files housed in the EXAMPLE.SAS directory. However, all files found in this directory begin with HW.

Acknowledgments

The assistance of many people made this book possible. Both of us enjoyed the full support of our departments in this effort. In addition, our good friend and colleague Anthony C. Lea, of the Applied Geography Program at Ryerson Polytechnical Institute, in Toronto, provided a preliminary bibliographical database to which items were added as needed. Brigitte Waldorf, of Princeton University, provided a very thorough and invaluable review of the draft manuscript. Incorporating her comments and suggestions certainly considerably improved our final product. Irina Vasiliev, from Syracuse University, reviewed the final draft manuscript and checked our calculations. While she identified an embarrassing number of mistakes, some of which were quite silly, any remaining errors remain our responsibility. Michael Kirchoff, of the Cartography Laboratory at Syracuse University, provided a very high level of professional assistance with the illustrations and figures. David Andrews and Ruth Croxford, of the Statistics Department, University of Toronto, made many useful and helpful comments on the first four chapters. Others we would like to recognize and thank for their commentaries on earlier drafts of this text are Martin

Cadwallader (University of Wisconsin-Madison), Arthur Getis (University of Illinois), David Hodge (University of Washington), Thomas Leinbach (University of Kentucky), Richard Morrill (University of Washington), and Sent Visser (Southwest Texas State University). Finally, we owe a debt of gratitude to the students in and teaching assistants of our 1989-1990 courses who suffered through a year with a not-quite-complete text.

Daniel A. Griffith
Carl G. Amrhein
Joseph R. Desloges

ABOUT THE AUTHORS

Daniel A. Griffith has been Professor of Geography at Syracuse University since 1988, and was previously a faculty member in the Geography Department at SUNY/Buffalo for 10 years. He received his doctorate (1978) in geography from the University of Toronto. In addition to numerous scholarly publications, Griffith has served on the editorial boards of the *Annals* of the Association of American Geographers and *Geographical Analysis*, and as chair of the AAG's Mathematical Models and Quantitative Methods Specialty Group.

Carl G. Amrhein has been at the University of Toronto since 1986, and was previously a faculty member in the Geography Department at the University of Kentucky. His doctorate in 1984 from SUNY/Buffalo has been followed by many journal articles and book chapters. He has served as the software review editor for *The Professional Geographer*, newsletter editor for the AAG's Mathematical Models and Quantitative Methods Specialty Group, and is currently serving as the Executive Vice-President of the Canadian Regional Science Association.

Both authors are Nystrom Dissertation Award winners from the Association of American Geographers.

Joseph R. Desloges joined the Geography Department at the University of Toronto in 1987, after receiving his doctorate from the University of British Columbia. His research interests are in physical geography, in particular statistical-based modelling of fluvial and glacial processes in past environments. He has published many research articles, monographs, and field guides for mountainous areas in western Canada.

PART I

DESCRIPTIVE STATISTICS

Statistical Analysis for Geographers is divided into three parts. The first, Part I, comprises seven chapters and presents the basic terminology and concepts needed to describe either population or sample data. Information content is one theme that runs through these chapters, in terms of population versus sample-specific information, measurement-scale information, and redundant information. Another common theme underlying these chapters concerns discriminating among statistics in order to determine which ones are meaningful. This part of the book establishes a foundation for Part II, "The Foundations of Inferential Statistics," which introduces the theory behind statistical inference together with the more commonly used simple parametric and nonparametric measures.

Following an introductory chapter, Chapter 2 presents the key concepts of scale of analysis, populations, and samples. Chapters 3, 4, and 5 discuss simple methods of describing data, particularly sample data. Students are introduced to measures of central tendency, dispersion, and correlation, as well as graphical methods for displaying data. Of particular importance for geographers is Chapter 5 on spatial statistics. Chapter 6 presents a brief summary of the relevant aspects of probability theory needed for the inferential work that follows. Since an entire course could be devoted to probability theory, only the minimal essentials can be covered here. The final chapter in this part, Chapter 7, introduces the four key frequency distributions that are used throughout the remainder of this book; literally hundreds of others exist but will not be discussed here because they rarely are encountered in the spatial analysis literature.

All too often there is a tendency among analysts to downplay the importance of a thorough understanding of the data at a descriptive level before proceeding to an inferential analysis. The first part of this book should not be viewed as the *easy* or *unimportant* portion. Indeed, it is our belief that a comprehensive understanding of material in this part makes the more complex inferential techniques relatively *easier* to comprehend and work with than would be the case without such a prior understanding. Additional time spent learning the material (definitions, logical constructs, theory) in Chapters 1 through 7 will likely reduce the effort necessary to effectively and efficiently deal with material appearing in subsequent chapters.

CHAPTER 1

INTRODUCTION AND BACKGROUND

Two reports attest to the importance of studying statistical applications in geography. One, the Lord Chorley report from Great Britain, entitled *Handling Geographic Information* (1987), concludes that a vast majority of empirical analyses are conducted with geographically referenced numerical data. By establishing the National Center for Geographic Information and Analysis, the National Science Foundation of the United States also recognizes this fact, and hence has dedicated considerable resources to the management and analysis of geo-referenced data on computers. Thus, students of many disciplines should benefit from an exposure to at least elementary statistical applications in a geography course. Such an exposure should include the proper addressing and handling of complications and difficulties commonly encountered in statistical analyses of geographically referenced data not normally encountered in an analysis of nongeographic data. These complications and difficulties will be highlighted throughout this book.

Besides seeking to explore and emphasize nuances and idiosyncrasies of geographic data analysis, this text also aspires to the general objective of an introductory statistics book. The goal of this book is to cultivate critical and analytical thinking skills in the reader, rather than to produce highly skilled statisticians. Again, this sort of thinking will be couched in terms of geographical analysis.

1.1. The purpose of statistical analysis

One general educational objective of statistical analysis is to promote critical and analytical thinking, or thinking that allows one to reasonably go about deciding what to do or believe (Walsh, 1985). This primary purpose has several secondary goals affiliated with it. On the one hand, statistical analysis seeks to develop measures upon which to base critical and analytical thinking; in so doing, it defines desirable properties of these measures in order to distinguish between useful and questionable or worthless ones. As a result, figuratively if not literally speaking, there are an infinite number of possible measures, many of which are discussed in the statistics literature but whose discussion is beyond the scope of this book; a small number of well-behaved, traditional measures are presented here, providing a basis for the reader to pursue studies of more contemporary, controversial and/or esoteric measures at a later date. Standards from classical statistics will be employed to identify those measures deemed useful. Once measures are obtained, their trustworthiness comes into question. Another secondary goal has to do with attaching some level of trust to statistical measures. Realizing this goal requires an understanding of scales of measurement and sources of error in a data analysis, as well as their respective impacts on conclusions drawn from an analysis (this topic is discussed at some length in Chapter 2).

Especially within the context of statistics, critical and analytical thinking moves one away from the "one-right-answer" paradigm. While many different statistical measures can be relevant to a single problem, each tends to be tuned to subtly different peculiarities in the data. As a result, different answers to a given question can be obtained by using different statistical measures or by attaching different levels of trust to the same statistical measure. Because so many of these measures exist, again it is worth stressing that desirable properties of measures must be established in order to discriminate between those that do and those that do not consistently provide a sound basis for critical and analytical thinking.

Viewed in this manner, statistics enables one to integrate numerical knowledge with intellectual skills that allow one "to weigh, reconcile, and assess reasons, arguments, and points of view, and to analyze conclusions" (Walsh, 1985, p. 20). It fosters the development of intellectual inquiry and critical habits of thought by encouraging the searching out of underlying assumptions (especially about the source of error in an analysis), guiding reasoning (especially in terms of chances that something will occur), possible implications (especially in the context of no "one--right-answer"), and alternative criteria for decision making (especially with regard to different numerical measures and/or different levels of trust that can be used). The student is encouraged to develop an attitude of thoughtfulness and perceptiveness when considering problems, a knowledge of the scientific method of logical inquiry and reasoning, and a skill in applying the latter method.

1.2. Pitfalls to avoid in statistical analysis

Another objective of statistical analysis, one many would argue is the dominant objective of statistics, is to make decisions about a population based upon information obtained from a subset or sample. A **statistical population** is a set of objects, and we are interested in the numerical values corresponding to some feature of these objects. This set must be defined in such a way that one can clearly specify whether or not any item is a member of it. This population may be finite or infinite in size. We will use N to denote the population size. A **sample** is a subset of a particular size n that is taken from the population. A geographic data set results by attaching a location to these numerical values. Consider the set of objects {a, b, c}. These objects might be anything geographers would find of interest. For example, human geographers might collect samples of cities, houses, neighborhoods, ethnic groups, individuals, or stores. Physical geographers might collect samples of soil, air quality, water quality, trees, glaciers. There are many possible attributes of an object that might be measured. For example, a house might be measured for size, price, and number of rooms; a soil profile might be measured for color, texture, grain size, and moisture content (see Problems #3 and #4 in Section 1.5 for another example). Assume that some attribute of the object (perhaps the number of cars the garages of three different houses might hold) has corresponding numerical values of {1, 2, 3}; all samples of size (n =) 2 for this population would be

objects	a, a	a, b	a, c	numeric values	1, 1	1, 2	1, 3
	b, a	b, b	b, c		2, 1	2, 2	2, 3
	c, a	c, b	c, c		3, 1	3, 2	3, 3

Many of the statistical pitfalls that have been identified over the years revolve about this notion of sample, specifically with reference to characteristics of a sample. Important properties to look for in a sample have to do with (a) how representative it is, (b) how large it is, (c) whether or not it is random, and (d) how likely its resulting measures are. A sample is **representative** if the objects it contains reflects the set of objects comprising the population in some reasonable fashion. In other words, it is a sample from which every source of visible or invisible forms of partiality and/or prejudice has been removed. For example, if a sample of size n = 3 were selected from the foregoing population of {a, b, c}, then a perfectly representative sample would be (a, b, c), whereas one highly unrepresentative sample would be (a, a, a). For the n = 2 case mentioned above, there is no perfectly representative sample; however, the samples (a, b), (a, c), (b, a), (b, c), (c, a), and (c, b) are better representatives of the population in question than are the samples (a, a), (b, b), or (c, c). The latter situation characterizes most sampling problems, for usually the sample size is substantially less than the population size (see Chapter 2 for a further discussion of this point).

The notion of randomness requires the concept of probability; **probability** refers to the chances that something will occur, and at least empirically may be defined for an exhaustive set of occurrences as the ratio of occurrences of a specific type to the total number of possible occurrences. A **simple random sample** is one in which the probability of selection is the same for each of the possible samples of size n; the total number of possible occurrences is the set of all possible samples of a given size n, and the specific type of occurrence is each individual sample. This property is known as **equally likely**.

Finally, measures that correspond to samples are based upon the numerical values for objects contained in samples. When the measure is based upon a sample, it is called a **statistic**. The corresponding measure for the parent population is called a **parameter**. Whether or not a measure is likely refers to the probability that it will result from sampling. For instance, by the time they enter university, most students are aware of the **arithmetic mean** measure (the sum all of the numerical values in a set divided by the number of numerical values in the set). For the above enumerated samples of size n = 2 drawn from the population having numerical values of {1, 2, 3}, and hence a parameter arithmetic mean of 2, the corresponding possible sample arithmetic means are

sample	1,1	1,2	1,3	2,1	2,2	2,3	3,1	3,2	3,3
arithmetic mean	1	1.5	2	1.5	2	2.5	2	2.5	3.

So for these nine samples, each having a mean, the probability of observing each arithmetic mean is

arithmetic mean	1	1.5	2	2.5	3
probability	1/9	2/9	3/9	2/9	1/9.

The most likely arithmetic mean here is 2. Noteworthy is the fact that this most likely statistic may be obtained from a representative sample, either (a, c) or (c, a), as well as from an unrepresentative sample, (b, b). Clearly, then, the sample properties of representativeness and likeliness are not equivalent!

1.2.1. Illuminating examples of pitfalls

Two interesting publications illustrate the need for critical thinking and analytical skills that are couched in the subject matter of statistics. The first, which is quite dated but still relevant in its content, is Huff's *How to Lie With Statistics* (1954); the second is Brower's "The Pernicious Power of the Polls" (1988). Comparative extracts from these two items exemplify noteworthy pitfalls to avoid in one's own statistical analyses and to look for in other statistical analyses. They also build upon the fundamentals of statistics that were just outlined above.

Huff discusses the problem of a sample with a built-in bias (i. e., partiality and/or prejudice), or what Brower labels a bum sample. This category of sample may arise in a number of different ways. A common situation is that the sample is self-selecting rather than randomly selected. Voluntary responses to surveys, for any number of possible motives, result in self-selected samples; one popular example is the television solicitations for 900-number polls. But even when a scientific selection strategy is employed, it may yield a self-selected sample. For instance, n persons are randomly selected from a population, but not all these persons reply to a distributed questionnaire; or n persons are randomly selected, but not from the total population. There may well be a systematic bias introduced into the sampie by either the nonrespondents or the section of the population that has been overlooked.

Brower cites the Hite report (1976) (prepared to portray the sexual attitudes of women) as exemplifying this situation and labels its sample self-aggrandizing--only 4500 responses were received from 100,000 questionnaires distributed to a crosssection of women's organizations. Not only do members of women's organizations not constitute the population of all women, but individual women may belong to more than one of these organizations. The nonresponse rate is a critical issue in public surveying; rarely will a scientist obtain a 100% response to a survey, and refusals are increasing, today almost always being at least 20%. Psychologists have identified strategies that help minimize the number of nonresponses (using certain colors of paper for the questionnaire and envelope, implementing follow-up mailings, combining mailing and telephone contacting, providing rewards for responses, and the such). In part this problem stems from (a) what Brower calls the bogus poll, where "some telemarketers pretend to be taking a poll in order to lure [the respondent] into a sales pitch" (p. 145) and (b) the frequency of polling (about 1/3 of all U.S. residents are polled in a single year, with 24% of these being contacted once, 28% being contacted twice, 16% being contacted thrice, and 32% being contacted four or more times).

Another source of built-in bias is associated with the accuracy of the numerical values that are obtained with the sample. Huff cites the example of an income survey of graduates from a prestigious university, where the desire to give a pleasing answer may result in an income inflation when revealing one's actual income, whereas the fear of governmental reprisal for incorrectly reporting income may result in an income deflation when revealing one's actual income. The scientist has no reason to suspect that these two measurement errors (more is said about this type of error in Chapter 2) will cancel each other. Brower refers to this kind of situation as a madness for numbers, and chaos on the prowl. He cites the recent revision of television rating measures from ones based upon weekly diaries of a family's viewing, usually kept from memory, to infrared scanner monitoring of the room within which a television set is located. This measurement refinement has resulted in a dramatic shift in television show patronage ratings; recording errors were uncovered that indicated some families viewed shows other than the ones they claimed to have watched.

A third source of built-in sample bias is closely related to sample size and historically has widely existed in advertising. Huff discusses this problem within the context of small sample size, where random samples repeatedly are selected until one yielding an acceptable measure is obtained. As was noted in the previous section, an unrepresentative sample can be obtained, but its probability of selection is smaller than that for a representative sample (1/3 versus 2/3 in the preceding example). So if the outcome of a sample is undesirable, then it can be discarded, and a new sample can be drawn. This process is facilitated by using a very small sample size (advertisers often report $n = 10$). Brower notes that many scientific sample sizes may be as small as 500 but usually are at least 1500; much smaller but controlled numbers may prevail in medical research, though. Closely related to this notion of sample size is one of observed sample differences, and their misleading interpretations. A common advertising campaign involves two leading soft drink companies conducting taste tests; frequently these taste tests indicate that nearly 1/2 of the tasters choose the opposite soft drink. But, if a taster is unable to discriminate between the two drinks, then by mere chance one would expect this result! Given these concerns, then, another objective of statistics has to do with evaluating how accurately a sample can be considered to represent its parent population. This issue, together with the attaching of some level of trust to measures made on a sample, is emphasized throughout this book.

Even if a sample has been chosen properly, one still can be misled by the statistical analyses conducted on it. It was previously mentioned that numerous statistical measures can be made that refer to a single construct, with each tending to be tuned to subtly different peculiarities in data. It was also noted that a scientist wants a measure with the most desirable statistical properties (these attributes will be established in Chapter 2). Measures should not be chosen based upon the conclusion they support. A popular example of inadequate gauging of a norm, or typifying value, when measured by the arithmetic mean, has a human reclining in such a way that his or her head is on ice and his or her feet are in boiling water; figuratively speaking, the arithmetic mean temperature this person is being exposed to is comfortable. This illustration emphasizes that the arithmetic mean becomes a worthless measure when only extreme values are involved. Other statistical measures of norms are more relevant in this type of numerical situation. The presence of different measures of a norm means that (a) the one used always should be clearly identified and (b) objective criteria for selecting the different measures should be spelled out.

Huff also describes the issue of semi-attached figures, in which a change in subject occurs between analysis and interpretation. He cautions that one must watch for a shift in emphasis where a particular point was not proven, so a substitute point is demonstrated and then the two points are cast as equivalents--two things are picked that sound the same but are not the same. In part Brower characterizes this situation as the loaded question syndrome and illustrates it by showing how a subtle change in the wording of a question can cause a sizeable number of respondents to change their answer. He notes that this result often occurs with highly emotionally charged topics. In a preferable situation people feel so strongly that different wordings of a given question may be irrelevant. Brower labels a second area of the semi--attached figures problem as "divide and conquer." In this context the question posed is such that the undesirable response is spread over a range of possibilities. He illustrates this situation with the example of a survey carried out by a certain pharmaceutical company. The survey pitted its painkiller, with one type of active ingredient, against two other painkillers, both of which use an alternative type of active ingredient. Hence, those doctors preferring the second type of active ingredient were split between the latter two brands, seemingly demonstrating a higher preference for the first brand. Huff cautions that things are not always what they seem to be, semantics!

There are often many ways of presenting a figure, and the one chosen should not be the one that sounds best for the purpose at hand. Brower also points out that ignorance, error, or even fraud can corrupt a number. Banks exploit this weakness by interchanging their use of simple, compound (together with the method of compounding), and actual annual percentage interest rates. Government officials frequently exploit this weakness when they present historical crime and disease statistics--more reported cases of a crime or disease are not always the same as more occurrences of either of these events. One also should beware of definitional changes; early in Reagan's administration the consumer price index showed an abrupt drop, not because deflation occurred but because the index's definition was changed. Finally, Huff describes what he calls the "gee whiz graph"; the range of the data (the horizontal scale) or the magnitude of a graph (the vertical scale) is truncated to make the resulting graph look more dramatic. The result is a graph that shows considerable change but in which the magnitude of change is difficult to asses.

The last prominent warning put forth in these two publications is called post hoc pretentious nonsense by Huff, and the dubious assumption by Brower. Huff's view here is characterized by the argument that if B follows A, then A causes B. He stresses that "permitting statistical treatment and the hypnotic presence of numbers to befog causal relationships is little better than superstition . . . and often more seriously misleading" (p. 98). He goes on to caution that common sense needs to be exercised when these sorts of arguments are encountered. Brower illustrates his version of this type of situation with the result of a highly publicized poll claiming that a white, college-educated, never-married 40-year-old woman had a probability of 1 in 100 of ever getting married. The well-informed reader realized that this result was at odds with U.S. Census Bureau information. The inconsistency here arose because the poll study used the unwarranted assumption that all women would follow the marriage pattern of high-school educated women, but starting later in their life cycle; the U. S. Census Bureau assumed more realistically that college-educated women would set a new pattern of delayed marriage.

Walsh (1985) argues that ". . . without a knowledge base, critical [and analytical] thinking skills exist in a meaningless vacuum, if indeed they exist at all" (p. 20). This section provides compelling support for her argument when the medium of critical and analytical thinking is statistical analysis. Accordingly, the contents of this section furnish a background for the subsequent subject matter of this book.

1.3. Background material: terminology, notation, and elementary algebra

In distinguishing between a sample and a population, and hence a statistic and a parameter, Section 1.2 indicates a need to review relevant terminology before a formal study of statistics is undertaken. The depiction of samples and measures in Section 1.1 indicates a need to review certain basic notation before this study is embarked upon. The emphasis in both of these preceding sections on establishing properties of statistical measures indicates a need to review selected elementary algebra within a sampling context.

1.3.1. Selected terminology and notation

A number of algebraic terms require review here. A **variable** is something that varies, or a quantity that can assume any one of a set of values. Classical statistics is concerned with numerical variables. A variable is denoted by an upper-case letter from the last part of the English alphabet (e. g., X), whereas the values taken on by a variable are denoted by the lower-case counterpart of the English alphabet letter (e. g., x). In contrast, a **constant** is something that does not vary, or a quantity that can assume only a single value for a set of objects. Again, classical statistics is concerned with numerical constants. A constant is denoted by a lower-case letter from the first part of the English alphabet (e. g., a) when a statistic is calculated, or its counterpart from the Greek alphabet (e. g., α) when a parameter is calculated. An **equation** is a statement of equality denoted by a combination of variables and constants (e. g., $Y = a + bX$). A **graph** is a diagram of a collection of all points whose coordinates satisfy a given functional relation, such as an equation. If the relation is linear, then the plot of points forms a straight line; if the relation is nonlinear, then the plot of points does not form a straight line.

A **superscript** is an alphanumeric character written above and to the right of another alphanumeric character. A specialized superscript called an **exponent** indicates the number of times the base value is to be multiplied by itself. For example, $2^3 = 2 \bullet 2 \bullet 2$ (the exponent is 3), and $x^4 = x \bullet x \bullet x \bullet x$ (the exponent is 4). A **subscript** is an alphanumeric character written below and to the right of another alphanumeric character (e. g., x_i). The subscript is a name that is attached to an object, albeit a number. If the object in question has two names (e. g., a first and a last name), then there will be two subscripts (e. g., x_{ij}); if the object has three names (e. g., a first, a middle, and a last name), then there will be three subscripts (e. g., x_{ijk}). Objects can have any number of subscripts, but in this textbook, the reader will never encounter more than three subscripts. An **absolute value** is a numerical value that is recorded regardless of sign and is denoted by a number housed between two parallel vertical lines (e. g., $|-2| = 2$, or $|2| = 2$). Finally, a **distribution** is composed of two parts, a set of containers and an allocation mechanism for assigning objects to these containers; the distribution is the frequency of objects by containers (which can be intervals on a number line).

Next, we define several additional basic statistical concepts. An **observation**, whose numerical value is denoted by a lower-case letter from the latter part of the English alphabet that is subscripted usually with the lower-case English letter i (see the preceding example), is the numerical value of a single variable observed for an element from the population; for a sample size of n, i ranges from 1 to n. A **frequency distribution** is a distribution of values for a variable using consecutive integers, where the values are allocated to intervals on a number line whose range is set by the variable, and the number of times each interval receives an allocation is recorded, by interval.

A **sampling distribution** is a frequency distribution based upon all possible samples of size n, where the variable is a given statistic (measure for a sample) and each observation is the statistic calculated for each possible sample. In Section 2.0 the sampling distribution for the arithmetic mean for a sample size of n = 2 was presented for the statistical population {1, 2, 3}. A **sample sampling distribution** is a sampling distribution based upon a subset of all possible samples of size n.

A **statistical hypothesis** is a statement of what is anticipated about population parameters; it is the benchmark for evaluating a statistic. Finally, **expectation** is used to establish some of the important properties of a statistic, and is the arithmetic mean of the sampling distribution. When this mean is equal to the corresponding population parameter, then the statistic is called **unbiased**, one of the properties deemed desirable in classical statistics. The elementary calculus of expectations and the elementary algebra it utilizes are essential to an illumination of this idea.

1.3.2. Basic notation and sampling distribution notation

Probably the most extensively used notation in elementary statistics is the summation operator, denoted by the upper-case sigma of the Greek alphabet, Σ, which operates on subscripts. This notation indicates that a sequence of numbers is to be added, in the sequence defined by the subscripts involved. The elementary algebra that accompanies this operator is somewhat simple but important to an understanding of the calculations that are involved in much of elementary statistics. This algebra is characterized by the following basic rules:

$$\Sigma_{i=1}^{n} a = \underbrace{a + a + \ldots + a}_{\text{n times}} = n \bullet a$$

$$\Sigma_{i=1}^{n} x_i = x_1 + x_2 + \ldots + x_n$$

$$\Sigma_{i=1}^{n} x_i \bullet y_i = x_1 \bullet y_1 + x_2 \bullet y_2 + \ldots + x_n \bullet y_n$$

$$\Sigma_{i=1}^{n} x_i^m = x_1^m + x_2^m + \ldots + x_n^m$$

$$(\Sigma_{i=1}^{n} x_i)^2 = (x_1 + x_2 + \ldots + x_n)^2$$

$$(\Sigma_{i=1}^{n} x_i)(\Sigma_{i=1}^{n} y_i) = (x_1 + x_2 + \ldots + x_n)(y_1 + y_2 + \ldots + y_n)$$

$$\Sigma_{i=1}^{n} a\, x_i = a\, \Sigma_{i=1}^{n} x_i$$

$$\sum_{i=1}^{n} (a\, x_i + b) = (a \sum_{i=1}^{n} x_i) + n \bullet b \text{ and}$$

$$\sum_{i=1}^{n} (x_i + y_i) = \sum_{i=1}^{n} x_i + \sum_{i=1}^{n} y_i \;.$$

Two noteworthy relationships that we will be call upon later are

$$\sum_{i=1}^{n} x_i^2 \neq (\sum_{i=1}^{n} x_i)^2 \text{ and}$$

$$\sum_{i=1}^{n} x_i \bullet y_i \neq (\sum_{i=1}^{n} x_i)(\sum_{i=1}^{n} y_i).$$

When an observed value has more than one subscript, then it may have more than one summation operator; a separate summation operator is necessary for each subscript attached to an object. In the case of a two-subscripts observation,

$$\sum_{j=1}^{n} \sum_{i=1}^{n} x_{ij} = \sum_{i=1}^{n} (x_{i1} + x_{i2} + \ldots + x_{in})$$

$$= (x_{11} + x_{12} + \ldots + x_{1n}) + (x_{21} + x_{22} + \ldots + x_{2n})$$

$$+ \ldots + (x_{n1} + x_{n2} + \ldots + x_{nn}).$$

In the case of a three-subscripts observation,

$$\sum_{i=1}^{n} \sum_{j=1}^{n} \sum_{k=1}^{n} x_{ijk} = \sum_{i=1}^{n} \sum_{j=1}^{n} (x_{ij1} + x_{ij2} + \ldots + x_{ijn})$$

$$= \sum_{i=1}^{n} [(x_{i11} + x_{i12} + \ldots + x_{i1n}) + (x_{i21} + x_{i22} + \ldots + x_{i2n})$$

$$+ \ldots + (x_{in1} + x_{in2} + \ldots + x_{inn})]$$

$$= (x_{111} + x_{112} + \ldots + x_{11n}) + (x_{121} + x_{122} + \ldots + x_{12n})$$

$$+ \ldots + (x_{1n1} + x_{1n2} + \ldots + x_{1nn}) + (x_{211} + x_{212} + \ldots + x_{21n})$$

$$+ (x_{221} + x_{222} + \ldots + x_{22n}) + \ldots + (x_{2n1} + x_{2n2} + \ldots + x_{2nn})$$

$$+ \ldots + (x_{n11} + x_{n12} + \ldots + x_{n1n}) + (x_{n21} + x_{n22} + \ldots + x_{n2n})$$

$$+ \ldots + (x_{nn1} + x_{nn2} + \ldots + x_{nnn}).$$

EXAMPLE 1.1. Using sigma notation with the selected ordered variables and constants of X = {3, 7, 4, 1, 9}, Y = {7, 5, 6, 1, 2}, a = 3, b = 1, n = 5, and m = 4.

$$\sum_{i=1}^{n} a = 3 + 3 + 3 + 3 + 3 = 15$$

$$\sum_{i=1}^{n} x_i = 3 + 7 + 4 + 1 + 9 = 24$$

$$\sum_{i=1}^{n} ax_i = (3 \cdot 3) + (3 \cdot 7) + (3 \cdot 4) + (3 \cdot 1) + (3 \cdot 9) = 3 \cdot (3 + 7 + 4 + 1 + 9) = 72$$

$$\sum_{i=1}^{n} x_i^m = 3^4 + 7^4 + 4^4 + 1^4 + 9^4 = 9300$$

$$\sum_{i=1}^{n} (ax_i + b) = \sum_{i=1}^{n} ax_i + \sum_{i=1}^{n} b = a \sum_{i=1}^{n} x_i + n \cdot b = [3 \cdot (3 + 7 + 4 + 1 + 9)] + (5 \cdot 1) = 77$$

$$\sum_{i=1}^{n} x_i \cdot y_i = (3 \cdot 7) + (7 \cdot 5) + (4 \cdot 6) + (1 \cdot 1) + (9 \cdot 2) = 99$$

$$(\sum_{i=1}^{n} x_i) \cdot (\sum_{i=1}^{n} y_i) = (3 + 7 + 4 + 1 + 9) \cdot (7 + 5 + 6 + 1 + 2)$$

$$= 24 \cdot 21 = 504$$

Next consider a sample that contains data arranged along two dimensions. These data can be represented in tabular form, where the rows (denoted by subscript i) of this table represent one dimension, and the columns (denoted by subscript j) the second dimension. Locations of numbers in this table are denoted by the pair (i, j). For instance,

column: 1 2 3 4
row 1: 2 3 6 7
row 2: 1 8 3 1.

Then

$$\sum_{i=1}^{2} \sum_{j=1}^{4} x_{ij} = (2 + 3 + 6 + 7) + (1 + 8 + 3 + 1) = 31.$$

Finally, expanding this dimensional viewpoint, suppose data are arranged in three dimensions, such as rows, columns, and sequences of the row-column tabulations (denoted by subscript k) for a third dimension (picture these data as shown in Figure 1.1). Locations of numbers in this tabular setting are denoted by the triplet (i, j, k).

$$\sum_{i=1}^{2} \sum_{j=1}^{3} \sum_{k=1}^{2} x_{ijk} = (2 + 7) + (3 + 1) + (8 + 9) + (1 + 5) + (3 + 4) + (6 + 4) = 53$$

A second, much less common notation used in elementary statistics is the product operator, denoted by upper-case pi of the Greek alphabet, Π, which also operates on subscripts. This notation indicates that a sequence of numbers is to be multiplied,

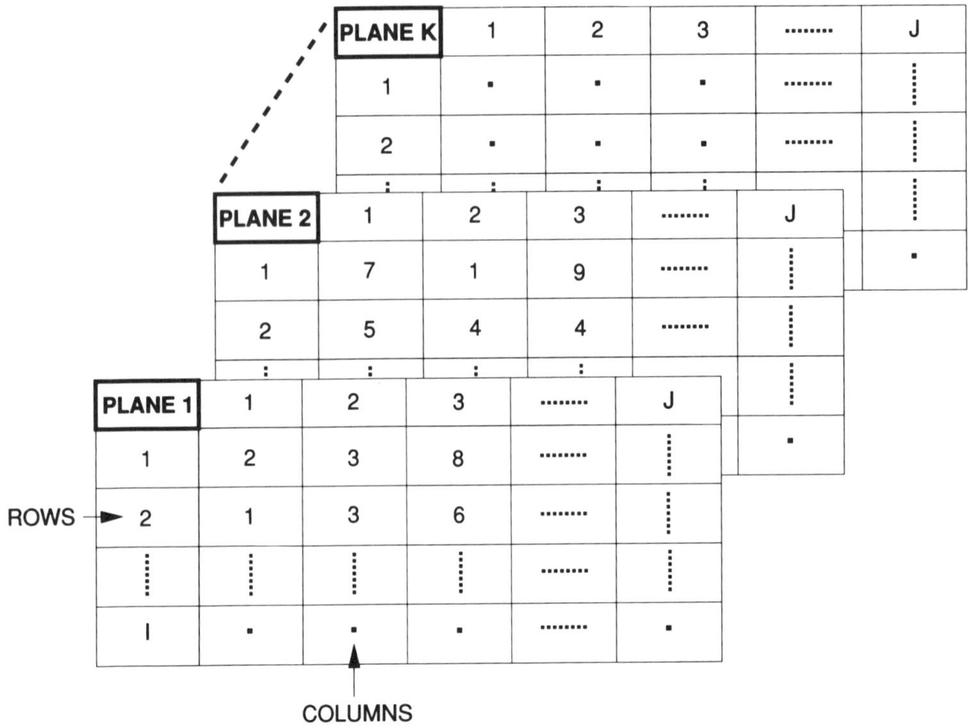

Figure 1.1: HYPOTHETICAL THREE-DIMENSIONAL DATA ARRAY USED IN EXAMPLE 1.1

with this sequence being defined by the subscripts involved. The elementary algebra that accompanies this operator is simple but important to an understanding of the calculations that are involved in some of the elementary probability theory that is associated with statistics. This algebra is characterized by the following basic rules:

$$\prod_{i=1}^{n} a = \underbrace{a \bullet a \bullet \ldots \bullet a}_{\text{n times}} = a^n$$

$$\prod_{i=1}^{n} x_i = (x_1) \bullet (x_2) \bullet \ldots \bullet (x_n)$$

$$\prod_{i=1}^{n} (x_i \bullet y_i) = (x_1 \bullet y_1) \bullet (x_2 \bullet y_2) \bullet \ldots \bullet (x_n \bullet y_n) \text{ and}$$

$$\prod_{i=1}^{n} x_i^m = (x_1^m) \bullet (x_2^m) \bullet \ldots \bullet (x_n^m) = [(x_1) \bullet (x_2) \bullet \ldots \bullet (x_n)]^m .$$

A third notation which commonly appears in probability work is the factorial calculations, denoted by an exclamation point, !. This is a multiplication operator that does not operate on subscripts. The notation indicates that a sequence of consecutive, descending nonnegative integer numbers is to be multiplied, with the sequence defined by the number before the exclamation point. Once again, the elementary algebra that accompanies this operator is somewhat simple and is characterized by the following basic rules:

$$n! = n(n - 1)(n - 2)(n - 3) \bullet \ldots \bullet (3)(2)(1) \text{ and}$$

$$\Pi_{i=1}^{k} (n_i!) = [(n_1)(n_1 - 1)(n_1 - 2)(n_1 - 3) \bullet \ldots \bullet (3)(2)(1)]$$

$$\bullet [(n_2)(n_2-1)(n_2-2)(n_2-3) \bullet \ldots \bullet (3)(2)(1)]$$

$$\bullet \ldots \bullet [(n_k)(n_k-1)(n_k-2)(n_k-3) \bullet \ldots \bullet (3)(2)(1)].$$

EXAMPLE 1.2. Using product and factorial operators with the variables and constants that were defined and used in Example 1.1.

$$\Pi_{i=1}^{n} a = 3 \bullet 3 \bullet 3 \bullet 3 \bullet 3 = 3^5 = 243$$

$$\Pi_{i=1}^{n} x_i = 3 \bullet 7 \bullet 4 \bullet 1 \bullet 9 = 756$$

$$\Pi_{i=1}^{n} x_i^m = 3^4 \bullet 7^4 \bullet 4^4 \bullet 1^4 \bullet 9^4 = 3.2665 \bullet 10^{11}$$

$$\Pi_{i=1}^{n} x_i \bullet y_i = (\Pi_{i=1}^{n} x_i)(\Pi_{i=1}^{n} y_i) = (3 \bullet 7 \bullet 4 \bullet 1 \bullet 9) \bullet (7 \bullet 5 \bullet 6 \bullet 1 \bullet 2)$$
$$= 317{,}520$$

$$n! = 5 \bullet 4 \bullet 3 \bullet 2 \bullet 1 = 120$$

$$\Pi_{i=1}^{n} x_i! = (3 \bullet 2 \bullet 1) \bullet (7 \bullet 6 \bullet 5 \bullet 4 \bullet 3 \bullet 2 \bullet 1) \bullet (4 \bullet 3 \bullet 2 \bullet 1) \bullet (1) \bullet$$
$$(9 \bullet 8 \bullet 7 \bullet 6 \bullet 5 \bullet 4 \bullet 3 \bullet 2 \bullet 1) = 2.6336 \times 10^{11}$$

A second class of frequently used notation in elementary statistics refers to the statistical measures in question as well as measures for their accompanying sampling distributions. For instance, the arithmetic mean is denoted by the statistic $\bar{x}$ (read as "x-bar") for a sample, and the Greek alphabet letter μ for the parent population (a parameter). This measure is defined, using summation notation, as

$$\Sigma_{i=1}^{n} x_i/n = \bar{x} \text{ for a given sample, and}$$

$$\Sigma_{i=1}^{N} x_i/N = \mu \text{ for the parent population.}$$

The arithmetic mean for the sampling distribution of sample means is denoted by $\mu_{\bar{x}}$ and is defined as

$$\sum_{j=1}^{m} \bar{x}_j/m = \mu_{\bar{x}} = \mu \quad \text{where m represents the total number of possible samples.}$$

Meanwhile, the arithmetic mean for a sample sampling distribution of sample means is denoted by $\bar{x}_{\bar{x}}$ and is defined as

$$\sum_{j=1}^{m} \bar{x}_j/m = \bar{x}_{\bar{x}} \quad \text{where m represents the total number of samples that have been drawn.}$$

In general, parameters and statistics for the sampling distribution of any statistic can be denoted in this same way. If Θ is a parameter, and t is its corresponding statistic, then the arithmetic mean of the sampling distribution of t is denoted by μ_t, while that for a sample sampling distribution is denoted by $\bar{t}$.

Before moving on from this discussion of statistical notation, the reader should note that the format for sampling distribution notation requires that either a statistic or a parameter be subscripted by a statistic. The subscript in this instance can never be a parameter. For example, μ_μ literally would mean that the entire population was used to calculate the arithmetic mean a specified number of times. But the population parameter is a constant, no matter how often it is calculated, and if a population parameter is obtained, there is no need to consider samples or sampling distributions. Similarly, $\bar{x}_\mu$ is just as nonsensical, for it literally means that the population arithmetic mean is calculated a subset of possible times. Again, the total number of times that a population parameter would need to be calculated is only one.

1.3.3. More notation and the calculus of expectations (optional)

The calculus of expectations often assumes that a given statistic has a certain theoretical sampling distribution form. This assumption has been criticized of late, with the advent of supercomputers and hence the ability to do large-scale numerical work in statistics that is based upon resampling techniques. Results by Efron (1982) and others indicate that this assumption may well be unwarranted, and its violation may be a source of serious error in statistical analysis (this topic will be discussed more thoroughly in Chapter 2). But since the discussion in this text is restricted to classical statistics, the calculus of expectations remains relevant.

The expectation, or expected value, is the arithmetic mean of a statistical distribution and is denoted by E(•). Suppose that a and b are constants, and X is a variable; then

$E(b) = b$ because every sample that is drawn will yield b,
$E(X) = \mu$ which is true by definition,

$E(aX) = aE(X) = a\mu$ because if all values of a variable are multiplied by a given constant, then its mean also must be multiplied by that same constant, and

$E(aX + b) = aE(X) + E(b) = a\mu + b.$

These simple rules for the calculus of expectations prove useful when dealing with sampling distributions.

But of interest here is the expectation with regard to statistics. For a sample arithmetic mean,

$E(\bar{x}) = \mu$ meaning that the sample mean is unbiased.

Next, consider a common measure of dispersion, the **variance**, which is defined as the average squared difference between an observed value and the arithmetic mean. The sample variance is denoted by s^2 (the population parameter is denoted by σ^2) and algebraically defined as

$$s^2 = \Sigma_{i=1}^{n} (x_i - \mu)^2/n.$$

Adding this statistic to the concepts and notation being developed in this text leads to the following sampling distribution notation, which parallels that for $\bar{x}$ presented in the previous section:

$\Sigma_{j=1}^{m} s_j^2/m = \mu_{s^2}$ the arithmetic mean of the sampling distribution of sample variances, where m represents the total number of possible samples,

$\Sigma_{j=1}^{m} (\bar{x}_j - \mu)^2/m = \sigma^2_{\bar{x}}$ the variance of the sampling distribution of sample means,

$\Sigma_{j=1}^{m} (s_j^2 - \sigma^2)^2/m = \sigma^2_{s^2}$ the variance of the sampling distribution of sample variances, where m represents the total number of possible samples,

$\Sigma_{j=1}^{m} s_j^2/m = \bar{x}_{s^2}$ the arithmetic mean of a sample sampling distribution of sample variances,

$\Sigma_{j=1}^{m} (\bar{x}_j - \bar{x}_{\bar{x}})^2/m = s^2_{\bar{x}}$ the variance of a sample sampling distribution of sample means, and

$\Sigma_{j=1}^{m} (s_j^2 - \bar{x}_{s^2})^2/m = s^2_{s^2}$ the variance of a sample sampling distribution of sample variances, where m represents the total number of samples that has been drawn.

The associated calculus of expectations problem is

$$E(s^2) = E[\sum_{i=1}^{n} (x_i - \mu)^2/n] = \sigma^2$$

meaning that the sample variance based upon the population mean is unbiased. But knowledge of the population arithmetic mean is unusual. If this parameter is unknown, then the best guess one has for it is $\bar{x}$, which yields an algebraic definition for the sample variance of

$$s^2 = \sum_{i=1}^{n} (x_i - \bar{x})^2/n$$

So the calculus of expectations problem becomes one of

$$E[\sum_{i=1}^{n} (x_i - \bar{x})^2/n] = n/(n-1)\,\sigma^2$$

meaning that the sample variance based upon the sample mean is biased. This result explains why the sample variance often is presented with (n - 1) rather than n in its denominator. The latter situation implies that sometimes a statistic is biased but can be simply adjusted or modified in order to become unbiased.

Finally, for illustrative purposes, consider a particular measure of linear relationship whose statistic is denoted by r and whose parameter is denoted by ρ (this measure is discussed at length in Chapter 4). In this last case,

$$E[\ln(1 + r) - \ln(1 - r)] = \ln(1 + \rho) - \ln(1 - \rho)$$

where ln denotes the natural logarithm, meaning that this measure of relationship is unbiased only when it is subjected to a rather sophisticated modification.

Adding this statistic measuring bivariate relationships to the concepts and notation being developed in this text leads to the following sampling distribution notation, in accordance with the previous section:

$\sum_{j=1}^{m} r_j/m = \mu_r$ — the arithmetic mean of the sampling distribution of the linear relationship measure, where m represents the total number of possible samples,

$\sum_{j=1}^{m} (r_j - \mu_r)^2/m = \sigma^2_r$ — the variance of the sampling distribution of the linear relationship measure,

$\sum_{j=1}^{m} r_j/m = \bar{x}_r$ — the arithmetic mean of a sample sampling distribution of the linear relationship measure, and

$\sum_{j=1}^{m} (r_j - \bar{x}_r)^2/m = s^2_r$ the variance of a sample sampling distribution of the linear relationship measure.

The examples presented in this section illustrate the usefulness of the calculus of expectations in providing a means for assessing bias in a statistic. One should realize, however, that there is no guarantee that a biased statistic can have its bias removed by either a simple or a complex modification; there is no way to adjust for bias present in many statistics, resulting in their being less acceptable as bases for analytical and critical thinking. These examples also illustrate the conventional notation used to denote statistics, parameters, sampling distribution statistics, and sample sampling distribution statistics.

1.4. Types of problems to which the reader will be exposed

Several types of problems will be presented in this text. However, first it is important to consider the sources of data that enable, as well as constrain, the spatial analysis. Data can be derived from many sources: Primary data are collected by the researcher personally. Examples of the collection of primary data include measurements and samples taken in the field by physical geographers, surveys, enumerations and, questionnaires administered by human geographers, and interviews and field studies conducted by cultural geographers. Secondary sources of information use information first collected by others. Examples include maps, national census publications, government statistics, and surveys and polls conducted by others. Tertiary data sources use a second party's summary of data originally collected by others. In other words, the spatial analyst is the third user in the chain. The relative merit of using each type of data is a function of time, resources available, and data reliability, that is, how much time an analyst has to complete the work, and the level of resources required to do the analysis. While primary data are usually preferred, time and cost constraints often force the use of secondary and tertiary data. This is especially true in human geography. More is said on this topic later in the book.

A second consideration for the spatial analyst concerns the manner in which location enters the problem. Several possibilities exist:

1. The data consist of observations that represent spatial entities. For example, a list of city characteristics represents spatial units (cities). The exact locations of these units, however, are not explicitly represented in space by specific locations. In this case, while it is necessary to partition space, the characteristics are the focus of study, not the individual city locations. A national census is an example of a data set that adopts this view of location.

2. Alternatively, the actual location of the city may be recorded by longi-

tude and latitude or some other referencing scheme. In this case, the actual location is the focus. For example, climatologists may be interested in the exact location of a place as a factor in pollution levels, rainfall, or any characteristic of place that is directly related to location. While the precise location of a place is now important, the location acts as a reference in space for the other characteristics.

For a situation falling between cases 1. and 2. a choice would be to record the relative location of places, perhaps by recording distances between places. In this case, the relative location is of interest, not the absolute location.

3. Finally, the location of a place may be the characteristic of interest. For example, in Chapter 16 trend surface analysis places primary importance on the location of a place, in terms of either the x- and y-coordinates in a Cartesian world or the longitude and latitude in a spherical coordinate system world. In this case, location in space is not relative, or a surrogate for organizing space, but rather is a focus of the study.

Initially the problems discussed examine a single variable (univariate). Later in the text, problems become more complex and examine several variables simultaneously (multivariate). Four sets of data are employed for illustrative purposes throughout this textbook. These data sets represent a range of topics geographers commonly address using statistical techniques. The first data set, presented in Table 1.1-A (the extension "-A" means that this table is located in the appendices), gives density of selected agricultural productivity in Puerto Rico for 1974 and has been selected because of the extensive experience Griffith has acquired over the years analyzing these data. Examples appearing in subsequent chapters that are based upon these data are designated by labels having the prefix PR. This data set has been compiled from the 1974 *U.S. Census of Agriculture* and the Commonwealth of Puerto Rico *Facts and Figures on Puerto Rico's Agriculture* and will be the primary data set for constructing examples in this book. The other three data sets described next are used to construct supplementary examples that either illustrate specific problems encountered in spatial analysis or important features of a particular statistical technique.

There are 73 areal units (observations), called municipios, in this data set (see Figure 1.2). For each municipio there are two labelling variables followed by three classification variables and eight agricultural variables, which historically have constituted the primary types of agricultural production on the island of Puerto Rico. Note that, unless otherwise indicated, all density measurements have been computed for a square-mile parcel. In addition, to facilitate certain illustrations used in this book, the 73 municipios have been grouped into their respective agricultural administrative regions and then arranged in alphabetical order, within these regions, by name. The following is a description of each variable together with an acronymlike

Figure 1.2: THE PARTITIONING OF PUERTO RICO INTO MUNICIPIOS AND AGRICULTURAL ADMINISTRATIVE REGIONS (used with permission of Artscan Datamap Limited, Toronto, Ontario)

mnemonic that appears in the Table 1.1-A results and frequently in ensuing chapter examples:

Label variable 1 number referring to the alphabetical order position of the municipio name

Label variable 2 name of the municipio

Classification variable 1 agricultural administrative region (ADM)

ADM = 1 if a municipio is located in the San Juan Region
ADM = 2 if a municipio is located in the Arecibo Region
ADM = 3 if a municipio is located in the Mayaguez Region
ADM = 4 if a municipio is located in the Ponce Region
ADM = 5 if a municipio is located in the Caguas Region

Classification variable 2 generic landform (LND)

LND = 0 if a municipio is located in coastal lowlands
LND = 1 if a municipio is located in interior highlands

Classification variable 3 rural/urban dichotomy (R-U)

R-U = 0 if ≤ 50% of a municipio's population is urban
R-U = 1 if > 50% of a municipio's population is urban

Agricultural variable 1 density of the number of farms (DFRM)

Agricultural variable 2 density of cuerdas (1 square mile = 658.94 cuerdas) of land occupied by farms reported in units of 10 (DCD)

Agricultural variable 3 density of quarts of milk sold reported in units of 4000 quarts (DMLK)

Agricultural variable 4 density of tons of sugarcane harvested reported in units of 200 tons (DSGR)

Agricultural variable 5 density of pounds of coffee harvested (DCF)

Agricultural variable 6 density of pounds of tobacco harvested reported in units of 200 pounds (DTB)

Agricultural variable 7	density of hundreds of bananas and plantains harvested reported in units of 1000 (DBN)
Agricultural variable 8	density of families living on farms (DFAM)

A univariate problem illustrated by the use of these Puerto Rico data might determine the average yield of milk for the 73 municipios. A multivariate problem might relate the milk production of a farm to farm density and landform classification.

The second data set, presented in Table 1.2-A, gives the location and manufacturing information for firms in the automotive parts industry in Ontario, Canada, for 1986 (see Figure 1.3) and has been selected because of the recent in-depth study Amrhein has undertaken of these data. This data set emphasizes economic and industrial geography issues. At times statistical concepts presented in subsequent chapters are demonstrated with these data; examples based upon these data are designated by labels having the prefix CA. This data set has been compiled from the 1986 *Canadian Auto Parts and Vehicle Manufacturers Directory*, published by the Automotive Directorate, Department of Regional Industrial Expansion, Ottawa.

There are 207 Ontario firms listed in the original data set from which 53 have been sampled as follows: first, observations with missing data have been deleted, and second, beginning with the first observation, every third observation has been selected from the remaining 163 firms. Because these observations are presented by location, these firms constitute the areal units of this data set. For each firm there is a single labelling variable, two locational variables (the first locational variable indicates that these data are drawn exclusively from Ontario), and 10 industry variables. In Table 1.2-A the data are organized by region. The following is a description of each variable with an acronymlike mnemonic that appears in the Table 1.2-A results and frequently in ensuing chapter examples:

Label variable 1	abbreviated name of the firm (NAME)
Location variable 1	first character in the postal code (PC1)

A = Newfoundland	M = Metropolitan Toronto
B = Nova Scotia	N = Southwestern Ontario
C = Prince Edward Island	P = Northern Ontario
E = New Brunswick	R = Manitoba
G = Eastern Quebec	S = Saskatchewan
H = Metropolitan Montreal	T = Alberta
J = Western Quebec	V = British Columbia
K = Eastern Ontario	X = Northwest Territories
L = Central Ontario	Y = Yukon

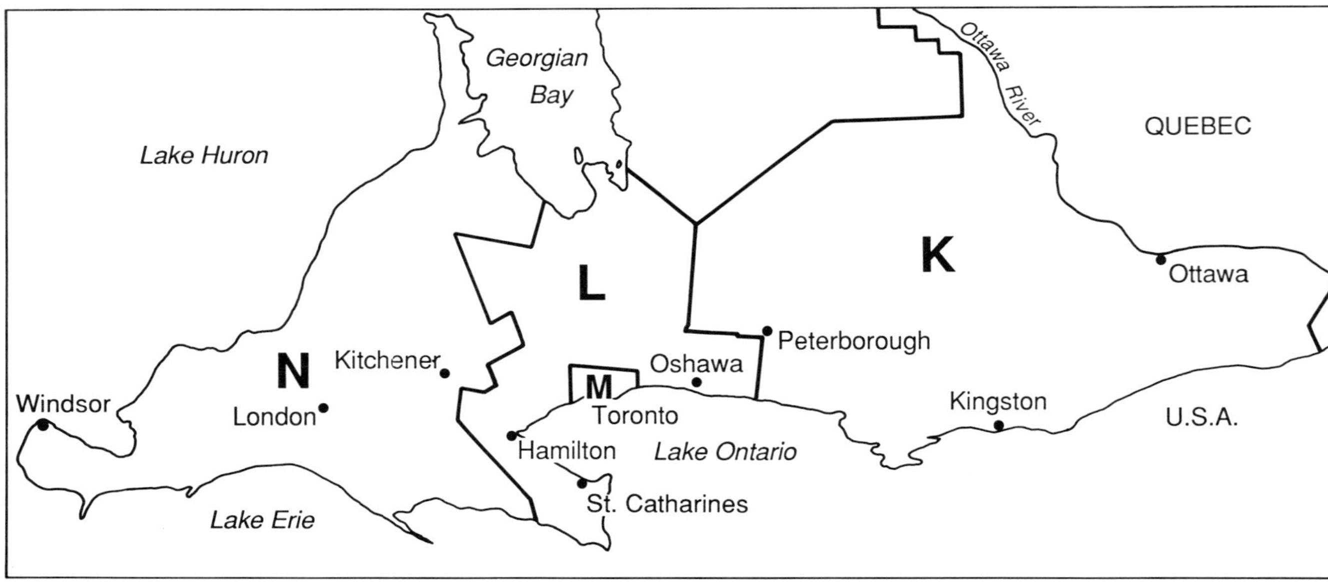

Figure 1.3: THE MAJOR POSTAL ZONES PARTITIONING OF SOUTHERN ONTARIO

Location variable 2 the last three characters of the postal code, locating the firm in a city, town, or region (PC2)

Auto-industry variable 1 foreign agent code (FAC)

FAC = 0 if the firm is an agent for a foreign government
FAC = 1 otherwise

Auto-industry variable 2 export code (EXP)

EXP = 0 if the firm exports primarily to the United States
EXP = 1 if the firm exports primarily to another foreign country

Auto-industry variable 3 market niche (MKT)

MKT = 0 if the firm's output is sold primarily to the original equipment market for the assembly of new cars
MKT = 1 if the firm's output is destined for spare parts for retail sales (after market) as well as original equipment
MKT = 2 if the firm's output is intended solely for the after market

Auto-industry variable 4 location of a Canadian firm's parent company code (PAR)

PAR = 0 if the parent company is located in Canada
PAR = 1 if the parent company is located in the United States
PAR = 2 if the parent company is located in Mexico
PAR = 3 if the parent company is located elsewhere

Auto-industry variable 5 Conservative view: categorized estimated sales reported in thousands of Canadian dollars
Alternative view: grouped interval estimate of sales reported in thousands of Canadian dollars (SALE)

Category 0:	no sales	Category 1000:	1000 - 2999
Category 1:	1 - 24	Category 3000:	3000 - 4999
Category 25:	25 - 99	Category 5000:	5000 - 24999
Category 100:	100 - 499	Category 25000:	25000 - 49999
Category 500:	500 - 999	Category 50000+:	50000+

Auto-industry variable 6 Conservative view: categorized estimated value of exports reported in thousands of Canadian

dollars
Alternative view: grouped interval estimate of the value of exports in thousands of Canadian dollars
Categories are the same as for SALE (EXP$)

Auto-industry variable 7 — Conservative view: categorized low estimate of the number of workers
Alternative view: grouped interval low estimate of the number of workers (LWK)

Category 1: 1 - 4	Category 100: 100 - 249
Category 5: 5 - 9	Category 250: 250 - 499
Category 10: 10 - 19	Category 500: 500 - 999
Category 20: 20 - 49	Category 1000+: 1000+
Category 50: 50 - 99	

Auto-industry variable 8 — Conservative view: categorized high estimate of the number of workers
Alternative view: grouped interval high estimate of the number of workers
Categories are the same as for LWK (for the largest firms HWK = LWK) (HWK)

Auto-industry variable 9 — product code for the group of products that forms the primary output of a firm (PPC)

0 = engine components, transmissions, drive trains, cooling systems, water pumps, fuel pumps [Standard Industrial Classification (SIC) codes 3251, 3259]
1 = suspension components, brake systems, shock absorbers, axles, steering assemblies (SIC codes 3254, 3255)
2 = electrical components, wiper motors, batteries, ignition systems, lights, wire assemblies (SIC code 3252)
3 = auto-body parts, interior and exterior assemblies (SIC codes 3253, 3256, 3257)
4 = rubber products, tires, belts, gaskets, flanges (SIC code 1511)
5 = fluids, chemicals, filters, paints (SIC code 3259 not in categories 0 or 6)
6 = accessories, trailer assemblies, alarms, radios (SIC code 3259 not in categories 0 or 5)
7 = basics, everything else such as nuts, bolts, bearings, and fittings

Auto-industry variable 10 — product code for the group of products that forms the secondary output of a firm using the same codes as for PPC (if a firm manufactures only one group of products, then that group is listed both as PPC and SPC) (SPC)

A range of univariate and multivariate problems can be examined with this data set, too, but of particular interest are applications of statistical techniques dealing with data that are categorical in nature.

The third data set, presented in Table 1.3-A, gives a set of physical geographic variables pertaining to hydrology and runoff for selected drainage basins in Ontario from 1960 to 1982, and has been selected because of Desloges's familiarity with it. The focus here is on problems addressed in the physical and environmental sciences. Examples appearing in succeeding chapters that are based upon these data are designated by labels having the prefix OR. This data set has been compiled from the 1960-1982 Environment Canada *Water Survey of Canadian* Surface Water Data, which is published annually. Latitude and longitude figures have been derived from topographic maps. Snowfall, temperature, and precipitation measures have been retrieved from the Environment Canada publication entitled *Atmospheric Environment Services Climate Summaries.*

There are 28 drainage basins (areal units) in this data set (see Figure 1.4). For each drainage basin there are two labelling variables, followed by eight attribute variables. The following is a description of each variable together with anacronym-like mnemonic that appears in the Table 1.3-A results and frequently in ensuing chapter examples:

Label variable 1 — number referring to the alphabetical order position of the drainage basin name

Label variable 2 — name of the drainage basin (NAME)

Physical variable 1 — area of the drainage basin measured in square kilometers (AREA)

Physical variable 2 — average spring discharge measured in cubic meters per second (Q)

Physical variable 3 — runoff, which is calculated for a given time interval t as Qt/AREA, where t is a 2-month time interval, Q is in cubic meters per second, and AREA is in square kilometers (RUN), runoff units are standardized to millimeters of water depth,

Physical variable 4 — latitude of the basin measured in degrees (LAT)

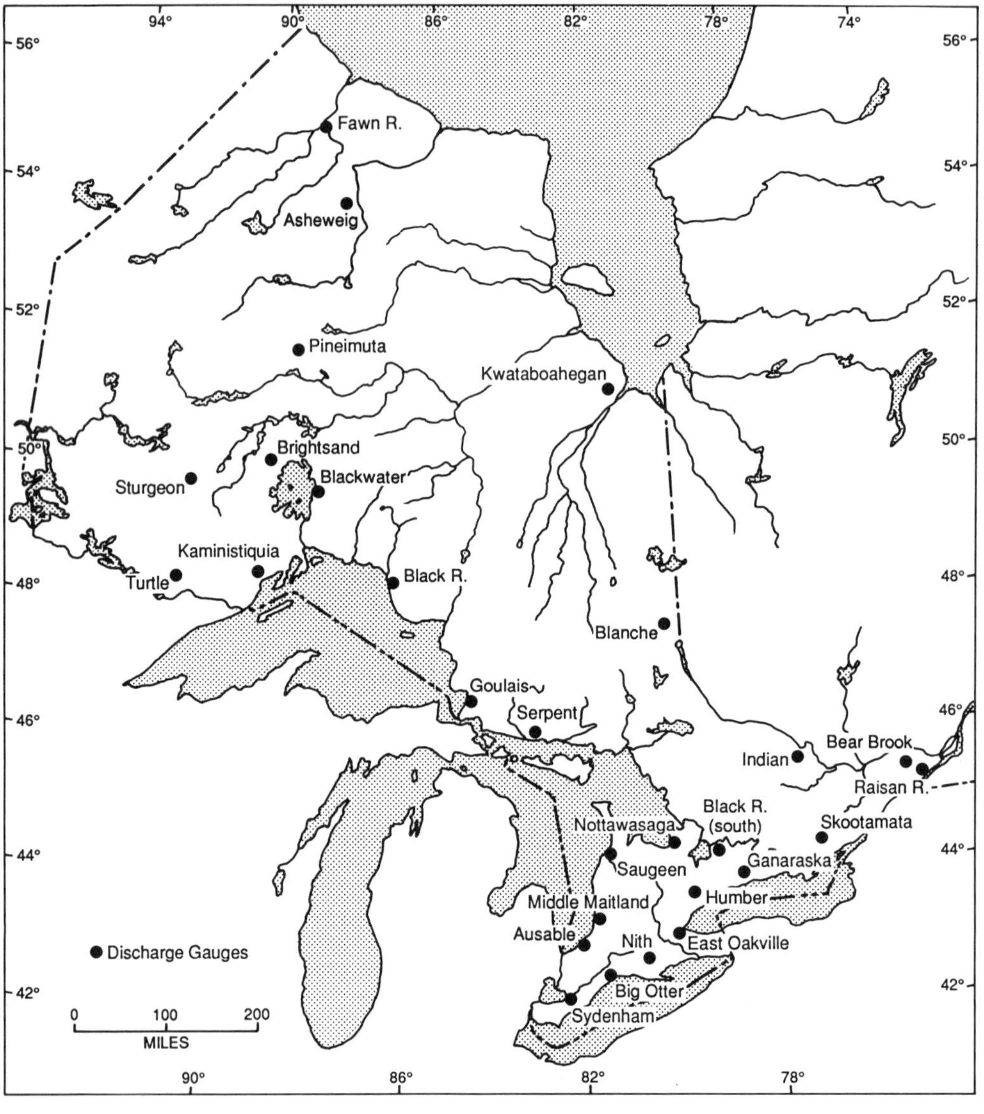

Figure 1.4: THE LOCATION OF SELECTED DRAINAGE BASINS IN ONTARIO

Physical variable 5	longitude of the basin measured in degrees (LONG)
Physical variable 6	mean annual snowfall measured in inches (SNOW)
Physical variable 7	mean annual precipitation measured in inches of water (PRECIP)
Physical variable 8	mean annual temperature measured in degrees Fahrenheit (TEMP)

The fourth data set, presented in Table 1.4-A, gives a set of glacial variables pertaining to selected physical geography characteristics of the Sentinel Glacier in British Columbia, Canada (just north of Vancouver); this data set has been selected because of Desloges's familiarity with it, too. Again the focus here is on problems addressed in the physical and environmental sciences. Examples appearing in succeeding chapters that are based upon these data are designated by labels having the prefix SG. This data set has been derived from published data compiled by various government agencies. Field measurements are from the Sentinel Glacier itself as well as ground station Alta Lake (located just to the north of Sentinel Glacier).

There are 19 observations in this data set. Each observation contains four variables in addition to the year of the measurement. The first two variables relate to physical characteristics of the glacier. The remaining two variables relate to the current environmental conditions around the glacier. One should note that although these observations represent a series of measurements in time, the data relating to net mass balance (BN) and equilibrium line altitude (ELA) can be treated as independent pairs. In other words, it is assumed that the previous year's climate does not influence BN/ELA observations in the current year. The following is a description of each variable together with an acronymlike mnemonic that appears in the Table 1.4-A results and frequently in ensuing chapter examples:

Classification variable 1	year in which the observation was obtained (DATE)
Physical variable 2	equilibrium line altitude defined as the elevation of the line that separates the zone of net mass (snow plus ice) accumulation from net mass loss (ELA)
Physical variable 3	net mass balance defined as the net gain (+) or loss (-) of glacier mass expressed as an average depth in meters of water equivalent (BN)
Physical variable 4	total precipitation accumulating during the winter months

of October through April (inclusive), measured in millimeters (PRECIP)

Physical variable 5 average temperature during the summer months of May through August (inclusive) measured in degrees Celsius (TEMP)

These four different data sets are used in distinct examples throughout the text. The choice of a data set for a particular illustration is dictated by the suitability of the data as well as the importance of the statistical technique application in a particular area of geographical inquiry.

1.5. Summary

Studying statistics is clearly one of the most difficult courses many undergraduate students in geography face. In part, the logic behind the construction of problems is very different from the logic learned in algebra and geometry courses, and in part, mastering the terminology is similar to learning a foreign language. There are rules and definitions that must be learned before the material can be mastered. Ultimately, exposure to many problems, and practice solving those problems, is the best learning environment. At the end of each chapter we list supplemental readings that contain additional examples; of particular interest are the SAGE and CATMOG series. Students who have taken our courses suggest that, generally these two series provide useful and (equally or more important) understandable examples.

1.6. Problems

1. Suppose that you are going to draw a sample of size 100 and that the source of the information is to be farms in Puerto Rico. The numerical population that you are interested in is the income earned from agricultural products for all commercial farms in a given year. Explain the pitfalls of sampling this population.

2. Identify some of the pitfalls associated with the sampling used to construct the Canadian automotive data set (CA) discussed in Section 1.4.

3. Five streets (street is the object) have been sampled and the number of houses counted. Let variable X represent the number of houses on each street as follows: {10, 9, 11, 6, 8}. Illustrate each of the following notational operations with this set of numbers. Let a = 10, b = 2, n = 5, and m = 3:

$\sum_{i=1}^{n} a$ $\quad \sum_{i=1}^{n} x_i$ $\quad (\sum_{i=1}^{n} x_i)^2$ $\quad \sum_{i=1}^{n} x_i^m$

$\sum_{i=1}^{n} a\, x_i = a \sum_{i=1}^{n} x_i$ $\quad \sum_{i=1}^{n} (a\, x_i + b) = (a \sum_{i=1}^{n} x_i) + n \bullet b$

$\prod_{i=1}^{n} x_i$ $\quad \prod_{i=1}^{n} a$ $\quad \prod_{i=1}^{n} x_i^m$

4. For the same streets sampled in Problem #3, variable Y represents the number of fire hydrants on the street as follows: {3, 3, 3, 1, 2}. Note that the object in this problem is the same as in Problem #3 (that is, the street). Using the values of X from Problem #3, illustrate each of the following notational operations with this set of numbers (n = 5). What information does each quantity provide?

$\sum_{i=1}^{n} x_i / y_i$ $\quad (\sum_{i=1}^{n} x_i)/(\sum_{i=1}^{n} y_i)$

5. Calculate the individual factorial terms for the variable X in Problem #4, and then calculate

$\prod_{i=1}^{n} x_i!$

6. Translate the following tabular data into the appropriate notation to depict row sums, column sums, the sum of all values in the table, the product of all of the row sums, and the product of each row sum times each column sum:

Row\column	1	2	3	4
1	10	5	7	14
2	6	6	6	7
3	9	5	11	8

7. Using the first 10 observations in the Ontario Drainage Basin data set, calculate the following terms:

a. the sum of runoff

b. the sum of the product AREA • Q

c. the sum of the square of PRECIP

1.7. References

Brower, B. (1988) "The Pernicious Power of the Polls." *Money*, 17 (March): 144-163.

Chorley, R. (Chairman) (1987) *Handling Geographic Information.* London: Her Majesty's Stationery Office, Report to the Select Committee on GIS.

Hite, S. (1976) *The Hite Report: A Nationwide Study on Female Sexuality.* New York: Macmillan.

Huff, D. (1954) *How to Lie With Statistics.* New York: W. W. Norton.

Walsh, D. (1985) "Socrates in the Classroom." *American Educator*, 9 (Summer): 20-25.

1.8. Additional readings

1.8.1. Review of basic mathematics and statistics

Baggaley, A. (1969) *Mathematics for Introductory Statistics.* New York: Wiley.

Blalock, H. (1979) *Social Statistics, 2nd revised ed.* New York: McGraw-Hill.

Ferguson, G. (1971) *Statistical Analysis in Psychology and Education.* New York: McGraw-Hill.

Johnson, R. (1976) *Elementary Statistics.* North Scituate, Mass.: Duxbury Press.

Summer, G. (1975) *Mathematics for Physical Geographers.* New York: Wiley.

Walker, H. (1964) *Mathematics Essential for Elementary Statistics*, rev. ed. New York: Holt, Rinehart & Winston.

Wilson, A., and Kirkby, M. (1980) *Mathematics for Geographers and Planners.* Oxford: Oxford University Press.

1.8.2. Statistics books for geographers

Barber, G. (1988) *Elementary Statistics for Geographers.* New York: Guilford Press.

Clark, W., and P. Hosking. (1986) *Statistical Methods for Geographers.* New York: Wiley.

Ebdon, D. (1977; 2nd ed. 1985) *Statistics in Geography. A Practical Approach.* Oxford: Basil Blackwell.

Hammond, R., and P. McCullagh. (1974) *Quantitative Techniques in Geography: An Introduction.* Oxford: Oxford University Press.

Krumbein, W., and F. Graybill. (1965) *Introduction to Statistical Methods in Geology.* New York: McGraw-Hill.

Norcliffe, G. (1977) *Inferential Statistics for Geographers.* New York: Halstead Press.

Shaw, G., and D. Wheeler. (1985) *Statistical Techniques in Geographical Analysis.* New York: Wiley.

Silk, J. (1979) *Statistical Concepts in Geography.* Winchester, Mass.: Allen & Unwin.

Taylor, P. (1977) *Quantitative Methods in Geography. An Introduction to Spatial Analysis.* Boston: Houghton-Mifflin.

Upton, J., and B. Fingelton. (1985, 1989) *Spatial Data Analysis by Example,* Vols. 1 & 2. New York: Wiley.

Unwin, D. (1981) *Introductory Spatial Analysis.* New York: Methuen.

CHAPTER 2

DATA AND MEASUREMENT

2.1. What are data?

Data (the plural of datum) come from observations made upon reality. They are gathered using commonly agreed upon standardized procedures, and hence are factual material or evidence. This evidence furnishes a basis for reasoning, decision-making, measuring, and/or calculating. Numbers of every kind have become the currency of modern life, and as such the discipline of statistics is concerned first with data in reference to these last two uses, from which derivative data are extracted for the first two uses. This numerical concern arises from the general belief that numbers suggest understanding, seem concrete and objective, and tend to be less ambiguous in meaning than words. Countless measurements (determining, for instance, how much, how many, how long, how often, how fast, where, or what kind) and calculations can be made on reality, with the possibilities limited only by one's imagination. The different attributes selected to measure are called variables (the values assumed by a given attribute are not necessarily constant from one measure to another; see Section 1.3.1), whereas the individual measurements made for a given attribute are called observations (i. e., an observation is made upon reality; see Section 1.3.1). But a sound data basis for reasoning and decision-making is obtained only when the statistics extracted with these calculations and from these measures exhibit certain desirable general properties, mostly with reference to error contained (or latent) in the data.

Expanding upon points mentioned in Section 1.4, we recall that data may be gathered in a number of different ways. Physical geographers often collect data from laboratory or field experiments, while human geographers sometimes collect data by surveying individuals or activity units (e. g., households, farms, stores). When data are secured in this fashion, their source labelled as primary; it comprises the originally collected information. If the primary source stems from an experiment, then considerable knowledge may be present about the resulting data. For example, idiosyncracies and nuances of technology used for making measurements can be recognized and corrected for (e. g., angle adjustments due to the relative position of the

Earth made to data acquired with satellites), various factors that influence the measurements being taken can be controlled to minimize their effects (e. g., road and driver conditions when determining new automobile gasoline mileage), and precision of the measurements (e. g., elevation measures are rounded to the nearest foot). In contrast, if the primary source stems from a survey, then probably less knowledge is present about the resulting data. For instance, even though shortcomings in selected characteristics of a survey instrument cannot be corrected, awareness of them becomes invaluable to an understanding and appreciation of the gathered data (many government census bureaus provide details about their survey instruments as a preface or epilogue to the published tabulated data). No control can be exercised over factors that may affect the statistical population in question. The best efforts seek to ensure a representative coverage of the population, and precision of the measurements often is compromised. For example, for various reasons and fears, most individuals refuse to provide precise income figures. Furthermore, experiments almost always yield data (faulty or inoperable equipment, or equipment malfunctions or breakdowns are an exception here). In contrast, survey returns frequently are fraught with failures, and the process of data collection itself becomes an issue. For instance, individuals selected to be surveyed may have others help them answer some or all questions, may refuse to respond, may respond to only part of a questionnaire, or may not be available at times when the survey is conducted; questions may be inappropriately worded or may be ordered improperly on the questionnaire, the wrong types or phrasings of questions may be employed, or the layout and design of the questionnaire may be inadequate.

Quite often, however, geographers and other spatial analysts acquire data from secondary sources, or sources that publish their own or sometimes someone else's primary data. Usually these data are published by government agencies, either in hard copy or digital form, and sometimes these data are unpublished but selectively and restrictively available through private enterprises (e. g., sales data for some retail chain). The key point here is that the researcher does not conduct an experiment or a survey in order to obtain data. For the most part these data are much cheaper and easier to obtain and make the accumulation of knowledge more efficient by removing redundancies in data collection efforts; the Puerto Rico agricultural data set described in Chapter 1 is an example of this sort of data, for it was collected by the U.S. Department of Agriculture. The major drawback with secondary data is the researcher's inability to ensure the quality of the resulting numbers. Published data may be altered by recording errors (e. g., in transcribing or editing). For example, in the Puerto Rico data described in Chapter 1, figures for the municipio of Guaynabo had been incorrectly included in the published totals for the agricultural administrative region of Arecibo rather than its true administrative region of San Juan. It cannot be emphasized too strongly that secondary data should be checked meticulously before being analyzed; supplemental data sets should be consulted for cross-checking purposes, so that gross recording errors can be identified and rectified. [Norcliffe (1977) illustrates the dilemma faced here. He points out that a recording of a person's being 8 feet 7.25 inches tall may be suspect because *The Guinness Book of Records* provides grounds for supposing this to be an error. "On the other hand a radio operator, on hearing of the Pearl Harbor attack on 7 December 1941, decided it was a preposterous hoax and did not bother to relay a message about it."] Selected data may be inadvertently omitted or suppressed. For instance, the U. S. Department of Agriculture suppressed data for certain Puerto Rican municipios in order to comply with the confidentiality laws it is bound by. Alternatively, questionable ad hoc procedures may distort the reported data, such as the cavalier rounding of figures, aggregation of reported results by combining individual attribute figures or amalgamating observations into areal units. Finally, ill-advised merging of several primary data sets also may distort data. Another major drawback with secondary sources is that data for specific attributes that a researcher may wish to measure may not have been collected. In this case the researcher is faced with either replicating at least part of the survey or experiment, which defeats a primary purpose of resorting to a secondary data source, or substituting an attribute for which data have been collected and hence introducing error into subsequent analyses; researchers frequently opt to follow this latter course of action. Of particular concern to geographers is assigning data distributed over space (such as data contained in individual census divisions) to a single surface point.

Regardless of whether a data set comes from a primary or a secondary source, though, measurement is involved.

2.2. Scales of measurement

Measurements result from a procedure that ascertains or appraises the dimensions, capacity, extent, or amount of something. When possible, it is preferable to use a universally recognized standard specified unit. For the most part, then, measurements are numerical in nature; any qualitative or categorical data can be easily quantified, as will be demonstrated in this chapter. But the resulting numbers may contain different amounts of information, depending upon their purpose or the generic method of measurement used to produce them. The generic method of measurement determines which arithmetic operations can be meaningfully performed and interpreted with given data. The method of measurement may be referred to as the measurement scale. There are four such scales, namely, nominal, ordinal, interval, and ratio, listed here in increasing order of sophistication and complexity. The information content of measures increases as one moves from a nominal scale to a ratio scale. The binary scale is a special dichotomous case of the nominal scale of measurement.

Nominal scale measurement is used to convert qualitative data to quantitative data; it is categorical in nature and scope. In essence, the numbers are nothing more than names of categories into which observations fall. Consider the simplest case of a dichotomy, which is binary in nature. For example, the Puerto Rico data have the three attributes of "agricultural administrative region" (ADM), "coastal lowlands/interior highlands" (LND), and "rural/urban dichotomy" (R-U) that have been measured using a nominal scale. Consider the second of these attributes. It involves two categories and thus is an example of the binary scale. LND takes on one of two values; the value of 0 for a coastal lowland municipio, and 1 for an interior highland municipio. Additional examples of nominal scale variables are contained in the Canadian Auto data set. Nine of the 14 variables are nominal scale. A dichotomous example from this data set is the foreign agent code (FAC), which assumes a value of 0 if the firm acts as an agent for a foreign corporation, and 1 otherwise. With regard to these variables from both data sets, the magnitude and difference of these numbers are arbitrary; conventionally 0 and 1 have been used partly for the sake of simplicity.

In the nominal measurement scale (1) the categories in question are **mutually exclusive** (a municipio can fall into one and only one category) and **collectively exhaustive** (a category exists for every municipio to fall into), (2) the dichotomous values involved are arbitrarily assigned to the prespecified categories (FAC could be acceptably defined to assume a value of 1 if the firm acts as an agent for a foreign corporation and 0 otherwise), (3) addition, subtraction, multiplication, and division are usually meaningless arithmetic operations with regard to the pair of categorical numbers, with their performance here yielding gross errors, and (4) the numbers can be accumulated in order to determine frequency of occurrence of a category (another reason to select 0 and 1), and hence sometimes are referred to as count data. Basically, the following question is being asked and answered with this measurement scale:

> Does observation i belong to Category A? If the answer to this question is yes, then measure a 1, and if the answer to this question is no, then measure a 0.

Clearly, since the two categories are mutually exclusive and collectively exhaustive, the affiliated question for Category B would yield a 0 everywhere a 1 appears for Category A, a 1 everywhere a 0 appears for Category A, and the sum of the two measures would always equal unity.

The ADM variable from the Puerto Rico data and the PC1 and PPC variables from the Canadian Auto data represent more general multichotomous situations involving nominal measurement. In the case of ADM, the integer numbers 1 through 5 are merely numerical names for the five classification categories involved, and just as easily could have been A, B, C, D, and E. They are just like the municipio numbers 1 through 73, which are simply numerical names that may be used in place of their longer alphabetical counterparts. In the case of PC1, the letters identify the appropriate postal zone in which the firm is located. Some numbering system, or the actual name of the postal zone, also could have been used. In the case of PPC, the numbers 0 through 7 identify the primary product produced by a firm. In this scheme, not only is the choice of numbers to identify product categories arbitrary, but the original groupings of parts into categories is subjective and may vary among researchers.

For purposes of statistical analysis, a multichotomous nominal variable always can be transformed into a number of binary variables. For example, the ADM variable could be transformed into five binary variables. In each case the question asked would be does municipio i belong to Category K? If the answer to this question is yes, then measure a 1, and if the answer to this question is no, then measure a 0. Again the sum of these five nominal measurements would always be unity, suggesting that actually only four of the nominal measurements are needed in any analysis (similar to only one of the two indicator variables for the LND variable being reported in the data set). Similar transformations can be constructed for variables, such as PPC, found in the Canadian Auto data set.

The **ordinal** measurement scale pertains to situations in which rankings of observations can be made utilizing an operationally defined attribute. Here, again for the sake of simplicity, the convention is to use an ascending or a descending sequence of consecutive positive integers. However, more substantive descriptive names also can be used, such as "poor," "middle class," and "wealthy." If ties occur, then an arithmetic average of the rankings that would have been assigned to the set of tied observations, if they had been distinct and separable, is assigned to each of the tied observations. Resulting numbers take on meaning in terms of their relative magnitudes with regard to the ordering involved; if descending order is used, then a rank of 1 is known to be greater than a rank of 2. But the distance between numbers has no precise meaning; a rank of 1 is not known to be twice as large as a rank of 2 or thrice as large as a rank of 3. The transitivity property is satisfied here, and addition have meaning. One way this measurement scale differs from the nominal scale can be exemplified by the ADM variable appearing in the Puerto Rico data set. There the consecutive integers of 1, 2, 3, 4, and 5 were used to label regions, but no ranking was implied by the assignment of this set of numbers, for not just San Juan but any of the five regions could have been named 1. In contrast, if these five regions were ranked ascendingly in accordance with the respective populations they embrace, then only San Juan could be assigned a value of 1, and Arecibo would no longer be assigned a value of 2.

There are four ordinal scale variables in the Canadian Auto data: SALE, EXP$, LWK, and HWK. Since the categories in each case have numerical limits, there is some information concerning how much larger one category is than another. Since every firm in a specific category is treated equally, no distinction among these firms is possible. For example, in the case of work force size (LWK or HWK), the categories could have been labelled "small," "medium," or "large," providing no information about the actual numbers of workers in a firm. Using numerical labels coincident with the range of each category, provides additional information. A specific firm, however, can be ranked only against firms in other categories. Although an approximate size difference can be determined, a precise comparison is not possible.

Interval measures build upon the ordinal scale by establishing a standard, meaningful unit of distance to separate numbers, which no longer are limited to integer values but are expanded in range to include all real numbers; they may be used in the operation of multiplication as well as those of addition and subtraction. Numbers located on this scale indicate neither quantities nor amounts; rather, distances are being measured. The most commonly cited example of interval measurement is that for temperature; both Fahrenheit and Celsius measures frequently are cited to illustrate this measurement scale. Such is the case in the TEMP variables from the Sentinel Glacier data and the Ontario Runoff data sets.

The fourth scale of measurement is **ratio**; it differs from that of interval in a subtle way. Here the arithmetic operation of division also can be performed; ratios are comparable because an absolute zero value (a zero with a precise meaning) exists, implying that if a measurement of zero is obtained, then there is a basis for saying that an observation has none of the attribute being measured. Interval measurement scales have no absolute zero value. Numbers located on this scale indicate actual amounts of an attribute that is being measured. In terms of temperature measures, then 0 on the Celsius scale (which is the freezing point of water, under standard atmospheric pressure) is equivalent to 32° on the Fahrenheit scale, and 0 on the Fahrenheit scale (which approximates the temperature produced by mixing equal quantities by weight of snow and salt, under standard atmospheric pressure) is equivalent to approximately -18° on the Celsius scale. In contrast, the Kelvin temperature scale has an absolute zero value, at which point no thermal energy exists. One example of a ratio measure is the DMLK variable from the Puerto Rico data set and another is the AREA variable from the Ontario Runoff data set. For all practical purposes, however, statistical analyses tend not to distinguish between interval and ratio measures.

The variables SALE, EXP$, LWK, and HWK provide examples of a special case of the ratio scale of measurement. These variables already have been presented as examples of ordinal scale measurements. However, the names of the categories of these variables provide additional information that could be used. Rather than simply labelling the categories using some scheme such as small, medium, or large, the category names provide numerical bounds on the categories. This information is not currently used in the ordinal classification. An alternative treatment that uses this information would classify these variables as grouped-ratio. In the absence of any information about the exact value for each of these variables for any observation, the least restrictive assumption that can be made is that the observations are evenly scattered within the interval. As a result, the value in the middle of each category is used for statistical calculations. Chapter 4 discusses several problems encountered in dealing with these variables as grouped ratio data.

EXAMPLE 2.1. Scales of measurement.

Table 2.1 presents sample variables at each scale of measurement for three quantities commonly measured in various fields of geography: income, location, and temperature. Note that in the case of income there is an increasing amount of information as the scale of measurement progresses from nominal to interval/ratio. For example, assuming that all income is from wages, then moving from a nominal to ordinal scale gives more information regarding one person's income relative to others. Further, moving from an ordinal to interval/ratio scale provides the exact level of wage income. Moving in the other direction, from interval/ratio down the scale to nominal, causes a loss of information. In general, while a higher order scale of measurement can always be reclassified into a lower order scale of measurement, the reverse process is not possible. *Therefore, data should always be collected at the highest scale of measurement possible.*

Also associated with measurement and the scale selected for measuring attributes are the nonsynonymous features of precision, accuracy, and validity. A discussion of these features naturally leads one into the topic of the next section, which has to do with error. The **precision** of a measure has to do with how exact the experiment or survey measurement instrument used is and hence how refined the measure yielded by this instrument is (the range of results produced when the same item is repeatedly measured). Historically a large part of the evolution of scientific technology has been preoccupied with the development of instruments that produce increasingly refined measures; mechanical devices and survey techniques are improving continuously. Refinement occurs as measures become increasingly closer to their counterparts in reality. The difference between what exists in the real world and what is measured by an instrument constitutes a class of error. If one instrument yields a measure that is closer to that value occurring in the real world than does a second instrument, then the first is said to be more precise than the second. Furthermore, since statistics also is concerned with calculations, calculation precision is of interest, too. There is no virtue over-generalizing results by excessive rounding of calculations; similarly, there is no virtue, and in fact there is error, in providing spurious calculation precision by including an unjustified number of places to the right of a decimal point.

The **accuracy** of a measure refers to its freedom from systematic bias so that distorted measures of reality are not obtained. For instance, within the context of experimental situations, if a mechanical apparatus is incorrectly calibrated, then its measures will always be incremented or decremented by a certain fixed amount. Within the context of survey situations, if one attempts to measure family income by the amount taxed by the government in a given year, then the amount of income removed by deduction adjustments always will be missing (a decrement always occurs). In contrast, precision would pertain to asking a family to indicate into which of a set of dollar ranges its income falls; as the dollar ranges decrease toward zero (i. e., become narrower), the precision of the family income measure should increase (such coarse ranges commonly are used because, as was mentioned earlier, people usually refuse to respond to detailed questions regarding their income).

TABLE 2.1

SAMPLE VARIABLES FOR INCOME, LOCATION, AND TEMPERATURE

Scale of Measurement			
Nominal	Ordinal	Interval	Ratio
(a) Examples of variables measuring income			
sources of income: wages, interest, dividends	lower class, middle class, upper class	taxable income, non-taxable income	actual income from each source
(b) Examples of variables measuring location			
desert, rain forest, urban, rural	interior, coastal, highlands, lowlands	latitude, longitude, UTM coordinates	distance from a fixed reference point
(c) Examples of variables measuring temperature			
comfortable or uncomfortable	hot, cold	degrees Fahrenheit, degrees celsius	kelvin

Finally, the **validity** of a measure pertains to its ability to measure what it purportedly is measuring. Normally this feature is not questionable when experimental data are used but becomes a serious issue when taking a survey or conducting a poll [In this context Budiansky (1988) mentions the use of "loaded" questions, encountering the "ignorance factor," and political "tactical polling"; Brower (1988) also notes the use of "loaded" questions as well as phrasing based upon dubious assumptions or "divide and conquer" strategies]. Mis-specified, imperfect, inappropriate, or ambiguous operational definitions for attributes are problematic. So too are improperly worded or phrased questions. The use of secondary data sources is one situation that gives rise to questions of validity. This is especially the case when attributes for which data have been collected are substituted for those for which data have not been collected. The use of surrogate variables and indices are plagued by this problem. A number of popular examples of data validity are summarized by Budiansky (1988). He notes, for instance, that in recent years the U.S. GNP has indicated economic growth dramatically slowing when in fact it has been accelerating, explaining that these estimates err in part because "the data that go into the figure are still heavily tilted toward industries that were dominant in the 1930s and 1940s; it is a poor measure of the newly ascendant service industries" (p. 47). Other similar examples include the Consumer Price Index and the various stock market indices.

2.3. Types of errors associated with data

When evaluating the quality of data or the selection of a measurement scale, five basic error sources are of interest here, namely, errors of calculation, measurement, specification, sampling, and random noise in the landscape. **Calculation error** refers to mistakes that arise from incorrect executions of arithmetic operations on numbers; for example, 2 + 2 = 3. With the widespread use of computers and hand calculators, this type of error should be nonexistent!

Measurement error refers to incorrect numbers in a data set. Many examples can be cited. For example, the undercounting of urban populations because the homeless and illegal aliens are so difficult to enumerate (the U.S. Census Bureau engaged in considerable litigation over this aspect of the 1980 census), the incorrect coding of numerical data (verification of unusual values is good practice, and in extreme-value cases common sense needs to prevail), the precision of a measuring instrument, (see Fuerbringer, 1989) and the use of surrogate variables such as trying to measure the demand for day care by indexing this demand with the number of households having women in the childbearing age cohort (this refers to the feature of measurement validity). Sometimes researchers strongly believe that measurement error dominates interval or ratio data and hence convert data measured at these scales to data measured at the nominal or ordinal scales. If they are wrong, then their conversion of data to a less sophisticated measurement scale results in a loss of information.

Specification error refers to the use of incorrect assumptions and/or the application of incorrect mathematical formulae or equations; for instance, if $y = ax + b$ is correct, but calculations use $y = ax$, then the calculation is mis-specified. This book will try to correct the incorrect assumption that geographic data can be statistically analyzed in the same way that nongeographic data are analyzed. One interesting and memorable example of specification error is supplied by behavioral research practices. Dr. Ronald Hart, a former Director of the U.S. Food and Drug Administration's National Center for Toxicological Research, once commented that

> The results clearly show that saccharin was ineffective either as a direct carcinogen or a bladder-tumor promoter. Although in rats saccharin was positive as a tumor promoter or carcinogen, in mice it was completely negative.

The issue is whether or not people are more like rats or more like mice. Federal legislation declared that any food additive that causes cancer in animals must be removed from the U. S. market, suggesting legislators assumed people are more like rats. This assumption certainly is a source of specification error in behavioral research.

Sampling error refers to one of the fundamental problems of statistical analysis, namely, a large collection of objects (the population) are to have measures taken on them, and these measures are to be analyzed, but for various reasons the measures are taken and analyzed for only a subset of these objects (a sample). Differences between results for this subset and the parent collection of objects that are due solely to the use of a subset rather than the entire collection constitute sampling error (that error introduced by virtue of the fact that samples vary from one to the next).

A particular category of sampling error is of interest here; to be of use in an analytical and critical thinking context, the subsets in question must display certain properties. The last class of error is labelled **random noise in the landscape**. In general this type of error represents a haphazard (stochastic) disturbance and is characteristic of ignorance or uncertainty about reality or is the outcome of a goodly number of factors that jointly behave in such a way that random disturbances appear to be introduced into the attribute measures. In other words, this error represents noise or a lack of understanding. For example, in analyzing a sequence of values for some attribute, say population of a region, that are observed at regular intervals in time, one finds that the population at time t tends to be very similar to that at its immediate past and future times, t - 1 and t + 1 (most people do not move into or out of regions in extremely short periods of time). But the change in population values over the time sequence in question is not purely systematic or predictable. This type of deviation from complete regularity may be viewed as a stochastic error component of population change over time. If one had complete knowledge about population changes (and hence error-free data), then perhaps one would know exactly what the population value would be for time t + 1, given its value at time t (the case of ignorance). Or this error may represent noise in the operation of the phenomenon that is being measured. This idea is exemplified by the static picked up by a radio (or television) receiver when no nearby station transmits at the tuned-in frequency. This type of deviation from a discernible radio signal (complete regularity) may be viewed as a stochastic error component. It comprises many atmospheric or various electrical disturbances that interactively and jointly behave as if random. One also may encounter static arising from measurement error that is attributable to the radio mechanism. If one has a strong radio signal being transmitted at the selected frequency, then almost always this static will not be detected. Similarly, in analyzing a sequence of values for some attribute, again say the population at a particular point in time over a set of geographic regions, one finds that the population located in adjacent regions tends to be very similar but does not change in a purely systematic manner over the geographic landscape. This deviation from systematic change will be called random noise in the landscape and viewed in a manner analogous to that described here for time sequences.

One goal of classical data collection and analysis is to judiciously compile figures and make calculations in such a way that all controllable sources of error are either eliminated or minimized. Useful insights and understanding cannot be gained when sampling error is reduced to the point that an analysis is overwhelmed or dominated by nonsampling errors.

EXAMPLE 2.2. Examples of types of errors (see Table 2.1 entries).

Table 2.2 presents examples of types of errors that can be encountered with variables measuring income, location, and temperature. Ideally, the spatial analyst wants to eliminate every source of error; a prudent use of resources accepts sampling and/or random noise, though. Dealing with this sampling error and random noise is the focus of statistics; managing random noise is often very difficult, if not impossible. The other three sources of error should be eliminated. Unfortunately, identifying the various ways in which error can appear in a data set is not always a straightforward task. In Table 2.2 several of the example sources of error might be debated. For example, is the improper or incomplete classification of sources of income better treated as a measurement problem or a specification problem? If this is a specification problem, then error minimization or elimination lies in identifying the missing sources of income. If there is a measurement problem as well, then identifying the income source is one problem, while quantifying its value is another. An illustration of such a situation would be an effort to include the value of a government-subsidized day-care program. The value of a subsidy per child would change in relation to local conditions, such as cost of living, as well as in relation to the family situation of the person being observed. A person with two children in the program likely experiences greater value from the service than a person with no children.

Detailed treatment of the identification and correction of various errors is presented in the third section of this book.

TABLE 2.2

SAMPLE ERRORS ENCOUNTERED WITH VARIABLES MEASURING INCOME, LOCATION, AND TEMPERATURE

Source of Error				
Calculation	Measurement	Specification	Sampling	Random Noise in the Landscape
(a) Errors in variables measuring income				
incorrect addition	using net versus gross or taxable versus non taxable income	excluding a source of income or the value of social services	national census estimates of income based on samples	the actual pecuniary value of social services varying across governments
(b) Errors in variables measuring location				
incorrect conversion of latitude and longitude to great-circle distances	incorrect determination of latitude or longitude	using straight-line versus great-circle distances	government topographic maps based on samples of points on soil maps	effect of cloud cover on location determined using remotely sensed data; the location of boundaries
(c) Errors in variables measuring temperature				
incorrect conversion of °F to °C	imprecise thermometer	using °F when °C should have been used	recording instruments (such as a Stevenson Screen)	microclimatic effects such as wind direction and amount of precipitation placed in a sample of locations

2.4. Populations and the necessity of drawing samples

In Chapter 1 the distinction was made between a population and samples drawn from that population. As the preceding section emphasizes, the drawing of a sample introduces error into data collection and analysis. Arguing for this course of action then raises several salient questions. First, why take a sample when one knows that it will introduce error into data? Second, what sorts of controls can be placed upon the corresponding sampling error?

The first of these two questions may be answered in a number of different ways. Probably the most frequently cited reason for sampling is the need to conserve on resources. Complete enumeration of a goodly number of attributes of every element in a sizeable population is very costly in terms of time required to complete the total enumeration, the monetary resources needed to properly survey each element of the population and then convert survey responses to tabulated and published numerical data, and the effort required to carry out the survey itself. In other words, sampling allows one to conserve on and satisfy binding constraints presented by available and/or finite resources. After all, one can obtain valid, reliable results from a good sample without having to resort to taking a census; sampling is efficient. Second, enumeration of the total population may be physically impossible, and so the only option available is to sample. For example, geographers sometimes talk about sampling points on a surface. But a surface has an infinite number of points, meaning that the total population of points cannot be enumerated. Or the population may be hypothetical and so cannot be totally enumerated (e. g., the flipping of a coin or the rolling of a die). Or the population may be unavailable and so cannot be totally enumerated (e. g., the number of customers that will patronize a store over the lifetime of that store). Third, data gathering may involve a destructive process; sampling allows data acquisition without sacrificing the entire population. For instance, to use radiocarbon dating techniques on an object, a sampled portion of that object must be destroyed (this example was popularized during the scientific dating of the Shroud of Turin). Quality control of manufactured items constitutes a similar problem; for example, in order to determined or verify the expected life of a brand of automobile tire, sample tires must be used in experimental road tests (clearly, all tires cannot be used by the manufacturer, or none would be available for sale). Taste testing is yet another form of this problem, for the food item is consumed in the course of sampling. Fourth, an analysis of some populations must be done in a timely manner so that either the population does not change or the data gathered do not become obsolete before a census is completed and its resulting data published; sampling provides a mechanism to do so. For instance, opinion polls often seek to assess the perceptions, beliefs, feelings, and the such, at particular points in time, and require data to be acquired and published within the context of a very brief time horizon, before opinions change or before the questions being addressed are no longer of public interest. Finally, in some circumstances the population is not wanted, while a sample is the focus of attention. For instance, in a lottery only the subset of "winners" is of interest; total enumeration of the population would be senseless.

The second question asks what sorts of controls can be placed upon sampling error and alludes to the establishment of desirable properties for a sample. The most prominent of these properties are (a) representativeness, (b) equally likely or random, and (c) likeliness of resulting sample statistics. Representativeness has been mentioned in Section 2.1. Here the most desirable composition of a sample is one that is derived from a good crosssection of the parent population and hence closely reflects the composition of this population. Unfortunately, without total enumeration of the population there is no way of knowing how well a sample represents its parent population. If a sample is drawn in such a way that it covers the population well, then data collected in this way will furnish a good representation of this population, and the affiliated sampling error will tend to be minimized. Many sampling strategies have been devised in order to improve the representativeness of a sample that is drawn; but there are important trade-offs between data interpretability in a statistical sense and the restrictiveness of such sampling strategy formulations. By placing constraints on the procedure for drawing a sample, certain samples are eliminated from the set of ones that are possible to be drawn. This situation is in conflict with the second property of randomness, in a random sample every element in the parent population has the same chance (i. e., equally likely), or probability, of being included. Drawing random samples helps to ensure that a representative sample is obtained. A quite unrepresentative random sample may well be selected, but the chances of this happening tend to be very small. In addition, the sampling error associated with random samples is well understood, so much so that probabilities can be attached to resulting statistics in order to qualify the extent to which one believes they are or are not misleading, that is, the likelihood of resulting sample statistics. One should recall that the main objective of sampling is to obtain accurate measures of population attributes without conducting a census. But sampling error is embedded in these measures. Drawing a random sample allows statistics to be calculated that on average are least distorted by this sampling error, for there are no systematic aspects of the sampling error associated with a random sample. In other words, samples drawn using well-designed procedures allow one to focus attention on other sources of error. Consequently, most random samples produce statistics with minimal deviation from the population attribute in question, meaning that there is a strong likelihood that the resulting statistic obtained with a random sample is a good approximation of its corresponding population parameter value.

One should note, however, that not all likely statistics come from representative samples, and occasionally, reasonably representative samples will fail to yield likely statistics. Furthermore, procedures implemented to increase the representativeness of a sample often deteriorate the randomness property of the sample, causing less to be known about the associated sampling error and hence raising questions about which statistics are most likely. But, as will be discussed in Chapter 8, some methods do exist that allow the representativeness of a sample to be enhanced without seriously compromising this randomness property.

2.5. *Information content of data*

The primary idea underlying sampling is that considerable information about a target population should be able to be gleaned from subsets of data taken from the parent population. Countless numbers of different statistics can be calculated using sample data, with the possibilities limited only by one's imagination. Thus, one needs to be able to discriminate between useful and essentially meaningless statistics and to guard against a blind faith in numbers. The truth behind numbers is too rarely questioned, and numbers are too easily manipulated. Therefore, properties are needed that evaluate the information a statistic contains about its corresponding population parameter value.

The first criterion that can be applied to a statistic in order to evaluate its information content has to do with the measurement scale involved. The least amount of information is captured by nominal measurement, which simply indicates to which of a set of qualitative categories an observation belongs. Ordinal measurement builds upon this concept by imposing a meaningful ordering on the numerical measures; hence additional information is obtained about observations in that one knows which attribute values are greater than or equal to others. Next, interval measurement expands upon the ordinal scale by establishing a serviceable standard unit of measure, allowing distances between numbers to take on precise meanings. Once more, additional information is obtained about observations, for now one is able to determine which observation values are closer and which are more distant. Finally, ratio measurement adds the notion of absolute zero to those features of an interval scale, allowing ratios to be compared. Therefore, as a measurement scale increases in sophistication and complexity, its information content increases. But one must recognize that employing an inappropriate measurement scale will do nothing more than yield spurious information when the scale is too sophisticated and discard valuable information when the scale is too simple. In many cases statistics have been derived that are applicable only to certain measurement scales, and become meaningless when calculated for data measured on other scales. For example, when measuring the sex of a person the nominal scale of measurement is not only the most appropriate scale it is the only scale that contains useful information.

The second criterion that can be applied to assess the information content of a statistic has to do with desirable properties of statistics calculated for random samples. Four of these properties emphasized in classical statistical theory are lack of bias, sufficiency, efficiency, and consistency. A statistic is said to be unbiased if the arithmetic mean of this statistic calculated for all possible samples of a given size n exactly equals its corresponding population parameter value (this concept was discussed in Section 1.3.1). The statistic is said to be **biased** if this arithmetic mean does not equal the corresponding population parameter. Sometimes when a statistic is biased, some simple modification can be made to it (perhaps increasing the sample size or multiplying by a certain constant; see Section 1.3.3) to produce an unbiased statistic. Many statistics, however, are not unbiased either in their raw or some modified form, and hence even though their calculation is possible, most of them should be avoided. One should note that establishing the unbiasedness of a statistic says nothing about individual sample results. But statistical theory demonstrates that an unbiased statistic calculated in the presence of random sampling tends to have its most likely value falling very close to the population parameter value, meaning that most often the sample statistic does not deviate dramatically from its corresponding population parameter. Attention is restricted in this book to statistics that are unbiased.

A **sufficient** statistic is one that summarizes all the *relevant* information about the parent population that is contained in a sample. Two kinds of information may be found in sample data, namely, some that is sample specific (this information changes from one sample to the next and emerges only when certain subsets of the population are treated) and some that is common to all random samples drawn from the parent population. Reducing sample data to a statistic, like moving to a lower scale of measurement, involves a loss of information. The goal is to lose only information that is sample-specific and not information that is pertinent to gaining knowledge about the population. A useful statistic will not entail a reduction of the sample data to such an extent that the information it captures is no longer a sufficient characterization of the population. It will distinguish between sample-specific information and information common to all samples and will exploit this latter type while overlooking the former. Without question one wishes to select statistics that are unbiased and sufficient.

Efficiency of a statistic refers to the degree to which the values for this statistic calculated for all possible samples of a given size n cluster about the corresponding population parameter value. If the statistic is unbiased, and random sampling was employed, then this clustering can be evaluated in terms of sampling error. Efficiency of statistics increases as sampling error decreases. But no statistic can be completely efficient, since such a state would require the absence of sampling error altogether, and hence analysis would be of the population rather than a sample. However, statistics can be compared, and hence relative efficiency can be established. Those statistics that are relatively more efficient are less susceptible to the impacts of sampling fluctuations and tend to be closer to the population parameter value than do their less efficient counterparts calculated for the same sample. Only the most efficient statistics are of interest here. There is little consolation in knowing that values of a statistic will average out to the population parameter value across all possible samples and that only information pertinent to the population is captured by the statistic, if for any given sample the statistic is likely to be far from the parameter value.

Finally, **consistency** has to do with the intuitive notion that the larger a sample size n is, the closer to its corresponding population parameter value a statistic should tend to be. As n increases, the set of statistic values for all possible samples of size n should increasingly concentrate around the population parameter value, and hence sampling error should decrease toward zero. In other words, as the sample size increases, probabilities calculated for various features from the set of all possible samples of size n will converge upon their theoretical probability counterparts. Generally speaking, this concept allows the likeliness of a statistic to be established. Clearly, consistency is a large-sample property and implies nothing about the concentration of statistic values around the population parameter for small sample sizes. Although some statistics are biased when n is small, if they are consistent this bias becomes negligible for large sample sizes; a statistic whose bias does not disappear as n increases needs to be modified in such a way that this bias is removed in order for the statistic to be consistent. Again the importance that a statistic be unbiased is highlighted.

Consequently, only measures made with an appropriate scale of measurement and a sufficiently large sample size render informative statistics although figuratively speaking an infinite variety of statistics can be calculated, only those that are unbiased, sufficient, relatively efficient, and consistent are of interest here. This class of statistics extracts the most information about a population from a sample.

2.6. Summary

An understanding of two of the topics in this chapter is absolutely essential before progressing to the next chapter. The scale of measurement at which observations are made ultimately constrains the range of statistical procedures that can be used with the data. Therefore, when choice is available, selecting the scale at which the data are collected is worth some time and careful consideration. While considering the scale of analysis, it is also well worth the effort to think very carefully about the population from which you think the sample is drawn. All too often, considerations about the identification of the population become lost in the numerical mechanics of the analysis. Equally all too often, the cost of this oversight can be seen in poor quality or inconclusive results.

2.7. Problems

1. Consider the Puerto Rico data set. Which measurement scales have been used for each of the variables DFRM, DCD, DMLK, DSGR, DCF, DTB, DBN, and DFAM? Explain why.

2. Given the multichotomous variable ADM in the Puerto Rico data set, create the appropriate set of indicator variables for the sample {Barranquitas, Camuy, Comerio, Guanica, Patillas, San Lorenzo, Villalba, Yauco}.

3. Consider the Canadian Auto data set. Which measurement scales have been used for each of the variables PC2, MKT, PAR, and SPC? Explain why.

4. Given the multichotomous variable PPC in the Canadian Auto data set, create the appropriate set of indicator variables for the sample {A&R, Bundy, Crown, GMA, HRuhl, Relmech, StndTube, TRW1}.

5. Consider the Sentinel Glacier data set. Which measurement scales have been used for each of the variables DATE, ELA, BN, and PRECIP? Explain why.

6. Consider the Ontario Runoff data set. Which measurement scales have been used for each of the variables NAME, AREA, Q, RUN, LAT, LONG, SNOW, and PRECIP? Explain why.

7. A geographer wishes to measure variation in household income in Syracuse for a given year and obtains figures from the Internal Revenue Service based upon tax returns for that year. Identify that source of error from the following set of five that best describes the anticipated error that occurs in his or her analysis of these data, and explain your choice:

calculation error	sampling error	measurement error
specification error	random noise in the landscape	

   (a) Error due to individuals not reporting income (do not report correct income on the tax return, receive nontaxable government benefits, earn less than the minimum amount that necessitates filing a tax return, and the like).

   (b) Error due to a power surge in the electrical lines to the computer which causes an incorrect summation.

   (c) Error due to the use of a subset of all possible households in Syracuse.

   (d) Error due to the inadvertent introduction of a program code mistake by the computer expert who composed the statistical software employed.

8. Consider the numerical population {0, 1, 1, 2, 2, 2, 3, 3, 4}. For a sample of size three, identify two dramatically different but clearly unrepresentative samples and two different but reasonably representative samples. Identify your criteria for making these choices.

9. Propose a geographic statistic (one capturing something about the geography of

a variable) that could be calculated for either the Puerto Rico, Canadian Auto, Sentinel Glacier, or Ontario Runoff data sets. Explain the properties of this statistic that need to be established or demonstrated before statisticians would find it to be an acceptable one.

10. Construct a table similar to Table 2.1 showing examples of variables that measure the attributes of regional economic health and soil structure.

11. Construct a table similar to Table 2.2 showing examples of errors that can arise in measurements of regional economic health and soil structure.

2.8. References

Brower, B. (1988) "The Pernicious Power of the Polls." *Money*, 17 (March): 144-163.

Budiansky, S. (1988) "The Numbers Racket: How Polls and Statistics Lie." *U. S. News and World Report*, 105 (#2; July 11): 44-47.

Fuerbringer, J. (1989) "Accuracy in Short Supply in Flood of U.S. Statistics." *The New York Times*, 139 (October 30, 1989): D1 and D4.

Norcliffe, G. (1977) *Inferential Statistics for Geographers*. New York: Halstead Press.

2.9. Additional readings

In addition to the texts listed in Chapter 1,

Johnston, R. *Classification in Geography*. CATMOG 6.

Taylor, P. (1977) *Quantitative Methods in Geography. An Introduction to Spatial Analysis*. Boston: Houghton-Mifflin.

Torgerson, W. (1958) *Theory and Methods of Scaling*. New York: Wiley.

CHAPTER 3

DESCRIPTIVE CLASSICAL STATISTICS: ELEMENTARY SUMMARIES

One troublesome consequence of the presence of error in data, regardless of its source, is that it tends to obscure both information content and obvious patterns. Single pieces of data are often difficult to understand, while grouped or aggregate data sacrifice detail for clarity. When describing data, the whole can be greater than the sum of its parts. Tabulating data may well camouflage pattern by diverting attention to individual numbers and hence compromise the overall intelligibility of the table. In an attempt to avoid these problems, a simple way of exploring data is to present them in pictorial, visual, or graphic forms.

3.1. Common calculations used to graphically portray spatial data

Observations grouped into categories can be displayed in a number of forms. The actual frequencies of observations in different categories can be displayed, or the actual frequencies can be compared first to the total number of observations and then displayed. This first approach produces **absolute frequency diagrams**. This second approach produces **relative frequency diagrams**. The transformation from the original data into a form required for relative frequency diagrams requires the calculation of a ratio, which may be expressed as a fraction, a decimal proportion, or a percentage.

Calculating a ratio that divides the frequency count of each category of a multinomial classification by the total frequency count, which is the sample size, produces a ratio. The sum of the ratios from all the categories in the classification scheme totals one. Therefore, a comparison of the ratios from different categories indicates the relative contribution of each category to the total. A representation of this ratio as a decimal number is called a proportion. Multiplying the proportion by 100 produces a percentage that describes the number of observations expected in each category if the total frequency equals 100.

EXAMPLE CA-3.1. Ratios, proportions, and percentages for PPC in the Canadian Auto data. The following table presents ratios, proportions and percentages for selected Canadian Auto data.

PPC	Frequency	Ratio	Proportion	Percentage
0	7	7/53	0.132	13.2
1	6	6/53	0.113	11.3
2	5	5/53	0.094	9.4
3	16	16/53	0.302	30.2
4	1	1/53	0.019	1.9
5	1	1/53	0.019	1.9
6	3	3/53	0.057	5.7
7	14	14/53	0.264	26.4
TOTAL	53	53/53	1.000	100.0

In this sample, for example, one firm manufactures rubber products as the primary output (PPC = 4). This represents 1 of 53 firms, or 0.019 of the total (all these firms manufacture parts only for new vehicles). More firms, 16 of 53, manufacture body parts than any other primary product. This group represents 0.302 of the total.

In the special case of a classification scheme comprising three categories, a distinctive visual representation of the proportion or percentage information can be useful. This diagram, called a ternary diagram, is particularly popular in physical and population geography. The ternary diagram is an equilateral triangle in which each side measures the percentage (or proportion) of observations in the relevant category. Lines parallel to one of the sides are drawn as shown in Figure 3.1. The sum of the percentages from the three categories must equal 100 since the three categories are mutually exclusive and collectively exhaustive. Following the appropriate line for each category produces some intersection point inside (or on the boundary) of the triangle. The triangle contains all possible combinations of the three percentage values, locating each sample in some "sample space." The location of several samples plotted on one diagram can be compared, or one sample can be compared to one or more ideal points. Figure 3.2 show the ternary diagrams for two applications. The top diagram in Figure 3.2 summarizes soil textures in terms of the sand, silt, and clay content. By dividing the body of the triangle into four smaller identical triangles, the relative concentration of the three grain sizes can be examined. For instance, samples that locate in subtriangle I are more than 50% sand, with the remainder silt and clay. Similarly, samples in subtriangle IV are mostly silt, whereas sub-triangle II samples are mostly clay. Samples in subtriangle III are more evenly divided among the three components. As an example, a soil that is 20% sand, 70% silt, and 10% clay would be located at point A in subtriangle IV. This is as expected given the dominance of silt. This diagram emphasizes that since the sum

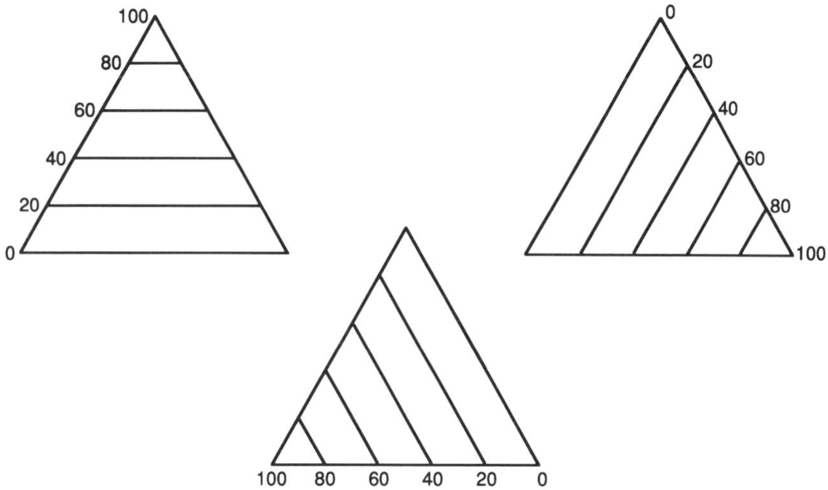

Figure 3.1: COMPONENTS OF A TERNARY DIAGRAM

of the three components must equal 100%, plotting two of the components automatically determines the third.

The bottom diagram in Figure 3.2 shows a ternary diagram for population studies. Here the three sides of the diagram are labelled by the age cohorts of dependents (say, less than 15 years of age), retired (perhaps greater than 64 years of age), and working (15 to 64 years old). A population that is 20% retired, 50% dependent, and 30% working would be located at point A, just on the boundary between subtriangles I and III. In this diagram, population in either subtriangles I or IV would have a minority of people supporting the majority, while subtriangle II represents the case of a majority of the population being workers.

3.2. Histograms

Data may be described in a useful way on the basis of their distribution by constructing a histogram. In fact, statisticians have identified and studied hundreds of different ideal variable distributions; four of these distributions will be examined in some detail in subsequent chapters of this book. With very small sample sizes ($n < 10$) a pattern frequently fails to materialize or is difficult to demonstrate when the observed values are organized in distributional form. However, a few statistics exist that are sensitive enough to detect information in sample sizes as small as three. Usually once n is beyond 20, this sort of graphic representation tends to be illuminating. The paramount question that is raised here asks how histograms can be constructed in an objective fashion, so that patterns that are uncovered are not merely an

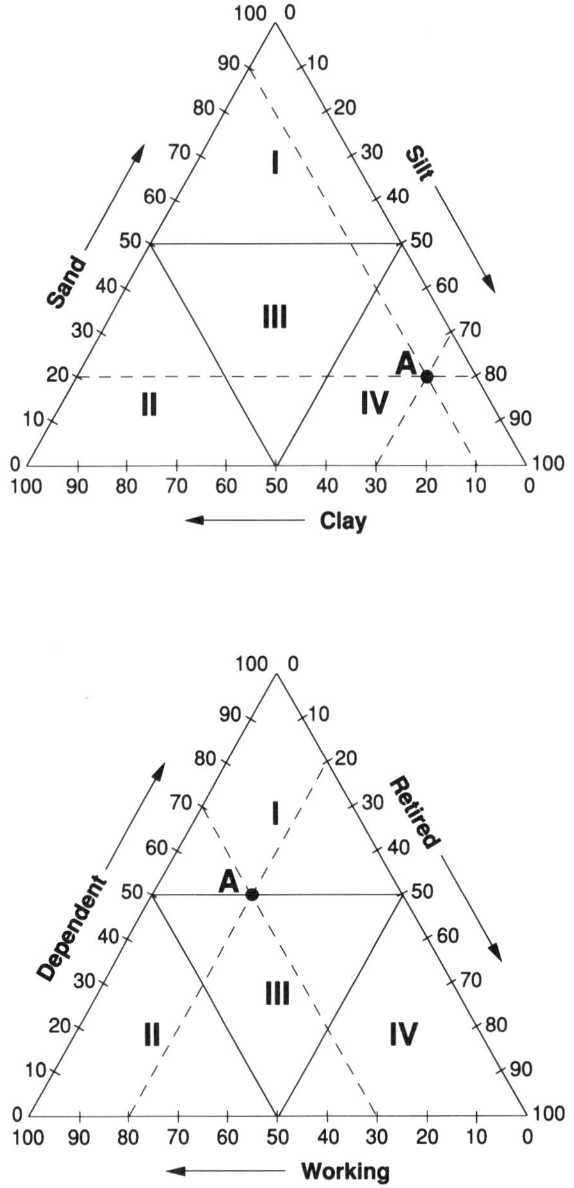

Figure 3.2: (TOP) TERNARY DIAGRAM OF SOIL TEXTURES; (BOTTOM) TERNARY DIAGRAM OF POPULATION AGE COHORTS

artifact of the construction technique. The resulting histogram is an indispensable statistical tool, for when it is based upon a representative sample, it will reveal a frequency distribution shape that is very similar to what would be found in the population from which the sample was drawn.

The **histogram** is either a dot or line graph, or a bar graph representation of an entire set of sample data. Some statisticians distinguish among the different graphic representations; we do not. In practice the choice of graphic form varies, although almost always a bar graph representation is published, while a dot or line graph representation is used on a day-to-day basis. Interval boundaries or important points (perhaps interval midpoints or the actual values for a discrete variable) along the X-axis are established, and the frequency of (or number of times) observations falling into these intervals at these points is measured along the Y-axis. Since discrete data and nominal scale data, especially the multinomial or polychotomous case, can assume only a limited number of values, the dot or line graph is a particularly useful means of portraying these data. In the case of a polychotomous classification, however, the ordering of categories along the X-axis is irrelevant. The bar graph is appropriate for data measured at the interval/ratio scale. Guidelines for the construction of this latter type of histogram follow.

1. Between 5 and 12 intervals or categories (some psychologists suggest seven plus or minus 2) should be adopted, with this number being denoted by k, although for very large data sets many more intervals may be desirable [usually, but not always, the most suitable number is given by the formula $k = 1 + 3.3 \log_{10}(n)$]. Except in the case of a variable having less than 5 unique values, fewer than 5 intervals could result in a loss of too much detail, while more than 12 intervals could conceal pattern and hamper understanding.

2. In the absence of more preferable criteria, interval width can be established by dividing the difference between the maximum and minimum variable values by the number of intervals selected, with some attention being paid to convenience by making numerical adjustments that facilitate handling of the resulting width (rule of thumb: interval boundaries should be measured to one decimal place beyond that of the data).

3. Intervals must be uniform in width in order to avoid a visual picture that can be very misleading. Open-ended intervals should be avoided.

4. Intervals must be mutually exclusive and collectively exhaustive, so that all observed values are assigned to one and only one of the intervals.

5. The vertical axis should display frequency, with the frequency of

interval/point i being denoted by f_i. The horizontal axis should display the interval boundaries/salient points.

6. Both axes should be labelled and the graph should bear a title.

Sometimes the bar graph that can be constructed with interval/ratio data is converted to a dot or line graph by locating frequencies at the midpoints of their respective intervals.

Two types of histogram can be constructed, absolute frequency and relative frequency. The first is based upon the absolute frequencies of intervals or category points, with the sum of the frequencies equaling the sample size:

$$\sum_{i=1}^{k} f_i = n.$$

The second is based on relative frequencies with the sum of the relative frequencies equal to 1.00.

EXAMPLE OR-3.1. Constructing a frequency distribution for AREA in the Ontario River Basin data.

The range for AREA is from approximately 199 to 6480. Using the formula for k to suggest an appropriate number of intervals (n = 28) yields:

$$k = 1 + 3.3 \log_{10}(28) = 5.8 \text{ or } 6.$$

The appropriate interval width for a histogram with six intervals is (6480 - 199)/6 or 6281/6 = 1046.83. However, an interval width of 1000 would be much more convenient. Adopting such an interval width requires seven intervals, rather than six. The following table presents the data required to produce the histogram in Figure 3.3. This shape can best be described as having two distinct peaks, although the magnitudes of these peaks are not equal (the first peak is 12 units high, while the second peak is 5 units high). Caution is necessary when examining histograms since secondary peaks in graphs drawn using sample data may be misleading; that is, the secondary peaks may not exist in the population from which the sample was drawn.

EXAMPLE PR-3.1. Constructing a frequency distribution for the density of sugarcane production (DSGR) in Puerto Rico.

In this case, the range of these data is 30.0 - 0.0 = 30.0. With n = 73, the suitable number of intervals to use for constructing a histogram is $k = 1 + 3.3 \log_{10}(73) = 7.1 \approx 7$. The appropriate interval size is 30.0/7 = 4.286 or 4,

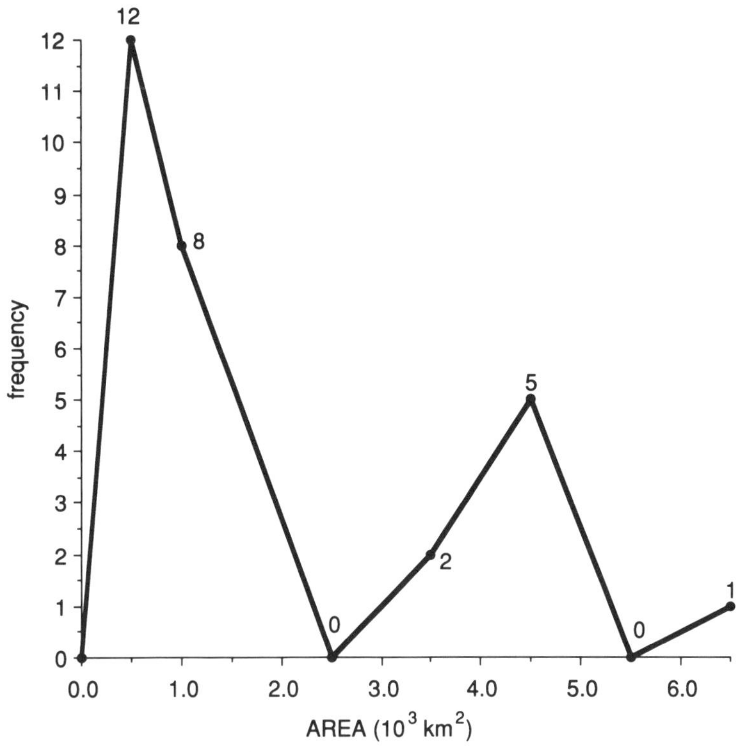

Figure 3.3: HISTOGRAM FOR AREA FROM THE OR DATA SET

Interval	Frequency	Range	Midpoint
0 to 999	12	(0, 999)	(999 + 1)/2 = 500
1000 to 1999	8	(1000,1999)	(2999 + 1)/2 = 1500
2000 to 2999	0	(2000,2999)	(4999 + 1)/2 = 2500
3000 to 3999	2	(3000,3999)	(6999 + 1)/2 = 3500
4000 to 4999	5	(4000,4999)	(8999 + 1)/2 = 4500
5000 to 5999	0	(5000,5999)	(10999 + 1)/2 = 5500
6000 to 6999	1	(6000,6999)	(12999 + 1)/2 = 6500

producing midpoints of {2, 6, 10, 14, 18, 22, 26, 30}. One additional interval emerges due to the error affiliated with rounding 4.287 to 4. The resulting histogram appears in Figure 3.4. Unlike Figure 3.3, this histogram shows high production (large values of DSGR) in a few places, with the majority of places (39 of 73) producing in the first interval. In other words, there is a very high frequency in the first interval with decreasing counts in the following intervals. This distribution resembles a reversed J

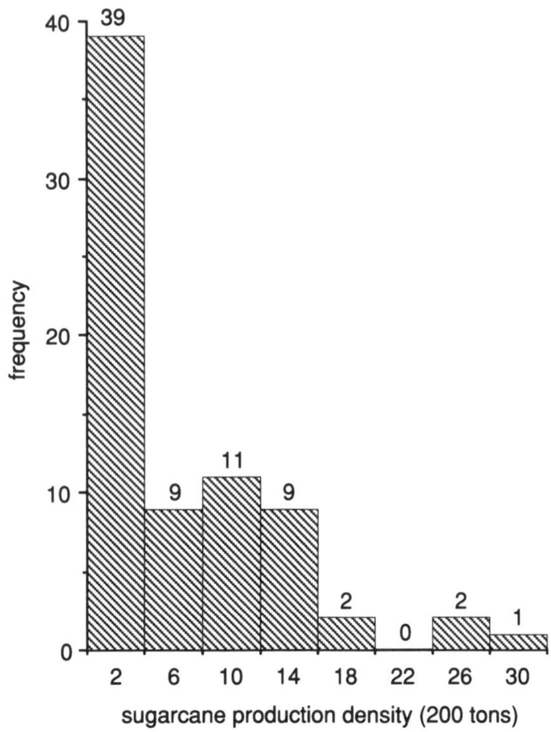

Figure 3.4: HISTOGRAM FOR DSGR FROM THE PR DATA SET

in shape (see Figure 3.12).

EXAMPLE CA-3.2. Constructing histograms for postal codes (PS1), production niches (PPC), and SALES for the Canadian Auto data.

Unlike Examples PR-3.1 and OR-3.1, these variables are measured at the nominal scale. Working first with PS1, let one category in the histogram represent each of the four postal codes represented in the data set (K, L, M, N). A count of the 53 firms shows 2 in Region K, 12 in Region L, 13 in Region M, and 26 in Region N. Since the data are nominal, the order in which the categories are displayed is irrelevant and there are no ranges or midpoints to calculate. The resulting histogram appears in Figure 3.5. One should note that with four categories the histogram indicates a marked concentration in Southwestern Ontario (N), and a noticeable absence of activity in Eastern Ontario (E). The other two regions together contain as many firms as Region N. Together with Figure 1.3 (map of Southern

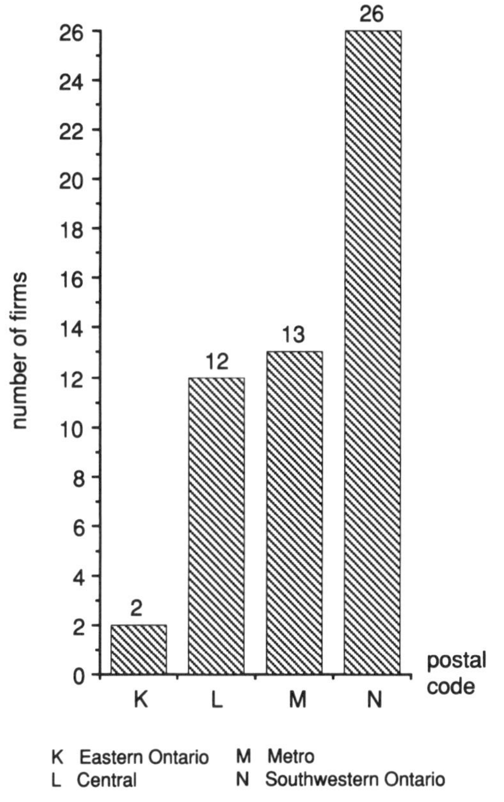

Figure 3.5: HISTOGRAM FOR PS1 FROM THE CA DATA SET

Ontario), the small number of firms in Region K is even more striking given the large area (however, it is largely a rural area). At the same time, the fact that Metropolitan Toronto (M), comprising a relatively small area, contains half as many firms as Region N suggests a different type of concentration than that suggested by the histogram alone.

Assume that rather than four regions, only two new regions ere of interest: metropolitan (comprised of Region M) and nonmetropolitan (comprised of Regions K + L + N). The frequencies in these two new regions are 40 for nonmetro and the original 13 for metro. Figure 3.6 displays this new histogram. The data now indicate a marked concentration in the nonmetropolitan Toronto portion of Southern Ontario of more than 3 to 1 (40 versus 13). However, there is also some important information contained in Figure 3.5 that is lost in the consolidation of categories. Most importantly, the near absence of firms in Region K is hidden. While reference to the

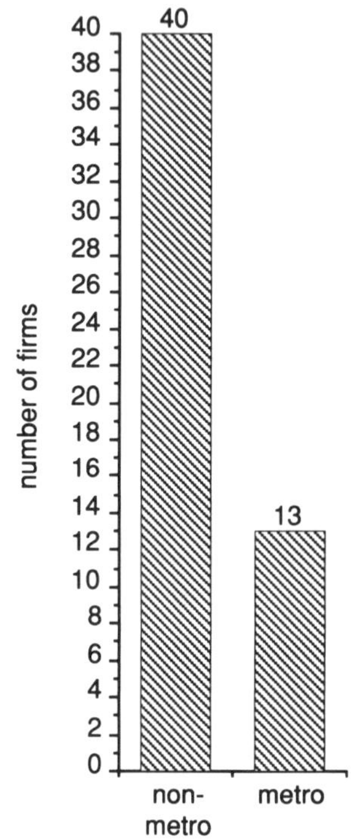

Figure 3.6: HISTOGRAM FOR PS1 FROM THE CA DATA SET BASED UPON CATEGORY CONSOLIDATIONS

map still suggests the importance of having almost one-third of the firms in the relatively small area of Metro Toronto, the distribution in the surrounding regions is lost to the reader of the histogram.

Using the market niche data from Example CA-3.1 produces the histogram in Figure 3.7. This histogram visually displays the frequency counts calculated before in Example CA-3.1. There is a distinct concentration of firms in categories three (body parts) and seven (other). The number of firms in the other niches ranges from one to seven. Next, reducing the number of categories from seven to three, where the three new categories are called "engine and mechanical parts" (Categories 0 + 1 + 2), "body parts" (Category 3), and "everything else" (Categories 4 + 5 + 6 + 7) (these three new groups will be labelled A, B, and C, respectively) results in the histogram appearing in Figure 3.8. The effect of the category consolidation is clear. Whereas Figure 3.7 shows a marked concentration in two of the seven groups, Figure 3.8 suggests an industry that is rather evenly

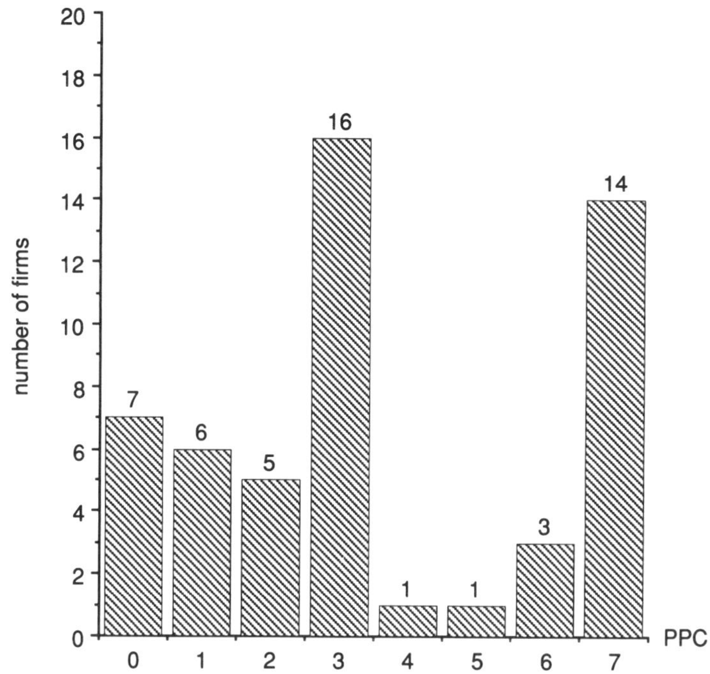

Figure 3.7: HISTOGRAM FOR PPC FROM THE CA DATA SET

distributed. The businesses of the firms certainly have not changed, only the manner in which the businesses have been grouped.

Finally, the variable SALE provides an example of a data set for which constructing the histogram proves difficult. In Chapter 2 the suggestion was made that SALE could be treated as either nominal or grouped-ratio. If SALE is treated as ratio, then intervals and midpoints should be constructed. However, since the data are grouped, it is not possible to construct intervals as in Examples PR-3.1 and OR-3.1; either the original groups can be used, or existing groups can be combined to form fewer groups. Suppose that the original groupings are adequate. Then two additional problems remain: the intervals are not of equal width, and the last interval is unbounded, preventing the selection of a midpoint in the usual fashion. The following table lists the intervals, the frequency in each interval, and the calculated midpoint. The midpoint for the last interval is chosen based upon information about the largest firms that is not contained in the original data; in other words, supplemental information from other

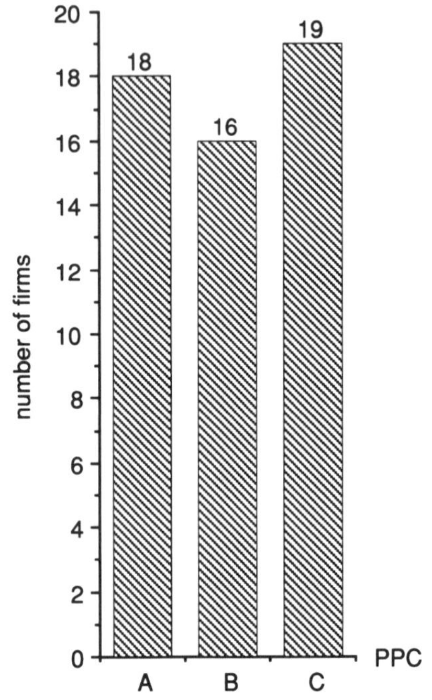

Figure 3.8: HISTOGRAM FOR PPC FROM THE CA DATA SET BASED UPON THE CATEGORY CONSOLIDATION CASE

sources has been used. The resulting histogram is shown in Figure 3.9. Note that given the problems encountered in treating SALE as a grouped-ratio variable, the histogram reverts to treating these data as being measured at the nominal scale using midpoints to identify the categories. As a result the varying interval width is not an issue. Since these data are discrete, the midpoint is calculated as [(low endpoint + high endpoint) + 1]/2, or (range + 1)/2.

Interval	Frequency	Range	Midpoint	
0	0	0		0
1 to 24	0	(1, 24)	(25 + 1)/2 =	13
25 to 99	0	(25, 99)	(124 + 1)/2 =	62.5
100 to 499	1	(100, 499)	(599 + 1)/2 =	300
500 to 999	2	(500, 999)	(1499 + 1)/2 =	750
1000 to 2999	3	(1000, 2999)	(3999 + 1)/2 =	2000
3000 to 4999	5	(3000, 4999)	(7999 + 1)/2 =	4000
5000 to 24999	21	(5000,24999)	(29999 + 1)/2 =	15000
25000 to 49999	7	(25000,49999)	(74999 + 1)/2 =	37500
50000 to ?	14	?	?	75000

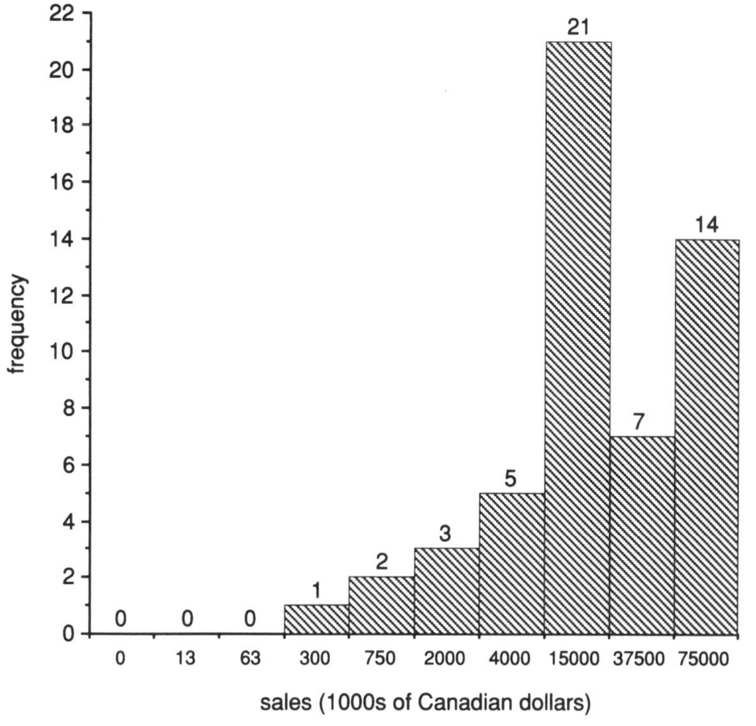

Figure 3.9: HISTOGRAM FOR SALE FROM THE CA DATA SET

One approach to condensing a histogram and thus making it more manageable is to convert the absolute frequencies to relative frequencies, simply by dividing all frequencies by n, so that the Y-axis is more compact. Now the Y-axis ranges from 0 to 1, rather than 0 to at most n. Moreover, the relative frequency is a proportional measure of the frequency of an occurrence. The histogram presented in Example PR-3.1 would appear the same in its relative form, but the Y-axis would range between 0 and 1. Hence, the frequency bar heights respectively would become 0.5342, 0.1233, 0.1507, 0.1233, 0.0274, 0.0, 0.0274, and 0.0137. These values sum to 1 (except for possible rounding error, which cancels out here).

Another functional form for summarizing the distribution of data is the **ogive**, or cumulative frequency distribution. Rather than recording the frequency of observations falling into each interval or occurring at each point, absolute or relative frequencies are replaced by subtotals for each interval/point, usually cumulating successive frequencies from left to right on the X-axis. The number assigned to each interval/point is the sum of the frequencies for that class plus all smaller-valued classes. The X-axis for an ogive is labelled with the true upper limits of intervals, rather than midpoints. Ogives are most useful when one needs to determine the number or percentage of cases falling below or above a certain value; most standard tables of ideal frequency distributions are tabulated in ogive form. Clearly, these sums must be nondecreasing if summing is from left to right, and they must be nonincreasing if summing is from right to left. Several statistics commonly calculated in applied spatial statistical analysis are based upon the ogive.

EXAMPLE OR-3.2. Constructing an ogive from a relative histogram for AREA in the Ontario Runoff data.

The following table presents the data required for constructing the cumulative frequencies curve shown in Figure 3.10. Note that the frequency of zero in the third and sixth intervals (recall Figure 3.3) appear in the cumulative frequency curve as flat areas. Note also that this curve, as required in the definition of an ogive, starts at zero and ends at a value of 1.00.

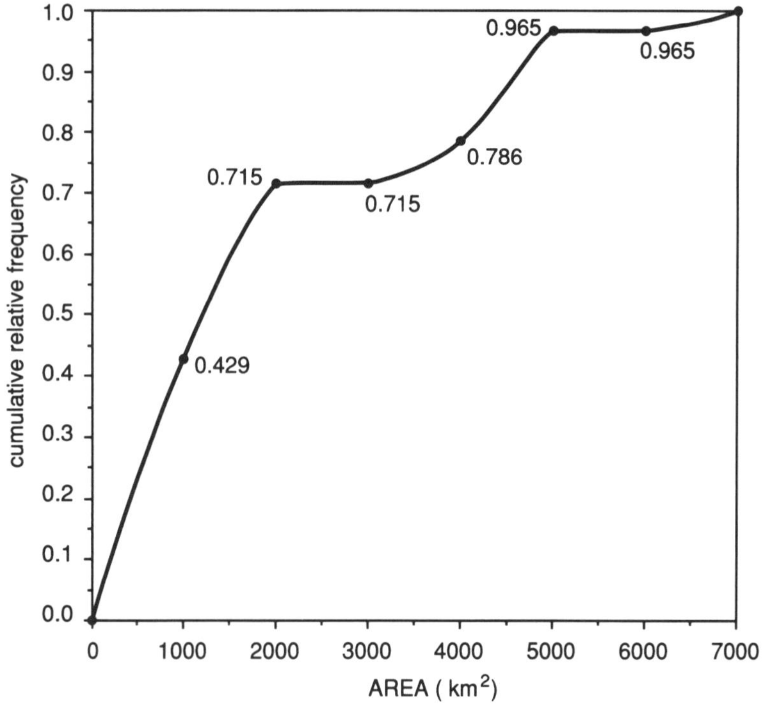

Figure 3.10: CUMULATIVE FREQUENCY CURVE (OGIVE) FOR DSGR FROM THE PR DATA SET

Ratio (n = 28)	Relative frequency	Cumulative relative frequency
12/28	0.429	0.429
8/28	0.286	0.715
0/28	0.000	0.715
2/28	0.071	0.786
5/28	0.179	0.965
0/28	0.000	0.965
1/28	0.035	1.000

EXAMPLE PR-3.2. Constructing an ogive from a relative histogram for the density of sugarcane production (DSGR) in Puerto Rico.

The following table presents the data required for producing the cumulative frequency curve shown in Figure 3.11. Note that, as before, the two cases with zero absolute frequencies (recall Figure 3.4) appear as flat areas. Note also that due to rounding error the cumulative frequency column sums to 0.999 rather than 1.000.

Ratio (n = 28)	Relative frequency	Cumulative relative frequency
39/73	0.534	0.534
9/73	0.123	0.657
11/73	0.151	0.808
9/73	0.123	0.931
2/73	0.027	0.958
0/73	0.000	0.958
2/73	0.027	0.985
1/73	0.014	0.999
0/73	0.000	0.999

Figure 3.12 shows the ogive or cumulative frequency distributions corresponding to five ideal frequency distributions encountered in geographic research. The cumulative frequency distribution corresponding to the normal frequency curve and sinusoidal frequency curve (note that the bimodal distribution encountered in EXAMPLE OR-3.1 is a sinusoidal type of curve) have been illustrated by example. In Chapter 4, Figure 4.2, the special cumulative frequency distribution called the **Lorenz curve**, is used to test the proportionality (or the uniformity) of the distribution of income over a population. Other graphic portrayals of data are discussed in standard elementary statistics textbooks. But, for the most part those commonly encountered in spatial analysis have been reviewed here.

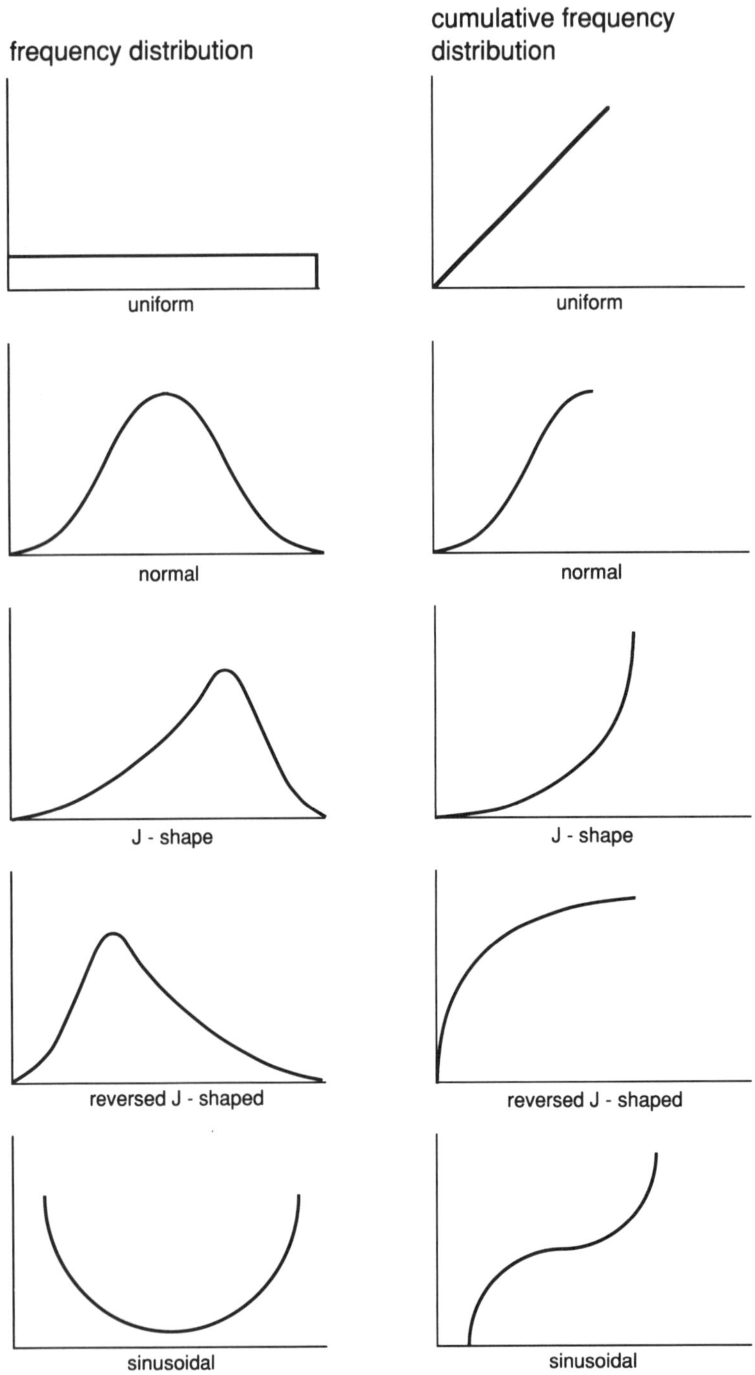

Figure 3.11: GENERALIZED SKEWED DISTRIBUTIONS

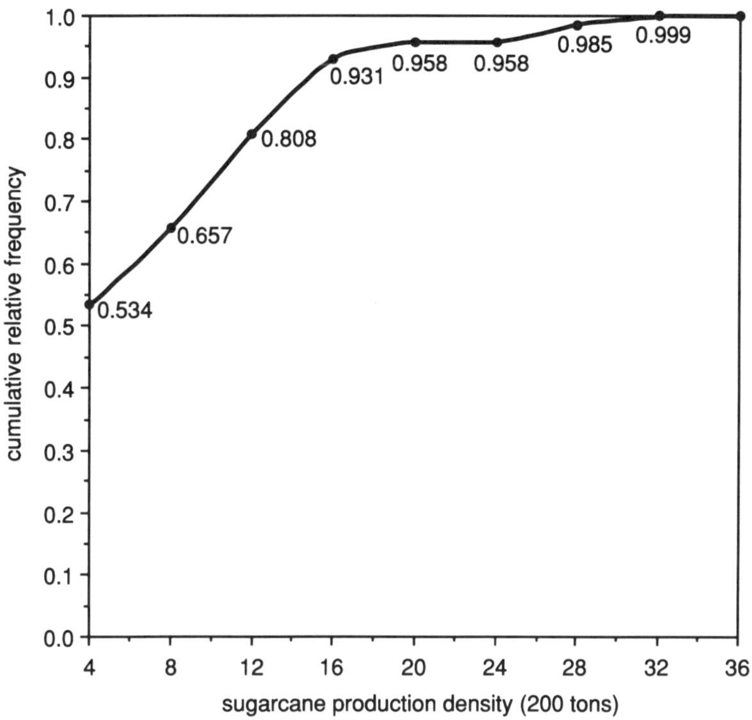

Figure 3.12: CUMULATIVE FREQUENCY CURVE (OGIVE) FOR AREA FROM THE OR DATA SET

3.3. The location quotient: one use of percentage data in spatial analysis

Ratio, proportion, and percentage data can be used to construct a measure of the degree of specialization/diversification or concentration of the categories of a multichotomous variable in an areal unit or region. Such a measure must refer to some assessment of the overall geographic distribution of these categorical data across areal units in the region. The proportions refer to relative frequencies of the categories of this variable, indicating which categories tend to locate predominantly in an areal unit and which tend to locate without regard to the underlying geographic distribution. This measure, which probably is the best known of those in this class, is called a **location quotient**, and may be defined as follows:

LQ = (% in category k in areal unit i)/(% in category k in all areal units),

which is a ratio involving percentages. This index shows the extent to which each of a set of areal units departs from the overall proportion, allowing for the comparison of some region's share of a particular category with the aggregate total for a map. A location quotient equal to 1 means the areal unit has exactly the same relative frequency for a category as is found across the entire map. A location quotient greater than 1 indicates a category that is overrepresented in the areal unit and hence has a relative concentration in the areal unit. The higher the LQ value, the greater the degree of concentration of a phenomenon in the areal unit in question. A location quotient less than 1 signifies underrepresentation of a category in the areal unit. This measure can be used to compare different categories for the same areal unit, or the same category for a given areal unit but at different points in time.

EXAMPLE PR-3.3. Calculating location quotients for the coastal lowlands category of the LND dichotomous variable across the agricultural administration regions for the Puerto Rico data.

Administrative region	Location quotient
San Juan	(12/19)/(42/73) = 1.10
Arecibo	(7/11)/(42/73) = 1.11
Mayaguez	(9/16)/(42/73) = 0.98
Ponce	(10/14)/(42/73) = 1.24
Caguas	(4/13)/(42/73) = 0.53

The results appearing in this example suggest that, in comparison with the ratio for the island as a whole, the percentage of coastal lowland municipios contained in the (a) Mayaguez Region is approximately equivalent, (b) Caguas Region is substantially less, (c) Ponce Region is noticeably greater, and (d) San Juan and Arecibo Regions is slightly more.

LQ seems to be quite an appealing measure, since it is independent of scale. It allows observations made for different situations to be compared, since they are standardized by their overall map incidences; departure from unity will mean the same, regardless of the phenomenon involved. Unfortunately, LQ does suffer from some weaknesses. Because this measure is based upon count data and is centered about unity, under-representation is compressed into the interval (0, 1), while overrepresentation is unrestricted in the interval (1, ∞). Studies also have found that LQ is highly sensitive to the size and shape of the areal units into which a surface has been partitioned. Except for special circumstances, this measure seems to be inferior to more traditional statistical manipulations (especially the z-score transformation, which will be discussed in Chapter 7).

3.4. Summary

It is important to realize that early in a project a small amount of time spent analyzing the data using techniques like those described in this chapter (and Chapter

4) can pay large dividends later on. All too frequently the urge to begin a project with the more sophisticated techniques described in Parts II and III proves to be irresistible. This is most unfortunate. The elementary material in the first three chapters is not first in the book only because it is easier to master than what follows. This material is first because it represents the logical sequence of steps a properly constructed analysis should follow. A spatial analyst must first completely understand the basic nature of the data before venturing into the techniques presented in the last two sections. Failure to gain this basic understanding can result in errors that otherwise would be avoided. In addition, from a practical standpoint, the results from an application of the elementary techniques covered in this first section (Chapters 1 through 5) may well provide adequate answers, saving the time, effort, and expense of additional analysis. A "good" statistical analysis is not necessarily the analysis that properly applies the most sophisticated techniques, but is rather the analysis that properly applies the most appropriate techniques.

3.5. Problems

1. Construct an absolute and a relative histogram for the polychotomous variable SPC from the Canadian Automotive data set in Table 1.1-A of the appendices.

2. Calculate ratios, proportions, and percentages for the three variables in the following table. Construct a ternary diagram using the variables quartz, feldspar, and lithic fragments.

TABLE I

NUMBER OF MINERAL GRAINS IN ALLUVIAL SAND SAMPLES FROM A BRAIDED RIVER THAT DRAINS VARIOUS ROCK TYPES

Sample	Number	Quartz	Feldspar	Lithic fragments
Plutonic rocks	1	66	128	26
	2	40	112	28
	3	65	76	47
	4	46	84	10
Volcanic rocks	5	2	32	156
	6	13	60	138
	7	3	43	85
	8	4	36	160

3. Construct an absolute histogram for the density of tobacco production and an absolute histogram for the density of farm land. Which of these two variables would benefit more from the use of a relative histogram?

4. Construct an ogive for the density of farm land.

5. Construct a relative histogram for the data presented in the following tables:

TABLE II

AVERAGE LENGTH (IN KILOMETERS)
OF JOURNEYS TO WORK IN CANADIAN CITIES

1.86	2.37	3.07	3.68	4.32	5.01	2.07	2.73
3.41	3.69	4.32	5.07	2.24	3.03	3.41	3.77
4.46	5.14	2.37	3.07	3.55	3.91	4.93	5.66

TABLE III

240 SAMPLES OF 8 HOUSES IN A SUBDIVISION
TO DETERMINE OWNERSHIP DURING A 13-YEAR PERIOD

Ownership composition: Number of houses occupied by the original owner	Ownership composition: Number of houses that have undergone at least one ownership change	Number of surveys that have the corresponding ownership composition
8	0	5
7	1	2
6	2	28
5	3	43
4	4	62
3	5	61
2	6	29
1	7	6
0	8	4

Note: The subdivision actually had 120 houses occupied by the original owners and 130 not occupied by the original owners.

TABLE IV

NUMBER OF BRAIN TUMORS
IN THE RZESZOW DISTRICT OF POLAND

County	Population (in 1000s)	Number of tumors per 100,000
Brzozow	70	0.000
Debica	110	3.636
Gorlica	83	10.843
Jaroslaw	102	5.882
Jaslo	101	3.960
Kolbuszowa	60	1.667
Krosno	101	1.980
Lesko	17	11.765
Lubaczow	40	2.500
Lancut	91	6.593
Mielec	73	5.479
Nisko	67	2.985
Przemysl	88	6.818
Przeworsk	57	3.509
Rzeszow	182	7.143
Sanko	56	7.143
Tarnobrzeg	67	4.478

6. Replicate the results for Example PR-3.3 but using the interior highlands category of the LND variable.

7. Calculate location quotients for the individual agricultural administration regions, based upon the U-R variable of the Puerto Rico data set.

3.6. Additional readings

In addition to the texts listed in Chapter 1,

Blalock, H. (1979) *Social Statistics*, 2nd rev. ed. New York: McGraw-Hill.

Chambers, J., W. Cleveland, B. Kleiner, and P. Tukey. (1983) *Graphical Methods for Data Analysis*. Belmont, Calif.: Wadsworth Publishing.

Cleveland, W. (1985) *The Elements of Graphing Data.* Monterey, Calif.: Wadsworth Publishing.

Reynolds, H. *Analysis of Nominal Data.* SAGE number 7.

Hildebrand, D., J. Laing, and H. Rosenthal. *Analysis of Ordinal Data.* SAGE number 8.

Tufte, E. (1983) *The Visual Display of Quantitative Information.* Cheshire, Conn.: Graphics Press.

CHAPTER 4

DESCRIPTIVE CLASSICAL STATISTICS: CENTRAL TENDENCY, DISPERSION, AND RELATIONSHIP

As was mentioned in Chapter 2, there are many possible statistics that one can calculate for a given data set, each having different properties, advantages, and disadvantages. Attention in this book is restricted to those classical statistics that researchers have found to be most useful when describing data by calculating summary data measures. Three types of statistics that often prove to be illuminating are ones that typify or characterize average, middle, or central tendency features of the data, ones that characterize the heterogeneity, scatter, spread, variability, or dispersion of the data, and ones that characterize relationships between pairs of variables in a data set. These specific types of measures furnish a compact description of data and are treated in this chapter.

4.1. Measures of central tendency

One of the important features of data when attempting to characterize them with some single summary statistic is to consider whether or not the associated frequency distribution is symmetric. In this context **symmetry** means that, on a two-dimensional diagram, a location exists somewhere along the horizontal (X-) axis of the frequency distribution such that pictorially the portion of this frequency distribution falling to the right of this location is a mirror image of that portion falling to the left of this location. Accordingly, this location is a midpoint where 50% of the frequency distribution falls to its right, and 50% falls to its left. A second important feature has to do with concentration of data points in the frequency distribution. Data will be most successfully typified by a single statistic if a sole point of considerable concentration, or "bunching up," exists in the frequency distribution, pictorially yielding a peak. When the frequency distribution in question is symmetric, the peak will be at the midpoint. Notable distributions which are symmetric, but do not "bunch up" at a single point, are the uniform and the sinusoidal frequency distributions. Usually then, the midpoint of a symmetric distribution represents a reasonable summary of the central tendency of the data. But how can this central tendency be measured, especially when a frequency distribution deviates from symmetry and/or a single point of concentration?

Three important measures of central tendency can be computed and have been studied extensively in classical statistics. They are the mean, the median, and the mode. Each of these three measures has certain advantages and disadvantages, has special mathematical properties, and is more appropriate in selected circumstances. The **arithmetic mean**, usually called the mean and denoted by $\bar{x}$ (x-bar) for a sample and μ (mu) for a population, is the most familiar, is the most commonly used, and historically has been the most important measure of central tendency. It is calculated by summing all the values of some variable in a data set and then dividing by the total number of observations in that data set. In other words, for a sample of size n,

$$\bar{x} = \sum_{i=1}^{n} x_i/n = (x_1 + x_2 + \ldots + x_n)/n$$

while for a population of size N,

$$\mu = \sum_{i=1}^{N} x_i/N = (x_1 + x_2 + \ldots + x_N)/N.$$

Suppose the number line is a seesaw. Let weights that are proportional to data values be placed on this number line at the corresponding positions of these data values. In this context the mean designates that point along this number line upon which these data point weights sit, below which the placement of a fulcrum would cause this number line to be parallel to the ground, or perfectly balanced.

Since the arithmetic operations of addition and division are involved in calculation of a mean, this measure is sensible only when data are at the interval/ratio scale of measurement. If the frequency distribution in question is perfectly symmetric, then the mean is an accurate typifying summary measure of this distribution's midpoint. The most appealing feature of this statistic is that it possesses all four of the desirable statistical properties mentioned earlier (sufficiency, efficiency, consistency, and unbiasedness). Furthermore, a well-developed body of sampling theory has been built on it. One algebraic property displayed by the mean is that the sum of the differences between each data value for a variable and the mean of that variable equals zero. A second property displayed is that the sum of the squared differences between each data value for a variable and the mean of that variable is a minimum compared to the sum of the squared differences from any other value. These two particular properties make the mean an ideal measure for calculating a number of more sophisticated statistics. Finally, the mean of the sampling distribution of any statistic is of special interest, especially when it equals the corresponding population parameter value (the statistical property of unbiasedness).

EXAMPLE PR-4.1. Calculating an arithmetic mean.

Figure 4.1: MAP OF THE ARECIBO AGRICULTURAL ADMINISTRATIVE REGION OF PUERTO RICO

In terms of the Puerto Rico data, the means of the different density variables for the Arecibo Agricultural Administrative Region (see Figure 4.1) are as follows:

$$\bar{x}_{DFRM} = (4.5 + 5.6 + 12.1 + 8.5 + 8.9 + 17.9 + 3.0 + 12.9 + 10.7 + 18.3 + 4.8)/11 = 107.20/11 = 9.75.$$

Variable	Sum	Mean	Variable	Sum	Mean
DCD	436.0	39.64	DCF	93105.3	8464.12
DMLK	742.9	67.54	DTB	58.5	5.32
DSGR	43.9	3.99	DBN	50.4	4.58
			DFAM	29.7	2.70

EXAMPLE OR-4.1. Means for four physical variables from the Ontario Runoff (AREA and RUN) and the Sentinel Glacier (ELA and TEMP) data sets.

Variable	Sum	Mean
AREA	53668.00	1916.71
RUN	8309.00	296.75
ELA	34550.00	1818.42
TEMP	242.60	12.77

If a frequency distribution is asymmetric then its mean tends to be a less useful, and sometimes misleading, measure of central tendency. As the peak becomes more and more extremely located to the right or left of the mean, the mean is increasingly less typifying, and the frequency distribution is said to be **skewed** or has one side (called the tail) that stretches out farther along the number line than does the side containing the peak. A curve is said to be skewed in the direction of this longer tail. Hence, if the peak falls to the right of the mean, then the frequency distribution is negatively skewed, whereas if the peak falls to the left of the mean, then the frequency distribution is positively skewed. Figure 4.2 shows skewed distributions for two variables, AREA and PRECIP from the Ontario River Basin data set. AREA (the top diagram of Figure 4.2) is positively skewed, whereas PRECIP (the bottom diagram of Figure 4.2) is negatively skewed.

In cases of substantial skewness, the median (MD) tends to be a better measure of central tendency than does the mean. The **median** is that numerical value above which half of the observations fall and below which half of the observations fall. This value is obtained by arranging the observations in either ascending or descending order according to the magnitude of values for the variable. If n is odd, then the median equals the value of the observation whose position is $(n + 1)/2$, or $MD = x_{(n+1)/2}$. If n is even, then the median equals the average of those two values occupying positions $n/2$ and $n/2 + 1$, or $MD = (x_{n/2} + x_{n/2+1})/2$.

When a frequency distribution is symmetric, then the mean and the median coincide. Unfortunately, sampling theory for the median historically is not as well developed as that for the mean, although **nonparametric statistical theory** (i. e., statistical theory in which a sampling distribution does not depend upon the frequency distribution of the population being sampled) tends to utilize the median rather than the mean. Furthermore, the median exhibits a poorer performance in terms of the four desirable statistical properties than does the mean. But a second discriminating factor, then, pertains to the measurement scale involved. If data are at the ordinal level of measurement, rather than the interval/ratio level, then the median is a sensible measure, and hence appropriate, whereas the mean no longer is a sensible measure of central tendency. One algebraic property displayed by the median is that the sum of the absolute differences between each data value for a variable and the median of that variable is a minimum. Another property is that the median is not influenced by extreme data values (while the mean is). This property is exploited in the construction of certain spatial statistics.

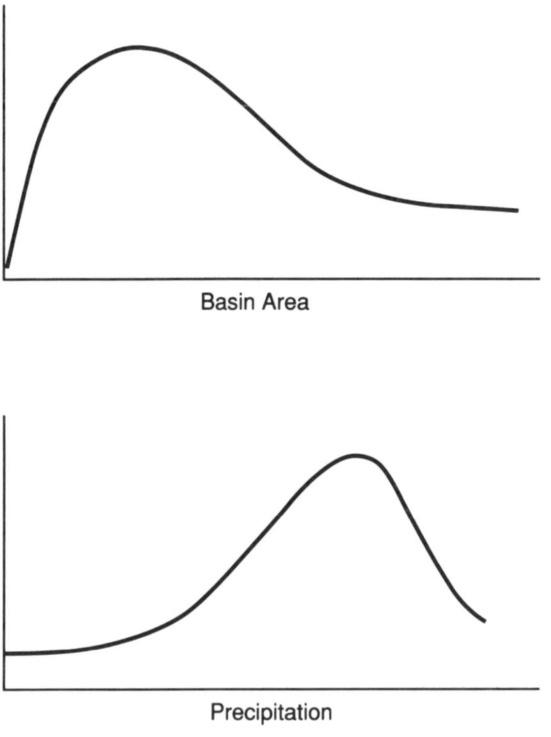

Figure 4.2: GENERALIZED SKEWED FREQUENCY DISTRIBUTION BASED UPON AREA AND PRECIP FROM THE OR DATA SET

EXAMPLE PR-4.2. Determining medians.

In terms of the Puerto Rico data set, the medians of the different density variables for the Arecibo Agricultural Administrative Region are as follows:

ordering of DFRM: 3.0 4.5 4.8 5.6 8.5 8.9 10.7 12.1 12.9 17.9 18.3.

The median is found in ordered position 6, having 5 municipio values greater than it and 5 municipio values less than it, and equals 8.9.

Variable	Median	Variable	Median	Variable	Median
DCD	36.9	DCF	610.4	DBN	0.6
DMLK	40.9	DTB	1.5	DFAM	1.9
DSGR	3.1				

EXAMPLE OR-4.2. Medians for four physical variables from the Ontario Runoff (AREA and RUN) and the Sentinel Glacier (ELA and TEMP) data sets.

Variable	Median	Variable	Median
AREA	1165	ELA	1797.5
RUN	282	TEMP	12.85

If a frequency distribution is measured at the nominal scale, then neither the mean nor the median is an appropriate measure of central tendency. Rather, the **mode**, or most frequently occurring value for discrete variables and most frequently occurring interval for continuous variables, is the appropriate statistic. As the most frequently observed value in the sample it is the most probable, and hence the typifying, measurement in a data set. The mode is the least used measure of central tendency and is severely limited for aggregated data based upon unequal interval widths. If either ordinal or interval/ratio data are studied, then for single-peaked, symmetric frequency distributions the mode coincides with the mean and the median. If the frequency distribution is positively skewed, then the mode falls to the left of the median; if the frequency distribution is negatively skewed, then the mode falls to the right of the median. For these higher levels of measurement, the mode becomes useful when multiple peaks occur. But in general the mode is the least important of the three measures of central tendency, primarily because it almost completely lacks the four desirable statistical properties.

EXAMPLE PR-4.3. Identifying modes.

In terms of the Puerto Rico data, the modes of the two classification variables for the Arecibo Agricultural Administrative Region are, respectively, coastal lowlands (7 of the 11 municipios fall into this category) and rural (7 of the 11 municipios fall into this category).

EXAMPLE OR-4.3. A comparison of means, medians, and modes.

The variables displayed in Figure 4.2 illustrate the effect of a skewed distribution on measures of central tendency. AREA has a mean value of 1916.714, a median of 1095, and a modal value of 350. For this positively skewed variable, the mean is to the right of the median value, which in turn is far to the right of the mode (located at the peak of the distribution). PRECIP has a mean of 31.554, a median of 32, and a modal value of 34. For this negatively skewed variable, the mean is pulled to the left of the median, which in turn falls to the left of the mode.

EXAMPLE 4.1. Relative location of the mean, median, and mode for ideal distributions.

Figure 4.3 shows the relative locations of the mean, median, and mode for several ideal distributions. For each curve, the left-hand distribution represents the situation in the absence of any skewness, while the right-hand distribution shows the effect of skewness. A comparison of these ideal curves with Figure 4.2 confirms, using AREA and PRECIP, the general case described in Figure 4.3. It is important to note the relative movement of the three measures of central tendency, first as the distribution deviates from a bell-shaped curve (compare the left-hand distributions) and second as the distribution becomes skewed (compare each pair of left-hand and right-hand curves).

In Part II of this book the effect on the mean as a distribution deviates from an assumed shape will be an important consideration in determining the usefulness of certain procedures.

Often data are tabulated in a shorthand form, either by displaying distinct values together with the frequency of their respective occurrences or by grouping similar values (those falling into designated intervals of equal size along a number line). These abridged versions of the data facilitate its scanning for pattern identification--sacrificing detail for the sake of clarity--and often allow useful modifications to be made to the formulae used to calculate measures of central tendency. For k distinct values, each having a frequency of f_k, the mean may be calculated as

$$\bar{x} = \Sigma_{i=1}^{k} f_i x_i / \Sigma_{i=1}^{k} f_i = (f_1 x_1 + f_2 x_2 + \dots + f_k x_k)/n$$

while the median and the mode remain the same. For grouped data, in the absence of information about how data values are positioned within intervals, the least restrictive assumption to make is that they are uniformly distributed (the frequency distribution is flat) within each interval. Accordingly, x_i becomes the midpoint of the interval, defined as half of the difference between an interval's endpoint values. In this context, then, the mean uses the foregoing aggregate data formula, the median is defined as

$$\begin{matrix}\text{lower class limit} \\ \text{of the interval m} \\ \text{containing the} \\ \text{median}\end{matrix} + \left[\frac{\text{interval width}}{\begin{matrix}\text{frequency of class interval m} \\ \text{which contains the median}\end{matrix}}\right](n/2 - \Sigma_{i=1}^{m-1} f_i)$$

and the mode score for the modal class is defined as

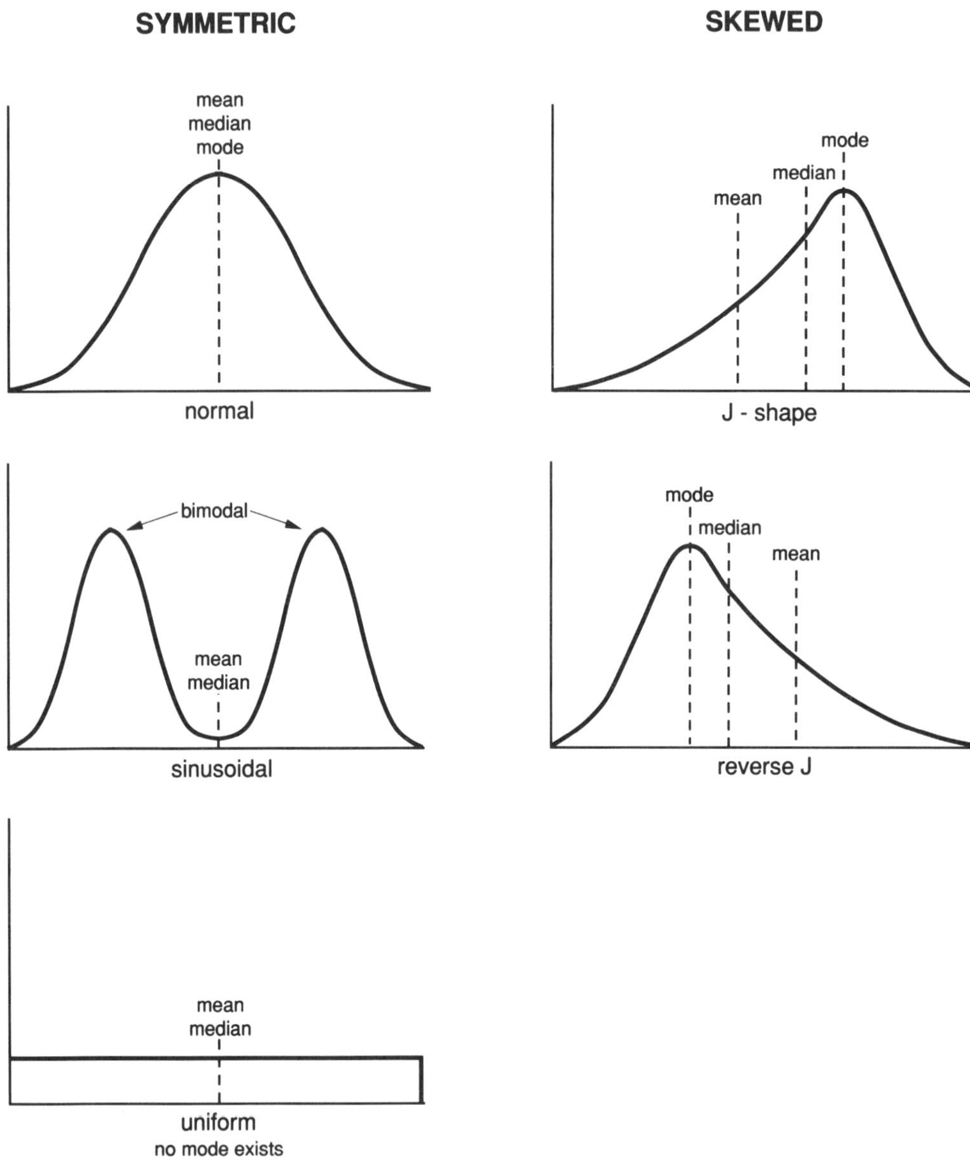

Figure 4.3: RELATIVE POSITIONING OF MEASURES OF CENTRAL TENDENCY FOR COMMONLY ENCOUNTERED IDEALIZED FREQUENCY DISTRIBUTIONS

$$\text{lower class limit of the interval m containing the mode} + \left[\frac{f_m - f_{m-1}}{(f_m - f_{m-1}) + (f_m - f_{m+1})}\right](\text{interval width})$$

Note that the assumption of a uniform distribution of data points within intervals can introduce measurement error into these latter calculations; hence, the use of raw data is preferable to the use of grouped data. But this sort of approximation tends to yield reasonable calculations and is heavily employed in the calculation of certain spatial statistics.

EXAMPLE CA-4.1 Calculation of the grouped mean for the variable SALE from the Canadian Automotive data set.

The problems encountered in treating certain variables from the CA data set as grouped-ratio measurements already have been discussed (recall Example CA-3.2). If SALE is treated as a grouped-ratio measurement, then a group mean can be calculated. The following table presents the calculations required to determine such a mean. The midpoint calculations are adopted from Example CA-3.2. The first three intervals with frequencies of zero are ignored.

Interval	Frequency	Midpoint	Product
100 to 499	1	300	300.00
500 to 999	2	750	1500.00
1000 to 2999	3	2000	6000.00
3000 to 4999	5	4000	20000.00
5000 to 24999	21	15000	315000.00
25000 to 49999	7	37500	262500.00
50000 to ?	14	75000	1050000.00

The column sum of the products is 1,655,300; the grouped mean is equal to 1,655,300/53 = 31,232.08. Recall that the midpoint of the last interval, since the interval is unbounded, is set at 75,000.00. This value was established using additional supplemental information not contained in the CA data set. Given the frequency in this interval, and the large value of the midpoint, this estimated value has a very large impact on the final value of the grouped mean. To illustrate this effect, first assume that the estimated midpoint for the last interval is 100,000, and then assume that it is 50,000. The resulting grouped means would be 37,835.85 and 24,628.30, respectively (the corresponding product sums would be 2,005,300 and 1,305,300). Clearly, given the uncertainty of the midpoint value, extreme caution is warranted in any subsequent analysis using this grouped mean. Since the interval widths of SALE are not equal, the grouped median and grouped mode should not be calculated using the formulae presented above. Such calculations are possible but unreliable and hence are not presented here.

4.2. Measures of dispersion

Another important feature of data to consider when attempting to summarize them is the degree of homogeneity displayed by the values. A perfectly homogeneous variable will have a single numerical value and hence be a constant. Perfectly heterogeneous data can have a flat frequency distribution (no peak exists) or perhaps be bimodal such that only two opposite extreme values exist that are positioned equidistant from the central tendency location. Consequently, additional valuable information (beyond that obtained with a measure of central tendency) about data can be gleaned from some measure of the spread of these data along a number line. Measures of dispersion capture this information. The most popular of these include the range, various quantiles, the variance, the standard deviation, and the mean deviation.

The **range** is a very simple and unsophisticated statistic that is defined as the difference between the largest and the smallest numerical values in a data set, or

$x_{max} - x_{min}$ for continuous data and $x_{max} - x_{min} + 1$ for discrete data.

Often this calculation distinction between continuous and discrete variables is not made. The range is quite limited in that it takes into account only the two extreme values of a data set, while ignoring all other values. More than likely neither of these two extreme values is near any peaks or concentrations in the frequency distribution, meaning that they are relatively rare values. Also, the numerous variation possibilities for the middle (n - 2) observations are overlooked. This measure can be useful when constructing sample frequency distributions or when extreme values are actually the focus of study (such as water levels for irrigation and flood-control projects).

EXAMPLE PR-4.4. Computing ranges. The ranges of the different density variables for the Arecibo Agricultural Administrative Region are

DFRM: $x_{max} - x_{min} = 18.3 - 3.0 = 15.3$ (see Example PR-4.2)

Variable	Min	Max	Range	Variable	Min	Max	Range
DCD	25.6	53.4	27.8	DCF	45.1	48874.9	48829.8
DMLK	0.0	368.6	368.6	DTB	0.0	23.7	23.7
DSGR	0.0	10.9	10.9	DBN	0.1	25.6	25.5
				DFAM	0.5	7.2	6.7

EXAMPLE OR-4.4. Ranges of four physical variables from the Ontario Runoff (AREA and RUN) and Sentinel Glacier (ELA and TEMP) data sets.

Variable	Min	Max	Range	Variable	Min	Max	Range
AREA	199.0	6480.0	6281.0	ELA	1600.0	2080.0	480.0
RUN	97.0	471.0	374.0	TEMP	11.5	14.2	2.7

The limited amount of information about dispersion extracted by a range can be enhanced by modifying this statistic in such a way that additional observations are used to calculate the measure. The next step is to move to quantiles, which are based upon selected partitionings of the rank ordering of the data values. Just as the median divides a data set into two equal sets of observations, terciles divide a data set into three equal sets, quartiles divide a data set into four equal sets, quintiles divide a data set into five equal sets, and percentiles divide a data set into 100 equal sets. In general, a **quantile** divides the data set into q equal subsets of observations. The fundamental notion here is that these quantile values are not the extremes and so should vary less from sample to sample. Although any quantile partitioning can be used, and any pair of quantiles compared, customarily the **quartile deviation** is selected. It may be defined as

(the value of the third quartile - the value of the first quartile)/2.

This measure of dispersion tends to be useful when comparing data from dramatically different frequency distributions; it has been used in geography to compare precipitation and temperature data from different weather stations. Unfortunately it is not suitable for interpreting the variability of a single frequency distribution. Thus, while more sensitive to data congeries than is the range, this measure of spread still fails to use all the information provided by each observation in a data set and so is inadequate in that it lacks the four desirable statistical properties.

EXAMPLE PR-4.5. Calculating an quartile deviation.

Consider the different density variables for the Arecibo Agricultural Administrative Region from the Puerto Rico data set. Recalling the ordering of DFRM in Example PR-4.2, the third quartile value would be set at position 9, having a value of 12.9, and the first quartile value would be set at position 3, having a value of 4.8. Hence, the quartile deviation for DFRM is (12.9 - 4.8)/2 = 4.05.

Variable	Rank 3	Rank 9	Quartile deviation		
DCD	32.7	47.5	(47.5 - 32.7)/2	=	7.40
DMLK	4.9	68.1	(68.1 - 4.9)/2	=	31.60
DSGR	0.1	9.1	(9.1 - 0.1)/2	=	4.50
DCF	227.9	12070.0	(12070.0 - 227.9)/2	=	5921.05
DTB	0.0	10.0	(10.0 - 0.0)/2	=	5.00
DBN	0.3	9.0	(9.0 - 0.3)/2	=	4.35
DFAM	1.1	4.5	(4.5 - 1.1)/2	=	1.70

Measures of dispersion that are based upon some average deviation of all data values better succeed in exploiting the total information contained in a data set (in a fashion parallel to the way the arithmetic mean achieves this goal). In Section 4.1 deviations about the mean summed to zero regardless of the dispersion present in data. In contrast, sums of squared deviations about the mean and absolute deviations about the median respectively produce minimum values when compared to the same calculations using values other than the mean or median. This is a useful property for measures of dispersion to have.

In the first case, the **variance**, denoted by s^2 for a sample and σ^2 (sigma squared) for a population, which is the most common measure of variability, may be defined as the average squared deviation about the arithmetic mean. This measure may be calculated as follows, for a sample of size n:

$$s^2 = \Sigma_{i=1}^{n} (x_i - \mu)^2/n = [(x_1 - \mu)^2 + (x_2 - \mu)^2 + \dots + (x_n - \mu)^2]/n$$

while for a population of size N,

$$\sigma^2 = \Sigma_{i=1}^{N} (x_i - \mu)^2/N = [(x_1 - \mu)^2 + (x_2 - \mu)^2 + \dots + (x_N - \mu)^2]/N.$$

The sample variance embraces the aforementioned four desirable statistical properties. But often the population mean is unknown, so it is replaced by the calculated sample mean. This substitution introduces measurement error into the calculation of the sample variance, modifying its formula to

$$s^2 = \Sigma_{i=1}^{n} (x_i - \bar{x})^2/n = [(x_1 - \bar{x})^2 + (x_2 - \bar{x})^2 + \dots + (x_n - \bar{x})^2]/n$$

which no longer is unbiased. But this statistic can be slightly modified in order to retain its unbiasedness simply by multiplying it by n/(n-1), yielding

$$s^2 = \Sigma_{i=1}^{n} (x_i - \bar{x})^2/(n-1) = [(x_1-\bar{x})^2 + (x_2-\bar{x})^2 + \dots + (x_n-\bar{x})^2]/(n - 1) .$$

This adjustment in the denominator (known as Bessel's correction) to compensate for systematic measurement error that is introduced by using $\bar{x}$ in place of μ is necessary because quite naturally the x_i values in the sample are positioned closer to their sample mean $\bar{x}$ than they are to their population mean μ ($\bar{x}$ is calculated from the set of x_i). Students should realize that these two formulae are used interchangeably in the literature, even though they should not be! Strange calculations sometimes are obtained because one formula is used when the other should have been used. If a student is suspicious about an answer for any reason, he or she should check to make sure the correct version of the variance formula has been used.

Two additional formulae merit attention at this point. First, for ease of calculation, the variance formula can be rewritten as follows:

$$s^2 = [\Sigma_{i=1}^{n} x_i^2/(n - 1)] - (\Sigma_{i=1}^{n} x_i)^2/[n(n - 1)].$$

This calculation formula requires summing both the values and the squared values of a data set and is reported in the literature; it is useful when using a hand calculator. The second formula is for aggregated data, for which the variance formula can be rewritten as follows:

$$s^2 = \Sigma_{i=1}^{k} f_i(x_i - \bar{x})^2/(\Sigma_{i=1}^{k} f_i - 1) = \Sigma_{i=1}^{k} f_i(x_i - \bar{x})^2/(n - 1).$$

Similar modified versions can be written for the formulae of both the population variance and the sample variance when μ is known.

The **standard deviation** is the positive square root of the variance. Accordingly the sample standard deviation is denoted by s, whereas the population standard deviation is denoted by σ. These values may be calculated as follows: for a sample of size n,

$$s = \sqrt{\Sigma_{i=1}^{n} (x_i - \mu)^2/n} = [\Sigma_{i=1}^{n} (x_i - \mu)^2/n]^{1/2}$$

for a sample where the population mean μ is unknown,

$$s = \sqrt{\Sigma_{i=1}^{n} (x_i - \bar{x})^2/(n - 1)} = [\Sigma_{i=1}^{n} (x_i - \bar{x})^2/(n - 1)]^{1/2}$$

while for a population of size N,

$$\sigma = \sqrt{\Sigma_{i=1}^{N} (x_i - \mu)^2/N} = [\Sigma_{i=1}^{N} (x_i - \mu)^2/N]^{1/2}.$$

For the sake of convenience the standard deviation usually is reported for data, since a sum of squared numbers tends to be cumbersome to work with and has units which are the square of the units in the original data. In addition, because the magnitude of the variance tends to be far from the range of the data, it is difficult to interpret, whereas the square root operation essentially removes effects of the squaring of the deviations and returns the standard deviation to the original units of the data. This latter value can be interpreted as a distance measure and as such is of special interest when calculated for the sampling distribution of any statistic. A customary round-off rule invoked here is to retain one more decimal place in the answer than is present in the original data. Although the standard deviation is more interpretable than is the variance, most sampling theory concerning dispersion rests upon the notion of variance.

EXAMPLE PR-4.6. Computing an unbiased standard deviation and variance.

In terms of the Puerto Rico data, the different density variables for the Arecibo Agricultural Administrative Region lead to the following computations:

DFRM: x_i	$\bar{x}$	$(x_i - \bar{x})$	$(x_i - \bar{x})^2$
4.5	9.75	-5.25	27.5625
5.6	9.75	-4.15	17.2225
12.1	9.75	2.35	5.5225
8.5	9.75	-1.25	1.5625
8.9	9.75	-0.85	0.7225
17.9	9.75	8.15	66.4225
3.0	9.75	-6.75	45.5625
12.9	9.75	3.15	9.9225
10.7	9.75	0.95	0.9025
18.3	9.75	8.55	73.1025
4.8	9.75	-4.95	24.5025

$$
\begin{aligned}
s^2_{DFRM} &= (27.5625 + 17.2225 + 5.5225 + 1.5625 + 0.7225 + 66.4225 \\
&\quad + 45.5625 + 9.9225 + 0.9025 + 73.1025 + 24.5025)/(11 - 1) \\
&= 273.0075/10 = 27.30
\end{aligned}
$$

$$
s_{DFRM} = \sqrt{273.0075/10} = 5.23
$$

Variable	Standard deviation	Variance
DCD	8.82	77.80
DMLK	103.58	10728.50
DSGR	4.28	18.33
DCF	15713.61	246917461.00
DTB	7.60	57.69
DBN	7.79	60.70
DFAM	2.04	4.18

EXAMPLE OR-4.5. Standard deviation and variance computations for four physical variables from the Ontario Runoff (AREA and RUN) and the Sentinel Glacier (ELA and TEMP) data sets.

Variable	Standard deviation	Variance
AREA	1822.80	3322519.80
RUN	106.62	11367.82
ELA	122.32	14962.18
TEMP	0.64	0.42

EXAMPLE CA-4.2. Calculation of the grouped variance for SALE from the Canadian Automotive data set.

Using the mean of 31,232.08 calculated in CA-4.1 (assume the midpoint of the last interval is 75,000), the following table presents the calculations necessary to obtain grouped variance and standard deviation values for SALE.

SALE: x_i (midpoint)	f_i	$(x_i - \bar{x})$	$(x_i - \bar{x})^2$	$f_i(x_i - \bar{x})^2$
300	1	-30932.08	$9.57 \cdot 10^8$	$9.57 \cdot 10^8$
750	2	-30482.08	$9.29 \cdot 10^8$	$18.58 \cdot 10^8$
2000	3	-29232.08	$8.55 \cdot 10^8$	$25.65 \cdot 10^8$
4000	5	-27232.08	$7.42 \cdot 10^8$	$37.10 \cdot 10^8$
15000	21	-16232.08	$2.63 \cdot 10^8$	$55.33 \cdot 10^8$
37500	7	6267.92	$3.93 \cdot 10^7$	$2.75 \cdot 10^8$
75000	14	43767.92	$1.92 \cdot 10^9$	$268.80 \cdot 10^8$

The sum of the frequencies is 53 (n = 53), the column sum of the product terms in the last column is 417.78 x 10^8. Thus the variance for SALE is s^2 = 419.78 x 10^8/(53-1) = 8.03 x 10^8. The standard deviation is the positive square root of the variance or, in this case, s = 2.83 x 10^4. Note once again that caution is necessary when dealing with results from a grouped variable such as SALE for which the last interval is unbounded in the original data.

One important advantage of the variance or standard deviation is that it equals zero for perfectly homogeneous data (a constant) and increases in value as heterogeneity increases in data. But one important disadvantage is that when extreme values are present, these two measures of variability may be misleading by suggesting that an unusually large dispersion exists. Moreover, they suffer from the same drawback as does the arithmetic mean, which is not surprising, since they are based upon this measure of central tendency. One attempt to circumvent this problem is to a measure based upon the median. As mentioned previously, the sum of the absolute deviations about the median is a minimum compared to the sum of the absolute differences from any other value, yielding the statistic

$$\sum_{i=1}^{n} |x_i - \text{median}|/n = [|x_1 - \text{median}| + |x_2 - \text{median}| + \ldots + |x_n - \text{median}|]/n.$$

However, since the mean ordinarily is the single most satisfactory measure of central tendency in a wide variety of situations, and since the mean and the median are equivalent for symmetric distributions, usually this statistic is written as

$$\text{mean deviation} = \sum_{i=1}^{n} |x_i - \bar{x}|/n = [|x_1 - \bar{x}| + |x_2 - \bar{x}| + \ldots + |x_n - \bar{x}|]/n.$$

The sum of absolute deviations from the mean is a more appealing statistic than is the sum of raw deviations from the mean, since, as was mentioned earlier, this second sum fails to summarize dispersion because it always equals zero. But for many situations use of the mean rather than the median in the mean deviation formula negates the property that it is a minimum, while the standard deviation or variance retains this property. Further, absolute values are not easy to manipulate algebraically, and the mean deviation fails to possess the four desirable statistical properties.

EXAMPLE PR-4.7. Computing a mean deviation.

In terms of the Puerto Rico data, the different density variables for the Arecibo Agricultural Administrative Region lead to the following calculations:

DFRM:	x_i	$\bar{x}$	$\|x_i - \bar{x}\|$	x_i	$\bar{x}$	$\|x_i - \bar{x}\|$
	4.5	9.75	|-5.25|	3.0	9.75	|-6.75|
	5.6	9.75	|-4.15|	12.9	9.75	| 3.15|
	12.1	9.75	| 2.35|	10.7	9.75	| 0.95|
	8.5	9.75	|-1.25|	18.3	9.75	| 8.55|
	8.9	9.75	|-0.85|	4.8	9.75	|-4.95|
	17.9	9.75	| 8.15|			

mean deviation = (5.25 + 4.15 + 2.35 + 1.25 + .85 + 8.15 + 6.75 + 3.15 + .95 + 8.55 + 4.95)/11 = 46.35/11 = 4.21

Variable	Mean deviation	Variable	Mean deviation
DCD	7.33	DCF	11314.50
DMLK	58.78	DTB	6.08
DSGR	3.62	DBN	5.50
		DFAM	1.56

Both the variance and the standard deviation are measures of dispersion that depend upon the units of measurement employed. A useful composite statistic that is constructed from the mean and the standard deviation of a data set is called the **coefficient of variation**. It may be defined as $CV = s/\bar{x}$ (or σ/μ for the population). One prominent advantage of this statistic is that it is a dimensionless number (the mean and the standard deviation both are measured in the same units, which cancel with the division operation), allowing it to be used to compare variables measured in different units or at different scales. This statistic is mentioned here because it is of some interest in spatial statistics. Because this statistic involves the construction of a ratio, it is meaningful only when data are measured at the ratio level. As dispersion decreases, so does CV; this statistic achieves its minimum of zero for a constant. Because statistical theory demonstrates that the same mean $\bar{x}$ and variance s^2 are independent, then CV can be shown to display the four desirable statistical properties.

EXAMPLE PR-4.8. Calculating a coefficient of variation.

In terms of the Puerto Rico data, the different density variables for the Arecibo Agricultural Administrative Region lead to the following computations, based upon the computations reported in Examples PR-4.1 and PR-4.6:

Variable	CV	Variable	CV
DFRM	5.23/ 9.75 = 0.536	DCF	15713.61/8464.12 = 1.856
DCD	8.82/39.64 = 0.223	DTB	7.60/5.32 = 1.429
DMLK	103.58/67.54 = 1.534	DBN	7.79/4.58 = 1.701
DSGR	4.28/3.99 = 1.073	DFAM	2.05/2.70 = 0.759

EXAMPLE OR-4.6. Calculation of the coefficient of variation for four physical variables from the Ontario Runoff (AREA and RUN) and Sentinel Glacier (ELA and TEMP) data sets based upon the results from Examples OR-4.1 and OR-4.5.

Variable	CV	Variable	CV
AREA	1822.80/1916.71 = 0.95	ELA	125.86/1818.42 = 0.07
RUN	106.62/296.75 = 0.36	TEMP	0.66/12.77 = 0.05

There are situations in which it is useful to be able directly to compare the distributions of two variables. This comparison can be made either by graphing the two cumulative relative frequency distributions on one graph (known as a quantile-quantile plot, or Lorenz curve) or by calculating a summary statistic (the Gini coefficient in this case, but others are available). One of the most common applications of the Lorenz curve in human geography addresses the distribution of income across a population. In this case, the two distributions of interest are population and income, which are used to construct cumulative percent columns. These cumulative percent columns form two cumulative relative frequency distributions that can be graphed by treating the two values for each group as a pair of Cartesian coordinates and plotting them. These points can be connected and then compared with a diagonal line drawn from the lower left-hand to the upper right-hand corner of the graph (at a 45° angle).

EXAMPLE 4.1. Constructing a Lorenz curve using hypothetical data.

Suppose that individuals in a population have been ranked by income and that this sequence has been divided into 10 equal parts (10% of the population fall into each group). The total income for each group of people is calculated by adding their individual incomes, yielding an aggregate group income. The proportion of total income, based upon summing the income of the entire population, has been calculated and entered into the following table. The population is ranked in decreasing order of income, from rich to poor.

	POPULATION		INCOME	
Group	Percent	Cumulative percent	Percent of the total	Cumulative percent
1 (rich)	10	10	30	30
2	10	20	20	50
3	10	30	10	60
4	10	40	10	70
5	10	50	10	80
6	10	60	5	85
7	10	70	5	90
8	10	80	4	94
9	10	90	3	97
10 (poor)	10	100	3	100

Figure 4.4 displays the data collected and organized in this example, with population measured on the X-axis, and income on the Y-axis.

The obvious question that this graph addresses concerns the relative distribution of income across the population. The diagonal line indicates a joint distribution in which each portion of the population earns an equal portion of the income. In other words, a situation in which the coordinates of the points are the same--(10, 10), (20, 20), (30, 30), . . . , (100, 100). The diagonal line thus describes a perfectly equitable distribution of income. The Lorenz curve describes the actual distribution. As a Lorenz curve deviates from the diagonal line, the distribution of income becomes increasingly inequitable. The extreme case would exist if 1% of the population earned 100% of the income, in which case the Lorenz curve would coincide with the axes of the graph. If the rank ordering of the population variable is reversed, the Lorenz curve switches sides of the diagonal. In this case, the Lorenz curve would fall below the diagonal line.

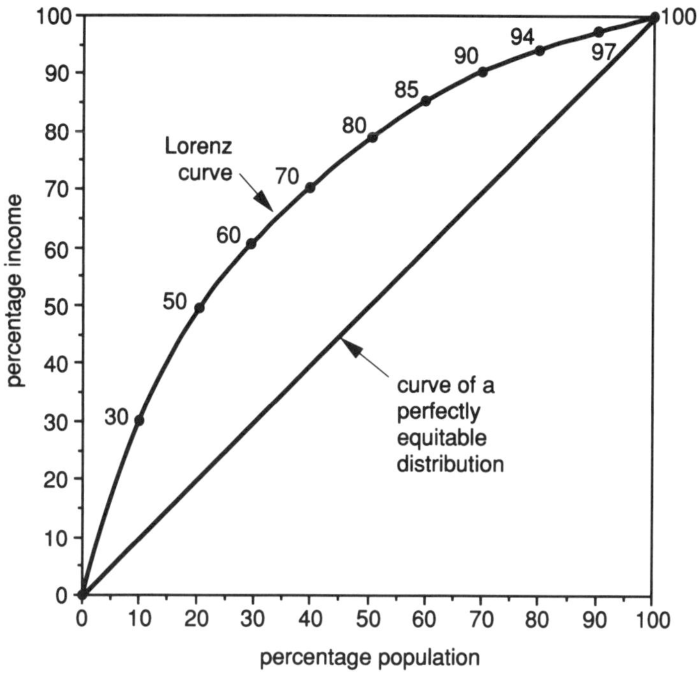

Figure 4.4: LORENZ CURVE FOR EXAMPLE 4.1

The **Gini coefficient** (G), also called an index of dissimilarity, calculates a summary measure of the deviation pictured in the Lorenz graph. The value of this coefficient equals the area of the space between the Lorenz curve and the diagonal line and thus increases as the maximum distance (measured perpendicular to the diagonal line) between the lines increases. There are several formulae in the literature for calculating G. To avoid having to keep track of the sign of the differences between the paired percentages, a version of G based on absolute values is adopted here; accordingly, G may be defined as

$$G = (0.5)(\Sigma_{i=1}^{k} |x_i - y_i|) \ ,$$

where $|x_i - y_i|$ represents the absolute value of the difference, k is the number of groups (10 in this case), and x_i and y_i are the relative frequencies (*not* the cumulative relative frequencies) of the population and income percentages for each group. The maximum value of G equals one-half the area of the graph, or the size of that triangle defined by the diagonal, the X-axis, and the Y-axis.

EXAMPLE 4.1 (continued).

In this case, the maximum value of G is 50. The calculated value is 30. A value of 30, compared to 50 (maximum inequality) and 0 (perfect equality) suggests a rather inequitable distribution of income. This result is not very surprising given that the richest 10% of the population earn 30% of the income, while the poorest 10% earn 3%, and the poorest 30% of the population earns 10% of the income.

4.3. Measures of relationships

There is a third important feature of data to consider when attempting to portray them in summary form, namely, (a) most data sets comprise more than one variable, and (b) there is information redundancy can result from the presence of multiple variables. Very few modern data sets that are collected consist of a single variable; if they do, then the sorts of summary statistics outlined in Sections 4.1 and 4.2 would suffice when characterizations of the data are sought. But what happens when a data set contains several variables? Recalling that one of the important goals of statistics is to assess the information content of data, this question can be reworded to ask how much information is held in common by two variables, or what is the rule of correspondence between two variables (how two variables are related)? Attention is restricted here to pairwise or bivariate relationships; more comprehensive relationship analyses are the topic of multivariate statistical analysis.

The relationship existing between two variables, say X and Y, may be described as Y = F(X). A question naturally arises about the function F in this sort of equation. Classical statistics restricts attention to forms of F that are **linear** or quasi-linear (linear after a sequence of algebraic manipulations has been performed), meaning that the equation to be analyzed is that of a straight line, and has the form Y = a + bX. Here the intercept of the equation is a, and the slope of the line in question is b. When addressing the notion of a relationship between two variables, attention is restricted to the slope of the line, b. In addition, since error (preferably due solely to sampling rather than measurement or arithmetic manipulation) is involved in statistical analysis, then each observation has its own equation of the form

$$y_i = a + bx_i + e_i$$

where e_i denotes the error that is associated with observation i and is represented by the variable E. If E is zero, then the information contained in variable Y is identical to that contained in variable X. If b is zero, then there is no information that is common to both variables X and Y, for variable Y is simply a constant plus error.

Two features of the linear equation are relevant to the notion of a relationship between variables X and Y. First, the **nature of the relationship** is identified by the slope of the line, b. If this slope is positive, then the values of Y increase as the values of X increase. If this slope is negative, then the values of Y increase as the values of X decrease. If this slope is zero, then the values of Y change in an unpredictable way as the values of X increase or decrease. Second, the **degree of the relationship** is identified by the magnitude of the error term E. These two features can be viewed by inspection of a plot of the set of data points $\{(x_i, y_i)\}$; such a graph is called a **scatter diagram**, since the points usually are scattered across it. As E goes to zero, this scatter of points collapses onto a line. The plot of variables DFRM versus DFAM, from the Puerto Rico data set, appearing in Figure 4.5 is an example of such a plot. In Figure 4.5b the set of points corresponding to this error-free situation is represented by the solid line. If this line slopes upwards to the right, then the relationship is labelled positive or direct (Figure 4.5b has such a slope); if this line slopes upwards to the left, then the relationship is labelled negative or inverse. If this line is either parallel or perpendicular to the X-axis, then no relationship exists. But E will not be zero, since error is present, causing the points to scatter rather than systematically align. Hence, the scatter of points must be approximated by a line (for instance, the solid line appearing in Figure 4.5b). If the slope of this approximated line is positive, then the values of Y tend to increase as the values of X increase. If the slope of this approximated line is negative, then the values of Y tend to increase as the values of X decrease. And if the slope of this approximated line is zero, then the relationship between variables Y and X is called random. The degree of the relationship refers to how close the scatter of points is to this approximation line. In keeping with the summary approach taken in Sections 4.1 and 4.2, a single numerical measure of the nature and degree of a relationship would be useful here. Such a measure is provided by the correlation coefficient.

EXAMPLE SG-4.1. Scatter diagram of temperature (TEMP) and precipitation (PRECIP) against the mass balance (BN) from the Sentinel Glacier data set.

In Figure 4.6, the values for TEMP are plotted along the left-hand vertical axis against the values for BN along the horizontal axis. These points are identified by the symbol •. The values for PRECIP are plotted along the righthand vertical axis against the values of BN along the horizontal axis. These points are identified by the symbol "+". An examination of this figure shows that the straight line relating TEMP and BN has a slope close to zero, suggesting that, in relation to BN, TEMP behaves in an unpredictable fashion. The straight line relating PRECIP and BN, however, has a distinct rise from the lower-left corner to the upper-right corner (indicating a positive slope), suggesting that there is some systematic change in the values of PRECIP in response to a change in the values of BN. There is no information at this stage in the analysis to indicate whether this relationship is important. This question is addressed in the next section.

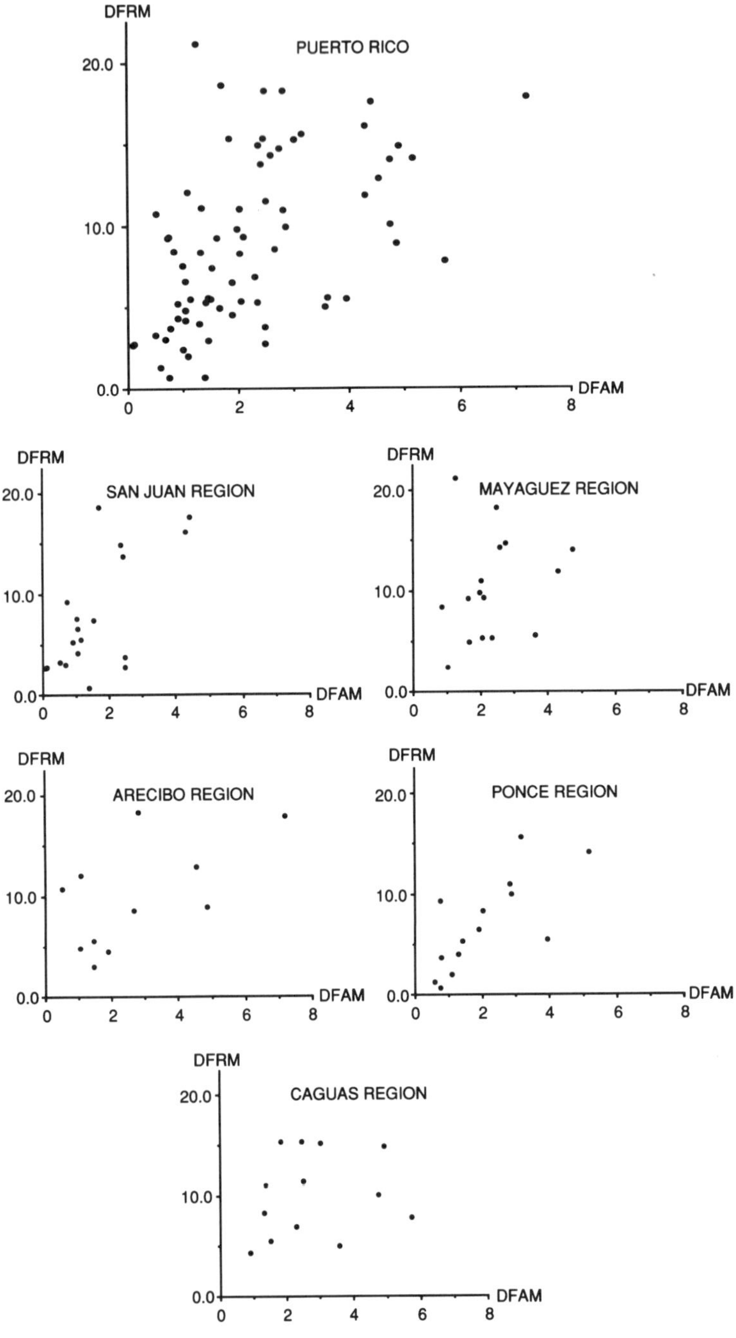

Figure 4.5: (a) ILLUSTRATIVE SCATTER DIAGRAM FOR THE DENSITY OF FARM FAMILIES AND DENSITY OF FARMLAND FROM THE PR DATA SET

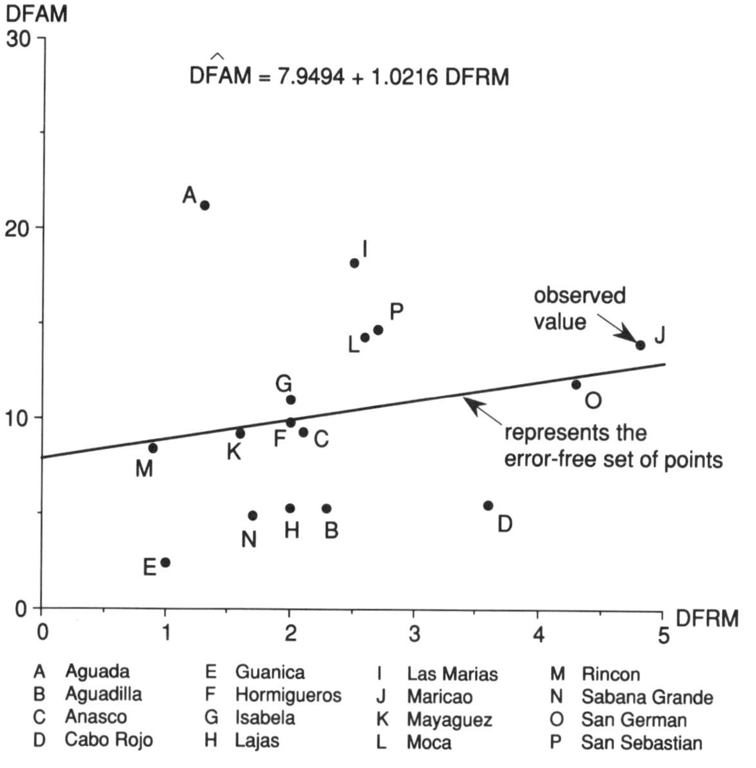

Figure 4.5: (b) A DETAILED ANALYSIS FOR THE MAYAGUEZ REGION

The **correlation coefficient** measures information content that is common to two variables by summarizing the nature and degree of a relationship prevailing between these variables. Naturally one would expect the correlation coefficient to be sensitive to the measurement scale used for these two variables, since the different measurement scales capture different amounts of information. In addition, as was seen with the coefficient of variation, because different variables do not have to be measured with a common unit, the correlation coefficient needs to be devised in such a way that it is dimensionless. Statisticians have formulated a set of correlation coefficients that satisfy these restrictions. All these coefficients are restricted to the interval [-1, 1], with -1 denoting a perfect negative relationship, 0 a random relationship, and 1 a perfect positive relationship between variables X and Y. As this coefficient approaches 0 (increasingly less information is held in common by the two variables), the described relationship is weak. As this coefficient approaches ±1 (increasingly more information is held in common by the two variables), the describ-

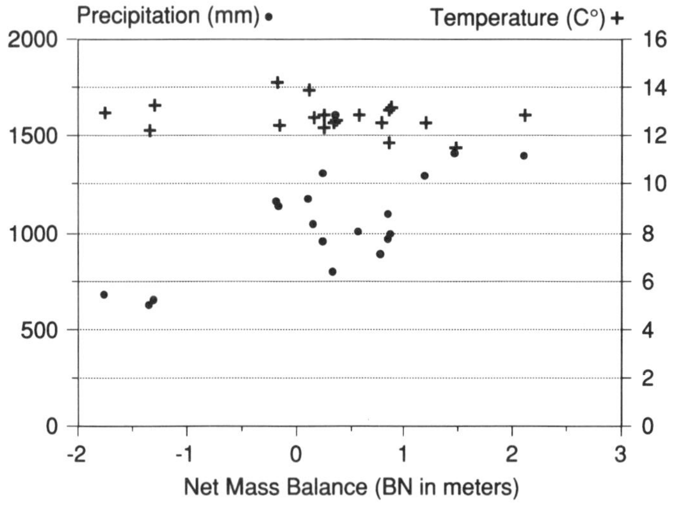

Figure 4.6: OVERLAID SCATTER DIAGRAMS FOR TEMP VERSUS BN AND PRECIP VERSUS BN FROM THE SG DATA SET

ed relationship is strong. And as this coefficient approaches approximately ±0.7, this described linear relationship is moderate. Illustrations of these target values appear in Figure 4.7; note the increased scatter of points as a value of 0 is approached, either from its left- or from its right-hand side.

A typology for correlation coefficients appears in Figure 4.8. Noteworthy at this point is that the phi coefficient, the point biserial, and the Spearman's rank all are special cases of Pearson's product moment coefficient. In other words, if a scientific hand calculator is being used, then its single correlation function is applicable to each of these sorts of data. Aside from purely historical reasons, these different coefficients need to be acknowledged because the different measurement scales involved dramatically affect the respective underlying sampling distributions.

The **phi coefficient** is appropriate for nominal data, specifically when both variables are dichotomies or binary in nature. In this context there are four number pair possibilities, namely, (0, 0), (0, 1), (1, 0), and (1, 1), with corresponding frequencies denoted here by A, B, C, and D. Which variable is labelled X and which is labelled Y is arbitrary. It is customary to tabulate the frequencies of these four number pairs in the following way:

level of measurement for variable Y

level of measurement for variable X	nominal	ordinal	interval / ratio
nominal	phi coefficient	*	point biserial
ordinal	*	Spearman's rank	*
interval / ratio	point biserial	*	Pearson's product moment

Figure 4.7: TYPOLOGY OF CORRELATION COEFFICIENTS

```
variable    X   Y
            0   1
   0        A   B
   1        C   D
```

The phi coefficient, denoted by ϕ, is defined as

$$\phi = [AD - BC]/[(A + B)(C + D)(A + C)(B + D)]^{1/2}.$$

This coefficient is positive when AD > BC, negative when AD < BC, and zero when AD = BC, if B = C = 0, then $\phi = 1$ [a perfect direct relationship; only (0,0) and (1,1) pairs are present], and if A = D = 0, then $\phi = -1$ [a perfect negative relationship; only (0, 1) and (1, 0) pairs are present]. A positive coefficient means that 0 values for X tend to coincide with 0 values for Y, and unity values for X tend to coincide with unity values for Y; a negative coefficient means that 0 values for X tend to coincide with unity values for Y, and unity values for X tend to coincide with 0 values for Y. The Pearson product moment equation reduces to this Phi coefficient equation when both X and Y are binary.

EXAMPLE CA-4.3. Computing a phi coefficient.

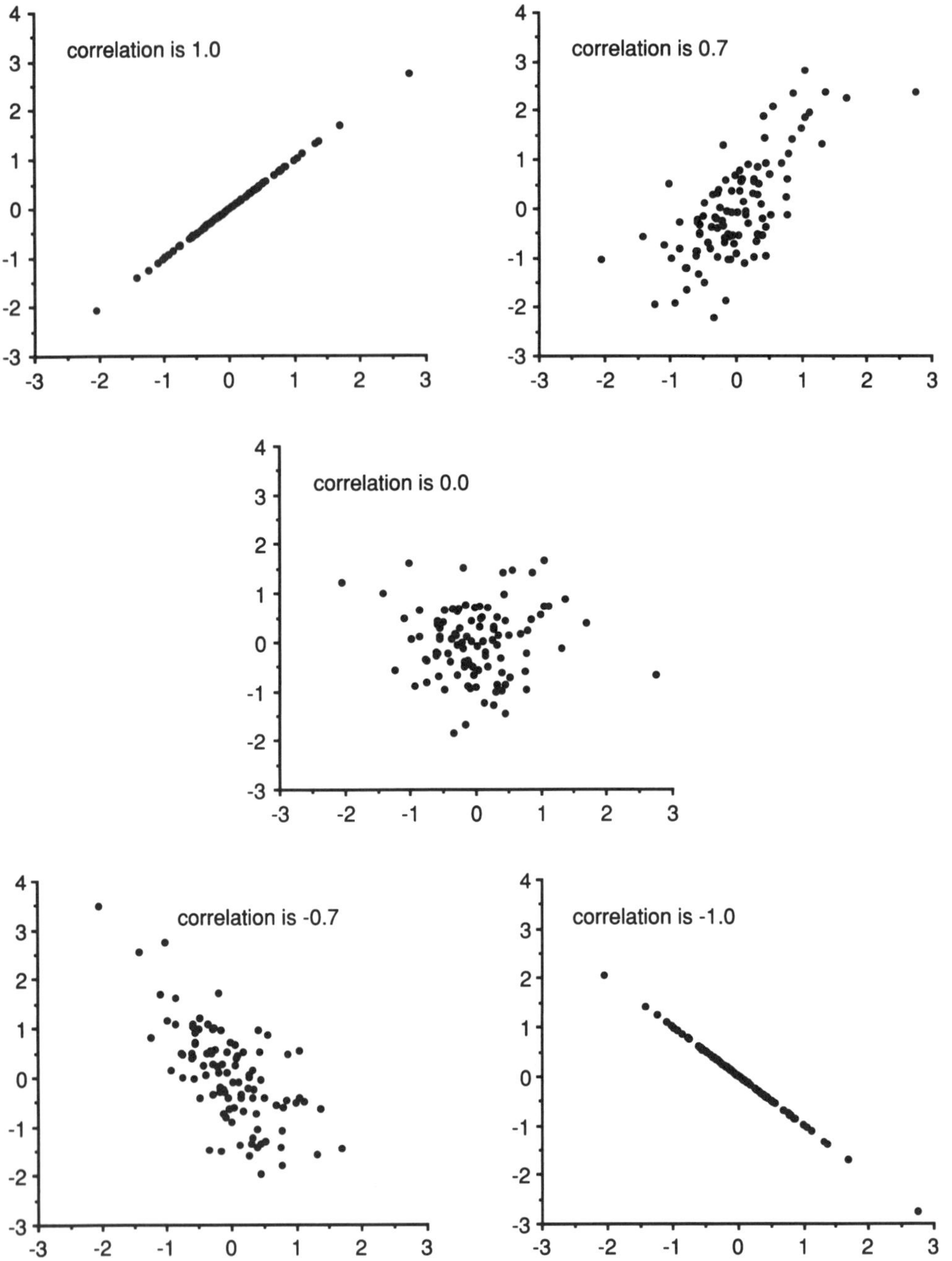

Figure 4.8: ILLUSTRATIONS OF SCATTER DIAGRAMS DEPICTING DIFFERENT NATURES AND DEGREES OF CORRELATION BETWEEN Y AND X

Consider the Canadian Auto data set. Two dichotomous variables measured at the nominal scale are FAC and EXP. Suppose that X is FAC, while Y is EXP. Let X = 0 if the firm acts as an agent for a foreign organization, and X = 1 otherwise. Similarly, let EXP = 0 if a firm exports primarily to the United States, and EXP = 1 otherwise. A tabulation of the results follows:

FAC	EXP	
	0 (U.S.)	1 (other)
0 (foreign agent)	18	2
1 (not an agent)	27	6

$$\phi = (18 \bullet 6 - 27 \bullet 2)/[(18 + 2)(27 + 6)(18 + 27)(2 + 6)]^{1/2} = 0.111.$$

This particular phi coefficient may be interpreted as indicating that no noticeable relationship exists between FAC and EXP. While the sign of the coefficient is positive, the value is rather close to zero. As a result, there appears to be no relationship between a firm's allegiance to a foreign organization, and its export behavior.

But what happens to this coefficient if the variable definitions are changed? The values of X can be switched, the values of Y can be switched, or the values of both X and Y can be switched. These switches cause only the row and/or column labels of the table to be switched, for by convention the table must be constructed so that the top left-hand cell is for (0, 0) frequencies, whereas the bottom right-hand cell is for (1,1) frequencies. This rearrangement of values does not affect the magnitude of the resulting numerical value but can change its sign. The denominator of the phi coefficient is the product of the row and column sums. Regardless of how these data are arranged in the table, the same four numbers are multiplied; the main difference from one set of variable definitions to the next is the order of addition or the order of multiplication, neither of which changes the final product (both addition and multiplication are commutative). But the numerator does change, the same two products are differenced, but in a different order. Therefore, the magnitude of the numerator remains unchanged, while the sign changes. Although the sign of the relationship may change across the possible tabulations, the substantive interpretation of the resulting coefficient remains unchanged (see Problem #10 at the end of this chapter). This consequence stresses the importance of clearly revealing variable definitions to a reader, even though they are arbitrary.

This correlation setup can be generalized to polychotomous classification schemes by analyzing **contingency tables**, which are arrangements of data into two-way tabulations. The (i, j) cell entry is usually a frequency count, o_{ij}. The relationship between two such categorical variables, where the first consists of k categories and the second consists of l categories, may be measured by a chi square statistic, which is defined as

$$\chi^2 = \Sigma_{i=1}^{k} \Sigma_{j=1}^{l} (e_{ij} - o_{ij})^2/e_{ij}$$

where e_{ij} denotes the proportion of the total sample size one would anticipate being classified into cell (i, j) given the percentage of a sample that appears in row i, say r_i/n, coupled with the percentage of this sample that appears in column j, say c_j/n. In other words,

$$e_{ij} = nr_jc_j/n^2 = n(r_i/n)(c_j/n) = (r_ic_j)/n.$$

Clearly the minimum chi square value is 0; this minimum value indicates that no relationship exists between the two categorical variables in question. As an aside, this chi square equation is an approximation, with the specification error associated with this approximation being negligible as long as every $e_{ij} \geq 5$; this issue is discussed, in Chapter 10.

EXAMPLE PR-4.9. Calculating a chi square statistic.

Consider the Puerto Rico data. The ADM polychotomous variable has five categories, while the LND dichotomous variable has two categories. The contingency table for this pair of categorical variables is

ADM	LND: 0 (coastal lowlands)	1 (interior highlands)	total
1 (San Juan)	12	7	19
2 (Arecibo)	7	4	11
3 (Mayaguez)	9	7	16
4 (Ponce)	10	4	14
5 (Caguas)	4	9	13
total:	42	31	73

χ^2 calculation

Cell	Observed frequency	Expected frequency	Squared difference	Chi square term
(1,0)	12	$73 \cdot 19 \cdot 42/73^2$	$(12-10.93)^2$	0.1047
(2,0)	7	$73 \cdot 11 \cdot 42/73^2$	$(7-6.33)^2$	0.0709
(3,0)	9	$73 \cdot 16 \cdot 42/73^2$	$(9-9.21)^2$	0.0048
(4,0)	10	$73 \cdot 14 \cdot 42/73^2$	$(10-8.05)^2$	0.4724
(5,0)	4	$73 \cdot 13 \cdot 42/73^2$	$(4-7.48)^2$	1.6190
(1,1)	7	$73 \cdot 19 \cdot 31/73^2$	$(7-8.07)^2$	0.1419
(2,1)	4	$73 \cdot 11 \cdot 31/73^2$	$(4-4.67)^2$	0.0961
(3,1)	7	$73 \cdot 16 \cdot 31/73^2$	$(7-6.79)^2$	0.0065
(4,1)	4	$73 \cdot 14 \cdot 31/73^2$	$(4-5.95)^2$	0.6391
(5,1)	9	$73 \cdot 13 \cdot 31/73^2$	$(9-5.52)^2$	2.1939

$$\chi^2 = 0.1047 + 0.0709 + 0.0048 + 0.4724 + 1.6190 + 0.1419 + 0.0961 + 0.0065 + 0.6391 + 2.1939 = 5.349$$

The **Spearman's rank coefficient** is appropriate for ordinal data, specifically when both variables are ranked without ties. It is customary to make both rankings either ascending or descending and to use the consecutive positive integers. Once more, the choice of which variable to label X and which to label Y is arbitrary. This statistic is denoted by r_s (ρ_s for the population) and is defined as

$$r_s = 1 - \{6 \sum_{i=1}^{n} d_i^2/[n(n^2 - 1)]\}$$

where d_i denotes the difference between the ranking of X and the ranking of Y for observation i. If the rank for X and the rank for Y are identical, then their difference is 0; if all of the observations have the exact same ranking on both variables, then d_i always is 0, resulting in

$$r_s = 1 - \{6 \bullet 0/[n(n^2 - 1)]\} = 1.$$

If all of the observations have the exact opposite rankings on the two variables, then the sum of the squared differences equals $n(n - 1)(n + 1)/3$, resulting in

$$r_s = 1 - \{6[n(n - 1)(n + 1)/3]/[n(n^2 - 1)]\} = -1.$$

As long as both variables are treated the same, ranking in ascending or descending order will not alter the sum of the squared differenced ranks, and the denominator will strictly be a function of sample (or population) size. Hence the choice of ascending or descending ranks is arbitrary. However, if one variable is ranked in ascending order and the other in descending order, then although the substantive interpretation of the resulting numerical value will not change, the sign of the calculated coefficient will change; this is the same type of definitional problem illustrated in Example CA-4.3. This outcome stresses the importance, once more, of informing the reader of how variables have been defined. Again, it is useful to note that this coefficient is a special case of the Pearson product moment coefficient if ties in rankings occur or if a scientific hand calculator is being used.

EXAMPLE PR-4.10. Calculating a Spearman's rank correlation coefficient.

Consider the Arecibo Agricultural Administrative Region data for Puerto Rico. If a researcher is concerned that the measurement error in these density data is neither negligible nor random, then he or she could convert each variable from a ratio scale to an ordinal scale. Upon completing this conversion, where in this instance rankings are in ascending order, the researcher would obtain the following Spearman's rank coefficients:

DFRM		DCD		DMLK		DCF		DMLK/DCF
Value	Rank	Value	Rank	Value	Rank	Value	Rank	Difference
4.5	2	25.6	1	46.7	7	678.7	7	7-7 = 0
5.6	4	32.5	2	32.3	4	268.6	4	4-4 = 0
12.1	8	39.9	7	68.1	9	413.1	5	9-5 = 4
8.5	5	35.0	5	4.9	3	12070.0	9	3-9 = -6
8.9	6	53.4	11	368.6	11	45.1	1	11-1 = 10
17.9	10	47.5	9	0.0	1	48874.9	11	1-11 =-10
3.0	1	34.8	4	89.2	10	80.5	2	10-2 = 8
12.9	9	50.9	10	50.4	8	3158.9	8	8-8 = 0
10.7	7	36.9	6	40.9	6	610.4	6	6-6 = 0
18.3	11	46.8	8	2.7	2	26677.2	10	2-10 = -8
4.8	3	32.7	3	39.1	5	227.9	3	5-3 = 2

DMLK & DCF:

$$r_s = 1 - \{6[0^2 + 0^2 + 4^2 + (-6)^2 + 10^2 + (-10)^2 + 8^2 + 0^2 + 0^2 + (-8)^2 + 2^2]/[11(11^2 - 1)]\} = -0.7455$$

DCD & DMLK: $r_s = 0.1273$ and DFRM & DCD: $r_s = 0.7636$.

The first coefficient suggests that there is a moderate to strong tendency for milk production density rank to increase as coffee production density rank decreases. The second coefficient suggests that there is a near random (weak positive) relationship between the ranks of density of farm land and density of milk production. The third coefficient suggests that there is a moderate to strong tendency for the rank of the density of number of farms to increase as the rank of the density of amount of farm land increases.

EXAMPLE SG-4.2. Calculation of the Spearman rank correlation coefficient for physical variables BN, TEMP, and PRECIP from the Sentinel Glacier data set.

In Example SG-4.1, Figure 4.6 showed the scatter diagram relating TEMP and PRECIP with BN. In order to assess the nature and degree of any suspected relationship, a correlation coefficient may be calculated. Further, if there is some doubt regarding the accuracy of the interval measurements of these variables, the scale of measurement can be reduced to an ordinal one, allowing Spearman's coefficient to be calculated. The following table contains calculations necessary to obtain r_S for BN and PRECIP. Since there are ties in the case of TEMP, the Spearman coefficient cannot be calculated using the formula just presented; the handling of this complication is addressed later in this chapter.

BN Value	BN Rank	PRECIP Value	PRECIP Rank	Difference	BN Value	BN Rank	PRECIP Value	PRECIP Rank	Difference
-0.15	5	1136.8	12	-7	1.47	18	1407.8	18	0
-0.18	4	1161.6	13	-9	-1.33	2	626.1	1	1
0.38	11	1608.7	19	-8	0.36	10	796.1	4	6
0.12	6	1178.3	14	-8	-1.74	1	685.7	3	-2
-1.30	3	657.3	2	1	0.27	9	953.4	6	3
0.57	12	1007.3	9	3	0.17	7	1043.8	10	-3
0.26	8	1308.0	16	-8	0.86	15	1103.9	11	4
0.78	13	884.5	5	8	1.20	17	1300.0	15	2
2.11	19	1397.4	17	2	0.85	14	971.1	7	7
0.88	16	994.5	8	8					

The sum of the squared differences is 592, so

$$r_s = 1 - \{(6 \bullet 592)/[19(19^2 - 1)]\} = 1 - (3552/6840) = 1 - 0.52 = 0.48.$$

As suggested by Figure 4.6, a positive relationship exists, indicating that as the value of BN increases, the value of PRECIP tends to increase. Given the target value of 0.7 stated earlier, a value of 0.48 suggests a moderately strong relationship.

A **point biserial correlation coefficient** is appropriate to use the variable designated by X is measured at an interval/ratio level, and the variable designated by Y is measured at a binary level (recall binary is a special case of nominal scale). Now which variable is labelled X and which is labelled Y is quite relevant. This coefficient, denoted r_{pbi}, is defined as

$$r_{pbi} = (\bar{x}_p - \bar{x}_{1-p})[p(1 - p)]^{1/2}/s_x$$

where p denotes the proportion of observations contained in the dichotomous group assigned a value of 1, $\bar{x}_p$ is the mean of variable X for this subset of observations, (1 - p) is the proportion of observations contained in the remaining dichotomous group assigned a value of 0, $\bar{x}_{1-p}$ is the mean of variable X for this subset of observations, and s_x is the **biased** standard deviation of variable X. Here the coefficient is 0 if all observations are in either of the two dichotomous groups [either p or (1 - p) is 0 then], or if the means of the two groups are the same ($\bar{x}_p = \bar{x}_{1-p}$). When all observations fall into one group, then Y is a constant, and the coefficient is meaningless; when the two group means are equal, then the relationship between X and Y is random. If the assignment of 0s and 1s in Y is switched, then the standard deviation remains the same, p and (1 - p) are switched but their product remains the same, and

$\bar{x}_p$ and $\bar{x}_{1-p}$ are switched, resulting in the same magnitude of difference but a different sign; again, then, the definition of Y affects the sign but not the magnitude of the coefficient.

EXAMPLE PR-4.11. Computing a point biserial correlation coefficient.

Consider the Arecibo Agricultural Administrative Region data for Puerto Rico. The LND variable (coastal lowlands = 0 and interior highlands = 1) is dichotomous, and is treated as the Y variable. Selected point biserial correlation coefficients using this variable include (where in turn DFRM, DSGR, and DMLK is the X variable):

DFRM & LND:

$r_{pbi} = \{(57.6/4 - 49.6/7)[(4/11)(7/11)]^{1/2}\}/(273.0075/11)^{1/2} = 0.7063$

DSGR & LND: $r_{pbi} = -0.7252$ and DMLK & LND: $r_{pbi} = -0.4060$

using calculations appearing in Example PR-4.6 This first coefficient suggests that there is a moderate tendency for density of farms to be high in the interior highlands and low in the coastal lowlands. The second coefficient suggests that there is a moderate tendency for dense sugarcane production to be located in the coastal lowlands and sparse sugarcane production to be located in the interior highlands. The third coefficient suggests that there is a weak-to-moderate tendency for density of milk production to be located in the coastal lowlands and sparse milk production to be located in the interior highlands. As stressed above, one should note that switching the definition of Y so that coastal lowlands is indicated by 1 and interior highlands is indicated by 0 would change the signs of these three example coefficients, but would neither change their magnitudes nor alter their substantive interpretations.

Pearson's product moment correlation coefficient is appropriate for interval/ratio data and is the general coefficient from which the three preceding special cases are derived. Again which variable is labelled X and which is labelled Y is arbitrary. This coefficient, denoted by r for sample values (denoted by ρ for population values), is defined as

$$r = \sum_{i=1}^{n} (x_i - \bar{x})(y_i - \bar{y})/(ns_x s_y)$$

when the biased sample variances are used, and

$$r = \sum_{i=1}^{n} (x_i - \bar{x})(y_i - \bar{y})/[(n - 1)s_x s_y]$$

when the unbiased sample variances are used. The numerator of this coefficient represents the product of the joint deviation, or the **covariation**, of the two variables about their respective means. The distribution of deviations around a mean is sometimes referred to as the first moment of a distribution. The phrase "product moment correlation coefficient" thus refers to the product of the first moments in the numerator of the coefficient. Note that the denominator of Pearson's r can never be negative, so the sign of the coefficient is derived from the numerator. Note also that either all biased or all unbiased statistics must be used to calculate this coefficient; a mixture of these types of statistics could result in $|r| > 1$, which is physically impossible. This formula has been rewritten in various reduced versions that ease or simplify the calculations involved. These same reduced formulae are obtained regardless of whether or not adjustment for bias is used in the numerator and the denominator. Perhaps the most convenient of these reduced forms defines r as

$$[\sum_{i=1}^{n} x_i y_i - (\sum_{i=1}^{n} x_i)(\sum_{i=1}^{n} y_i)/n]/\{[\sum_{i=1}^{n} x_i^2 - (\sum_{i=1}^{n} x_i)^2/n]^{1/2}[\sum_{i=1}^{n} y_i^2 - (\sum_{i=1}^{n} y_i)^2/n]^{1/2}\}$$

where the calculations are limited to sums of the X and Y values, the squared X and Y values and the cross-product of the X and Y values.

EXAMPLE PR-4.12. Calculating a product moment correlation coefficient.

Consider the Arecibo Agricultural Administrative Region data for Puerto Rico. Selecting the three most illustrative correlations from amongst those for the pairwise density variables yields the following Pearson's product moment coefficients:

		Mean deviation		Product moment
DSGR	DFAM	DSGR - 3.99	DFAM - 2.70	(DSGR-3.99)(DFAM-2.70)
6.1	1.9	2.11	-0.8	-1.688
10.9	1.5	6.91	-1.2	-8.292
3.6	1.1	-0.39	-1.6	0.624
0.1	2.7	-3.89	0.0	0.000
1.0	4.9	-2.99	2.2	-6.578
0.2	7.2	-3.79	4.5	-17.055
3.1	1.5	-0.89	-1.2	1.068
0.0	4.5	-3.99	1.8	-7.182
9.8	0.5	5.81	-2.2	-12.782
0.0	2.8	-3.99	0.1	-0.399
9.1	1.1	5.11	-1.6	-8.176

$$\begin{aligned} r &= [(-1.688 - 8.292 + 0.624 + 0.000 - 6.578 - 17.055 + 1.068 - 7.182 \\ &\quad - 12.782 - 0.399 - 8.176)/10]/(4.28 \cdot 2.04) \\ &= [-60.460/10]/(4.28 \cdot 2.04) = -0.6925 \end{aligned}$$

using the standard deviations presented in Example PR-4.6 (one should note that if the biased standard deviations were computed in Example PR-4.6, then the numerator of this calculation would be divided by 11; furthermore, noticeable rounding error is present here, for the actual correlation coefficient is -0.6906). Furthermore,

$$\text{DTB \& DBN: } r = 0.0047 \quad , \quad \text{DCF \& DBN: } r = 0.9718$$

This first coefficient suggests that there is a moderate tendency for sugarcane production density to increase as density of farm families decreases. The second coefficient suggests that there is essentially a random relationship between the tobacco and banana/plantain production densities. The third coefficient suggests that there is a strong tendency for coffee production density to increase as banana/plantain production density increases.

EXAMPLE SG-4.3. Calculation of Pearson product moment correlation coefficients for BN with TEMP, and then with PRECIP from the Sentinel Glacier data set.

Recall that in Example SG-4.2 it was not possible to calculate r_S for the relationship between BN and TEMP because of tied observations for TEMP. Using Pearson's r, such a calculation is possible. Pearson's r also is calculated here for the relationship between BN and PRECIP. The required calculations, using the original formula for r, appear in the following table (the unbiased standard deviations are BN = 0.96, TEMP = 0.66, and PRECIP = 269.19). Let BN' = (BN - 0.29), TEMP' = (TEMP - 12.77), and PRECIP' = (PRECIP - 1064.33). Then,

			Mean deviation			Product moment	
BN	TEMP	PRECIP	BN'	TEMP'	PRECIP'	BN' • TEMP'	BN' • PRECIP'
-0.15	12.4	1136.8	-0.44	-0.37	72.47	0.16	-31.89
-0.18	14.2	1161.6	-0.47	1.43	97.27	-0.67	-45.72
0.38	12.7	1608.7	0.09	-0.07	544.37	-0.01	48.99
0.12	13.9	1178.3	-0.17	1.13	113.97	-0.19	-19.37
-1.30	13.3	657.3	-1.59	0.53	-407.03	-0.84	647.18
0.57	12.9	1007.3	0.28	0.13	-57.03	0.04	-15.97
0.26	12.9	1308.0	-0.03	0.13	243.67	0.00	-7.31
0.78	12.6	884.5	0.49	-0.17	-179.83	-0.08	-88.12
2.11	12.9	1397.4	1.82	0.13	333.07	0.24	606.19
0.88	13.2	994.5	0.59	0.43	-69.83	0.25	-41.20
1.47	11.5	1407.8	1.18	-1.27	343.47	-1.50	405.29
-1.33	12.2	626.1	-1.62	-0.57	-438.23	0.92	709.93

BN	TEMP	PRECIP	Mean deviation BN'	TEMP'	PRECIP'	Product moment BN' • TEMP'	BN' • PRECIP'
0.36	12.5	796.1	0.07	-0.27	-268.23	-0.02	-18.78
-1.74	13.0	685.7	-2.03	0.23	-378.63	-0.47	768.62
0.27	12.3	953.4	-0.02	-0.47	-110.93	0.01	2.22
0.17	12.8	1043.8	-0.12	0.03	-20.53	0.00	2.46
0.86	13.1	1103.9	0.57	0.33	39.57	0.19	22.55
1.20	12.5	1300.0	0.91	-0.27	235.67	-0.25	214.46
0.85	11.7	971.1	0.56	-1.07	-93.23	-0.60	-52.21

In the preceding example, the column sum of the product moments for the BN' • TEMP' column is -2.82, and thus $r_{BN,TEMP}$= -2.82/(19 • 0.96 • 0.66) = -0.23. The column sum of the product moments for BN' • PRECIP' is 3107.32. Therefore, $r_{BN,PRECIP}$ = 3107.32/(19 • 0.96 • 269.19) = 0.63. Recalling Figure 4.6, the first correlation between BN and TEMP supports the earlier observation that the ideal straight line for these data is relatively flat, or the slope of this line is quite close to zero; $r_{BN,TEMP}$ = -0.23 suggests a very slight negative relationship. Given the relatively small number of points in the sample, a value of -0.23 suggests at best a weak or possibly no relationship. The second correlation, 0.63, also supports the earlier observation of a strong positive relationship. Recall from Example SG-4.2 that r_S = 0.48 was interpreted as suggesting a reasonably strong positive relationship. A value of $r_{BN,PRECIP}$ = 0.63 certainly reinforces this impression.

4.3.1. Obtaining specialized coefficients from the product moment formula

As stated before, the product moment coefficient represents the general correlation coefficient case. Each of the other three coefficients discussed can be derived as special cases of the Pearson coefficient. First, consider the phi coefficient. Both X and Y are binary variables, each taking on a value of either 0 or 1. Accordingly, each variable follows a binomial distribution, having a mean of (# ones)/n = n_p/n = p, and a variance of [(# ones)/n][(# zeroes)/n] = (n_p/n)(n_{1-p}/n) = p(1-p). For variable X the number of ones equals (C + D), and for variable Y the number of ones equals (B + D), with n = A + B + C + D (see the preceding discussion of the phi coefficient). The cross-product term $\sum_{i=1}^{n} x_i y_i$ equals 0 when either variable X or variable Y is 0, and equals 1 when both variable X and variable Y are 1; hence, this sum equals D. The variance terms become n[(C + D)/n][1 -(C+D)/n] = (A + B)(C + D)/n, and n[(B + D)/n][1 - (B + D)/n] = (A + C)(B + D)/n. Substituting these values into the product moment correlation coefficient formula yields

$$\{D/n - [(C + D)/n][(B + D)/n]\}/\{[(A + B)(C + D)/n][(A + C)(B + D)/n]\}^{1/2}$$

$$= [nD - (C + D)(B + D)]/[(A + B)(C + D)(A + C)(B + D)]^{1/2}$$

$$= (AD - BC)/[(A + B)(C + D)(A + C)(B + D)]^{1/2}$$

which is the previously stated phi coefficient. Therefore, this phi coefficient can be computed on a hand calculator by entering the paired binary values of X and Y.

Next, as emphasized before, Spearman's rank correlation coefficient is based upon untied, ranked data. Because both variable X and variable Y are ranked using consecutive integers, they both have the same mean, namely, n(n + 1)/(2n), and the same biased variance, namely, $[(n + 1)(2n + 1)/6 - (n + 1)^2/4]$; these two expressions come from the mathematical theory of difference equations and summation operators. The cross-product term, or product moment term, equals (where r_{xi} signifies the ith ranking of variable X, and r_{yi} signifies the ith ranking of variable Y)

$$\sum_{i=1}^{n} [r_{xi} - (n + 1)/2][r_{yi} - (n + 1)/2]/n$$

$$= [-\sum_{i=1}^{n} d_i^2/2 + n(n + 1)(2n + 1)/6 - n(n + 1)^2/4]/n.$$

Therefore, r_s equals

$$\{[-\sum_{i=1}^{n} d_i^2/2 + n(n + 1)(2n + 1)/6 - n(n + 1)^2/4]/n\}/[(n + 1)(2n + 1)/6 - (n + 1)^2/4]$$

$$= -6\sum_{i=1}^{n} d_i^2/[2n(n + 1)(2n + 1) - 3n(n + 1)^2] + 1$$

$$= 1 - 6\sum_{i=1}^{n} d_i^2/[(n - 1)n(n + 1)]$$

which is Spearman's rank coefficient. r_s also can be computed on a hand calculator by entering the ranked binary values of X and Y. In addition, this derivation illustrates why Spearman's rank formula does not apply when ties are present; the formulae for summed consecutive integers and summed squared consecutive integers would be incorrect (introducing specification error).

Finally, consider the point biserial coefficient where Y is the binary variable and X is the interval/ratio variable, and the cross-product term $\sum_{i=1}^{n} x_i y_i = \sum_{i=1}^{n} x_{pi}$. Those values of variable X whose corresponding value of Y is 0 would become 0 in this summation. From the earlier discussion of nominal variables, the mean of variable Y is p, the variance of variable Y is p(1 - p), and the sample mean of variable X becomes $\bar{x} = (\sum_{i=1}^{n} x_{pi} + \sum_{i=1}^{n} x_{(1-p)i})/n$, since it includes both those observations whose value of Y is 0 and those whose value of Y is 1. Making the appropriate substitutions into the product moment coefficient formula renders

$$\{\sum_{i=1}^{n} x_{pi}/n - [(\sum_{i=1}^{n} x_{pi} + \sum_{i=1}^{n} x_{(1-p)i})/n]p\}/[s_x^2 p(1 - p)]^{1/2}$$

$$= [\sum_{i=1}^{n} x_{pi}(1 - p)/n - \sum_{i=1}^{n} x_{(1-p)i}p/n]/\{s_x[p(1 - p)]^{1/2}\}$$

$= [n_p\bar{x}_p(1 - p)/n - n_{1-p}\bar{x}_{1-p}p/n]/\{s_x[p(1 - p)]^{1/2}\}$

$= [p\bar{x}_p(1 - p) - (1 - p)\bar{x}_{1-p}p]/\{s_x[p(1 - p)]^{1/2}\}$

$= (\bar{x}_p - \bar{x}_{1-p})[p(1 - p)]^{1/2}/s_x$

$= r_{pbi}$

which is the point biserial coefficient. r_{pbi} also can be computed on a hand calculator by inputing the paired values of X and Y. In addition, this algebraic manipulation demonstrates why the biased sample standard deviation must be used to calculate the point biserial coefficient and why the group denoted by unity for variable Y is represented by p.

4.4. Summary

The discussion of correlation coefficients concludes the introduction of elementary statistical concepts. Chapter 5 concludes a parallel treatment of spatial statistics. Following Chapter 5, Part II of this book begins an examination of inferential statistics. A thorough understanding of the descriptive statistics in this section is necessary to understand inferential statistics. The foundation built here not only supports the more complicated techniques to follow, but also should act as a guide for pointing the direction that might be followed in ensuing analyses. A complete statistical analysis must start with the basics presented here because ignoring the basics can result in serious consequences for geographic data analysis.

4.5. Problems

1. Construct a scatter diagram for the following pairs of variables for the different data sets presented in Chapter 1:

    a. DFAM versus DFRM from the Mayaguez Region of the Puerto Rico data set (see Figure 4.5).
    b. LWK versus SALE from the Canadian Auto data set (first 20 observations only).
    c. LAT versus SNOW from the Ontario River Basin data set.
    d. ELA versus PRECIP from the Sentinel Glacier data set.

2. For the Arecibo Region, verify the arithmetic mean calculations presented in Example PR-4.1.

3. For the Arecibo Region, verify the medians identified in Example PR-4.2.

4. For the Arecibo Region, verify the modes identified in Example PR-4.3.

5. For the Arecibo Region, verify the range calculations presented in Example PR-4.4.

6. For the Arecibo Region, verify the quartile deviation calculations presented in Example PR-4.5.

7. For the Arecibo Region, verify the standard deviation and variance calculations presented in Example PR-4.6.

8. For the Arecibo Region, verify the mean deviation calculations presented in Example PR-4.7.

9. For the Arecibo Region, verify the coefficient of variation calculations presented in Example PR-4.8.

10. Using the data presented in Example CA-4.3, show that only a sign change occurs when variables are redefined and that neither the magnitude nor the interpretation of the correlation coefficient changes (there are three additional definitional possibilities).

11. Calculate and interpret the coefficient of variation for DFRM for the Mayaguez Region.

12. Construct a contingency table and calculate the associated chi square statistic for U/R versus LND of the Mayaguez Agricultural Administrative Region from the Puerto Rico data set.

13. Verify the Spearman's rank correlation coefficient calculations presented in Example PR-4.10; then calculate and interpret the Spearman's rank correlation coefficient for DTB versus DCF of the Mayaguez Agricultural Administrative Region from the Puerto Rico data set. [Note: Recall how ties are to be handled.]

14. For the Arecibo Region, verify the point biserial correlation coefficient calculations presented in Example PR-4.11; then calculate and interpret the point biserial correlation coefficient for LND versus DMLK of the Mayaguez Agricultural Administrative Region from the Puerto Rico data set.

15. For the Arecibo Region, verify the Pearson's product moment correlation coefficient calculations presented in Example PR-4.12. Next show that the same results for Problem #13 and Problem #14 are obtained with this general correlation coefficient formula. Then calculate and interpret the Pearson's product moment correlation coefficient for DSGR versus DMLK of the Mayaguez Agricultural Administrative Region from the Puerto Rico data set.

16. For the Metropolitan Toronto region in the CA data set, calculate the grouped mean, standard deviation, and variance for the EXP$ (value of export sales) variable. Use a value of $75,000.00 for the midpoint value of the last interval (> $50,000.00).

17. Verify the calculations in Examples OR-4.1 and OR-4.5 of the mean, standard deviation, and variance for variable TEMP from the Sentinel Glacier data set.

4.6. Additional readings

In addition to the texts listed in Chapter 1,

Gardiner, V., and G. Gardner. *Analysis of Frequency Distributions*. CATMOG 19.

Haggett, P., A. Cliff, and A. Frey. (1977) *Locational Methods*, Vol. 2 of 2nd ed. of *Location Analysis in Human Geography*. New York: Wiley, Halstead Press.

Reynolds, H. *Analysis of Nominal Data*. Sage number 7.

Hildebrand, D., J. Laing, and H. Rosenthal. *Analysis of Ordinal Data*. Sage number 8.

CHAPTER 5

DESCRIPTIVE SPATIAL STATISTICS: CENTRAL TENDENCY, DISPERSION, AND RELATIONSHIP

Four desirable properties of statistics were outlined in Chapter 2. One of these, sufficiency, refers to the ability of a statistic to capture all the information contained in a sample that pertains to the population parameter in question. Consider the four values {0, 1, 1, 2} and a simple map that is partitioned into four areal units. These four values can be arranged on this map in the following 12 different ways:

0	1	0	1	1	0	1	0	1	1	1	1
1	2	2	1	1	2	2	1	0	2	2	0

1	2	2	1	1	2	2	1	0	2	2	0
0	1	0	1	1	0	1	0	1	1	1	1

In each of these 12 cases the classical mean statistic $\bar{x}$ equals 1 (addition is commutative), indicating that the mean is unable to distinguish among these 12 maps. The mean has failed to capture any locational information regarding the geographic arrangement of data values on a map because locational information is ignored. In other words, the classical mean is not a sufficient statistic for spatial data.

In response to this shortcoming, spatial analysts began exploring ways of "adjusting" classical statistics (such as the mean and standard deviation presented in Chapter 4) in an effort to enable these traditional statistics to capture locational information. One of the first steps in this effort was an exploitation of the two-dimensional bell-shaped (called a normal) distribution, since geographic data are located in two dimensions. Measures developed using this physical analogy reasoning are called **geostatistics**. Obviously expecting the real world to conform to a two-dimensional normal distribution is not realistic. In fact, inspection of the geographic distribution of population in North America reveals almost a completely opposite sort of distribution. Rather than a bell shape having a single peak in the center of the continent, this specific population distribution more closely resembles a bowl. The resulting dissatisfaction with geostatistics produced efforts that advanced a complementary set of statistics whose sole purpose was to capture locational information. Measures developed using this line of reasoning fall under the heading of point pattern analysis. Here the quantitative attributes of geographic distributions are ignored, while the accompanying locational patterns are analyzed. In point pattern analysis, treatment of geography is restricted to punctated surfaces (that is, point patterns); partitioned surfaces are of little interest. These statistics are a function of regional demarcations and thus are very sensitive to the definition of regions and the location of boundaries. The resulting calculated values are highly unstable. For these and other reasons point pattern analysis has been severely criticized. Abandoning an approach that focuses strictly on the locational information latent in a geographic data set, attention has returned to classical statistical endeavors. But now the theme concerns reformulation of these classical statistics so that locational information is uncovered by them. This brand of research falls under the heading of spatial autocorrelation and is akin to the time series analysis avenue of research that is conducted in econometrics.

Each of these three groups of techniques, namely, geostatistics, point pattern analysis, and spatial autocorrelation, is examined next. Each of the groups is best suited to a particular type of application. Although there currently is no single set of applications that works in every situation, recent advances in the use of spatial autocorrelation provide some hope for the future.

5.1. Popular geostatistics

Common geostatistics include the spatial mean, spatial median, standard distance, and areal frequency distribution. In practice these are the spatial statistics that rely heavily upon aggregate data formulae, such as those presented in Chapter 4. In theory each item has a geographic reference, usually represented by a Cartesian coordinate of the form (x, y). But most surfaces are partitioned into areal units, with data for these aforementioned coordinates being aggregated and then reported for their affiliated areal units. Here, as in classical statistics, if one is ignorant of the geographic distribution of these data within individual areal units, then the best assumption to invoke is a uniform distribution; this assumption leads to choropleth maps. In conjunction, geostatistics calculated for these areal units assign the aggregated data to their corresponding areal unit centroids and then use the coordinates of these centroids for calculation purposes. Clearly this assumption introduces measurement error into an analysis, for the true coordinates of individual pieces comprising an aggregated attribute are being measured as the corresponding areal unit centroid coordinates.

Areal unit centroids are relatively easy to calculate. First, a sequence of well-placed areal unit boundary coordinates [(k + 1) points] need to be obtained such that the first and last points in this sequence are the same. This set of points overlays a polygon on the areal unit in question, and hence provides a closed polygon (the first and last points are identical) approximation of this areal unit; the accuracy of this approximation is determined in part by the size and the positioning of this set of points (see Figure 5.1, for example). The resulting area and centroid may be defined as follows, where an arbitrary unit of measure is used:

$$\text{Area} = |\textstyle\sum_{i=1}^{k} x_i(y_{i+1} - y_{i-1})/2| \quad \text{where } y_o \equiv y_k$$

$$x_c = |\textstyle\sum_{i=1}^{k} (y_i - y_{i+1})(x_i^2 + x_i x_{i+1} + x_{i+1}^2)|/(6 \bullet \text{Area})$$

$$y_c = |\textstyle\sum_{i=1}^{k} (x_i - x_{i+1})(y_i^2 + y_i y_{i+1} + y_{i+1}^2)|/(6 \bullet \text{Area})$$

EXAMPLE PR-5.1. Calculating the area and centroid of Rincon.

In Tables 5.1a and 5.1b (found at the end of this chapter), 48 digitized boundary coordinates for the Rincon municipio from the Mayaguez Agricultural Administrative Region are presented, together with those calculations necessary for computing the area and centroid of this municipio. The relevant summations of these data are as follows:

$$\text{Area} = |23.670/2| = 11.840$$
$$x_c = |\,1213.49|/(6 \bullet 11.840) = 17.082 \text{ and}$$
$$y_c = |-3756.26|/(6 \bullet 11.840) = 52.876.$$

An accompanying plot of the digitized points together with this centroid appear in Figure 5.1.

Once these centroids have been obtained, various geostatistics can be calculated.

One measure of central tendency in a geographic distribution of data is the **spatial mean**. This statistic exploits the orthogonal axes property of a Cartesian coordinate system; accordingly the X-axis and the Y-axis can be treated separately. The x_is and the y_is of a geographic data set, respectively, become the Cartesian coordinate X-axis and Y-axis values. The aggregate data for an areal unit become the frequency (f_i) for each of a unit's centroid coordinate axes values. Accordingly, the spatial mean is defined as the Cartesian coordinate point

$$(\bar{x}, \bar{y}) = (\textstyle\sum_{i=1}^{n} f_i x_i / \sum_{i=1}^{n} f_i,\ \sum_{i=1}^{n} f_i y_i / \sum_{i=1}^{n} f_i)$$

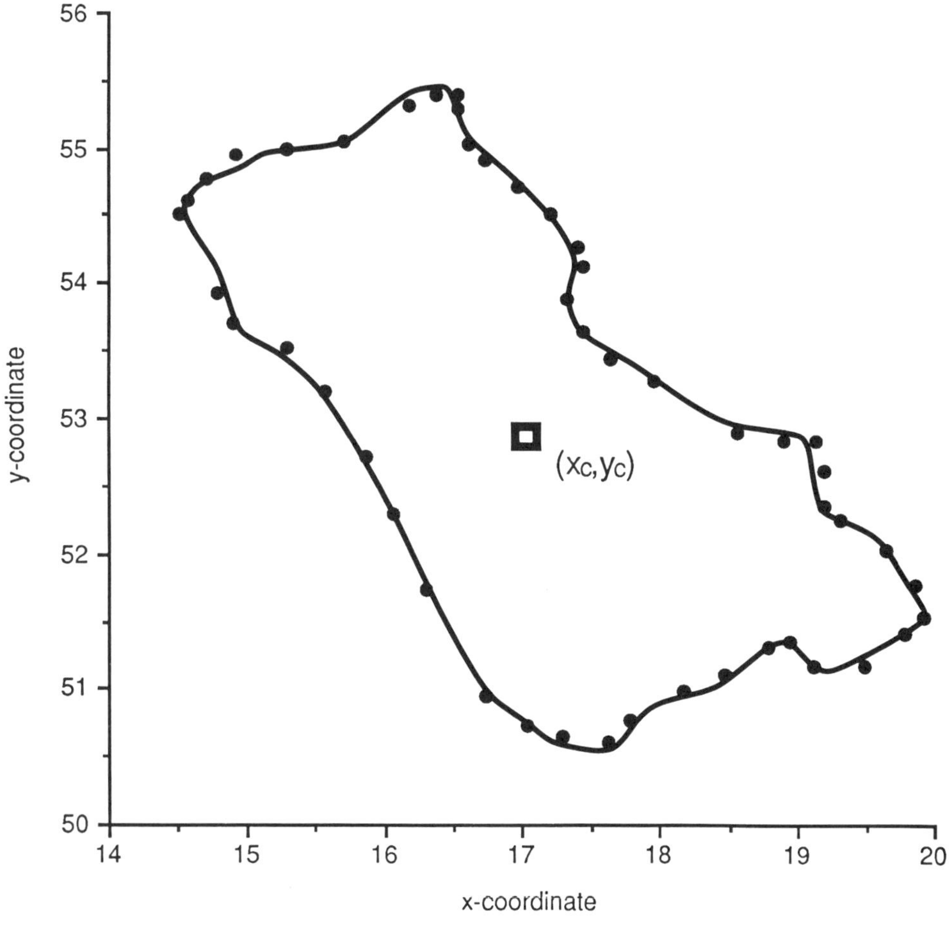

Figure 5.1: PLOT OF THE BOUNDARY AND CENTROID POINTS FOR RINCON

where n is the number of areal units. This location is a center of gravity. Probably the most publicized spatial mean is the population centroid calculated by the U.S. Bureau of the Census after the surveys from each decennial census are tabulated; a physical marker is placed in this location. With the passing of time this location has been moving from the northeast toward the southwest.

EXAMPLE PR-5.2. Calculating spatial means for the Mayaguez Agricultural Administration Region.

Using the milk production density (DMLK) data presented in Table 1.1-A (found in the appendices) and the municipio centroids presented in Table 5.2, the spatial mean for this particular geographic distribution may be calculated as follows:

		Centroid coordinates			
Municipio	f_i	x_i	y_i	f_ix_i	f_iy_i
1	0.1	20.0314	54.6399	2.003	5.464
2	5.6	23.1881	61.1065	132.172	348.307
3	0.0	23.8571	50.1807	0.000	0.000
4	22.5	22.6322	34.4176	509.225	74.396
5	4.4	36.6477	30.8942	161.250	135.934
6	0.1	24.7229	40.4245	2.472	4.042
7	39.6	30.1030	60.8053	1192.079	2407.890
8	6.1	29.4894	32.8150	179.885	200.172
9	0.0	32.1001	47.2462	0.000	0.000
10	0.0	34.9884	43.1383	0.000	0.000
11	3.8	24.8361	45.0923	94.377	171.351
12	2.1	25.8709	56.0336	54.329	117.671
13	0.0	17.0815	52.8758	0.000	0.000
14	3.5	35.0721	37.4654	122.752	131.129
15	0.5	29.2389	39.0352	14.619	19.518
16	6.9	32.6112	53.1334	225.017	366.620
column sum	95.3			2690.180	4682.494

$$\bar{x} = 2690.180/95.3 = 28.229$$

$$\bar{y} = 4682.494/95.3 = 49.134$$

An accompanying plot of the municipio centroids together with this spatial mean appear in Figure 5.2.

The spatial means for the remaining seven density variables for this particular region are as follows:

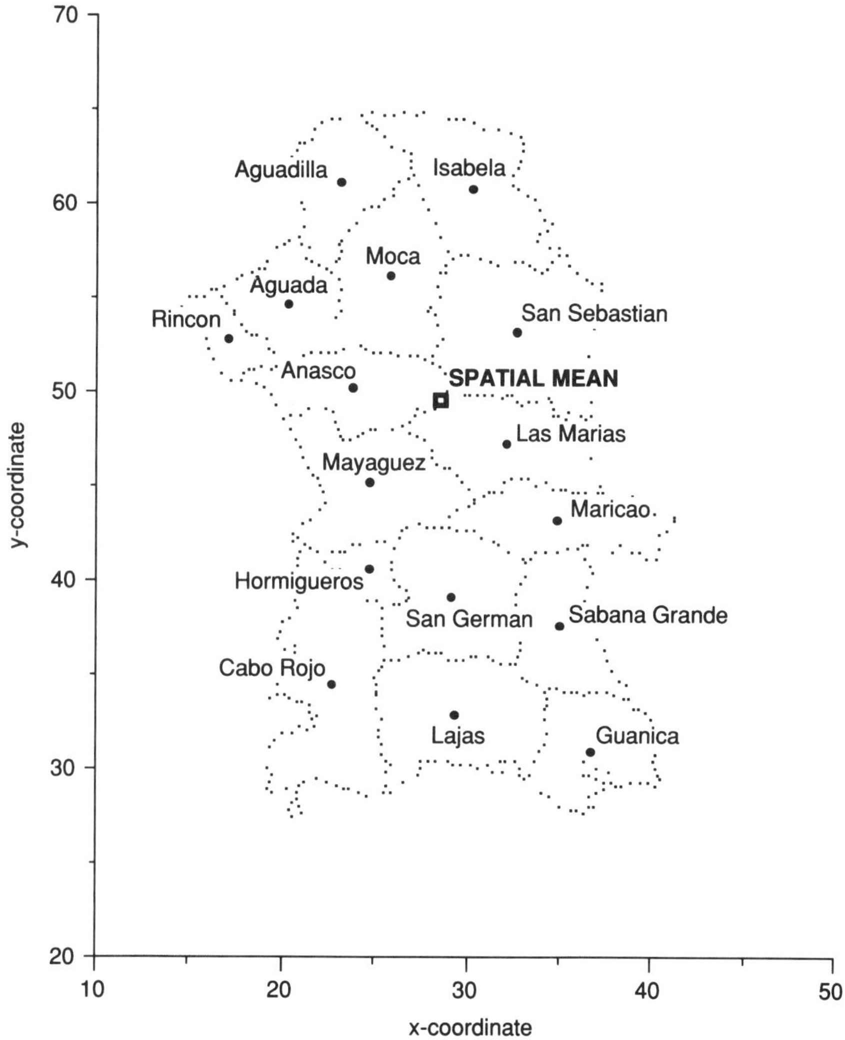

Figure 5.2: PLOT OF THE MUNICIPIO CENTROIDS AND THE SPATIAL MEAN FOR MILK PRODUCTION FOR THE MAYAGUEZ AGRICULTURAL ADMINISTRATIVE REGION

Variable	x-coordinate	y-coordinate
DFRM	27.3685	48.3951
DCD	27.9610	45.0466
DSGR	25.9592	46.6379
DCF	30.3448	46.4375
DTB	30.0195	51.4321
DBN	30.6537	46.2339
DFAM	28.3475	45.6002

The spatial mean can be approximated manually by superimposing a sparse grid upon a map. Although those lines of this grid that are perpendicular to the X-axis must be uniformly spaced, as do the grid lines that are perpendicular to the Y-axis, these two sets of spacings are not required to be equal. Once a grid is superimposed, then intervals along both the X-axis and the Y-axis may be numbered with consecutive odd integers, beginning with 1. These numbers actually are associated with the midpoints of their intervals, implying that even integers are associated with the grid lines themselves. Next, since the aggregate areal unit data are treated as though they cluster at areal unit centroids, the aggregate data for all centroids falling between a pair of parallel grid lines are summed and used as a frequency to be multiplied by the corresponding axis interval integer. At this point the preceding formula for a spatial mean is employed. It is worth noting here that (a) this approximation method introduces additional measurement error into the spatial mean calculation, (b) this sort of changing of the units of measure is irrelevant, since the location of the spatial mean is independent of measurement units, (c) at least two intervals are needed on each axis, and (d) as the grid lines become increasingly denser, the numerical outcome moves closer and closer to that obtained with a continuous Cartesian coordinate system. Therefore, not only does this approximation method furnish a quick and reasonable guess, but it also allows one to evaluate the sensitivity of the spatial mean to measurement error.

EXAMPLE PR-5.3. Calculating the approximate spatial mean of DMLK for the Mayaguez Agricultural Administration Region.

Since the simple approximation is for a scale that has just two equal size intervals, the X-axis can be divided into two such intervals by determining the midpoint of the range of x-coordinate values:

$x_{min} + (x_{max} - x_{min})/2 = 17.0815 + (36.6477 - 17.0815)/2 = 26.8646.$

All x-coordinate values between 17.0815 and 26.8646 are assigned a value of 1, and all values between 26.8646 and 36.6477 are assigned a value of 3.

TABLE 5.2

CALCULATED MUNICIPIO AREAS AND CENTROIDS
FOR THE MAYAGUEZ REGION

Number	Municipio	Actual area	Estimated area	x_c	y_c
2	Aguada	30.21	26.8713	20.0314	54.6399
3	Aguadilla	35.57	31.0115	23.1881	61.1065
6	Anasco	40.05	33.2918	23.8571	50.1807
12	Cabo Rojo	72.35	59.6684	22.6322	34.4176
25	Guanica	36.52	34.5940	36.6477	30.8942
31	Hormigueros	11.16	9.3191	24.7229	40.4245
33	Isabela	55.47	46.5532	29.1030	60.8053
37	Lajas	60.23	51.0072	29.4894	32.8150
39	Las Marias	47.03	39.6011	32.1001	47.2462
44	Maricao	36.85	32.4462	34.9884	43.1383
46	Mayaguez	56.95	45.8654	24.8361	45.0923
47	Moca	50.43	43.0506	25.8709	56.0336
56	Rincon	14.14	11.8399	17.0815	52.8758
58	Sabana Grande	35.24	30.9182	35.0721	37.4654
60	San German	53.90	45.6620	29.2389	39.0352
63	San Sebastian	70.80	60.6780	32.6112	53.1334

The sums of all municipio centroids that fall within these two intervals are, respectively, 34.3 and 61.0 (these sums can be determined by superimposing these interval lines on a map). Similarly, for the Y-axis,

$$y_{min} + (y_{max} - y_{min})/2 = 30.8942 + (61.1065 - 30.8942)/2 = 46.0004.$$

All y-coordinate values between 30.8942 and 46.0004 are assigned a value of 1, and all values between 46.0004 and 61.1065 are assigned a value of 3. The sums of all municipio centroids that fall within these two intervals are, respectively, 40.9 and 54.4. Therefore, the approximate spatial mean equals

$$([34.3 \cdot 1 + 61.0 \cdot 3]/95.3, [40.9 \cdot 1 + 54.4 \cdot 3]/95.3) = (2.2802, 2.1417)$$

which translates into (28.2351,47.0703) on the measurement scale used in Table 5.2 (the x-coordinates of 21.9731 and 31.7562 are mapped into the numbers 1 and 3, and the y-coordinates of 38.4473 and 53.5534 are mapped into the numbers 1 and 3). Consequently, by aggregating the areal unit attributes into four geographic groups defined by the midpoints of the municipio centroid x- and y- ranges, the spatial mean is erroneously located about 2.6 units south and 0.007 units east of its actual location. This mistake is attributable to measurement error (incorrectly measuring the municipio centroids).

The **spatial median** is that point (U, V) minimizing the expression

$$\sum_{i=1}^{n} f_i[(x_i - U)^2 + (y_i - V)^2]^{1/2}$$

where $[(x_i - U)^2 + (y_i - V)^2]^{1/2}$ is the Euclidean distance separating areal unit centroid (x_i, y_i) and the spatial median (U, V). Calculation of this point is far more complicated than the simple midpoint determination of classical statistics. In fact this expression must be minimized in an iterative fashion. In the first step of this iteration, let

$$U_0 = \sum_{i=1}^{n} f_i x_i / \sum_{i=1}^{n} f_i \text{ and}$$

$$V_0 = \sum_{i=1}^{n} f_i y_i / \sum_{i=1}^{n} f_i).$$

These are the first guesses for the spatial median (U, V). Next, for step τ the centroid coordinate approximations are defined as

$$U_\tau = \{\sum_{i=1}^{n} f_i x_i / [(x_i - U_{\tau-1})^2 + (y_i - V_{\tau-1})^2]^{1/2}\} / \{\sum_{i=1}^{n} f_i / [(x_i - U_{\tau-1})^2 + (y_i - V_{\tau-1})^2]^{1/2}\}$$

$$V_\tau = \{\sum_{i=1}^{n} f_i y_i / [(x_i - U_{\tau-1})^2 + (y_i - V_{\tau-1})^2]^{1/2}\} / \{\sum_{i=1}^{n} f_i / [(x_i - U_{\tau-1})^2 + (y_i - V_{\tau-1})^2]^{1/2}\}.$$

Thus for $\tau = 1$, $\tau - 1 = 0$, and the first guesses for the spatial median are used in the right-hand expression. In the second step, $\tau = 2$, $\tau - 1 = 1$ and the (U, V) coordinate calculated for $\tau = 1$ are used in the left-hand expression. The step counter τ continues to be incremented until changes from step to step in the spatial median approximations become negligible. The spatial median obtained in this final step is that point on a surface having the minimum distance to all areal unit centroids.

EXAMPLE PR-5.4. Calculating spatial medians for the Mayaguez Agricultural Administration Region.

A convergence criterion of 10^{-8} was employed in order to terminate the estimation algorithm. The initial guess is the spatial mean. Because of relatively rapid convergence, results for DFRM are reported here.

τ	x-coordinate	y-coordinate	Coordinate name
0	27.3729000	48.3914000	(U_0, V_0)
1	27.4837151	48.5748596	(U_1, V_1)
2	27.5237598	48.6526680	(U_2, V_2)
3	27.5359764	48.6869812	(U_3, V_3)
4	27.5385685	48.7026215	(U_4, V_4)
5	27.5382366	48.7100601	(U_5, V_5)
6	27.5372829	48.7137680	(U_6, V_6)
7	27.5364094	48.7156982	(U_7, V_7)
8	27.5357609	48.7167473	(U_8, V_8)
9	27.5353165	48.7173233	(U_9, V_9)
10	27.5350304	48.7176590	(U_{10}, V_{10})
11	27.5348492	48.7178497	(U_{11}, V_{11})
12	27.5347424	48.7179680	(U_{12}, V_{12})
13	27.5346756	48.7180367	(U_{13}, V_{13})

The remaining seven spatial medians for the Mayaguez region are:

Variable	τ_{max}	x-coordinate	y-coordinate
DCD	17	28.0396938	44.3157997
DMLK	41	29.2233601	55.5407181
DSGR	35	25.2879391	46.8061638
DCF	171	32.0910721	47.2215538
DTB	22	30.9503345	52.8197708
DBN	74	32.0770836	46.8628311
DFAM	16	29.0744553	44.5761108

One measure of dispersion for a geographic distribution is the standard distance. This geostatistic also relies heavily upon the aggregate formulae presented in Chapter 4. Again the orthogonal property of Cartesian coordinate system axes is exploited; once more the X-axis and the Y-axis can be treated separately. As before, the x_is and the y_is of a geographic data set, respectively, become the Cartesian coordinate X-axis and Y-axis values, and the aggregate data for an areal unit become the frequency for each of a unit's centroid coordinate axes values. Accordingly, the **standard distance** may be defined as

$$SD = \{\sum_{i=1}^{n} f_i(x_i - \bar{x})^2/\sum_{i=1}^{n} f_i + \sum_{i=1}^{n} f_i(y_i - \bar{y})^2/\sum_{i=1}^{n} f_i\}^{1/2}$$

where n is the number of areal units, and $(\bar{x}, \bar{y})$ is the spatial mean; one should note that this definition is in terms of a biased sample variance, which usually makes little difference, since the sum of the frequencies often is quite large. Moreover, it is the square root of the sum of the variance along the X-axis and the variance along the Y-axis [i. e., $(s_x^2 + s_y^2)^{1/2}$]. This measure of geographic dispersion is measurement-scale specific, unlike the spatial mean, where scale becomes irrelevant once this mean's position is located on a map. It also can be used to construct contour lines of equidistant deviation about the spatial mean; these contour lines can be elliptical in shape.

EXAMPLE PR-5.5. Calculating standard distances for the Mayaguez Agricultural Administration Region.

Using the DMLK variable from the Puerto Rico data set and the municipio centroids presented in Table 5.2, the standard distance for this particular geographic distribution may be calculated as follows ($\bar{x}$ = 28.229 and $\bar{y}$ = 49.134 from Example PR-5.2):

Municipio	f_i	$(x_i - \bar{x})^2$	$(y_i - \bar{y})^2$	$f_i(x_i - \bar{x})^2$	$f_i(y_i - \bar{y})^2$
Aguada	0.1	67.201	30.315	6.720	3.032
Aguadilla	5.7	25.411	143.341	144.841	817.042
Anasco	0.0	19.114	1.096	0.000	0.000
Cabo Rojo	22.5	31.324	216.572	704.794	4872.880
Guanica	4.4	70.875	332.690	311.848	1463.837
Hormigueros	0.1	12.293	75.855	1.229	7.536
Isabela	39.6	3.512	136.219	139.070	5394.285
Lajas	6.1	1.589	266.310	9.691	1624.490
Las Marias	0.0	14.985	3.564	0.000	0.000
Maricao	0.0	45.689	35.948	0.000	0.000
Mayaguez	3.8	11.512	16.335	43.745	62.074
Moca	2.1	5.861	47.604	11.677	99.969
Rincon	0.0	124.267	14.001	0.000	0.000
Sabana Grande	3.5	46.828	136.156	163.898	476.647
San German	0.5	1.020	101.986	0.510	50.993
San Sebastian	6.9	19.204	15.995	132.505	110.367
column sum	95.3			1670.528	149983.101

$$SD = [(1670.528/95.3) + (14983.101/95.3)]^{1/2} = 13.2193$$

The standard distances for the remaining seven density variables for this particular region are as follows:

Variable	s_x^2	s_y^2	Standard distance
DFRM	28.9453	64.0891	9.6454
DCD	27.3779	90.2799	10.8470
DSGR	23.9422	92.3293	10.7825
DCF	25.0872	25.7462	7.1298
DTB	3.7588	88.5081	9.6056
DBN	23.4104	23.6725	6.8962
DFAM	25.0591	81.2739	10.3053

The standard distance also can be approximated manually by superimposing a sparse grid upon a map, although this approximation is of little use unless the spatial mean has been approximated, too. The same grid lines used for the spatial mean approximation must be used in this calculation. The frequencies, which are the summed aggregate data for all centroids falling between a pair of parallel grid lines, are multiplied by the corresponding axis interval integer. At this point the preceding formula for a standard distance measure is employed. This approximation method introduces measurement error into the standard distance measure and is almost completely dependent upon the grid unit scale employed; standard distances for the same data but different grid unit scales will be proportional, however.

EXAMPLE PR-5.6. Calculating the approximate standard distance of variable DMLK for the Mayaguez Agricultural Administration Region.

The same scale is used here as was used in Example PR-5.3. Therefore,

$$\begin{aligned} SD &= \{[34.3(1 - 2.2802)^2 + 61.0(3 - 2.2802)^2]/95.3 \\ &\quad + [40.9 \cdot (1 - 2.1417)^2 + 54.4 \cdot (3 - 2.1417)^2]/95.3\}^{1/2} \\ &= (0.9215 + 0.9799)^{1/2} = (1.9014)^{1/2} = 1.3789. \end{aligned}$$

which translates [$1 = (21.9731 - 17.0815)/4.8916$, so $s_x^2 = 4.8916^2 \cdot 0.9215$; $1 = (38.4473 - 30.8942)/7.5531$, so $s_y^2 = 7.5531^2 \cdot 0.9799$] into a value of 8.8283 on the measurement scale used in Table 5.2.

Example PR-5.6 demonstrates that this approximation method more or less yields unsatisfactory results for standard distances, primarily because the measurement error introduced by using fewer common centroids can substantially suppress the actual geographic variation.

Areal frequency distributions are nothing more than a three-dimensional map of that geographic distribution under study. Usually these frequency distributions are based upon at least 20, and as many as 125 areal units and may utilize an oblique perspective for this three-dimensional map. The map presented in Figure 5.3 shows the areal frequency distribution for DMLK in the Mayaguez Agricultural Administration Region.

5.2. *Popular point pattern analysis statistics*

Point pattern analysis restricts attention to the locational information that is overlooked by classical statistics, in turn ignoring the attribute information captured by classical statistics. This brand of analysis has addressed two related location problem themes, namely, (a) describing the geographic distribution of a set of points

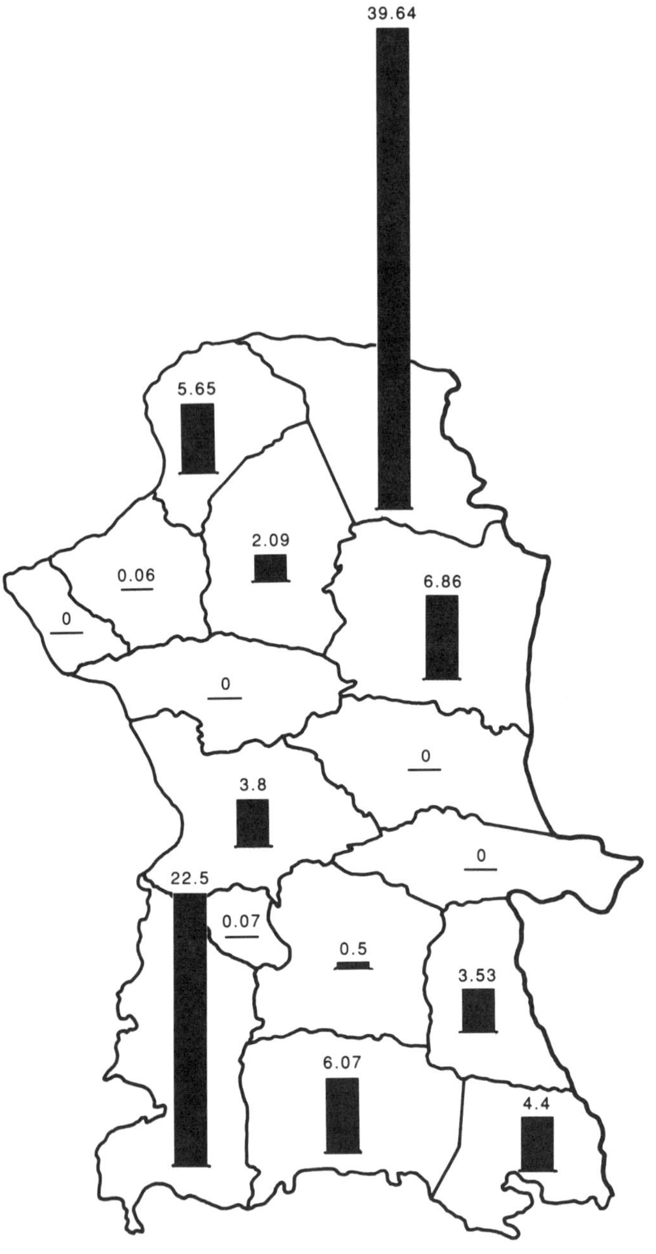

Figure 5.3: AN AREAL FREQUENCY DISTRIBUTION FOR MILK PRODUCTION DENSITY IN THE MAYAGUEZ AGRICULTURAL ADMINISTRATIVE REGION

according to their spacing and (b) describing the geographic distribution of points according to their density. That measure formulated for this first theme is the **nearest neighbor statistic**, which is denoted by R. The statistic R indexes the distance between closest, or nearest neighbor, points on a surface. In this case the population in question is that set of all points on a surface; the sample is that set of points where the attributes of interest are located. This perspective implies that the population is infinite in size and that the observance of any given location should be rare. This nearest neighbor statistic may be defined as follows:

$$R = 2\bar{d}(n/\text{area})^{1/2}$$

where n is the sample size, $\bar{d}$ (which is calculated by dividing the sum of the distances from each point to its nearest neighbor--one distance for each point--by n) is the average nearest neighbor distance, and the area refers to the size of the region within which the points are located. The interpretation of R is analogous to the notion of correlation presented in Chapter 4. Intuitively speaking, if points cluster on a surface, one on top of another in a clump, then the distance between nearest neighbors will be 0, and hence R will be 0. A nearest neighbor statistic value of 0 thus describes a clustered point pattern. If there is no relationship between point locations on a surface, then their pattern will be random and R will be equal to 1 (this numerical value is obtained mathematically by resorting to the notion of random noise in the geographic landscape). A nearest neighbor statistic value of 1 thus describes a random point pattern. If nearby points are positioned as far apart from each other as is physically possible, then they will form a uniform geographic distribution (such a distribution is reminiscent of central place patterns discussed in economic geography), with an accompanying R value of nearly 2.15. A nearest neighbor statistic value of close to 2.15 thus describes a uniform point pattern. A sample of the range of values and accompanying illustrative point patterns is presented in Figure 5.4.

EXAMPLE PR-5.7. Calculating the nearest neighbor statistic of the municipio centroids for the Mayaguez Agricultural Administration Region.

Consider the geographic distribution of municipio centroids presented in Figure 5.2, together with the centroid coordinates presented in Table 5.2. The nearest neighbor centroids and their accompanying nearest neighbor distances are in the following table below.

Here $\bar{d} = (84.94374/16) = 5.3090$, and summing the areas for the digitized municipios produces area = 602.378. Hence,

$$R = 2(5.3090)(16/602.378)^{1/2} = 1.7305.$$

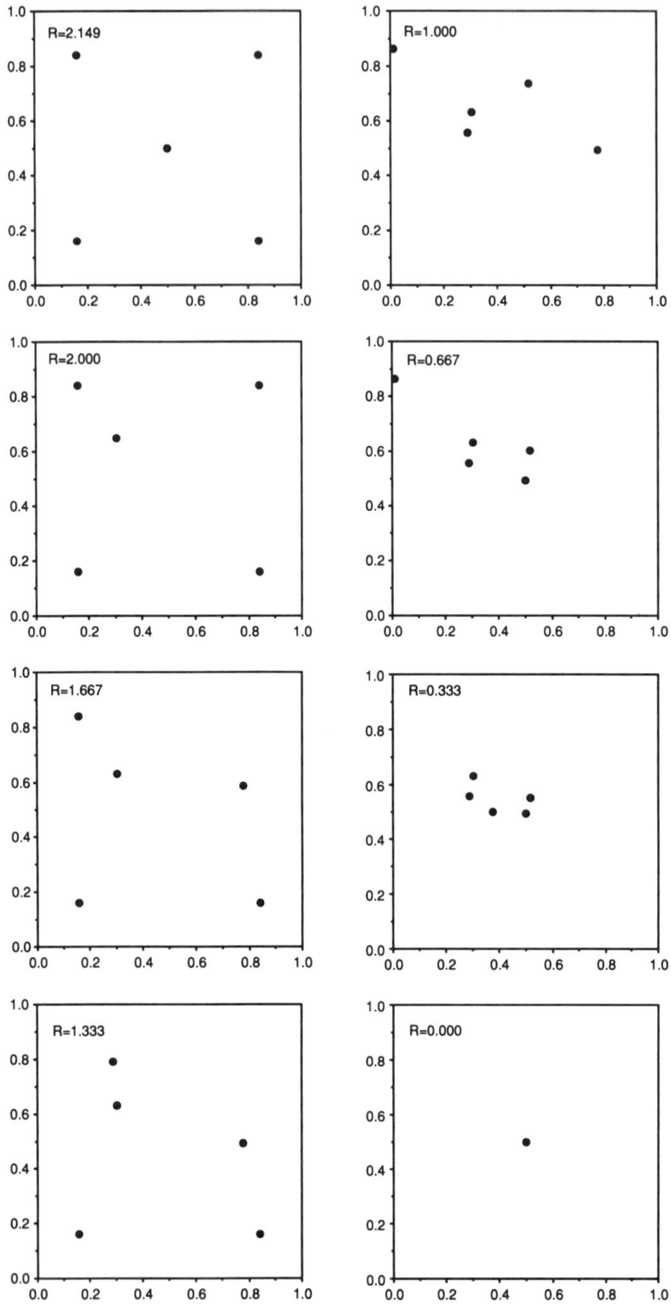

Figure 5.4: EXAMPLE POINT PATTERNS FOR SELECTED NEAREST NEIGHBOR STATISTICS COVERING THE RANGE OF POSSIBLE VALUES, FOR N = 5

This nearest neighbor statistic value suggests that the geographic distribution of municipio centroids may be described as a point pattern whose spacing has a moderate tendency toward being uniform. Intuitively speaking this conclusion makes sense.

Municipio	Nearest neighbor	Distance
Aguada	Rincon	3.43714
Aguadilla	Moca	5.73862
Anasco	Mayaguez	5.18172
Cabo Rojo	Hormigueros	6.36034
Guanica	Sabana Grande	6.75746
Hormigueros	Mayaguez	4.66917
Isabela	Moca	6.37807
Lajas	San German	6.22524
Las Marias	Maricao	5.02166
Maricao	Las Marias	5.02166
Mayaguez	Hormigueros	4.66917
Moca	Aguadilla	5.73862
Rincon	Aguada	3.43714
Sabana Grande	Maricao	5.67352
San German	Hormigueros	4.72487
San Sebastian	Las Marias	5.90934
Column sum		84.94374

Two interesting and noteworthy distance relations appear in Example PR-5.7. First, two points may be reciprocal nearest neighbors (such as Aguada and Rincon), or two points may not be reciprocal nearest neighbors (such as Anasco and Mayaguez). When a reciprocal pair occurs, then one must keep in mind that since the average distance is based upon n distance measures, the reciprocal nearest neighbor distance must appear twice (one nearest neighbor distance for each of the two points involved) in the calculation of $\bar{d}$. Second, a point may be the nearest neighbor to more than one point (such as Hormigueros, which is the nearest neighbor to Cabo Rojo, Mayaguez, and San German), or a point may be the nearest neighbor to no other points (such as Anasco, Cabo Rojo, Guanica, Isabela, Lajas, and San Sebastian).

One major drawback of interpreting the statistic R is that it fails to distinguish between homogeneous point patterns, like those presented in Figure 5.4, and heterogeneous point patterns exhibiting distance relations in addition to nearest neighbor ones, such as those appearing in Figure 5.5. In an attempt to circumvent this particular problem, the nearest neighbor statistic has been modified in such a way that different categories of nearest neighbors are identified. These categories are established by choosing a number of sectors into which a map is divided and then identifying the nearest neighbor point in each sector. The same map partitioning is used for each point and may be visualized as etching a set of lines radiating from a central point on a piece of acetate, with these lines being uniformly spaced (in order to maintain equiprobable possibilities) around this central point. The number of sectors equals the number of sections into which these lines partition a surface. Next, the piece of acetate is superimposed upon each point in a point pattern, subject to the constraint that all superimpositions must maintain the same orientation of the radials, with the central point of the radial lines coinciding with the point on the map. Distance is measured to the closest point in each of the sectors (now a point will have many nearest neighbors), and the average of this much larger collection of distances is calculated. Given this setting, the formula for the statistic R becomes

$$R_k = 2(\bar{d}/\sqrt{k})(n/\text{area})^{1/2}$$

where k is the number of sectors (absence of a subscript on R implies that only one sector has been employed). Noteworthy is the reduction of this version of R to the earlier version; if no lines radiate from the point on the acetate, then there is one sector, $k = 1$, and when the point on the acetate is superimposed upon each map point only a single nearest neighbor is identified.

A second major drawback, which was mentioned in the introduction to this chapter, derives from the sensitivity of R to the demarcation of the appropriate region. In the case of Example PR-5.7, most of the regional boundary is determined by the Caribbean Sea, so this issue is far less problematic. But in the case of, say, retail establishments in an urban area, this issue becomes critical. Consider a given set of points, such as any of those presented in Figure 5.4. Let this set of points be held constant, so that as the region changes in size, no new points appear, and none of the original points disappear. Then, regardless of the value of R presented in Figure 5.4, as the area increases toward infinity, R will decrease to 0. In other words, if a sufficiently large region is defined, even the uniform point pattern displayed in Figure 5.4 ($R = 2.149$) will render an R value of 0. In contrast to this scenario, suppose the region continually shrinks until it is a set of tentacles extending from the most central position. Now R can actually take on values that exceed 2.15. But point patterns cannot be more regular than a uniform spacing, implying that values beyond 2.15 are physically meaningless. Theoretically these limiting cases are avoided by assuming an infinite surface over which a homogeneous geographic distribution of points occurs; but the real world does not consist of infinite surfaces!

Besides spacing of points, one can analyze the density of points. This approach to point pattern analysis is called **quadrat analysis**, which involves the following steps:

Step 1: Construct a set of square grids, called quadrats

Step 2: Superimpose these quadrats, as a grid pattern, over the study area

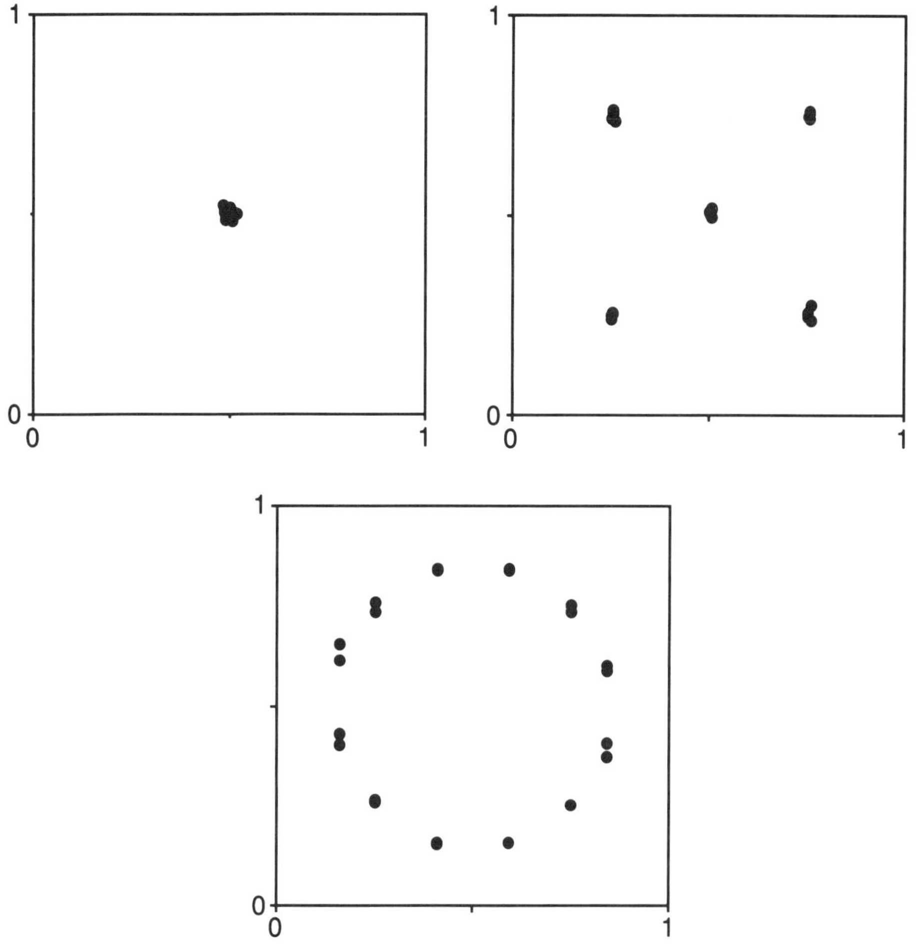

Figure 5.5: THREE POINT PATTERNS HAVING A SIMILAR NEAREST NEIGHBOR STATISTIC

Step 3: Construct a frequency distribution whose horizontal axis is the number of points, and whose vertical axis is the number of quadrats.

Operational problems encountered here include (1) positioning of the grid pattern on the study area and (2) choosing an appropriate grid size. Usually one tries to choose an orientation and placement that minimizes the area of quadrats that extends beyond the study area. Statistically, however, one should choose a random orientation and placement in order to eliminate sampling bias. Empirical experience has found that often the optimal size of quadrat is 2 • area/n, suggesting that the appropriate grid interval width is $(2 \cdot \text{area}/n)^{1/2}$ but there is no formal proof that this size is always optimal. Once the frequency distribution has been calculated, then the coefficient of variation (the CV statistic discussed in Chapter 4) needs to be calculated. A CV value of unity is consistent with a random distribution of points. On the one hand, in the limit, a clustered point pattern will have all points falling into a single quadrat, resulting in a frequency distribution having a very large frequency (all but one of the quadrats) at 0, and a frequency of 1 at n. A CV value substantially greater than 1 will be consistent with this point pattern. On the other hand, in the limit, a uniform distribution of points will have the same number of points falling into each quadrat, resulting in a frequency distribution consisting of a single spike, with this spike occurring at the number of points present in each quadrat. A CV value of 0 will be consistent with this point pattern.

EXAMPLE 5.1. Quadrat analysis of a simulated (randomly generated) point pattern (see Figure 5.6).

Point	Coordinates X	Y	Point	Coordinates X	Y	Point	Coordinates X	Y
A	0.9715	0.0921	J	0.9779	0.3508	R	0.2031	0.4658
B	0.4362	0.5167	K	0.2388	0.8501	S	0.3914	0.2248
C	0.5298	0.5917	L	0.8500	0.2611	T	0.9190	0.0951
D	0.2299	0.9623	M	0.2540	0.6383	U	0.8770	0.8829
E	0.7377	0.2828	N	0.7554	0.7903	V	0.6289	0.3625
F	0.2151	0.3499	O	0.4286	0.7874	W	0.6166	0.3090
G	0.8813	0.7432	P	0.5735	0.4306	X	0.0706	0.6234
H	0.1593	0.9059	Q	0.6816	0.8277	Y	0.8287	0.9198
I	0.9118	0.2428						

The frequency distribution that accompanies this set of 25 points, using the optimal grid size of a 0.2-by-0.2 square, is as follows:

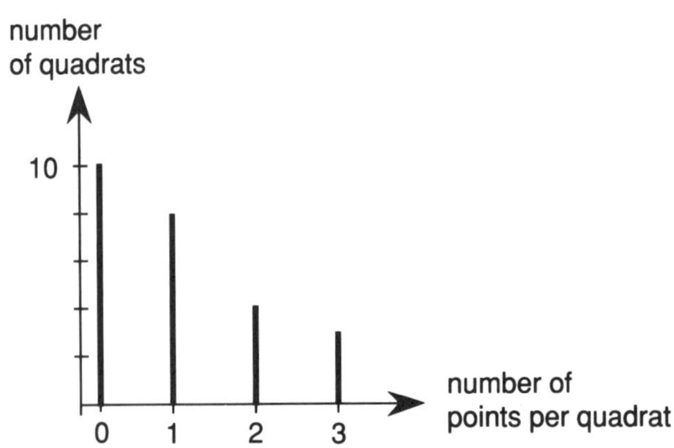

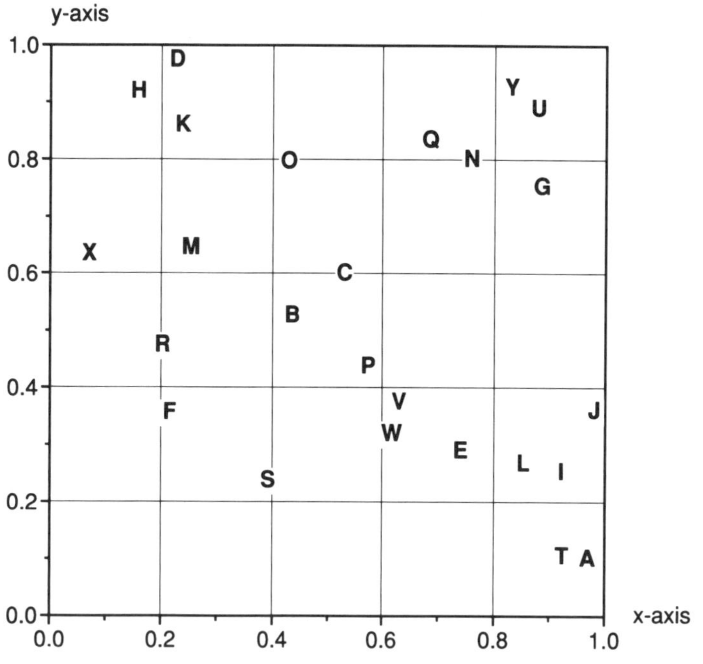

Figure 5.6: SIMULATED RANDOM POINT PATTERN

In the last example the CV statistic essentially equals unity.

5.3. An introduction to popular spatial autocorrelation statistics

After the limited success enjoyed by techniques of point pattern analysis, the initial focus of spatial autocorrelation analysis was a return to classical statistics, but with the thought of reformulating these statistics in order for their new versions to capture locational information. In part, the impetus for this approach came from successes with parallel developments in time series analysis. To the surprise and dismay of many analysts, though, complexities of spatial data series have proven to be far more problematic than those of their time series counterparts. The prominent issue being addressed here is how to handle these complications that lie dormant in traditional data sets but emerge when geographic data sets are analyzed. These difficulties are not visible to an analyst until an attempt is made to capture location information. They disappear in classical statistics by invoking appropriate assumptions.

Spatial autocorrelation is a direct extension of classical correlation, and hence, as the phrase indicates, its accompanying spatial statistics measure the nature and degree of relationships. But the prefix "auto" (literally translated as "self") means that pairs of values of a single variable X are correlated; this notion is in distinct contrast with the classical statistics idea of correlation between two variables. Dependence giving rise to autocorrelation stems from an ordering or arrangement of values. Thus, **spatial autocorrelation** may be defined as the relationship among values of a single variable that is attributable to the geographic arrangement of areal units on a map. This concept is at odds with classical statistical analysis, which assumes that such an ordering does not exist, and hence observations always are pairwise independent. Spatial autocorrelation statistics equivalent to a correlation coefficient approaching unity imply that similar values of variable X tend to cluster, those approaching negative unity imply that dissimilar X values tend to cluster, and those approaching 0 imply that the X values tend to be randomly scattered across a map. These three limiting cases are illustrated with hypothetical data in Figure 5.7.

The arrangement of areal units on a surface can be described as a configuration for the n observed values of some variable X and may be depicted by an n-by-n table. The rows of this table, denoted by subscript i, and the columns of this table, denoted by subscript j, are the same ordered sequence of areal units; entries in this table may be represented by c_{ij}. 0 entries in this table ($c_{ij} = 0$) indicate that the corresponding row and column areal units are not side-by-side or juxtaposed in the geographic configuration. Unity entries in this table ($c_{ij} = 1$) indicate that the corresponding row and column areal units are juxtaposed. Moreover, relative geographic location is measured at the nominal scale.

EXAMPLE PR-5.2. Constructing an areal unit configuration table.

Recalling the 12 hypothetical maps presented at the beginning of this chapter, we can be depict their configuration of areal units as follows:

A	B
C	D

A	E	I	M
B	F	J	N
C	G	K	O
D	H	L	P

geographic configuration of areal units

19	83	84	13
38	55	58	26
50	41	38	75
16	78	23	27

random pattern (no spatial autocorrelation)

84	83	58	38
78	75	50	27
55	41	26	19
38	23	16	13

similar values clustering (positive spatial autocorrelation)

50	23	58	38
19	84	16	55
78	13	83	27
38	75	26	41

dissimilar values clustering (negative spatial autocorrelation)

Figure 5.7: THREE LIMITING CASES OF SPATIAL AUTOCORRELATION FOR A CONTRIVED GEOGRAPHIC LANDSCAPE

The corresponding configuration table is

	A	B	C	D
A	0	1	1	0
B	1	0	0	1
C	1	0	0	1
D	0	1	1	0

This configuration table is used to capture locational information. Note that the binary configuration table adopts the most simplistic view of spatial structure. Each value in the table is either 0 or 1, a place is either adjacent to another, or not. Other methods for constructing configuration tables exist that attempt to measure the importance of one pair of areal unit connections relative to others. These techniques, however, further complicate already elaborate calculations.

As was seen with classical correlation coefficients, understanding spatial autocorrelation measures requires a knowledge of the measurement scale for variable X. If a nominal scale is used, then spatial autocorrelation is measured with the **join count statistic**. If an interval/ratio scale is used, then spatial autocorrelation is measured with either the Moran Coefficient or the Geary Ratio.

If data are measured with a nominal scale, then an areal unit either has or does not have the characteristic in question. Let an areal unit be coded with a 1 if the characteristic in question is present, and a 0 otherwise. Then n_1 areal units are coded 1, and n_2 areal units are coded 0, where $n_1 + n_2 = n$. Imagine a game-board composed of white and black squares. WW denotes the number of juxtaposed areal units for which both units in a pair have been coded with 1, BB denotes the number of cases in which both units in a pair have been coded 0, and BW denotes the number of cases in which one areal unit in a pair has been coded with 1 while the other has been coded with 0. Accordingly,

$$\sum_{j=1}^{m} \sum_{i=1}^{n} c_{ij} = 2(WW + BW + BB) = 2J$$

where J is multiplied by 2 because the table contains a record of areal unit A beside B as well as B beside A; hence, J equals one-half of the sum of the 1s in the configuration table. In keeping with the permutation perspective employed in enumerating all possible maps in the introduction of this chapter, the expected values of these three statistics, when no spatial autocorrelation is present, are

$$E(WW) = (WW + BW + BB)n_1(n_1 - 1)/[n(n - 1)] = Jn_1(n_1 - 1)/[n(n - 1)]$$
$$E(BB) = (WW + BW + BB)n_2(n_2 - 1)/[n(n - 1)] = Jn_2(n_2 - 1)/[n(n - 1)] \text{ and}$$
$$E(BW) = 2(WW + BW + BB) \bullet n_2 n_1/[n(n - 1)] = 2Jn_2 n_1/[n(n - 1)] .$$

This last statistic is multiplied by two because BW takes into account a 1 beside a "0" as well as a 0 beside a 1.

As similarly coded areal units increasingly are juxtaposed BW goes to 0, and either BB increases, WW increases, or both, depending upon the magnitudes of n_1 and n_2 (positive spatial autocorrelation). On the other hand, as dissimilarly coded areal units tend to be juxtaposed then BB goes to 0, WW goes to 0, and BW goes to J, one-half of the total number of 1s in the configuration table (negative spatial autocorrelation).

EXAMPLE 5.2. Calculating join count statistics for the geographic distribution of urban propensity in the Mayaguez Agricultural Administration Region.

First, the areal unit configuration table for the Mayaguez Region is as follows:

		A	B	C	D	E	F	G	H	I	J	K	L	M	N	O	P
A:	Aguada	0	1	1	0	0	0	0	0	0	0	0	1	1	0	0	0
B:	Aguadilla	1	0	0	0	0	0	1	0	0	0	0	1	0	0	0	0
C:	Anasco	1	0	0	0	0	0	0	0	1	0	1	1	1	0	0	1
D:	Cabo Rojo	0	0	0	0	0	1	0	1	0	0	1	0	0	0	1	0
E:	Guanica	0	0	0	0	0	0	0	1	0	0	0	0	0	1	0	0
F:	Hormigueros	0	0	0	1	0	0	0	0	0	0	1	0	0	0	1	0
G:	Isabela	0	1	0	0	0	0	0	0	0	0	0	1	0	0	0	1
H:	Lajas	0	0	0	1	1	0	0	0	0	0	0	0	0	1	1	0
I:	Las Marias	0	0	1	0	0	0	0	0	0	1	1	0	0	0	0	1
J:	Maricao	0	0	0	0	0	0	0	0	1	0	1	0	0	1	1	0
K:	Mayaguez	0	0	1	1	0	1	0	0	1	1	0	0	0	0	1	0
L:	Moca	1	1	1	0	0	0	1	0	0	0	0	0	0	0	0	1
M:	Rincon	1	0	1	0	0	0	0	0	0	0	0	0	0	0	0	0
N:	Sabana Grande	0	0	0	0	1	0	0	1	0	1	0	0	0	0	1	0
O:	San German	0	0	0	1	0	1	0	1	0	1	1	0	0	1	0	0
P:	San Sebastian	0	0	1	0	0	0	1	0	1	0	0	1	0	0	0	0

Let urban propensity be defined such that 1 denotes a municipio having at least 50% of its population in urban areas, and 0 denotes its having less than 50% of its population in urban areas (see Table 1.1-A). Then $n_1 = 4$, the number of 1s, and $n_2 = 12$, the number of 0s. The sum of the numbers in the preceding geographic configuration table equals 64, implying that J = 32. The observed spatial autocorrelation statistics for this example are

$$WW = 1,\ BW = 12,\ \text{and}\ BB = 19,$$

with their respective expected values being

$$\begin{aligned} E(WW) &= 32 \bullet 4 \bullet 3/(16 \bullet 15) &&= 1.6, \\ E(BW) &= 2 \bullet 32 \bullet 12 \bullet 4/(16 \bullet 15) &&= 12.8, \text{ and} \\ E(BB) &= 32 \bullet 12 \bullet 11/(16 \bullet 15) &&= 17.6. \end{aligned}$$

Since the observed statistics appear to be very close to what one would find for a random geographic distribution, the spatial distribution of the urban propensity classification across the Mayaguez Region seems to be random.

Note that in Example PR-5.8 the sum of the three join count statistics equals the sum of their three expected values, which equals J.

When interval/ratio measurements are of concern, then either the **Moran Coefficient** or the **Geary Ratio** measures of spatial autocorrelation are appropriate. This first statistic, denoted here by MC, is written in terms of the type of cross-products found in a classical correlation coefficient formula, but using juxtaposed areal unit values here. MC is defined as

$$MC = [n/(\Sigma_{i=1}^{n} \Sigma_{j=1}^{n} c_{ij})][\Sigma_{i=1}^{n} \Sigma_{j=1}^{n} c_{ij}(x_i - \bar{x})(x_j - \bar{x})/\Sigma_{i=1}^{n} (x_i - \bar{x})^2] \ ,$$

where the denominator of the first term is the sum of the 1s appearing in the configuration table, the denominator of the second term is the sum of the squared deviations used to calculate a classical sample variance, and the numerator of the second term is the cross-product of deviations from the mean for those juxtaposed areal units (there will be one of these cross-products for each 1 present in the configuration table, since all of the remaining cross-products are multiplied by $c_{ij} = 0$). Because calculating MC is similar to calculating a classical correlation coefficient, the values it takes on tend to range between -1 and 1; they are not restricted to this range, though, as is the traditional correlation coefficient. Accordingly, MC values approaching unity imply that similar values of variable X tend to cluster on a map, indicating positive spatial autocorrelation. MC values approaching -1 imply that dissimilar X values tend to cluster, indicating negative spatial autocorrelation. MC values approaching the expected value of $-1/(n - 1)$ (which converges to 0 as n goes to infinity) imply that the X values tend to be randomly scattered, indicating an absence of or 0 spatial autocorrelation.

The Geary Ratio, denoted here by GR, is written in terms of paired comparisons of juxtaposed areal unit values and may be defined as

$$GR = [(n - 1)/(2 \Sigma_{i=1}^{n} \Sigma_{i=1}^{n} c_{ij})][\Sigma_{i=1}^{n} \Sigma_{i=1}^{n} c_{ij}(x_i - x_j)^2/\Sigma_{i=1}^{n} (x_i - \bar{x})^2]$$

where the denominator of the first term is twice the sum of the 1s appearing in the configuration table, the denominator of the second term is the sum of the squared deviations used to calculate a classical sample variance, and the numerator of the second term is the squared differences between values for juxtaposed areal units (there will be one of these squared differences for each 1 present in the configuration table). The values taken on by GR tend to range between 0 and 2; they are not restricted to this range, though. Obviously if juxtaposed values are identical, then their difference is 0, establishing an absolute lower limit of 0. Accordingly, GR values approaching 0 imply that similar values of variable X tend to cluster on a map, indicating positive spatial autocorrelation is present. GR values approaching 2 imply that dissimilar X values tend to cluster, indicating negative spatial autocorrelation. GR values approaching the expected value of 1 imply that the X values tend to be randomly scattered, indicating an absence of or 0 spatial autocorrelation.

These two spatial autocorrelation statistics often imply the same results, but they are inversely related, meaning that as MC increases, GR will decrease. Since MC exploits cross-products whereas GR exploits paired comparisons information, they sometimes uncover nuances of data, that suggest something different about the observed map pattern.

EXAMPLE PR-5.9. Calculating the Moran Coefficient and the Geary Ratio of DMLK for the Mayaguez Agricultural Administration Region.

$$n = 16\ ,\ \sum_{j=1}^{m} \sum_{i=1}^{n} c_{ij} = 64\ ,\ \sum_{i=1}^{n} (x_i - \bar{x})^2 = 1674.8192\ ,$$

$$\sum_{j=1}^{m} \sum_{i=1}^{n} c_{ij}(x_i - \bar{x})(x_j - \bar{x}) = 80.3312$$

$$\sum_{j=1}^{m} \sum_{i=1}^{n} c_{ij}(x_i - x_j)^2 = 11063.0984$$

and so MC = (16/64)(80.3312/1674.8192) = 0.0120, which is quite close to the expected value of -0.0667 for 0 spatial autocorrelation, and therefore GR = (15/128)(11063.0984/1674.8192) = 0.7741, which is close to the expected value of 1 for zero spatial autocorrelation, implying that there is a very weak but slightly positive tendency for similar values of DMLK to cluster on the map.

Spatial autocorrelation statistics for the remaining seven density variables for this particular region are as follows:

Variable	MC	GR	Variable	MC	GR
DFRM	0.1173	0.7496	DCF	0.2438	0.6671
DCD	-0.0583	0.9584	DTB	-0.1344	1.0371
DSGR	0.1116	0.7649	DBN	0.1973	0.7256
			DFAM	0.0540	0.9003

These index values suggest that the levels of spatial autocorrelation present range from weakly negative through weak-to-moderately positive.

5.4. Summary

This chapter has introduced commonly applied techniques in each of the groups of spatial statistics. It is probably clear that explicitly incorporating spatial structure into any study greatly complicates the analysis over what would be encountered if space were ignored (that is, the world of classical statistics). Each of these groups of techniques provides a range of more sensitive and sophisticated procedures. This is particularly true of point pattern analysis and spatial autocorrelation techniques. The latter group especially is experiencing rapid growth as the power of microcomputers permits the development of techniques with enormous computational requirements. Clearly only the most trivial problems in spatial statistics can be solved by hand. This will become increasingly obvious in following chapters.

5.5. Problems

1. Verify the area and centroid calculations of Table 5.2 for Hormigueros using the following 31 digitized (x,y) boundary points (keep in mind that the first and last points are identical):

(26.09,41.67)	(25.88,41.74)	(25.54,42.04)	(25.06,42.10)	(24.24,41.92)
(23.80,41.76)	(23.42,41.56)	(23.06,41.50)	(22.70,41.54)	(22.26,41.46)
(22.00,41.48)	(22.02,41.24)	(22.56,40.70)	(23.00,40.54)	(23.48,40.32)
(23.60,39.84)	(23.74,39.52)	(24.10,39.24)	(24.94,38.84)	(25.36,38.60)
(25.62,38.64)	(26.00,38.66)	(26.38,38.66)	(26.78,38.88)	(26.90,39.14)
(26.72,39.46)	(26.34,39.62)	(26.10,40.00)	(25.96,40.48)	(25.98,40.90)
(26.09,41.67)				

Construct a plot of these digitized points and the resulting centroid.

2. Verify the area and centroid calculations of Table 5.2 for Sabana Grande using the following 43 digitized (x,y) boundary points (keep in mind that the first and last points are identical):

(36.10,41.67)	(35.78,41.62)	(35.38,41.68)	(34.96,41.62)	(34.50,41.38)
(33.92,41.34)	(33.64,41.50)	(33.28,41.18)	(33.12,40.86)	(33.18,40.16)
(32.82,39.86)	(32.66,39.28)	(32.62,38.62)	(32.60,37.88)	(32.70,37.36)
(32.84,36.74)	(32.80,35.84)	(32.80,35.38)	(32.98,34.96)	(33.28,34.58)
(33.90,34.44)	(34.34,34.06)	(34.68,34.18)	(35.44,34.04)	(35.88,34.06)
(36.32,33.96)	(36.92,33.96)	(37.70,33.98)	(38.44,33.74)	(38.72,33.82)
(38.64,34.46)	(38.26,34.66)	(37.80,35.20)	(37.60,35.68)	(37.38,36.26)
(36.96,36.88)	(36.72,37.62)	(36.62,38.28)	(36.42,38.92)	(36.50,39.58)
(36.70,40.28)	(36.60,40.68)	(36.42,41.10)	(36.10,41.67)	

Construct a plot of these digitized points and the resulting centroid.

3. Consider the following subset of well-positioned boundary points from Problem #1 (points 1, 6, 11, 16, 21 and 26):

(26.09,41.67)	(23.80,41.76)	(22.00,41.48)	(23.60,39.84)	(25.62,38.64)
(26.72,39.46)	(26.09,41.67)			

Recalculate the area and the centroid to illustrate the magnitude of measurement error involved when fewer well-positioned digitized points are employed.

4. Consider the following subset of ill-positioned boundary points from Problem #1 (the first five points plus the last point):

   (26.09,41.67) (25.88,41.74) (25.54,42.04) (25.06,42.10) (24.24,41.92) (26.09,41.67)

   Recalculate the area and the centroid to illustrate the magnitude of measurement error involved when ill-positioned digitized points are employed.

5. Verify the spatial means presented in Example PR-5.2, using those data appearing in Tables 1.1-A and 5.2.

6. Calculate the corresponding approximations to the spatial means presented in Example PR-5.2, using just two intervals on each of the X- and Y- axes.

7. Verify the spatial medians for density of farm land and density of farm families presented in Example PR-5.4, using those data appearing in Tables 1.1-A and 5.2.

8. Verify the standard distance measures presented in Example PR-5.5, using those data appearing in Table 1.1-A and Table 5.2.

9. Calculate the corresponding approximations to the standard distances presented in Example PR-5.5, following the setup employed in Problem #6 and adhering to the guidelines established in Example PR-5.6.

10. Using the set of random numbers presented in Table 7.1-A and outlining a 25-by-25 square region on a sheet of graph paper, select 15 random (x, y) co--ordinate pairs, and locate these pairs in the study region (state the positions within the random numbers table from which you extracted your sample). Next, calculate the traditional as well as the modified nearest neighbor statistic for this distribution of points, and interpret the resulting R_k value. Third, conduct a quadrat analysis of this distribution of points.

11. Interpret each of the following 13 nearest neighbor statistics presented by Silk (1979). For each of the phenomena described, state whether or not you think it is statistically independent or dependent over space. Briefly explain each of your answers.

Service	Nearest neighbor index	Number of outlets
post offices	1.589	51
fresh fish stores	1.207	20
laundromats	1.171	35
funeral homes	1.076	5
newspaper stands	1.063	109
barber shops	1.051	41
beauty shops	1.036	79
drug stores	1.029	34
card shops	0.964	118
dry cleaners	0.913	32
women's clothing stores	0.492	100
furniture stores	0.199	19
savings & loan banks	0.090	27

12. Calculate the join count statistics for each of the following map patterns, and then describe the nature and degree of spatial autocorrelation suggested by these statistics:

1	1	1	0
1	1	0	0
1	0	0	0

1	0	1	0
0	1	0	1
1	0	1	0

1	0	1	1
0	0	0	1
1	1	0	0

13. Verify the Moran Coefficient and the Geary Ratio measures presented in Example PR-5.9, using those data appearing in Table 1.1-A.

5.6. References

Silk, J. (1979) *Statistical Concepts in Geography*. Winchester, Mass.: Allen & Unwin.

5.7. Additional readings

Bartlett, M. (1976) *The Statistical Analysis of Spatial Pattern.* London: Chapman and Hall.

Thomas, R. *An Introduction to Quadrat Analysis.* CATMOG 12.

Kellerman, A. *Centrographic Measures in Geography.* CATMOG 32.

Aplin, G. *Order-neighbor Analysis.* CATMOG 36.

Cliff, A., and J. Ord. (1973) *Spatial Autocorrelation.* London: Pion.

Cliff, A., and J. Ord. (1981) *Spatial Processes, Models and Applications.* London: Pion.

Cliff, A., P. Haggett, J. Ord, K. Bassett, and R. Davies. (1975) *Elements of Spatial Structure: A Quantitative Approach.* Cambridge: Cambridge University Press.

Duncan, O., R. Cuzzort, and B. Duncan. (1961) *Statistical Geography, Problems in Analyzing Areal Data.* Glencoe, Il.: The Free Press.

Getis, A., and B. Boots. (1978) *Models of Spatial Processes: An Approach to the study of Point, Line and Area Patterns.* Cambridge: Cambridge University Press.

Griffith, D. (1987) *Spatial Autocorrelation: A Primer.* Washington, D.C.: The Association of American Geographers.

Haggett, P., A. Cliff, and A. Frey. (1977) *Locational Methods*, Vol. 2 of 2nd ed. of *Location Analysis in Human Geography.* New York: Wiley, Halstead Press.

Lewis, P. (1977) *Maps and Statistics.* London: Methuen.

Odland, J. (1988) *Spatial Autocorrelation.* Beverly Hills, Calif.: Sage (Volume 9 in the Scientific Geography Series).

Ripley, B. (1981) *Spatial Statistics.* New York: Wiley.

Rogers, A. (1974) *Statistical Analysis of Spatial Dispersion.* London: Pion.

TABLE 5.1

CENTROID CALCULATION TERMS FROM 48 DIGITIZED BOUNDARY POINTS FOR RINCON

Table 5.1a: Digitized Points Sequence

Point	y_i	x_i	y_{i-1}	y_{i+1}	x_{i+1}	Point	y_i	x_i	y_{i-1}	y_{i+1}	x_{i+1}
1	16.53	55.40	16.38	16.54	55.30	26	18.78	51.30	18.94	18.46	51.10
2	16.54	55.30	16.53	16.62	55.04	27	18.46	51.10	18.78	18.18	50.98
3	16.62	55.04	16.54	16.74	54.92	28	18.18	50.98	18.46	17.78	50.76
4	16.74	54.92	16.62	16.98	54.72	29	17.78	50.76	18.18	17.62	50.60
5	16.98	54.72	16.74	17.20	54.50	30	17.62	50.60	17.78	17.28	50.64
6	17.20	54.50	16.98	17.40	54.26	31	17.28	50.64	17.62	17.02	50.72
7	17.40	54.26	17.20	17.44	54.12	32	17.02	50.72	17.28	16.74	50.94
8	17.44	54.12	17.40	17.32	53.88	33	16.74	50.94	17.02	16.30	51.74
9	17.32	53.88	17.44	17.44	53.64	34	16.30	51.74	16.74	16.06	52.30
10	17.44	53.64	17.32	17.64	53.44	35	16.06	52.30	16.30	15.86	52.72
11	17.64	53.44	17.44	17.96	53.28	36	15.86	52.72	16.06	15.58	53.20
12	17.96	53.28	17.64	18.56	52.90	37	15.58	53.20	15.86	15.30	53.52
13	18.56	52.90	17.96	18.90	52.84	38	15.30	53.52	15.58	14.90	53.70
14	18.90	52.84	18.56	19.14	52.84	39	14.90	53.70	15.30	14.78	53.92
15	19.14	52.84	18.90	19.20	52.62	40	14.78	53.92	14.90	14.52	54.50
16	19.20	52.62	19.14	19.20	52.36	41	14.52	54.50	14.78	14.58	54.62
17	19.20	52.36	19.20	19.32	52.26	42	14.58	54.62	14.52	14.70	54.78
18	19.32	52.26	19.20	19.64	52.04	43	14.70	54.78	14.58	14.92	54.96
19	19.64	52.04	19.32	19.86	51.78	44	14.92	54.96	14.70	15.30	55.00
20	19.86	51.78	19.64	19.92	51.54	45	15.30	55.00	14.92	15.72	55.06
21	19.92	51.54	19.86	19.78	51.40	46	15.72	55.06	15.30	16.18	55.32
22	19.78	51.40	19.92	19.48	51.16	47	16.18	55.32	15.72	16.38	55.40
23	19.48	51.16	19.78	19.12	51.16	48	16.38	55.40	16.18	16.53	55.40
24	19.12	51.16	19.48	18.94	51.34	49	16.53	55.40			
25	18.94	51.34	19.12	18.78	51.30						

Note: $(x_1, y_1) \equiv (x_{49}, y_{49})$.

TABLE 5.1b: Computation Terms

Point	$y_{i+1}-y_{i-1}$	$x_i(y_{i+1}-y_{i-1})$	Point	$y_{i+1}-y_{i-1}$	$x_i(y_{i+1}-y_{i-1})$	Point	$y_{i+1}-y_{i-1}$	$x_i(y_{i+1}-y_{i-1})$
1	0.16	8.86	17	0.12	6.28	33	-0.72	-36.68
2	0.09	4.98	18	0.44	22.99	34	-0.68	-35.18
3	0.20	11.01	19	0.54	28.10	35	-0.44	-23.01
4	0.36	19.77	20	0.28	14.50	36	-0.48	-25.31
5	0.46	25.17	21	-0.08	- 4.12	37	-0.56	-29.79
6	0.42	22.89	22	-0.44	-22.62	38	-0.68	-36.39
7	0.24	13.02	23	-0.66	-33.77	39	-0.52	-27.92
8	-0.08	- 4.33	24	-0.54	-27.63	40	-0.38	-20.49
9	0.00	0.00	25	-0.34	-17.46	41	-0.20	-10.90
10	0.32	17.16	26	-0.48	-24.62	42	0.18	9.83
11	0.52	27.79	27	-0.60	-30.66	43	0.34	18.63
12	0.92	49.02	28	-0.68	-34.67	44	0.60	32.98
13	0.94	49.73	29	-0.56	-28.43	45	0.80	44.00
14	0.58	30.65	30	-0.50	-25.30	46	0.88	48.45
15	0.30	15.85	31	-0.60	-30.38	47	0.66	36.51
16	0.06	3.16	32	-0.54	-27.39	48	0.35	19.39

Point	y_i-y_{i+1}	$x_i^2+x_ix_{i+1}+x_{i+1}^2$	$(y_i-y_{i+1})(x_i^2+x_ix_{i+1}+x_{i+1}^2)$	Point	y_i-y_{i+1}	$x_i^2+x_ix_{i+1}+x_{i+1}^2$	$(y_i-y_{i+1})(x_i^2+x_ix_{i+1}+x_{i+1}^2)$
1	-0.01	9190.87	- 91.91	25	0.16	7901.23	1264.20
2	-0.08	9131.20	- 730.50	26	0.32	7864.33	2516.60
3	-0.12	9068.40	-1088.20	27	0.28	7815.25	2188.26
4	-0.24	9015.71	-2163.77	28	0.40	7763.28	3105.31
5	-0.22	8946.77	-1968.30	29	0.16	7705.39	1232.86
6	-0.20	8871.57	-1774.30	30	0.34	7687.15	2613.63
7	-0.04	8809.67	- 352.39	31	0.26	7705.39	2003.40
8	0.12	8748.01	1049.77	32	0.28	7751.08	2170.31
9	-0.12	8670.43	-1040.46	33	0.44	7907.55	3479.32
10	-0.20	8599.60	-1719.91	34	0.24	8118.32	1948.39
11	-0.32	8541.87	-2733.40	35	0.20	8271.94	1654.39
12	-0.60	8455.68	-5073.41	36	0.28	8414.34	2356.01
13	-0.34	8385.71	-2851.14	37	0.28	8541.89	2391.73
14	-0.24	8376.20	-2010.29	38	0.40	8622.10	3448.85
15	-0.06	8341.37	- 500.49	39	0.12	8686.56	1042.39
16	0.00	8265.62	0.00	40	0.26	8816.26	2292.22
17	-0.12	8209.01	- 985.07	41	-0.06	8930.38	- 535.82
18	-0.32	8158.88	-2610.84	42	-0.12	8976.28	-1077.15
19	-0.22	8083.96	-1778.48	43	-0.22	9032.16	-1987.08
20	-0.06	8006.28	- 480.37	44	-0.38	9068.40	-3445.99
21	0.14	7947.49	1112.64	45	-0.42	9084.90	-3815.66
22	0.30	7888.93	2366.69	46	-0.46	9137.83	-4203.40
23	0.36	7852.04	2826.72	47	-0.20	9194.19	-1838.83
24	0.18	7879.70	1418.35	48	-0.15	9207.48	-1381.14

TABLE 5.1b: Computation Terms, Continued

Point	$x_i - x_{i+1}$	$y_i^2 + y_i y_{i+1} + y_{i+1}^2$	$(x_i - x_{i+1})(y_i^2 + y_i y_{i+1} + y_{i+1}^2)$	Point	$x_i - x_{i+1}$	$y_i^2 + y_i y_{i+1} + y_{i+1}^2$	$(x_i - x_{i+1})(y_i^2 + y_i y_{i+1} + y_{i+1}^2)$
1	0.10	820.22	82.02	25	0.04	1067.11	42.69
2	0.26	824.69	214.42	26	0.20	1040.14	208.03
3	0.12	834.67	100.16	27	0.12	1006.89	120.83
4	0.20	852.79	170.56	28	0.22	969.88	213.38
5	0.22	876.22	192.77	29	0.16	939.88	150.38
6	0.24	897.88	215.49	30	-0.04	913.54	- 36.54
7	0.14	910.37	127.45	31	-0.08	882.38	- 70.59
8	0.24	906.20	217.49	32	-0.22	854.82	-188.06
9	0.24	906.20	217.49	33	-0.80	818.78	-655.03
10	0.20	922.96	184.59	34	-0.56	785.39	-439.82
11	0.16	950.55	152.09	35	-0.42	764.17	-320.95
12	0.38	1000.37	380.14	36	-0.48	741.37	-355.86
13	0.06	1052.47	63.15	37	-0.32	715.20	-228.86
14	0.00	1085.30	0.00	38	-0.18	684.07	-123.13
15	0.22	1102.47	242.54	39	-0.22	660.68	-145.35
16	0.26	1105.92	287.54	40	-0.58	643.88	-373.45
17	0.10	1112.85	111.29	41	-0.12	635.11	- 76.21
18	0.22	1138.44	250.45	42	-0.16	642.99	-102.88
19	0.26	1170.20	304.25	43	-0.18	658.02	-118.44
20	0.24	1186.84	284.84	44	-0.04	684.97	- 27.40
21	0.14	1182.07	165.49	45	-0.06	721.72	- 43.30
22	0.24	1156.03	277.45	46	-0.26	763.26	-198.45
23	0.00	1117.50	0.00	47	-0.08	795.13	- 63.61
24	-0.18	1086.43	-195.56	48	0.00	812.31	0.00

CHAPTER 6

AN EXCURSION INTO ELEMENTARY PROBABILITY THEORY

In previous chapters we identified the objectives of statistical analysis and surveyed informative classical as well as spatial statistics. We also commented on and referred to sampling. The question that we address at this point, then, asks whether or not a sample statistic is likely to be observed, given the assumed parent population from which its sample was selected. Answers to this question are at the heart of inferential statistics. Probability and statistics are two separate, but related, fields of the mathematical sciences, with the laws of probability enabling the theory of statistics to answer this question. To establish such answers, though, we need to examine probability theory. Probability is the engine driving scientific declarations about the likelihood of what will occur when samples are drawn from a specified population.

6.1. What is a probability?

A probability assigns to each possible outcome of an activity some number giving a precise measure of the chance that individual outcomes will occur. To obtain this measure, the term probability may be defined in three subtly different ways that cast the results in terms of a numerical value derived from a relative frequency. The first definition, which describes **empirical probabilities**, states that a probability is merely the observed relative frequency with which an outcome occurs, or can be expected to occur. When a sample is selected, the proportion of times an outcome occurs in that sample is an empirical probability. For example, in the Puerto Rico data set, a sample of size 25 might be selected, with 14 municipios classified as coastal lowlands and 11 classified as interior highlands, yielding an empirical probability of 14/25 for randomly selecting a coastal lowlands municipio (since there are 73 municipios in the population, and 31 of them are classified as interior highland municipios, the actual coastal lowlands probability is 42/73). Alternatively, when a sample is drawn, this measure can be calculated as the expected proportion of times an outcome actually occurs relative to the possible number of times it could occur. This definition of probability is based upon deductive logic and exploits arguments of symmetry or geometry. For instance, since there are only two possible municipio LND classes, the expected probability may be posited as being 1/2.

The second definition of probability is built upon a more comprehensive concept of sampling. Consider all possible samples of a given size n that can be drawn from a population; this set is called a **sample space**. If each sample contained in this sample space is equally likely to be drawn, then probability is defined to be a ratio between the total number of samples displaying a certain attribute (perhaps a statistic that is calculated for each sample) in the numerator and the total number of samples contained in this sample space in the denominator. This is a **theoretical probability**. While probability in this case still is a relative frequency or proportion, the fraction is constructed in relation to the list of all possibilities. Because this definition requires the existence of an equally likely sample space (random sampling), it is advantageous to construct equally likely sample spaces whenever possible.

The third definition of probability integrates the two preceding definitions in that **probability** is defined to be a limit of the ratio of an outcome to the total number of possible outcomes. This empirical probability is defined as a proportion over the long run and is calculated for each sample size n. The value of the probability converges on a theoretical probability as n increases to infinity in either an asymptotic (that is, smooth and nondecreasing) or oscillatory (that is, flipping back and forth on each side) fashion. This concept of probability is cast within the context of the **law of large numbers**, which states that the ratio of the number of outcomes of interest to the number of sampling repetitions approaches the elusive theoretical probability of the outcome for an individual sample selection as the number of samples drawn increases. Because the meaning of probability is based upon the concept of a limiting relative frequency, its applicability is restricted to replicable and repeatable situations yielding the outcome in question.

Given these three somewhat distinctive views of probability, general mathematical properties of this notion may be established. First, denote the probability of some item as P(x) or of a value taken by some random variable as $P(X = x)$ or $P(a \leq X \leq b)$. Since all three definitions render ratios containing counts in both the numerator and the denominator, a probability must be non-negative. In other words, $P(x) \geq 0$. Second, since each definition of probability refers to a proportion of some total, a probability must have an upper bound of 1. In other words, $P(x) \leq 1$. Combining these two rules requires that $0 \leq P(x) \leq 1$. Third, if the set of possible outcomes is mutually exclusive (the occurrence of a single, arbitrary outcome precludes occurrence of any of the other outcomes) and collectively exhaustive (all possible outcomes are included), then

$$\sum_{i=1}^{K} P(x_i) = 1$$

where K is the total number of possible outcomes, and x_i is the number of occurrences with outcome i. If outcomes are not mutually exclusive, then they can be reconstructed in such a way that their modified forms are mutually exclusive. Note that these properties do not determine an assignment of probabilities to outcomes; rather, they function merely to rule out assignments inconsistent with the foregoing definitions of probability. Establishing a better understanding of these three properties of probabilities, as well as expanding them into a larger arsenal of helpful probability attributes, requires a reconnaissance into elementary set theory. It also requires exploring the methods of counting all possible samples.

6.2. Counting all possible samples of size n

The principles of counting all possible samples of a given size n, without resorting to enumerating all of them, are predicated upon selected characteristics of a sampling procedure. These characteristics interact with the proposition, emphasized in the previous section, that all outcomes should be equally likely. Strategies affiliated with random sampling tend to yield samples representative of the parent population. The theme of this chapter is to ascertain which statistics computed for samples tend to be likely. The two remaining prominent sample characteristics developed later in this chapter pertain to whether or not sampling is conducted with replacement, and whether or not the order of observation selection in drawing a sample is important. Classical sampling theory assumes that the procedure involves replacement, with order being important. This situation serves as the starting point of the discussion.

The initial proposition is that if each of a sequence of K procedures can be performed in n_k different ways, where the subscript k denotes the kth procedure, then the number of possible ways a given sequence of procedures can be performed is

$$\prod_{k=1}^{K} n_k.$$

In other words, if there are three procedures to be executed, say A, B, and C, and each respectively can be performed in (a_1, a_2, a_3), (b_1, b_2), and (c_1, c_2, c_3, c_4) ways, then the total number of possible execution sequences is 3 • 2 • 4 = 24. For example, imagine three bins A, B, and C. Bin A has three balls (a_1, a_2, a_3), bin B has two balls (b_1, b_2), and bin C has four balls (c_1, c_2, c_3, c_4). Thus, there are 24 possible outcomes from a process that selects one ball from each bin, namely,

$a_1 - b_1 - c_1$, $a_1 - b_1 - c_2$, $a_1 - b_1 - c_3$, $a_1 - b_1 - c_4$
$a_1 - b_2 - c_1$, $a_1 - b_2 - c_2$, $a_1 - b_2 - c_3$, $a_1 - b_2 - c_4$
$a_2 - b_1 - c_1$, $a_2 - b_1 - c_2$, $a_2 - b_1 - c_3$, $a_2 - b_1 - c_4$
$a_2 - b_2 - c_1$, $a_2 - b_2 - c_2$, $a_2 - b_2 - c_3$, $a_2 - b_2 - c_4$
$a_3 - b_1 - c_1$, $a_3 - b_1 - c_2$, $a_3 - b_1 - c_3$, $a_3 - b_1 - c_4$
$a_3 - b_2 - c_1$, $a_3 - b_2 - c_2$, $a_3 - b_2 - c_3$, $a_3 - b_2 - c_4$.

This total enumeration of all possibilities can be summarized conveniently in a tree diagram, as is done in Figure 6.1. This proposition proclaims that there are three possibilities for executing the first procedure, and for each of these three there are two possibilities for subsequently executing the second procedure; hence, there are 3 • 2 = 6 total possibilities for sequentially executing the first two procedures. Next, for each of these six possibilities for executing A and B jointly, there are four possibilities for subsequently executing the third procedure; thus, there are 6 • 4 = 24 total possibilities for sequentially executing the three procedures. This structure is apparent in the preceding tabulation, where the subscript of a is 1, then 2, then 3, followed by the subscript of b alternating between 1 and 2, followed by the subscript of c being either 1, or 2, or 3, or 4. Consequently, the total number of outcomes in a sampling situation can be computed as the product of the number of times each individual outcome can materialize.

The characteristic of replacement refers to sampling from a population, observing the element that is drawn, and then replacing that element before making the next sample selection. For a population of size N, then, a random sample means that every element has the same probability of being selected on the first draw, with this probability equaling 1/N (a uniform distribution). Because replacement occurs, all subsequent element selections also will have a probability of 1/N. Invoking the foregoing counting proposition for sequences, there are N possibilities for the first selection, N possibilities for the second selection, and so forth, through to the nth selection; thus, there are

$$\prod_{i=1}^{n} N = N^n$$

sequential possibilities. For example, consider a population whose composition is {a, b, c, d}, and consider the drawing of samples of size n, where replacement occurs and order is important. Then since n = 2, there are a total of $4^2 = 16$ possible samples, with the enumeration of these possibilities creating a corresponding sample space of

a a	b b	c c	d d		
a b	a c	a d	b c	b d	c d
b a	c a	d a	c b	d b	d c

The probability of selecting any one of these samples, with a random drawing, and employing this perspective is $1/N^n$.

In contrast, the characteristic of no replacement refers to sampling from a population in such a way that once an element is drawn, it is not returned to the population prior to the next sample selection. For a population of size N, then, the probability of being selected on the first draw is 1/N, the probability of being selected on the second draw is 1/(N - 1), the probability of being selected on the third

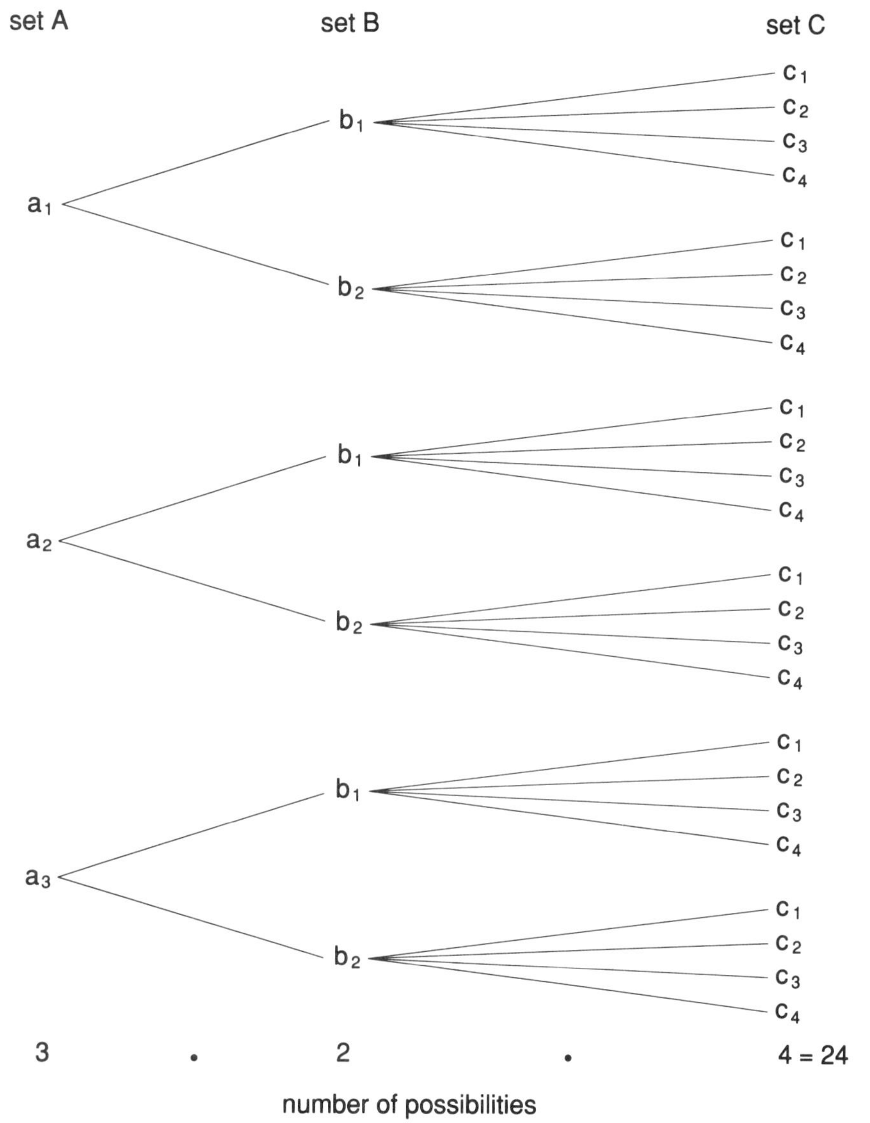

Figure 6.1: TREE DIAGRAM FOR THREE PROCEDURES, A, B, AND C, THAT RESPECTIVELY CAN BE PERFORMED IN SEQUENCE

draw is 1/(N - 2), and so forth, through to the nth selection, whose probability is 1/(N - n + 1). Therefore, without replacement, there are

$$N \bullet (N - 1) \bullet (N - 2) \bullet \ldots \bullet (N - n + 1)$$

sequential possibilities; this notation was presented in Chapter 1 under the heading of the **factorial** arithmetic operation. In the nomenclature of counting principles, this particular category of multiplication is referred to as a **permutation**; it counts the number of ways any ordered sequence of n distinct objects can be taken from a set of N distinct objects, and may be denoted as P(N, n). In shorthand form, it may be written as

$$P(N, n) = N!/(N - n)!.$$

Returning to the preceding population and sampling example, there now are a total of 4 • 3 = 12 possible samples, with a corresponding sample space of

a b	a c	a d	b c	b d	c d
b a	c a	d a	c b	d b	d c.

In other words, the first row of the initial sample space has been eliminated by the constraint of no replacement. This sequential selection procedure without replacement is identical to sampling all n elements at the same time. Although the individual selections have a varying probability, all sequences of size n have an equal probability of selection of 1/P(N, n). This quality of permutations often allows their use to be construed as random sampling. The magnitude of possibilities here is such that $P(N, n) < N^n$, for $n > 1$. Finally, the sample space constructed with permutations is contained in the sample space constructed with replacement. Those samples in the latter sample space that are not in the former sample space actually appear with a probability of zero (it is impossible for any single element to repeat itself in a sample), and the remaining samples experiencing a uniform increment in their probabilities. This simple rule for permutations becomes more complicated if there are identical, or repeated, elements in the population. Treatment of this topic is left to more advanced texts.

As an important aside, 0! takes on a special meaning within this context. It counts the number of ways no items can be arranged, which is one--nothing cannot be arranged. Thus, 0! = 1.

The two categories of sampling with or without replacement both entail selection sequences. Maintaining sequential ordering suggests that the order of selection, or the arrangement of elements in a sample, is important. However, arrangement may not be important. If order is deemed unimportant, then distinguishable arrangements must be identified, and redundant arrangements must be removed from the sample space. This reduction of the sample space is accomplished quite easily when sampling is done without replacement, and order is unimportant. This restrictive sampling perspective takes all arrangements of a set of elements and treats them as being equivalent. From the previous discussion on permutations, then, each set of n distinct sampled elements has P(n, n) = n! possible arrangements. Since all n! of these arrangements are to be treated as a single sample, their count must be reduced to 1. This calculation is achieved by dividing n! by n!; n!/n! = 1. But there is a total of P(N, n) of these arrangements. Therefore, the elimination of redundant samples when order is judged unimportant means the total number of samples reduces to P(N, n)/n!. In the nomenclature of counting principles, this particular category of multiplication and division is referred to as a combination. **Combinations** count the number of ways any unordered set of n distinct objects can be taken from a set of N distinct objects, and may be denoted as C(N, n). In shorthand form, it may be written as

$$C(N, n) = N!/[n!(N - n)!].$$

Again returning to the preceding population and sampling example, there now are a total of 4 • 3/2 = 6 possible samples, with a corresponding sample space of

b a   c a   d a   c b   d b   d c.

In other words, both the first and the second rows of the initial sample space have been eliminated by the constraints of no replacement and order unimportant. This selection without replacement and order unimportant is identical to sampling all n elements at the same time. Although the individual selections again have a varying probability, all groups of size n have an equal probability of selection of 1/C(N, n). The magnitude of possibilities here is such that $C(N, n) < P(N, n) < N^n$, for n > 1, and the sample space constructed without replacement and order unimportant is contained in the two sample spaces constructed with order being important.

The final category is that of replacement and order unimportant. Counting the total number of possibilities associated with this perspective does not reduce to a simple formula as in the previous three cases. Now one has to know what the composition of a sample is. The probability of each sample composition may well be different, causing this perspective to deviate from the notion of random sampling. Accordingly, one rarely sees this sampling perspective utilized. In order to account for all possible samples within this context, the idea of a permutation must be generalized to indistinct elements. In other words, arrangements of n elements need to be counted in such a way that redundancies attributable to repetitive entries can be removed. Suppose that the n items consist of k distinct elements, each repeating itself n_j times. Then combining a generalization of the permutation and combination concepts implies that there are n! total arrangements, with each element introducing $n_j!$ redundancies. Since the n! total needs to be divided, in turn, by each of these $n_j!$ repetitions in order to remove redundancies, the total number of arrangements when indistinct items occur is given by

$$n!/[\prod_{j=1}^{k} n_j!]$$

which reduces to P(n, n) when all $n_j = 1$ (all entries are distinct). This term must be multiplied by the number of combinations that can lead to the elements contained in this sample composition, which is given by C(N, k), as well as any combination terms associated with differential element repeats in each sample composition. The products of these permutation and combination expressions yield the probabilities of various sample compositions, whereas the sum of the individual combination products yields the total number of possible samples. Once more returning to the initial example, there are a total of 10 {4 from C(4, 1)[2!/2!], and 6 from C(4, 2)[2!/(1!1!)] rendering C(4, 1) + C(4, 2) = 10} possible samples, a corresponding sample space of

a a b b c c d d
a b a c a d b c b d c d.

In other words, only the third row of the initial sample space has been eliminated by the single constraint of order unimportant. The probability of each sample in this first row is 1/4, while the probability of each in this second row is 1/6; these are not equally likely samples.

These four sampling perspectives may be summarized in tabular form as follows:

		Replacement: YES	Replacement: NO
Order important	YES	N^n	P (N,n)
	NO	sum of combination products	C (N,n)

They all involve enumeration of the full set of N^n possibilities, where order is important and replacement occurs, followed by various partitionings of this original sample space into feasible and infeasible samples and then an assigning of probabilities to the resulting feasible sample. Any sample classified as being infeasible actually has a probability of 0 assigned to it.

EXAMPLE 6.1. Counting the possible samples for the four different sampling perspectives.

Consider a population of distinct elements for which N = 100 together with the drawing of all samples of size n = 5 from it.

a. Sampling with replacement and order important means that there are N^n or 100^5 = 10,000,000,000 possible samples associated with this perspective.

b. Sampling without replacement and order important means that there are P(100, 5)

= 100!/(100 - 5)! = 100!/95! = 100 • 99 • 98 • 97 • 96
= 9,034,502,400

possible samples associated with this perspective.

c. Sampling without replacement and order unimportant means that there are C(100,5)

= 100!/[5!(100 - 5)!] = 100!/(5! • 95!)
= (100 • 99 • 98 • 97 • 96)/(5 • 4 • 3 • 2 • 1) = 75,287,520

possible samples associated with this perspective.

d. Finally, sampling with replacement and order unimportant leads to

Number of distinct entries	Combination terms	Permutation terms	Composition probability
5 same	C(100,1)	5!/5!	1/10000000
4 same, 1	C(100,2) • C(2,1)	5!/4!	99/20000000
3 same, 2 same	C(100,2) • C(2,1)	5!/(3!2!)	99/10000000
3 same, 1, 1	C(100,3) • C(3,1)	5!/3!	4851/5000000
2 same, 2 same, 1	C(100,3) • C(3,2)	5!/(2!2!)	14553/10000000
2 same, 1, 1, 1	C(100,4) • C(4,1)	5!/2!	470547/5000000
1, 1, 1, 1, 1	C(100,5)	5!	1411641/1562500

meaning that the number of possible samples associated with this perspective

C(100, 1) + C(100, 2)•C(2, 1) + C(100, 2)•C(2, 1) + C(100, 3)•
C(3, 1) + C(100, 3)•C(3, 2) + C(100, 4)•C(4, 1) + C(100, 5)
= 91,962,520 .

6.3. Elementary set theory

Now we return to those fundamentals of set theory necessary for explaining the concept of and introducing the calculus of probability. Much of set theory can be graphically portrayed in Venn diagrams; a **Venn diagram** consists of a rectangle whose interior represents the size of the sample space in question, and within which circles are inscribed representing the size of groups of selected samples. The sample space encompassed by the rectangle is called a **universe** in set theory parlance. Each grouping of samples embraced by a circle is called a **set**. Those samples obtained by combining two or more sets constitute a **union**, with this set operation being denoted by $\cup$. The union of two sets, say A and B, means that a given sample is contained in either set A, or set B, or both; the word "or" designates union. If two circles overlap, then the area of the rectangle that they hold in common is called an **intersection**, with this set operation being denoted by $\cap$. The intersection of two sets, say A and B, means that a given sample is contained in both set A and set B; the word "and" designates intersection. If none of the circles overlap, then they are mutually exclusive or **disjoint**, with no pair of sets having any samples in common; this "intersection" of two nonoverlapping circles is termed the **empty set**, and is denoted by $\varnothing$. If all sets jointly capture the entire sample space, then they are collectively exhaustive; at least the universe is collectively exhaustive. That part of the universe falling outside of a circle is called the **complement** of the set demarcated by this circle, and is denoted by an apostrophe ('). A **subset** is a set that is completely contained in another set, and for two sets, say A and B, is denoted by $A \subset B$

Set theory operations adhere to the following mathematical rules, given the sets A, B, and C:

1. commutative: $A \cup B = B \cup A$; $A \cap B = B \cap A$
2. associative: $A \cup (B \cup C) = (A \cup B) \cup C$; $A \cap (B \cap C) = (A \cap B) \cap C$
3. distributive: $A \cap (B \cup C) = (A \cap B) \cup (A \cap C)$;
   $A \cup (B \cap C) = (A \cup B) \cap (A \cup C)$

EXAMPLE 6.2. Illustrating the basic terminology of set theory.

Consider the population {a, b, c, d} together with all possible samples of size n = 2, drawn with the perspective of replacement and order important. Let set A = {(x, y) ∍ x ≡ a} = {(a, a), (a, b), (a, c), (a, d)} (note: the symbol ∍ denotes "such that"), set B = {(x, y) ∍ y ≡ a} = {(a, a), (b, a), (c, a), (d, a)}, and set C = {(x, y) ∍ x or y ≡ b} = {(a, b), (b, b), (c, b), (d, b), (b, a), (b, c), (b, d)}. The Venn diagram for this illustration appears in Figure 6.2. Now $A \cap B$ = {(a, a)}, $A \cup B$ = {(a, a), (a, b), (a, c), (a, d), (b, a), (c, a), (d, a)}. $A \cap B \cap C = \varnothing$. C' = {(a, a), (a, c), (a, d), (c, a), (d, a), (c, c), (c, d), (d, c), (d, d)}.

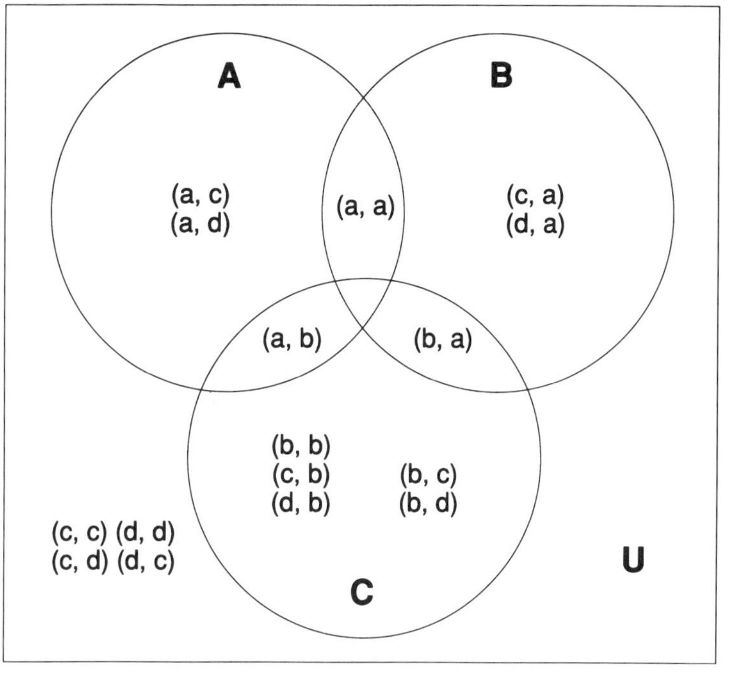

Population: { a, b, c, d }   Sets: A = { (a, a), (a, b), (a, c), (a, d) }
B = { (a, a), (b, a), (c, a), (d, a) }
C = { (a, b), (b, a), (b, b), (b, c), (c, b), (b, d), (d, b) }

Figure 6.2: VENN DIAGRAM FOR EXAMPLE 6.2

Since probability refers to relative frequencies, then within the context of Example 6.2 the following probabilities can be calculated. Since 4 of the 16 samples contain the letter a as the first element, P(A) = 4/16 = 1/4. In a similar fashion, since 4 of the 16 samples contain a as the second letter, P(B) = 4/16 = 1/4, and since 7 of the 16 samples contain b in either position, P(C) = 7/16, $P(A \cap B) = 1/16$, $P(A \cup B) = 7/16$, $P(A \cap B \cap C) = 0$, and P(C') = 9/16. Set theory is used in this fashion throughout this textbook when assigning numerical values to probability measures.

6.4. Determining probabilities

Calculating the probability of elementary sets (individual samples of size n) tends not to be an extremely tedious task. Calculating the probability of compound sets (collections of individual samples of size n) tends to be more problematic. Several basic probability propositions facilitate the computation of these latter probabilities. The first is the **addition rule**, which states that

$$P(A \cup B) = P(A) + P(B) - P(A \cap B).$$

The probability $P(A \cap B)$ is subtracted because it is contained in both P(A) and P(B) and hence is counted twice when these two probabilities are added. If $A \cap B = \varnothing$, then this proposition reduces to $P(A \cup B) = P(A) + P(B)$, since $P(\varnothing) = 0$. However, one cannot assume that the intersection is empty; rather, one must demonstrate that it is empty. Assuming an empty intersection when in fact it is nonempty introduces specification error into the calculation and could lead to a violation of the axiom that $P(x) \leq 1$. Returning to Example 6.2 and recalling how probabilities are computed for sampling with replacement and order unimportant, $P(A) = P(a, \bullet) = (4/4) \bullet (1/4) = 1/4$, $P(B) = P(\bullet, a) = (4/4) \bullet (1/4)$, $P(A \cap B) = P(a, a) = (1/4) \bullet (1/4) = 1/16$, and hence $P(A \cup B) = 1/4 + 1/4 - 1/16 = 7/16$, which is the value obtained in the preceding section through total enumeration. This addition rule proposition extends to more than two sets.

The second proposition is the **complementation rule**, which states that

$$P(A) = 1 - P(A'), \text{ or } P(A') = 1 - P(A)$$

since $A \cup A'$ is the universe, which by construction is collectively exhaustive. This proposition is useful when the set of interest can be described with an expression of the form "at least . . ." or "at most . . . ," since then the complement may be expressed respectively as "less than . . ." or "more than . . . ," for which probabilities often are easier to calculate.

The third proposition relates to sampling situations in which constraints are applied, such as no replacement or order unimportant. When information becomes available that affects the probability assignments being made, then a conditional probability must be assigned. The **conditional probability** of A given that B has occurred is denoted by $P(A|B)$ and defined as

$$P(A|B) = P(A \cap B)/P(B), \ P(B) > 0;$$

similarly,

$$P(B|A) = P(B \cap A)/P(A) = P(A \cap B)/P(A), \ P(A) > 0.$$

This formula yields the **multiplication rule**, which states that

$$P(A \cap B) = P(A|B) \bullet P(B) = P(B|A) \bullet P(A).$$

If $P(A|B) = P(A)$, then the probability of A is unconditional; similarly, if the probability of B is unconditional, then $P(B|A) = P(B)$. In other words, the probability of A given that B has occurred, or the probability of B given that A has occurred, remains unchanged; the occurrence of A does not influence B, and vice versa. This reduction can occur only when $P(A \cap B) = P(A) \cdot P(B)$, resulting in $P(A) \cdot P(B)/P(B) = P(A)$, or $P(A) \cdot P(B)/P(A) = P(B)$. Therefore, **statistical independence** may be defined as $P(A \cap B) = P(A) \cdot P(B)$; otherwise A and B are dependent. This is an extremely important property, since independence furnishes far simpler probabilities to work with than does dependence. Random sampling is intentionally structured in a way that allows this property to be exploited.

EXAMPLE PR-6.1. Sampling probabilities associated with the classification variables in the Puerto Rico data set.

Let the set A denote the collection of municipios contained in the San Juan Agricultural Administrative Region, set B denote the collection in the Arecibo Region, set C the collection in the Mayaguez Region, set D the collection in the Ponce Region, and set E the collection in the Caguas Region (see Figure 1.2). The following probabilities can then be calculated.

Set	Region	Probability
A	San Juan	P(A) = 19/73
B	Arecibo	P(B) = 11/73
C	Mayaguez	P(C) = 16/73
D	Ponce	P(D) = 14/73
E	Caguas	P(E) = 13/73

Now, let set X denote the collection of municipios located in the coastal lowlands, and set Y the collection in the interior highlands. $P(X) = 42/73$ and $P(Y) = 31/73$.

Now,

$P(A') = 1.0 - P(A) = 1.0 - 19/73 = 54/73$
$P(A \cup B) = P(A) + P(B) - P(A \cap B) = 19/73 + 11/73 - 0 = 30/73$
$P(A \cap X) = P(A|X) \cdot P(X) = (12/42) \cdot (42/73) = 12/73$.
(Note that 12 municipios in San Juan province are coastal lowlands, yielding $P(A|X) = 12/42$; see Appendix A for Chapter 1.)

For A and X to be independent, $P(A) \cdot P(X)$ must equal 12/73. Since $P(A) \cdot P(X) = (19/73) \cdot (42/73) \neq 12/73 \neq P(A \cap X)$, A and X are not statistically independent outcomes. In other words, using the definition of conditional probability, P(A|X) = (12/73)/(42/73) = 12/42. P(X|A) = P(12/73)/(19/73) = 12/19.

EXAMPLE CA-6.1. Sampling probabilities associated with two classification variables in the Canadian Auto data.

Use the postal code (PC1) and primary product code (PPC) variables as described in Figures 3.5 and 3.8. In other words, let set K denote the firms in Eastern Ontario, L the firms in Central Ontario, M the firms in Metropolitan Toronto, and N the firms in Southwestern Ontario. Regroup the PPC variables so the set X represents firms with PPC equal to 0, 1, or 2, Y the firms with PPC equal to 3, and Z the firms with PPC equal to 4, 5, 6, 7. Following is a table with rows representing location (sets K, L, M, N) and columns representing primary products (sets X, Y, Z). Each element in the table represents the number of firms in a region described by the row, producing products described by the column.

Location (PC1)	Primary product (PPC)			
	X	Y	Z	Total
K	0	0	2	2
L	6	3	3	12
M	5	5	3	13
N	7	8	11	26
Total	18	16	19	53

Dividing each row total by 53 produces the probability of selecting a firm in the row location; P(K) = 2/53, P(L) = 12/53, P(M) = 13/53, and P(N) = 26/53. Similarly, dividing each column total by 53 produces the probability of selecting a firm producing in the column category; P(X) = 18/53, P(Y) = 16/53, and P(Z) = 19/53. Dividing each entry in the table by either the row or column total produces the two conditional probabilities as follows: P(column set|row set) = (table entry/row total) and P(row set|column set) = (table entry/column total). For example, P(N|Y) = 8/16, but P(Y|N) = 8/26. Similarly, P(M|Z) = 3/19 and P(Z|M) = 3/13. The independence of location and production niche for M and Z can be tested by determining if $P(M \cap Z)$

= P(M|Z) • P(Z) = (3/19)(19/53) = 3/53 in fact equals P(M) • P(Z) = (13/53)(19/53). Clearly, these terms are not equal, so M and Z are not independent, and thus location and production niche are not independent.

EXAMPLE 6.3. Example 6.1 revisited.

When sampling with replacement and order is important, then the probability of the first selection is 1/100, the probability of the second selection is 1/100, of the third is 1/100, of the fourth is 1/100, and of the fifth is 1/100. Sampling done at any point in this sequence fails to influence the selection probability at any other point in the sequence. Further, $(1/100)(1/100)(1/100)(1/100)(1/100) = 1/100^5$. Therefore, sample drawings with replacement and order important are statistically independent. When sampling without replacement and order is important, then the probability of the first selection is 1/100, the probability of the second selection is 1/99, of the third is 1/98, of the fourth is 1/97, and of the fifth is 1/96. Now sampling done at any point in this sequence influences the selection probability at all subsequent points in the sequence. Therefore, sample drawings without replacement and order important are statistically dependent. When sampling without replacement and order is unimportant, then the probability of the first selection is 1/100, the probability of the second selection is 2/99, of the third is 3/98, of the fourth is 4/97, and of the fifth is 5/96. Again sampling done at any point in this sequence quite differentially influences the selection probability at all subsequent points in the sequence. Therefore, sample drawings without replacement and order unimportant also are statistically dependent. Finally, sampling with replacement and order unimportant assigns different probabilities to different sample compositions; thus, the initial sequence of drawings must influence the latter probabilities, since the first selection in all cases has a probability of 1/100.

This notion of statistical independence is especially relevant to the contingency table analysis presented in Chapter 4. In this context both the row and column sums of a contingency table can be converted into relative frequencies, which are called **marginal probabilities** because the quantities are totals that appear in the margins of a table. Meanwhile, each of the two classification variables labelling the rows and the columns of a contingency table produces a mutually exclusive and collectively exhaustive partitioning of the sample into sets. This particular partitioning allows addition over the sets of a given classification variable, in accordance with the addition proposition of probabilities. The cross-classifications, or cells identified by row and column intersections, are joint probabilities, namely, $P(C_{1j} \cap C_{2k})$ for set j of classification variable C_1 and set k of classification variable C_2. Within this context of probabilities, then, contingency tables address the issue of whether or not the two classification variables are independent, asking whether or not $P(C_{1j} \cap C_{2k})$ equals $P(C_{1j}) \cdot P(C_{2k})$, for all j and k. This connection to classical descriptive statistics furnishes a worthwhile application of conditional probabilities and a connection between the chi square statistic (developed in Chapter 4) and the notion of statistical independence.

EXAMPLE PR-6.2. An application of the multiplication rule to a contingency table for the cross-classifications produced by ADM and LND.

ADM LND:	Coastal lowlands	Interior highlands	Marginal probability
San Juan	12/73	7/73	19/73
Arecibo	7/73	4/73	11/73
Mayaguez	9/73	7/73	16/73
Ponce	10/73	4/73	14/73
Caguas	4/73	9/73	13/73
Marginal probability	42/73	31/73	73/73

The probabilities to be compared are (42/73)(19/73) with 12/73, (42/73)(11/73) with 7/73, (42/73)(16/73) with 9/73, (42/73)(14/73) with 10/73, (42/73)(13/73) with 4/73, (31/73)(19/73) with 7/73, (31/73)(11/73) with 4/73, (31/73)(16/73) with 7/73, (31/73)(14/73) with 4/73, and (31/73)(13/73) with 9/73. Technically only one of these comparisons has to bear a nonzero value, that is, only one has to produce a difference <u>not</u> equal to zero. For statistical independence to hold, every comparison of this sort in the population must bear a nonzero value.

EXAMPLE CA-6.2. Example CA-6.1 revisited.

Note that the probabilities calculated after the table in example CA-6.1 were, in fact, the marginal probabilities. A complete table of probabilities for example CA-6.1 follows.

One comparison for dependence was already performed in the previous example and found to produce a nonzero difference.

6.5. Bayes' Theorem

The notion of probability discussed thus far in this chapter deals with numerical value assignments before samples are selected; these are known as **prior probabilities**. Another matter to consider is the likelihood that a particular sampling perspec-

Location (PC1)	Primary product (PPC) X	Y	Z	Marginal probability
K	0	0	2/53	2/53
L	6/53	3/53	3/53	12/53
M	5/53	5/53	3/53	13/53
N	7/53	8/53	11/53	26/53
Marginal probability	18/53	16/53	19/53	53/53

tive had been engaged in order to draw a sample once this drawn sample is presented for analysis; these are known as **posterior probabilities**. One means of integrating these two classes of probability, while simultaneously exploiting the concept of statistical dependence, is found in Bayes' theorem. This theorem describes how to compute posterior probabilities from given prior probabilities and conditional probabilities.

Suppose the universe is partitioned into n mutually exclusive and collectively exhaustive sets, denoted by A_i, i = 1, 2, . . . , n, so that

$$S = A_1 \cup A_2 \cup A_3 \cup \ldots \cup A_n.$$

Consider some other set, say B, that is superimposed on this same universe. Thus,

$$B = S \cap B = (A_1 \cup A_2 \cup \ldots \cup A_n) \cap B$$
$$= (A_1 \cap B) \cup (A_2 \cap B) \cup \ldots \cup (A_n \cap B).$$

Accordingly, using the addition rule as it applies to mutually exclusive sets,

$$P(B) = \Sigma_{i=1}^{n} P(A_i \cap B) = \Sigma_{i=1}^{n} P(B|A_i) \cdot P(A_i).$$

Further, since $P(A_i|B) = P(A_i \cap B)/P(B)$,

$$P(A_i|B) = P(B|A_i) \cdot P(A_i)/\Sigma_{i=1}^{n} P(B|A_i) \cdot P(A_i).$$

In other words, one is trying to assign probabilities to the sets A_i given knowledge about set B.

EXAMPLE PR-6.3. Bayes' theorem applied to assessing which sampling perspective was engaged, given a sample already drawn.

Suppose a sample of size n = 3 has been drawn from the Puerto Rico data set, namely {Ceiba, Ceiba, Lares}. What is the probability that this sample was selected implementing the perspective of replacement and order important? In this situation, set B is the sample that has been drawn, and set A_i refers to the four possible sampling perspectives (i = 1, 2, 3, 4). P(B|replacement, order important) = $1/73^3$, P(B|replacement, order unimportant) = $216/73^2$, P(B|no replacement, order important) = 0 (Ceiba appears twice in the sample), and P(B|no replacement, order unimportant) = 0. Let the probability that a sampling perspective was employed be such that P(replacement, order important) = p_1, P(replacement, order unimportant) = p_2, P(no replacement, order important) = p_3, and P(no replacement, order unimportant) = p_4. Then the posterior probability of using the classical sampling perspective, given that set B was obtained, is P(replacement, order important|B) = $p_1(1/73^3)/[p_1(1/73^3) + p_2(216/73^2) + 0 + 0]$ = $p_1/(p_1 + 15768 \bullet p_2)$.

This last example affords an opportunity to mention yet another kind of probability. All the probabilities reviewed in this chapter may be classified as **objective probabilities**. They are based upon counting principles, relative frequencies, and sampling experiments and as such are void of the influence of individual opinions or claims to intuitive knowledge. But what happens if nothing is known about the relative frequency of occurrence of an outcome? Usually in these circumstances an individual defaults to his or her "degree of belief" about how often some outcome might occur; these measures are known as **subjective probabilities**. They are strongly influenced by the information that is available to an individual, as well as the manner in which the individual evaluates this information. Consequently, subjective probabilities for the same outcome can vary quite dramatically from person to person. With regard to Example PR-6.3, little information is available describing the relative frequency with which the four sampling perspectives are utilized by the scientific community. In the presence of pure ignorance about this matter, the least restrictive assumption is that these four perspectives are equally likely, resulting in $p_i = 1/4$. Substituting this probability into the solution presented in Example PR-6.3 yields P(replacement, order important|B) = 1/15769. In contrast, since the classical perspective often is promoted by statisticians, one might suspect that it is utilized 99% of the time; now the solution becomes 11/1763. In other words, these two sets of subjective probabilities return noticeably different posterior probabilities for a single problem.

6.6. Summary

Probability theory is perhaps the most difficult set of concepts for students to master on the first encounter. Entire courses taught by statistics departments are devoted to probability theory alone. We have extracted the minimum set of topics required to permit reasonable coverage of materials that follow. Of particular importance are the concepts contained in conditional probability and statistical independence. In addition, the tabular format for the organization and presentation of the data in Example CA-6.1 will prove very useful in the several chapters to follow.

6.7. Problems

1. Consider a sample of size n = 3 and a population {a, b, c, d}. Determine the number of samples for this situation, using the appropriate counting principles, that is associated with each of the four sampling perspectives. Then verify your results by enumerating the individual sample spaces.

2. Calculate the probability of drawing an individual sample that accompanies each of the four perspectives for Problem #1.

3. Construct a contingency table for the classificatory variables ADM and U-R appearing in the Puerto Rico data set. Then compute the joint probabilities and the marginal probabilities for this table. Are these classification variables statistically independent?

4. Construct a Venn diagram for DTB using LND, from the Arecibo Agricultural Administrative Region of the Puerto Rico data set, to divide this collection of numbers into two sets, say A and B. Apply the addition rule and the multiplication rule to this result.

5. Using DFRM of the Puerto Rico data set and the three sets A = {municipios $\ni$ DFRM $\in$ [0.0,7.5]}, B = {municipios $\ni$ DFRM $\in$ [6.0, 10.15]}, and C = {municipios $\ni$ DFRM $\in$ [7.0,17.5]}, calculate the following probabilities:

   (a) $P(A)$
   (b) $P(B)$
   (c) $P(C)$
   (d) $P(A')$
   (e) $P(B')$
   (f) $P(C')$
   (g) $P(A \cup B)$
   (h) $P(B \cup A)$
   (i) $P(A \cap B)$
   (j) $P(A \cup B \cup C)$
   (k) $P(A \cap B \cap C)$
   (l) $P[A \cap (B \cup C)]$
   (m) $P[(A \cup B) \cap (A \cup C)]$
   (n) $1 - P(A \cap B)$
   (o) $P(A|B)$
   (p) $P(A|C)$
   (q) $P(B|C)$
   (r) $P(B|A)$
   (s) $P(C|A)$
   (t) $P(C|B)$
   (u) $P(C|A \cap B)$

6. Given the sample {Ciales, Luquillo, Moca, Arroyo, Gurabo}, which was selected from the Puerto Rico data set, use Bayes' theorem to determine the posterior probabilities that each of the four sampling perspectives was used to draw it. First assume complete ignorance about sampling perspective preferences. Next, assume that engaging sampling (a) with replacement and order important has a probability of 0.30, (b) without replacement and order important has a probability of 0.15, (c) without replacement and order unimportant has a probability of 0.50, and (d) with replacement and order unimportant has a probability of 0.05.

7. Given a retail firm that operates three spatially separated outlets, say A, B and C, consider the following situation. Suppose that these three outlets respectively account for 50%, 30%, and 20% of the total merchandize sold by this firm. Further, let the associated percentage of returned merchandise be 3%, 4%, and 5%. If a piece of merchandise is selected at random, calculate the probability that it will be returned. If a piece of returned merchandise is selected at random, calculate the probability that it came from outlet B.

8. Construct two tables of the conditional probabilities similar to the table in Example CA-6.1. Test two nonzero entries in each table for independence.

6.8. Additional readings

In addition to texts listed in Chapter 1,

Lindgren, B. (1968) *Statistical Theory*, 3rd ed. New York: Macmillan.

Lipschutz, S. (1965) *Schaum's Outline Of Theory and Problems of Probability*. New York: McGraw-Hill.

CHAPTER 7

SELECTED THEORETICAL FREQUENCY DISTRIBUTIONS

There are, literally, hundreds of different ideal frequency distributions; their relative counterparts constitute important probability distributions. Historically these distributions have been a focus of study primarily because they greatly simplify and often make tractable the mathematical and numerical work involved in statistical analysis. This motivation for using ideal frequency distributions is reinforced by the preponderance of empirical evidence that has accumulated over the years demonstrating that they indeed do an extremely good job in representing various aspects of reality.

Four of the possible theoretical frequency distributions are presented in this chapter, strictly because these are the most common ones found in the literature pertaining to spatial data analysis. These four distributions are the uniform, the binomial, the normal, and the Poisson. These distributions are described in terms of the probabilities they yield (their relative versions), the expectations they render (their absolute versions), and their parameters.

7.1. The uniform frequency distribution

The simplest of the four frequency distributions discussed here is known as the **uniform distribution**. This distribution is defined for a given interval, say [a, b], such that the probability of observing any value within this interval is p, whereas the probability of observing any value outside of this interval is zero. Let X (x_i is observation i of variable X) be a **random variable**, or a variable whose observed values of x_i, i = 1, ..., N are determined by chance. If X is continuous, then its probability density function (that is, the equation that generates the characteristic curve associated with the distribution) is defined as

$$f(x) = \begin{cases} 1/(b - a) & \text{if } a \leq x \leq b \\ 0 & \text{otherwise,} \end{cases}$$

so that the probability density is spread evenly or uniformly (hence its name) in a continuous fashion over the interval of possible values. If the random variable X in question is discrete, however, then its probability mass function consists of aggregations or congeries of probability mass of a single magnitude that are assigned to the discrete points along the x-axis contained in the interval [a, b]. If there are k discrete points contained in this interval, then p = 1/k, where k = (b - a + 1). Graphs of these two forms of the uniform frequency distribution appear in Figure 7.1.

The mean of a continuous uniform frequency distribution is defined as

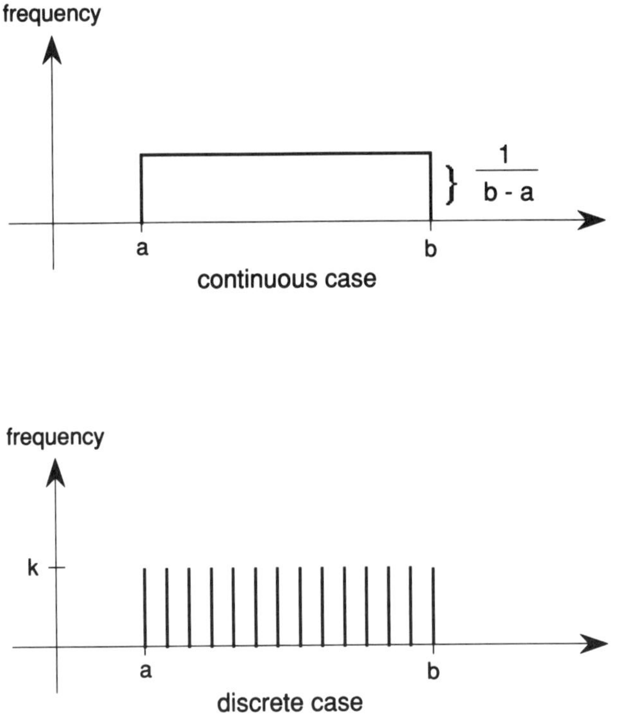

Figure 7.1: (TOP) AN EXAMPLE OF A CONTINUOUS UNIFORM FREQUENCY DISTRIBUTION; (BOTTOM) AN EXAMPLE OF A DISCRETE UNIFORM FREQUENCY DISTRIBUTION

$$\mu = (b + a)/2$$

whereas the variance of this continuous form is defined as

$$\sigma^2 = (b - a)^2/12.$$

Because the probability of a continuous random variable refers to the area under the curve for this variable, then $P(X = x) = 0$ for a continuous uniform distribution; only interval probabilities are nonzero here.

The mean of a discrete uniform frequency distribution is defined as

$$\mu = [b(b + 1) - (a - 1)a]/(2k)$$

which reduces to $(k + 1)/2$ if the first k positive integers are used, whereas the variance of this discrete form is defined as

$$\sigma^2 = (b - a)(b - a + 2)/12$$

which reduces to $(k^2 - 1)/12$ if the first k positive integers are used.

This distribution usually is employed in statistical analyses in geography for one of two purposes. First, it provides the probability basis for classical sampling experiments (replacement and order important). To select a sample scientifically from a population, every element in this population must have an equal probability of being chosen; if constrained sampling is involved, then once the constraints have been taken into account, individual elements still must have an equal chance of being sampled. These and other specifics of sampling are discussed in Chapter 8. This sampling notion leads to what are called random samples; such samples often are drawn with the aid of a random numbers table, or a table whose single digits are drawn from the population {1, 2, 3, 4, 5, 6, 7, 8, 9, 0} (where 0 denotes the number 10) using the uniform distribution. An example of a random numbers table is presented in Table 7.1-A (the extension -A means that this table is located in the appendices), which has been constructed by randomly selecting 1800 single-digit integers from the preceding population using the MINITAB pseudo-random number generator.

EXAMPLE 7.1. The collection of random digits reported in Table 7.1-A.

These numbers result in a sample mean of 5.4939 and a sample standard deviation of 2.9180. The frequency distribution for this set of numbers is as follows (recalling that 0 in the table denotes 10):

Digit	1	2	3	4	5	6	7	8	9	0
Observed frequency	190	186	182	169	168	178	190	155	193	189
Expected frequency	180	180	180	180	180	180	180	180	180	180

Diagnostics support this set as a likely and representative sample from the population {1, 2, 3, 4, 5, 6, 7, 8, 9, 0}. Here a = 1 and b = 10 (hence k = 10, since there are 10 distinct integers), and X is discrete, so that the population parameters are

$p = 1/10, \mu = 10 \bullet 11/(2 \bullet 10) = 5.5$
$\sigma^2 = 9 \bullet 11/12 = 8.25$, and hence $\sigma = 2.87$.

Clearly the sample mean and standard deviation are not very different from these population parameter values.

Saying that Table 7.1-A is a random numbers tables means that one can (1) enter into this table at any point, (2) follow any arbitrary path along the horizontal or the vertical axis in either direction as long as backtracking is avoided, and (3) select single or juxtaposed multiple digits (e. g., double digits denote a drawing from the population {01, 02, . . . , 99, 00}) from the entries, and the selected sequence of numbers will be a random drawing based upon the uniform probability distribution. The expected frequency values that accompany those observed frequency values appearing in Example 7.1 result from multiplying the sample size (n = 1800) by the probability (p = 1/10) [in other words, $np_i = 1800(1/10) = 180$, i = 1, 2, . . . , 10]. All of these expected values are the same because a uniform distribution is being used.

Besides being used for drawing random samples, the uniform distribution is employed in certain statistical analyses in geography when ignorance prevails. If one has no knowledge of how some variate is distributed, the least prejudicial assumption to make is that of a uniform distribution. The use of this perspective already has been encountered in classical descriptive statistics dealing with aggregated data (see Chapter 4) and in descriptive spatial statistics concerned with aggregate areal unit data (see Chapter 5). Mathematical statistics shows that it is the most likely distribution in the presence of ignorance.

7.2. The binomial frequency distribution

Another theoretical frequency distribution of interest refers to dichotomous situations, or ones entailing only two possible outcomes, and is known as the **binomial distribution**. The random variable in question is discrete, the probability that the first outcome will materialize is denoted by p, and the probability that the second outcome will materialize is denoted by (1 - p) (often this quantity is denoted by q). The random variable X is a count of the number of times this first outcome materializes (of course, the assignment of the labels first and second to two such outcomes is arbitrary). The total number of each outcome is of primary interest, rather than exactly which observations yield this first outcome. In other words, order is unimportant. The sample consists of n independent outcomes of this random variable. Recall that independent observations are ones in which the result of a particular outcome in no way affects the outcome that next materializes, or any other outcome that is observed. Moreover, the probability p remains constant in value throughout the entire sample. If more than two outcomes are possible, a binomial random variable can be created by dividing the set of possible outcomes into two distinct groups.

Three graphs of this binomial frequency distribution appear in Figure 7.2. The first of these examples, Figure 7.2a, depicts a binomial situation for which p = 0.05. When the probability of the first outcome is very small (as it approaches zero), then the frequency distribution becomes increasingly skewed to the right, or positively skewed. Figure 7.2b depicts a binomial situation for which p = 0.5. When the probability of the first outcome exactly equals the probability of the second outcome, then the frequency distribution is perfectly symmetrical. The third example here, Figure 7.2c, depicts a binomial situation for which p = 0.95. When the probability of the first outcome is very large (as it approaches unity), then the frequency distribution becomes increasingly skewed to the left, or negatively skewed. The most useful binomial analyses tend to be for cases where p approaches one-half and n is not too large. As p approaches zero, then one helpful approximation to the binomial may be adopted, whereas as n becomes very large, another helpful approximation may be adopted (in ideal circumstances these approximations introduce negligible specification error). These approximations will be treated subsequently.

The different categories of a binomial frequency distribution result from a count of the number of times the first outcome materializes. If the sample size is n, then the possibilities are 0, 1, 2, . . . , n, or (n + 1) different categories. As was mentioned earlier, the probability for each of these categories refers to the combination of first and second outcomes that materializes, with the order of these occurrences being unimportant. These probabilities are obtained by constructing a polynomial expansion of the term $[p + (1 - p)]^n = 1^n$. The algebra involved here highlights why (1) the two outcomes must be mutually exclusive and collectively exhaustive (so that they sum to unity), (2) p must be constant across all n materializations (so that the same p appears in each of the n expressions being multiplied), and (3) the n outcomes must be independent (so that they can be multiplied). The result of this polynomial expansion is a set of (n + 1) terms, which in general define the probability mass function of a binomial random variable as

$$P(X = x) = \{n!/[x!(n - x)!]\}p^x(1 - p)^{n-x}$$

where the combinatorial term n!/[x!(n - x)!] is called the binomial coefficient and counts the number of different sequences of x first outcomes and (n - x) second outcomes when order is not important.

The mean of a binomial frequency distribution is defined as

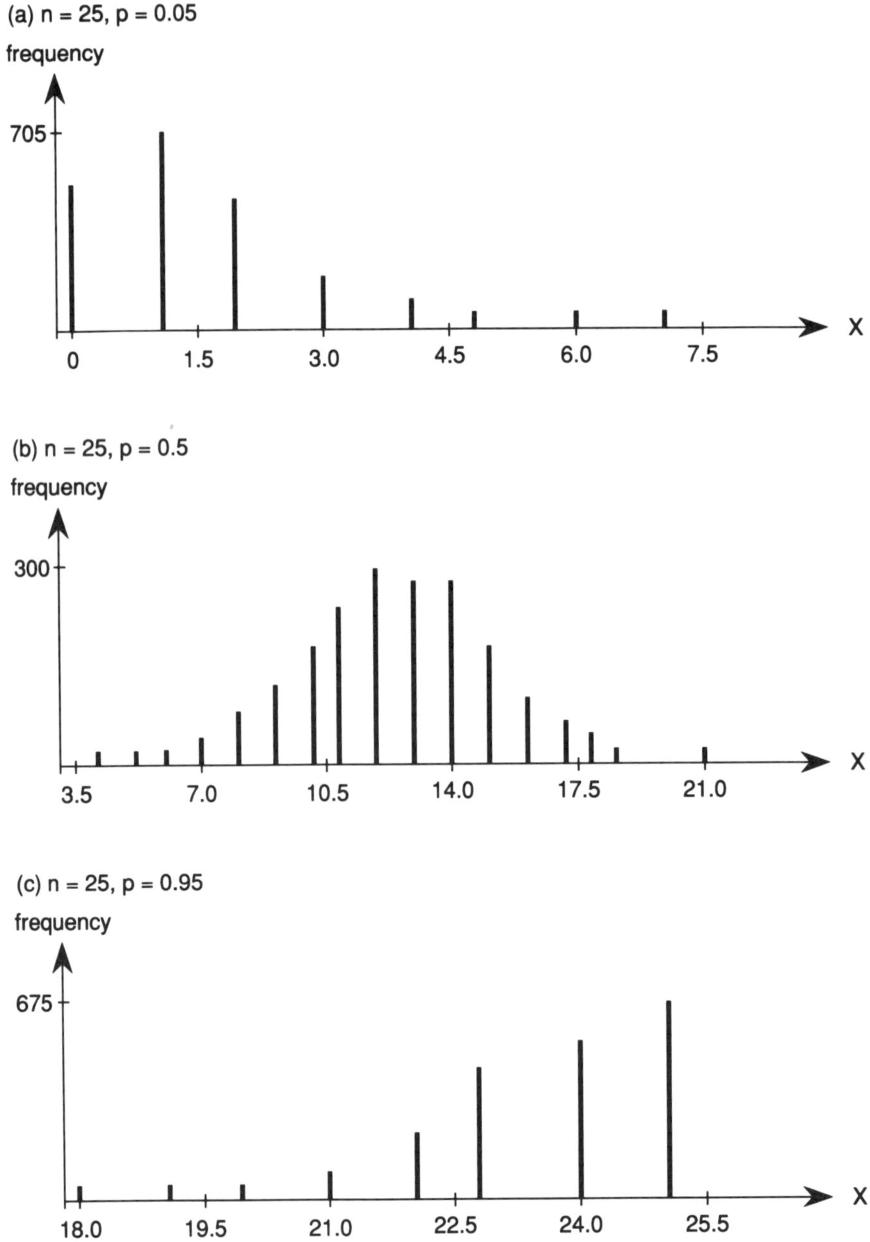

Figure 7.2: THREE EXTREME CASES OF THE BINOMIAL FREQUENCY DISTRIBUTION (1800 RANDOM SAMPLES DRAWN)

μ = np for the first outcome, and

μ = n(1 - p) for the second outcome,

whereas the variance of this discrete random variable is defined as

σ^2 = np(1 - p), which is the same for either outcome.

Summing the probabilities of all outcomes yields

$\sum_{x=0}^{n} \{n!/[x!(n - x)!]\}p^x(1 - p)^{n-x} = 1.$

This distribution is employed regularly in statistical analyses in geography for the evaluation of sample composition representativeness. It also has been used in modelling spatial diffusion processes. Recall that random sampling usually occurs when little is known about the population characteristics and, as a result, the representativeness of a sample remains unknown. The binomial distribution allows information from some sample (either the one selected or a pilot study sample) to be utilized in order to ascertain whether or not the selected sample can be viewed as having a sizeable probability of being representative (whether or not it actually is representative remains concealed as long as the parent population characteristics are unknown).

EXAMPLE PR-7.1. Evaluation of sample compositions.

The LND variable of the Puerto Rico data set is dichotomous, having 42 municipios classified as coastal lowlands and 31 municipios classified as interior highlands. Consider the outcome of coastal lowlands; p = 42/73 = 0.57534. Each of 100 students enrolled in a quantitative geography class was given the assignment of drawing a sample of size 10 from this population. These 100 samples yielded the following set of sample compositions:

Count of outcomes	0	1	2	3	4	5	6	7	8	9	10
Coastal lowlands	1	0	3	4	10	22	26	18	10	5	1

Here $\mu = 10(42/73) = 5.7534$, whereas $\bar{x} = 5.8500$, and $\sigma = \sqrt{10(42/73)(31/73)} = 1.5631$, whereas $s = 1.7371$. Because a dichotomy is involved, designating interior highlands as the first outcome would yield a mirror image of the foregoing distribution

Count of outcomes	0	1	2	3	4	5	6	7	8	9	10
Interior highlands	1	5	10	18	26	22	10	4	3	0	1

Here $\mu = 10(31/73) = 4.2466$, whereas $\bar{x} = 4.1500$; σ remains 1.5631 and s remains 1.7371.

The expected frequency values that accompany the coastal lowlands observed frequency values presented in Example PR-7.1 can be calculated again by multiplying the sample size times the appropriate probability--specifically, np_i, where n is the sample size of 10 and p_i is the probability that is calculated from the binomial probability mass function, i = 0, 1, 2, . . . , 10. Here the expected values are not a constant, as was true for the uniform distribution, since the binomial probabilities differ across the set of 11 categories.

EXAMPLE PR-7.2. Calculating the expected values for a binomial frequency distribution for coastal lowlands.

In the following table the first column presents the possible counts, the second column is the set of expected frequencies if p is known (5.7534 calculated from the parent population), and the third column is the set of expected frequencies if p is not known ($10\hat{p} = \bar{x} = 5.85$; hence $\hat{p} = 5.85/10 = 0.585$).

x	100{10!/[x!(10 - x)!]} • $(42/73)^x(31/73)^{10-x}$	100{10!/[x!(10 - x)!]} • $0.585^x(0.415)^{10-x}$
0	0.0191	0.0152
1	0.2584	0.2136
2	1.5753	1.3549
3	5.6916	5.0931
4	13.4945	12.5641
5	21.9395	21.2530
6	24.7704	24.9659
7	19.1771	20.1102
8	9.7432	10.6305
9	2.9334	3.3300
10	0.3974	0.4694

For the illustration summarized in Examples PR-7.1 and PR-7.2, sample compositions of 4 coastal lowlands and 6 interior highlands, 5 coastal lowlands and 5 interior highlands, 6 coastal lowlands and 4 interior highlands, 7 coastal lowlands and 3 interior highlands, and 8 coastal lowlands and 2 interior highlands are most likely (about 90% of the samples have one of these five compositions) and have an X value that is not too different from the mean or a composition that is not too different from the population. As Example PR-7.2 suggests, sampling error can somewhat distort an analysis of sample composition, but this distortion tends not to be too severe. All in all, then, the binomial distribution furnishes a valuable tool for diagnosing sample representativeness.

7.3. The Poisson frequency distribution

A third discrete theoretical frequency distribution, labelled the **Poisson distribution**, specifically addresses the situation of a rare event, where the probability of an outcome materializing is very low (usually any value less than 0.05). The probability in question can be labelled either p or q = 1 - p (so watching one's ps and qs becomes necessary). For example, a scientist at the National Center for Atmospheric Research in Boulder, Colorado, found what appears to be a counter-example to the maxim stating "No two snowflakes are alike" (Knight, 1988). This was the first such discovery from an enormous number of research flights; clearly her find is a rare event. At the same time, many events that appear to be rare are almost to be expected. For example, when Evelyn Adams won a New Jersey lottery twice in four months during 1986, her feat was reported as an amazing coincidence with odds of one in 17 trillion. In fact, the Law of Large Numbers (see Chapter 2) suggests that her odds actually were closer to one in thirty (Kolata, 1990).

The Poisson distribution deals with count data, and so the values X can assume must be nonnegative. In fact, one rationale for its use derives from a limiting situation for the binomial random variable: letting n go to infinity and p go to zero, in such a way that np remains fixed at some positive value, results in the binomial probability mass function converging to a Poisson probability mass function. Within this context, the mean occurrence of an outcome must be relatively small in comparison with the total possible range of occurrences. Another rationale for applying the Poisson distribution arises in connection with the occurrence of outcomes of a particular type within a specified continuous interval; this perspective is particularly relevant to the point pattern analyses discussed in Chapter 5. Accordingly, occurrences of the outcome in mutually exclusive intervals are independent, the probability that an outcome will occur is proportional to the length of the interval, and the probability of multiple occurrences of an outcome in a small interval is near zero. One pertinent and useful mathematical statistics finding is that the joint probability distribution for two uniform random variables is a Poisson random variable; this property enabled Figure 5.6 to be constructed. Although attention here will be restricted to the point pattern context of Poisson applications, the student should realize that other geo-referenced rare events also have been analyzed within the framework of Poisson probabilities. Some examples include spatial interaction modelling, frequency of thunderstorms, and queuing for retail purchases.

Three graphs of the Poisson frequency distribution appear in Figure 7.3. The first of these examples, Figure 7.3a, depicts a Poisson situation for which np = 50. This figure exemplifies the tendency toward symmetry as np increases; the relationship between the binomial and the Poisson is strongly suggested here. Figure 7.3b depicts a Poisson situation for which np = 1; now p is considerably smaller than in

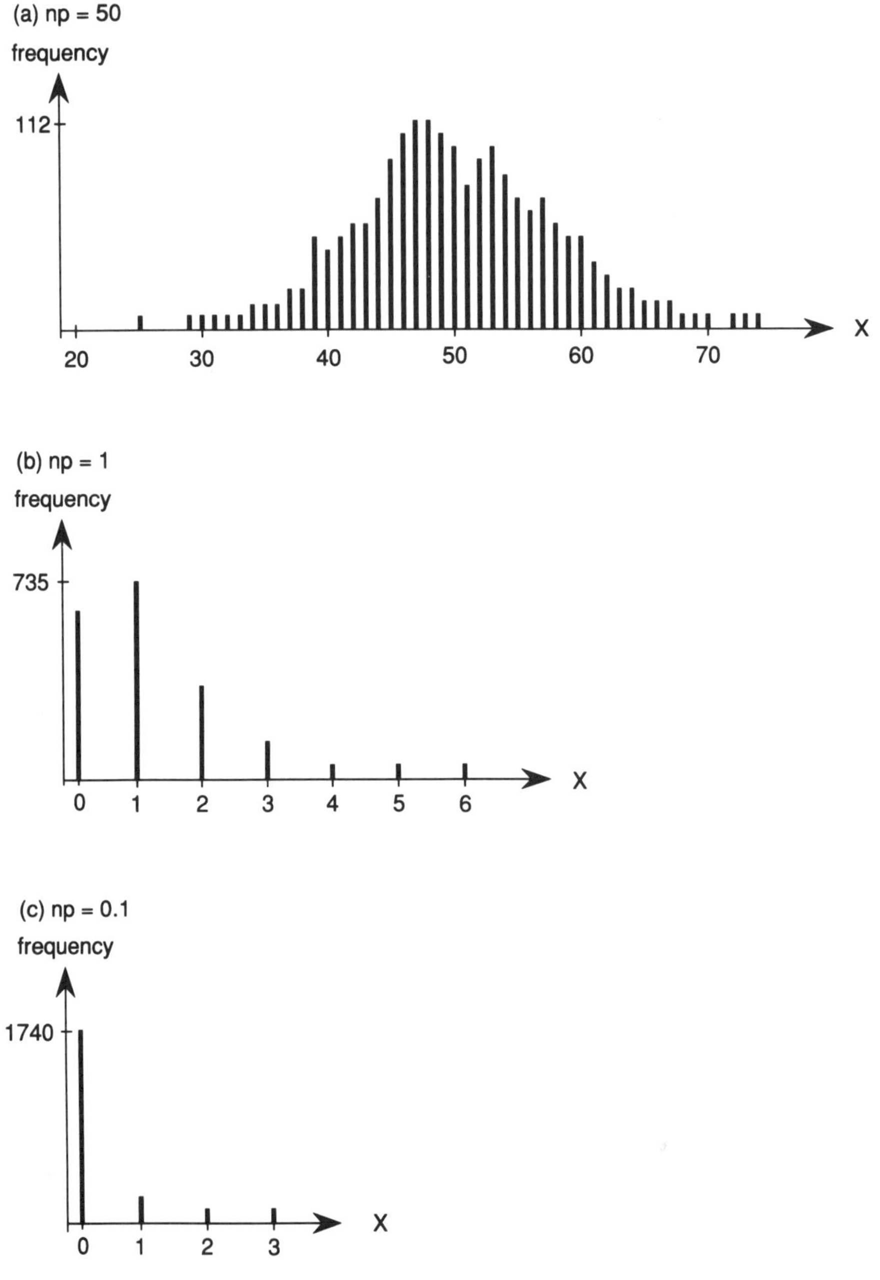

Figure 7.3: THREE SYMBOLIC CASES OF THE POISSON FREQUENCY DISTRIBUTION (1800 RANDOM SAMPLES DRAWN)

the preceding case. The skewness exhibited in this figure is reminiscent of that found in Figure 7.2a. The third example, Figure 7.3c, depicts a Poisson situation for which np = 0.1; now p is very small. This frequency distribution is nearly a spike at 0, emphasizing the rare-event notion of the Poisson random variable. These three cases demonstrate that the frequency distribution for a Poisson random variable increasingly spreads out over the potential values of X as np increases. Consequently, the Poisson distribution provides a worthwhile approximation to the binomial frequency distribution as p approaches zero.

The probability mass function of a Poisson random variable may be defined as

$$P(X = x) = \mu^x \exp(-\mu)/x!, \ x = 0, 1, 2, \ldots$$

where for the discrete binomial type of perspective $\mu = \sigma^2 = np$, and for the continuous interval viewpoint $\mu = \sigma^2$ = average density of outcomes. A fundamental feature of the Poisson random variable is the equality of its mean and its variance. This property allows the coefficient of variation to be used in statistical analyses involving Poisson frequency distributions. In other words, if a phenomenon is Poisson distributed, then the coefficient of variation equals one. By construction,

$$\Sigma_{x=0}^{\infty} \mu^x \exp(-\mu)/x! = 1.$$

Note that although the Poisson random variable is defined for all nonnegative values of X, its frequency distribution usually is truncated (see Figure 7.3) because of the near-zero probabilities that are affiliated with even moderate magnitudes of X.

As mentioned previously, this distribution commonly is utilized in statistical analyses in geography for undertaking point pattern analyses. If points are being sampled from a two-dimensional surface, then the population under study is infinite in size. The resulting probability of selecting any single point on a surface is nearly zero (it cannot be exactly zero or no points would appear in a sample). Clearly the outcome is a rare event.

EXAMPLE 7.2. Example 5.1 revisited.

The simulated point pattern was generated by sampling jointly from a pair of independent, continuous uniform distributions, where a = 0 and b = 1. The resulting frequency distribution, namely,

Number of points/quadrat	0	1	2	3
Number of quadrats	10	8	4	3

visually resembles a Poisson frequency distribution. Here the estimate $\bar{x}$ = 25/25 = 1, whereas $s = \sqrt{26/(25 - 1)} = 1.0408$; thus, CV = 1.0408/1 = 1.0408, which is very close to unity. The corresponding expected frequency values here are (np_i, i = 0, 1, 2, . . .).

Number of points/quadrat	1^xexp(-1)/x!	25 • 1^xexp(-1)/x!
0	0.367879	9.19699
1	0.367879	9.19699
2	0.183940	4.59849
3	0.061313	1.53283
4	0.015328	0.38321
5	0.003066	0.07664
6	0.000511	0.01277
7	0.000073	0.00182
8	0.000009	0.00023
9	0.000001	0.00003
10	0.000001	0.00002

Truncation occurs at X = 10 because the probabilities beyond this value are negligible. The differences between these expected and observed frequencies are attributable to sampling error, subject to the condition that indeed $\mu = 1$.

As before, the expected values in Example 7.2 are not constant, since p_i varies across the possible values of X.

Features of the outcome to be described by a selected random variable always should be contrasted with the theoretical properties of this variable. In the case of the Poisson random variable, one may question whether or not n could go to infinity; in Example 5.8, this possibility would be dictated by what the points are in reality. Suppose that the binomial is considered a more appropriate probability model for this example (this contention cannot be true for this example population, since the point pattern was generated with a pair of uniform random variates). Then, since 3 is the maximum value on the x-axis (i. e., the greatest number of points observed in any quadrat) in Example 5.8, n in the binomial formula would need to be 3, and hence $3\hat{p} = \bar{x} = 1$, resulting in $\hat{p} = 1/3 = 0.3333$, which is greater than 0.05. The foregoing Poisson calculations should fail to approximate closely the corresponding binomial computations, since p is relatively small, as is seen in the next example.

EXAMPLE 7.3. Poisson (shown above) and binomial calculations for the same problem.

x	{3!/[x!(3 - x)!]} • $(0.3333)^x(.06667)^{3-x}$	25{3!/[x!(3 - x)!]} • $(0.3333)^x(0.6667)^{3-x}$
0	0.296341	7.4085
1	0.444444	11.1111
2	0.222189	5.5547
3	0.037026	0.9256

The difference between the two probabilities for $P(X = 0)$ is 0.071538 (0.367879 - 0.296341), between the two for $P(X = 1)$ is 0.076565, between the two for $P(X = 2)$ is 0.038249, between the two for $P(X = 3)$ is 0.024287, and between the two for $P(X > 3)$ is 0.018989. These differences are attributable to specification error. As can be seen from this example, for n as small as 3 and p as large as 0.3333, the specification error involved in replacing a Poisson computation with a binomial one is marked. This notion is more generally illustrated by the histograms appearing in Figures 7.2 and 7.3; sometimes a Poisson distribution visually but erroneously may appear to conform to a binomial distribution, and vice versa.

7.4. The normal frequency distribution

The final theoretical frequency distribution to be entertained here for analytical purposes is for normal random variables and is called the **normal distribution**. This distribution is continuous; as such, its probabilities refer to areas under a curve and pertain only to values falling between two points along the x-axis (the probability of a single point is zero, since a line has no area). This distribution visually resembles a bell in shape. The normal curve is symmetric, smooth, and based upon an indefinitely large population size; hence, it can only be approximated by empirical frequency distributions. The affiliated probability density function is given by an equation that is far more complex than any of the other three presented in this chapter. For routine applications standard tables exist (see Table 7.1-A, for example) so that students never need to use this formula for computational purposes (many scientific hand calculators are programmable, allowing exact values to be calculated). The ordinate of a point on the graph of the normal distribution for $X = x$ may be defined as follows:

$$f(x) = \{1/[\sigma(2\pi)^{1/2}]\}\exp\{-[(x - \mu)^2]/(2\sigma^2)\}.$$

The area between two values of x that falls under the curve described by this formula is obtained by integrating the function f(x); since calculus is not a prerequisite for a student of elementary statistics, this procedure is described here. This area (which can be obtained from a normal curve table) is denoted by $P(a < X < b)$. Since lines have no area, using $<$ instead of $\leq$ or $>$ instead of $\geq$ is an issue of conceptual rather than arithmetic concern.

The mean of a normal frequency distribution is defined as μ, and the variance of this distribution is defined as σ^2. The mean can take on any value on the real number line. Large values of σ yield frequency distributions that are relatively shallow and quite spread out about μ, whereas small values of σ yield frequency distributions that are relatively peaked and quite condensed about μ. The parameters μ and σ are independent.

Three graphs of this normal frequency distribution appear in Figure 7.4. The

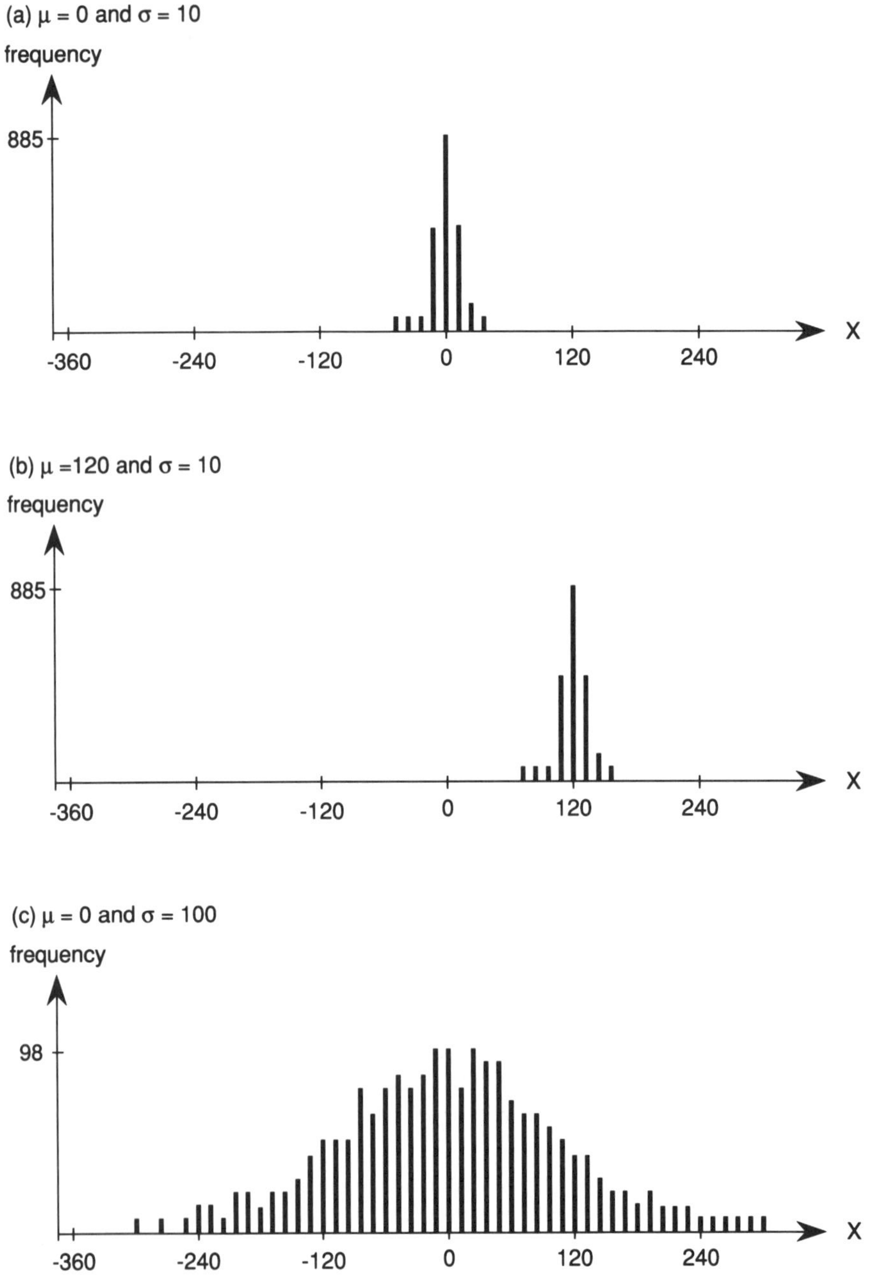

Figure 7.4: THREE SYMBOLIC CASES OF THE NORMAL FREQUENCY DISTRIBUTION (1800 RANDOM SAMPLES DRAWN)

first of these examples, Figure 7.4a, depicts a normal distribution situation for which $\mu = 0$ and $\sigma = 10$. Figure 7.4b uses the same sample as Figure 7.4a, except now $\mu = 120$ and $\sigma = 10$. Note that the form of this distribution does not change when the population mean is altered; the mean and the variance are independent parameters. The third example here, Figure 7.4c, depicts a normal distribution situation for which $\mu = 0$ and $\sigma = 100$. All three graphs were constructed using the same x-axis in order to demonstrate that as σ decreases, the frequency distribution becomes more peaked. The respective sample means (n = 1800) for these three sets of data are -0.128, 119.87, and -1.79; the respective sample standard deviations are 9.744, 9.744, and 96.90. All these values are reasonably close to their corresponding population parameter values.

The normal distribution is perhaps the most important probability model in all of statistics. Many numerical populations have distributions that can be described very accurately by a normal curve with appropriate values of μ and σ. When the underlying distribution is discrete rather than continuous, the normal curve often gives a superb approximation to the discrete frequency distribution. It furnishes a good characterization of error in scientific studies, especially sampling error (the pattern of random errors formed by a statistic around its population parameter value); as such, it forms the basis for much of inferential statistics. Furthermore, although individual variables may deviate dramatically from a normal distribution, under suitable conditions sums and averages of these variables will have approximately a normal distribution.

EXAMPLE PR-7.3. Constructing a normal frequency distribution.

Consider the variable DCD from the Puerto Rico data set. The sample mean of this variable is 36.17, and the unbiased sample standard deviation (see Chapter 4) is 11.96. A histogram can be constructed for DCD by establishing intervals about the mean that are 11.96 units wide. For example, the first two intervals on either side of the mean are one standard deviation wide or, 36.17 + 11.96 and 36.17 - 11.96. The intervals are 36.17 to 48.13, and 24.21 to 36.17. In this way the sample frequency distribution can be related easily to a table of normal curve probabilities. One such construction yields

Interval boundaries	Observed frequency	Normal probability	Expected frequency
60.09 - ∞	2	0.0228	1.6608
48.13 - 60.09	9	0.1359	9.9211
36.17 - 48.13	19	0.3413	24.9182
24.21 - 36.17	31	0.3413	24.9182
12.25 - 24.21	10	0.1359	9.9211
-∞ - 12.25	2	0.0228	1.6608

Again the expected frequency values are obtained by computing the product np_i, where n is the sample size of 73 and p_i is the normal probability for each interval calculated using the formula for the normal curve density function (this can be approximated with Table 7.2-A). Whereas p_i varies across the intervals, the top sequence of probabilities is a mirror image of the bottom sequence of probabilities, since the normal curve is symmetric. The two open-ended intervals are problematic in that they violate the property of uniform interval size. However, the probability that an outcome will occur in either of these intervals is sufficiently small that this violation introduces negligible measurement error. Once more, it is hoped that the differences between observed and expected frequencies will be attributed to sampling error.

EXAMPLE OR-7.1. Constructing a normal frequency distribution.

Consider the variable SNOW from the Ontario Runoff data set. The sample mean of this variable is 86.25, and the unbiased sample standard deviation (see Chapter 4) is 18.78. A histogram can be constructed for SNOW by establishing intervals about the mean that are 18.78 units wide. For example, the first two intervals on either side of the mean are one standard deviation wide or, 86.25 + 18.78, and 86.25 - 18.78. The intervals are 86.25 to 105.03 and 67.47 to 86.25. In this way the sample frequency distribution can be related easily to a table of normal curve probabilities. One such construction yields

Interval boundaries	Observed frequency	Normal probability	Expected frequency
123.81 - ∞	1	0.0228	0.6384
105.03 -123.81	4	0.1359	3.8052
86.25 -105.03	11	0.3413	9.5564
67.47 - 86.25	8	0.3413	9.5564
48.69 - 67.47	3	0.1359	3.8052
-∞ - 48.68	1	0.0228	0.6384

Again, the expected frequency values are obtained by computing the product np_i, where n is the sample size of 28 and p_i is the normal probability for each interval calculated using the formula for the normal curve density function (this can be approximated with Table 7.2-A). Whereas p_i varies across the intervals, the top sequence of probabilities is a mirror image of the bottom sequence of probabilities, since the normal curve is symmetric. As in the previous example, the two open-ended intervals are problematic in that they violate the property of uniform interval size. However, here too the probability that an outcome will occur in either of these intervals is sufficiently small that this violation introduces negligible measurement error. Once more, it is hoped that the differences between observed and expected frequencies will be attributed to sampling error.

But how are these normal probabilities retrieved from tabulated normal curve areas? This topic is treated in the next section.

7.5. The standard normal curve

All normal random variables adhere to the same probability density function, taking on various bell-shaped forms in accordance with the specific numerical values of μ and σ. Therefore, the key to working with normal frequency distributions is to standardize all of them into a single, universal reference normal curve. A tabulated version of this curve is presented in Table 7.2-A. In other words, the measures of probability affiliated with some variable X are determined by the relative position of X with respect to its population mean and population standard deviation. Accordingly, the standardization procedure involves converting each x to a **z-score** (the random variable represented by a z-score is denoted by Z), which is defined as

$$z = (x - \mu)/\sigma, \text{ or}$$

$$z = (x - \bar{x})/s \text{ if population parameter values are unavailable.}$$

This z-score is known as a standardized variable because its units are standard deviations, and it is dimensionless. Subtraction of the mean shifts the position of the original normal curve so that it is centered at zero. Dividing by the standard deviation is equivalent to changing units along the horizontal axis so that a distance of σ along the x-axis corresponds to a distance of 1 along the z-axis. The normal probability distribution associated with standard scores is called the **standard normal distribution**. The probabilities presented in Table 7.2-A are governed by all the rules outlined in Chapter 6. Converting variable X to variable Z causes μ and σ respectively always to be transformed to 0 and 1, and the probability density function simplifies to

$$f(z) = [1/(2\pi)^{1/2}]\exp[-(z^2)/2].$$

This particular probability distribution infrequently serves as a model for a naturally arising random variable. Instead, it is a reference distribution from which information about other normal distributions can be obtained.

Because the normal distribution is symmetric, only areas under the curve in one direction need to be tabulated. The areas in the opposite direction are a mirror image of this set of calculations.

EXAMPLE PR-7.4. Transforming the DCD variable appearing in Example PR-7.3 to a z-score variable.

The original density data are converted to z-scores using the sample mean and sample standard deviation as follows:

$z_i = (x_i - 36.1742)/11.9587.$

This yields

DCD	z-score	DCD	z-score	DCD	z-score	DCD	z-score
38.0	0.1527	25.6	-0.8842	62.5	2.2014	37.5	0.1109
9.4	-2.2389	32.5	-0.3072	40.3	0.3450	51.7	1.2983
22.7	-1.1267	39.9	0.3116	42.0	0.4872	48.6	1.0391
13.2	-1.9211	35.0	-0.0982	26.4	-0.8173	39.4	0.2698
52.6	1.3736	53.4	1.4405	37.9	0.1443	39.1	0.2447
40.7	0.3785	47.5	0.9471	21.6	-1.2187	30.4	-0.4828
26.2	-0.8340	34.8	-0.1149	24.3	-0.9929	37.3	0.0942
24.5	-0.9762	50.9	1.2314	49.4	1.1060	36.5	0.0273
15.0	-1.7706	36.9	0.0607	45.9	0.8133	43.1	0.5792
23.7	-1.0431	46.8	0.8886	41.1	0.4119	36.2	0.0022
23.3	-1.0765	32.7	-0.2905	41.4	0.4370	39.5	0.2781
34.2	-0.1651	41.7	0.4621	43.6	0.6210	18.5	-1.4779
53.6	1.4572	28.1	-0.6752	36.8	0.0523	39.8	0.3032
14.3	-1.8291	37.0	0.0691	62.9	2.2349	36.8	0.0523
2.9	-2.7824	48.3	1.0140	57.1	1.7499	37.5	0.1109
27.7	-0.7086	39.0	0.2363	49.5	1.1143	27.2	-0.7504
28.2	-0.6668	44.5	0.6962	26.7	-0.7922	35.4	-0.0647
20.4	-1.3190	32.3	-0.3239	44.6	0.7046	43.0	0.5708
31.7	-0.3741						

The mean of these z-scores is 0.0000, and the standard deviation is 1.0000. If these z-scores are ranked in ascending order, one obtains, reading left to right,

-2.7824	-2.2389	-1.9211	-1.8291	-1.7706	-1.4779
-1.3190	-1.2187	-1.1267	-1.0765	-1.0431	-0.9929
-0.9762	-0.8842	-0.8340	-0.8173	-0.7922	-0.7504
-0.7086	-0.6752	-0.6668	-0.4828	-0.3741	-0.3239
-0.3072	-0.2905	-0.1651	-0.1149	-0.0982	-0.0647
0.0022	0.0273	0.0523	0.0523	0.0607	0.0691
0.0942	0.1109	0.1109	0.1443	0.1527	0.2363
0.2447	0.2698	0.2781	0.3032	0.3116	0.3450
0.3785	0.4119	0.4370	0.4621	0.4872	0.5708
0.5792	0.6210	0.6962	0.7046	0.8133	0.8886
0.9471	1.0140	1.0391	1.1060	1.1143	1.2314
1.2983	1.3736	1.4405	1.4572	1.7499	2.2014
2.2349					

EXAMPLE OR-7.2. Transforming the SNOW variable appearing in Example OR-7.1 to a z-score variable.

The original density data are converted to z-scores using the sample mean and sample standard deviation as follows:

$z_i = (x_i - 86.2500)/18.7747.$

This yields

SNOW	z-score	SNOW	z-score	SNOW	z-score
76.0	-0.5459	48.0	-2.0373	72.0	-0.7590
88.0	0.0932	88.0	0.0932	88.0	0.0932
87.0	0.0399	68.0	-0.9721	91.0	0.2530
60.0	-1.3982	132.0	2.4368	120.0	1.7976
92.0	0.3063	80.0	-0.3329	85.0	-0.0666
110.0	1.2650	88.0	0.0932	80.0	-0.3329
96.0	0.5193	112.0	1.3715	90.0	0.1997
90.0	0.1997	108.0	1.1585	64.0	-1.1851
92.0	0.3063	80.0	-0.3329	70.0	-0.8655

The mean of these z-scores is 0.0000, and the standard deviation is 1.0000. If these z-scores are ranked in ascending order, one obtains, reading left to right,

```
-2.0373 -1.3982 -1.3982 -1.1851 -0.9721 -0.8655 -0.7590
-0.5459 -0.3329 -0.3329 -0.3329 -0.0666  0.0399  0.0932
 0.0932  0.0932  0.0932  0.1997  0.1977  0.2530  0.3063
 0.3063  0.5193  1.1585  1.2650  1.3715  1.7976  2.4368
```

Appealing interval boundaries for the ranked data appearing in Example PR-7.4 and OR-7.2 are easily identified; they occur at -2, -1, 0, 1, and 2.

In Table 7.2-A the tabulated standard normal curve probabilities are presented as area measures between the mean and values to the right of the mean; symmetry permits this condensation. The rows and columns of this table represent digits of a z-score; the row provides the first digit both to the right and to the left of the decimal point, while the column provides the second digit to the right of the decimal point. A z-score of 1.57 then directs attention to the row labelled 1.5 and the column labelled 0.07 of this table. The intersection of the relevant row and column contains the area under a standard normal curve between the mean of zero and the z-score defined by a sum of the row and column label values. The intersection of row 1.5 and column 0.07 identifies an area of 0.4418, which is $P(0.00 < Z < 1.57)$.

EXAMPLE PR-7.5. Calculating the normal probabilities for Example PR-7.3.

From Example PR-7.4 the normal probabilities being sought are

$P(2 < Z < \infty) = 0.5000 - P(0 < Z < 2) = 0.5000 - 0.4772 = 0.0228$
$P(1 < Z < 2) = 0.4772 - P(0 < Z < 1) = 0.4772 - 0.3413 = 0.1359$
$P(0 < Z < 1) = 0.3413$
$P(-1 < Z < 0) = P(0 < Z < 1) = 0.3413$
$P(-2 < Z < -1) = P(1 < Z < 2) = 0.1359$
$P(-\infty < Z < -2) = P(2 < Z < \infty) = 0.0228$

EXAMPLE OR-7.3. Calculating the normal probabilities for Example OR-7.1.

From Example OR-7.2 the normal probabilities being sought are as follows. Note that since we are now working with Z scores, given the same set of intervals, the calculations in Example PR-7.5 are the same as the calculations in this example.

$P(2 < Z < \infty) = 0.5000 - P(0 < Z < 2) = 0.5000 - 0.4772 = 0.0228$
$P(1 < Z < 2) = 0.4772 - P(0 < Z < 1) = 0.4772 - 0.3413 = 0.1359$
$P(0 < Z < 1) = 0.3413$
$P(-1 < Z < 0) = P(0 < Z < 1) = 0.3413$
$P(-2 < Z < -1) = P(1 < Z < 2) = 0.1359$
$P(-\infty < Z < -2) = P(2 < Z < \infty) = 0.0228$

The standard normal curve table can be used in a reverse fashion, too. If the probability $P(0 < Z < z)$ is known, say 0.4750, then the area can be found within the interior of the table, and the corresponding row and column labels identified (1.9 and 0.06, meaning $z = 1.96$). If additional z-score digits are available, then they must be either rounded off (introducing measurement error) or interpolated between adjacent z-score values (which often introduces specification error). Similarly, if a probability is given that does not appear in the standard normal curve table, then its tabular location must be approximated, followed by an estimate of the corresponding row and column labels.

7.5.1. Transforming the standard normal curve

There are applications in which a mean and standard deviation other than 0 and 1 may be desirable. While a mean of 0 and standard deviation of 1 may represent the most commonly used transformation, any pair of values (one for the mean, the other for the standard deviation) can be specified. Given a set of z-scores, or standard deviates, adding a constant value k_1 to each score produces a new set of scores with a mean equal to k_1. Similarly, multiplying each deviate by some constant value k_2 produces a new set of deviates with a standard deviation equal to k_2. Combining the two transformations by first multiplying each deviate by k_2 and then adding a constant k_1 produces a transformed set of deviates with a mean equal to k_1 and a standard deviation equal to k_2. Note that in this third, combined, transformation the *order* of operations *is important*; first multiply, and then add.

EXAMPLE CA-7.1. Transformations of standard scores using the sales variable (i. e., Sales of Auto Parts) from the Canadian Auto data for Central Ontario Firms.

In the following table, the first column shows the original value for each of the 12 firms in the region. The second column shows the accompanying z-scores. The third, fourth, and fifth columns show the deviates after transformations of (1) adding a constant of 10, (2) multiplying by a constant of 3, and (3) first multiplying by 3 then adding 10. The original values have a mean of 16,633.33 and a standard deviation of 21,138.60.

Original sales (thousands of $)	z-score	(Z + 10)	(Z • 3)	(Z • 3) + 10
25,000	0.396	10.396	1.188	11.188
100	-0.782	9.218	-2.346	7.654
50,000	1.578	11.578	4.734	14.734
50,000	1.578	11.578	4.734	14.734
5,000	-0.550	9.450	-1.650	8.350
5,000	-0.550	9.450	-1.650	8.350
5,000	-0.550	9.450	-1.650	8.350
500	-0.763	9.237	-2.289	7.711
100	-0.782	9.218	-2.346	7.654
3,000	-0.645	9.355	-1.935	8.065
50,000	1.578	11.578	4.734	14.734
5,000	-0.550	9.450	-1.650	8.350
Mean	-0.0035	9.9965	-0.0105	9.9895
Standard deviation	1.0026	1.0026	3.0079	3.0079

This table presents the values for the mean and standard deviation for each column of deviates. Except for rounding error, the means for the z-scores and (Z • 3) are 0, and for (Z + 10) and (Z • 3) + 10, they are 10. Similarly, the standard deviations for the columns z-score and (Z + 10) are 1, and for (Z • 3) and (Z • 3) + 10, they are 3.

7.6. Normal curve approximations to the binomial and the Poisson

The normal curve can supply a practical approximation to the binomial frequency distribution when n is very large. This approximation is particularly useful because the factorial terms n!, x!, and (n - x)! become cumbersome or impossible to calculate. Keep in mind that such an approximation introduces specification error, but this error becomes negligible as p goes to 0.5 and n goes to infinity. As a rule of thumb, the range that yields sufficiently close and acceptable approximations is delimited by both $np \geq 5$ and $n(1 - p) \geq 5$; whenever either of these values is less than 5, the binomial distribution is too skewed to be adequately approximated by the normal distribution. To compute this approximation, the difference between a binomial random variable, which is discrete, and a normal random variable, which is continuous must be accommodated. The necessary modification needed here, commonly referred to as the continuity correction factor, is a conversion of each binomial point value x to an interval (x - 0.5, x + 0.5). This interval now can be related to the normal curve. In essence, this conversion is equivalent to superimposing the probability mass function for a binomial frequency distribution on a normal curve. Because a discrete distribution is involved here, using the inequalities < instead of ≤ and > instead of ≥ makes a difference.

Let X be a binomial random variable. Then normal curve approximations to its probabilities are defined as

$$P(a < X < b) = P\{(a + 0.5 - np)/[np(1 - p)]^{1/2} < Z < (b - 0.5 - np)/[np(1 - p)]^{1/2}\}$$
$$P(a < X \leq b) = P\{(a + 0.5 - np)/[np(1 - p)]^{1/2} < Z < (b + 0.5 - np)/[np(1 - p)]^{1/2}\}$$
$$P(a \leq X < b) = P\{(a - 0.5 - np)/[np(1 - p)]^{1/2} < Z < (b - 0.5 - np)/[np(1 - p)]^{1/2}\}$$
$$P(a \leq X \leq b) = P\{(a - 0.5 - np)/[np(1 - p)]^{1/2} < Z < (b + 0.5 - np)/[np(1 - p)]^{1/2}\}$$

Parallel results can be written for the inequalities > and ≥.

EXAMPLE 7.4. Example PR-7.2 reexamined.

Consider the binomial probability $P(X = 5)$
$= P(4 < X < 6) = P(4.5 \leq X \leq 5.5) = \{10!/[5!(10 - 5)!]\}(0.60274 \bullet 0.39726)^5$
$= 0.1983$.

The normal approximation to this probability is given by
$P[(4.5 - 10 \bullet 0.60274)/(10 \bullet 0.60274 \bullet 0.39726)^{1/2} \leq Z \leq$
$(5.5 - 10 \bullet 0.60274)/(10 \bullet 0.60274 \bullet 0.39726)^{1/2}]$
$= P(-0.987 < Z < -0.341)$
$= 0.3365 + (7/10)(0.3389 - 0.3365) - [0.1331 + (1/10)(0.1368 - 0.1331)]$
$= 0.2047$.

The difference of 0.0064 between the binomial probability and the normal approximation probability is due to specification error. The normal curve table values are obtained, because of symmetry, as $P(0.341 < Z < 0.987) = P(Z < 0.987) - P(Z < 0.341)$. The terms (7/10)(0.3389 - 0.3365) and (1/10)(0.1368 - 0.1331) represent linear interpolations of the tabulated areas associated with the third digits to the right of the decimal found in the z-scores; calculation of exact values with the standard normal probability density function produced no difference between them and these interpolated values for the first four decimal places.

The normal curve also can supply a practical approximation to the Poisson frequency distribution when n and μ are very large, and hence the factorial and exponential terms $x!$, $\exp(-\mu)$, and μ^x become cumbersome or impossible to calculate. Again keep in mind that such an approximation introduces specification error, and no suitable general rule of thumb exists that establishes a range yielding sufficiently close and hence acceptable approximations. To compute this approximation, the difference between a Poisson random variable, which is discrete, and a normal random variable, which is continuous, must be accommodated by the continuity correction factor. In essence, this conversion is equivalent to superimposing the probability mass function for a Poisson frequency distribution on a normal curve.

Let X be a Poisson random variable. Then normal curve approximations to its probabilities are defined as

$$\begin{aligned}
P(a < X < b) &= P[(a + 0.5 - \mu)/\sqrt{\mu} < Z < (b - 0.5 - \mu)/\sqrt{\mu}] \\
P(a < X \le b) &= P[(a + 0.5 - \mu)/\sqrt{\mu} < Z < (b + 0.5 - \mu)/\sqrt{\mu}] \\
P(a \le X < b) &= P[(a - 0.5 - \mu)/\sqrt{\mu} < Z < (b - 0.5 - \mu)/\sqrt{\mu}] \\
P(a \le X \le b) &= P[(a - 0.5 - \mu)/\sqrt{\mu} < Z < (b + 0.5 - \mu)/\sqrt{\mu}]
\end{aligned}$$

Parallel results can be written for the inequalities > and ≥.

EXAMPLE 7.5. Example 7.2 reexamined.

Consider the Poisson probability $P(X = 3)$

$$= P(2 < X < 4) = P(2.5 \le X \le 3.5) = 1 \bullet \exp(-1)/3! = 0.061313.$$

The normal approximation to this probability is given by

$$\begin{aligned}
&P[(2.5 - 1)/1 \le Z \le (3.5 - 1)/1] \\
&= P(1.5 < Z < 2.5) = 0.4938 - 0.4332 = 0.0606.
\end{aligned}$$

The difference of 0.000713 between the Poisson probability and the normal approximation probability is due to specification error. This approximation is much better than the binomial one presented in Example 7.3.

7.7. The use of probability paper

Throughout this and previous chapters, the importance of determining the distributional characteristics of the selected sample has been emphasized. One method of determining the normality of the sample is to calculate a test statistic such as chi square (see Chapter 4). Another method is to plot the observations on probability paper. Using probability paper requires plotting the cumulative proportion of observations in a sample against the increasing (or decreasing) values of some variable. The spacing of cumulative probabilities is such that a straight line is obtained if the observations are normally distributed with respect to the variable measured.

EXAMPLE SG-7.1. Using probability paper when analyzing the Sentinel Glacier data set.

Using the PRECIP variable from the Sentinel Glacier, first rank the observations from smallest to largest. Each observation then represents one year out of 20. Note that although there are 19 observations, 20 years are represented, including the 6 months preceding the first observation and the 6 months following the last observation. Since each observation thus equals 5% of the total (p = 0.05), the cumulative probability is the number of observations ranked, including the current observation, multiplied by 0.05. The rearranged data follow:

Observed value	Cumulative probability	Observed value	Cumulative probability
626.1	0.05	1103.9	0.55
657.3	0.10	1136.8	0.60
685.7	0.15	1161.6	0.65
796.1	0.20	1178.3	0.70
884.5	0.25	1300.0	0.75
953.4	0.30	1308.0	0.80
971.1	0.35	1397.4	0.85
994.5	0.40	1407.8	0.90
1007.3	0.45	1608.7	0.95
1043.8	0.50		

The total cumulative probability ends at 0.95, since there are 19 observations for 20 years. The cumulative probability is plotted on the x-axis and the ranked precipitation values are on the y-axis. The resulting graph appears in Figure 7.5. As can be seen, the fit is not ideal; that is, the plotte-d points do not form a straight line. However, the trend is certainly linear, as shown by the straight line drawn through these points. The fit is not ideal because the plotted values are seasonal totals rather than annual

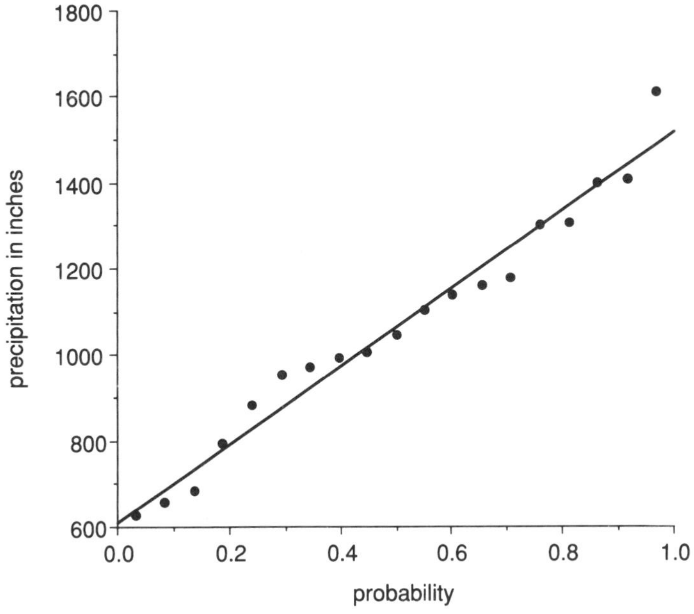

Figure 7.5: NORMAL PROBABILITY PAPER PLOT FOR EXAMPLE SG-7.1

extreme values, which might be expected to produce a trend that is more linear in nature.

There are a variety of probability papers that can be used in specific situations (for example, log-normal, Gumbel, and log-log paper).

7.8. Summary

A chapter on theoretical frequency distributions is always a difficult chapter to place in a book. The content does not follow directly the content of preceding chapters but is required for the chapters that follow. Frequency distributions appear in many statistical applications. At times there appears to be a different distribution for each application. However, the distributions described in this chapter are those that occupy the majority of applications in the geographic sciences. A few others (chi square, t, and F distributions) are used extensively in certain inferential applications, but these distributions relate more to the distribution of statistics, rather than populations and samples, and theoretically can be related to the normal distribution.

The inferential procedures that begin in Chapter 10 and occupy the entire third part of the book rely heavily on assumptions regarding the distributions of samples, populations, and sample statistics. A reasonable familiarity with the distributions presented in this chapter will greatly aid in understanding what is to come!

7.9. Problems

1. Select the first set of 50 single digits in the first row of Table 7.1-A, namely, {9, 7, 3, 9, 4, 5, 1, 6, 7, 0, 2, 2, 7, 4, 1, 2, 8, 2, 3, 3, 0, 7, 6, 9, 7, 1, 6, 9, 2, 9, 7, 4, 6, 7, 5, 2, 1, 5, 7, 3, 3, 3, 0, 6, 5, 4, 1, 1, 7, 1}. Determine the population mean and variance, as well as the sample mean and variance of this sample. Then construct the observed and the theoretical uniform frequency distributions for this sample.

2. Verify the probability calculations appearing in Example PR-7.2.

3. Construct the binomial frequency distribution for all possible samples of size 5, drawn from the Puerto Rico data, for the dichotomous variable U-R. Repeat this exercise for the ADM variable, distinguishing between the San Juan Agricultural Administrative Region and all other administrative regions.

4. Over a 100-year period, observers noted that a given river flooded 27 times. Construct the binomial probability and frequency distributions of all possible flooding scenarios for this river over the next 10-year period. Those data collected on 10-year intervals for all comparable rivers in a study region yielded the data appearing in the following table:

The Temporal Distribution of the Number of Rivers That Flooded in a Particular Study Region Over a 10-year Period

No. of years flooding occurred	0	1	2	3	4	5	6	7	8	9	10
No. of rivers that flooded	1	2	6	3	7	11	7	6	2	4	1

5. Verify the probability calculations appearing in Example 7.2.

6. The observed frequency distribution appearing in the following table resulted from the first stages of a quadrat analysis of the spatial distribution of dwellings on Rousay Island, off the coast of England. Construct the probability and theoretical frequency distributions for this set of data.

The Spatial Distribution of Dwellings on Rousay Island, by Quadrat

No. of dwellings	0	1	2	3	4	5	6	7	8	9	10	11
Quadrat frequency	21	5	0	6	2	2	1	0	2	0	1	1

7. The data appearing in the following table refer to a rare-event situation. Construct the probability and theoretical frequency distributions for this set of data.

Number of Brain Tumors in the Rzeszow District of Poland

County	Population (in 1000s)	Number of tumors per 100,000
Brzozow	70	0.000
Debica	110	3.636
Gorlica	83	10.843
Jaroslaw	102	5.882
Jaslo	101	3.960
Kolbuszowa	60	1.667
Krosno	101	1.980
Lesko	17	11.765
Lubaczow	40	2.500
Lancut	91	6.593
Mielec	73	5.479
Nisko	67	2.985
Przemysl	88	6.818
Przeworsk	57	3.509
Rzeszow	182	7.143
Sanko	56	7.143
Tarnobrzeg	67	4.478

8. Suppose the Poisson distribution is considered a more appropriate probability model for Problem #4. Construct the probability and frequency distributions for this set of data employing this assumption. Discuss the specification error affiliated with this particular example.

9. Compute the following normal curve probabilities:

(a) $P(0.0 < Z < 1.5)$
(b) $P(0.5 < Z < 1.6)$
(c) $P(-0.51 < Z < 1.92)$
(d) $P(-1.43 < Z < -0.77)$
(e) $P(-1.50 < Z < 0.55)$
(f) $P(0.377 < Z < 1.249)$

10. The data appearing in the following table refer to a variable that should be normally distributed. Construct the probability and frequency distributions for this set of data.

Average Length (in kilometers) of Journeys to Work in Canadian Cities

1.86	2.37	3.07	3.68	4.32	5.01	2.07	2.73	3.41	3.69	4.32	5.07
2.24	3.03	3.41	3.77	4.46	5.14	2.37	3.07	3.55	3.91	4.93	5.66

11. The set of 73 municipios of the Puerto Rico data set were rated according to the level of agricultural production occurring in each one. The corresponding ratings appear in the following table and should be normally distributed. Construct the probability and frequency distributions for this set of data.

1974 Agricultural Productivity Index for Municipios of Puerto Rico

22.82	6.87	19.15	10.40	20.12	28.31	23.74	11.85
11.74	14.86	19.60	21.69	29.72	8.15	2.27	31.97
27.36	17.49	24.63	20.71	20.11	33.39	19.71	93.85
33.37	31.80	36.60	24.64	30.78	21.26	22.51	13.38
18.96	25.63	16.42	18.46	24.57	27.90	28.88	30.68
11.88	21.28	11.71	13.62	26.44	25.87	29.08	18.87
22.84	16.18	29.30	32.96	21.11	15.17	22.69	17.21
20.71	15.16	23.30	22.73	20.47	20.01	36.08	24.28
21.56	32.49	12.48	47.16	33.39	14.89	24.87	28.62
25.90							

12. Construct the normal probability and frequency distributions for the DFRM and DTB agricultural variables appearing in the Puerto Rico data set.

13. Verify the probability calculations appearing in Examples 7.4 and 7.5.

14. Suppose the normal approximation is considered worth exploring for Problem #4. Construct the probability and frequency distributions for this set of data employing this assumption. Discuss the specification error affiliated with this particular example.

15. Suppose the normal approximation is considered worth exploring for Problem #6. Construct the probability and frequency distributions for this set of data employing this assumption. Discuss the specification error affiliated with this

particular example.

16. Duplicate the analysis of Example CA-7.1 using the variable SALE for the Metropolitan Toronto region in the Canadian Automotive data. Verify that the means and standard deviations for the transformed z-scores are as expected.

17. Using the Sentinel Glacier data set, construct the cumulative probability curve for the variable PRECIP on cumulative probability graph paper. From the graphed line estimate the mean annual rainfall and the standard deviation of annual rainfall. If a positive mass balance occurs only in years that receive more than 1000 mm of precipitation, what is the probability of a positive mass balance in any given year?

18. The distribution of grain sizes in soil samples determines many of the physical properties of the soil. Analysis of a sample yielded the following information. Note that ϕ units are a logarithmic transformation of millimeter units, where $\phi = (-\log d)/(\log 2)$ and d = the grain size in millimeters. Thus, positive ϕ units are small grain sizes and negative ϕ are large grain sizes.

Grain size (mm)	Grain size (ϕ)	Cumulative Probability
0.063	4.0	0.038
0.088	3.5	0.085
0.125	3.0	0.125
0.177	2.5	0.280
0.250	2.0	0.520
0.350	1.5	0.782
0.500	1.0	0.848
0.707	0.5	0.962
1.000	0.0	0.985
1.410	-0.5	0.995

Plot grain size in ϕ units against cumulative probability on cumulative probability paper. Do the grain sizes appear to be normally distributed?

7.10. References

Knight, N. (1988) "No Two Alike?" *Bulletin of the American Meteorological Society*, 69: 496.

Kolata, G. (1990) "1-in-a-trillion Coincidence, You Say? Not Really, Experts Find." *The New York Times*, 140 (February 27): C1 and C2.

7.11. Additional readings

In addition to the texts listed in Chapter 1,

Gardiner, V., and G. Gardner. *Analysis of Frequency Distributions*. CATMOG 19.

Getis, A., and B. Boots. (1978) *Models of Spatial Processes: An Approach to the study of Point, Line and Area Patterns*. Cambridge: Cambridge University Press.

Ord, J. (1972) *Families of Frequency Distributions*. London: Griffin Limited.

PART II

THE FOUNDATIONS OF INFERENTIAL STATISTICS

Chapters 8 through 12 comprise the second part of this text. After studying the preliminary material and descriptive techniques presented in Part I, an analyst should be anxious to make some type of statement that relates a sample, with its characteristics, to its parent population. A major theme underlying the chapters in this portion of the book is specification error (and diagnostics for it), with special attention being paid to how it arises from the presence of redundant information latent in data. This process begins with the sampling strategy, Chapter 8, and culminates in some form of inferential statement. Chapter 9 presents the theory behind an inferential test, both parametric and nonparametric. Chapter 10 applies material covered in the preceding chapter to simple cases regarding population parameters and nonparametric alternatives. Chapter 11 presents tests that permit the comparison of two (or more) populations and samples. In this latter chapter, the power and utility of a classical analysis of variance is presented. The final chapter in this section introduces the very powerful technique of linear regression, perhaps the most widely applied technique in the statistical arsenal of procedures. Unfortunately, the widespread use of a technique increases the chances of its misuse as well.

A complete mastery of the material in Chapters 8 and 9 should reduce the misuse of techniques presented in Chapters 10 through 16. As with Part I, extra time spent in these two critical chapters will surely pay handsome dividends later.

A constant source of puzzlement as we teach statistics courses is the frequency with which advanced students (graduates and seniors) working on research projects return with the following question: Well, here are the data I collected; what can I do with them? Such a question suggests that little time has been spent planning the data collection phase of a project in conjunction with the analysis stage. The fact that not all these students had the benefit of one of our courses is a small comfort. It is difficult to cover the many dimensions of sampling frameworks in a textbook with a limited size. Not only would the size of such a chapter grow very large, but more importantly, it is the type of topic that is better presented in tutorials or laboratory settings. The length of Chapter 8 is not at all proportional to the importance of its topic. Of all the chapters in the book (except perhaps Chapter 7, on probability theory), Chapter 8 provides just a minimal treatment. Only primary data collection is discussed in any detail. Discussions of the vagaries of survey sampling and the use of secondary data sources are left to other readings.

Chapter 12 introduces linear regression in an inferential context. As we stated before, this is probably the single most widely used inferential technique. It is equally widely misused. Misuse arises from a number of sources including poor data, violations of assumptions, and inappropriate interpretations. Again, the naive analyst may well go directly to regression analysis, assuming that all the necessary understanding will arrive once the results are at hand. The availability of computer packages with thick manuals may encourage such slipshod analysis. We cannot repeat too often, time spent learning the basics will reduce errors in application in the long run. Our motto is: *Learn the basics and the rest will follow!*

CHAPTER 8

SAMPLING DESIGNS

One theme woven through the first seven chapters stresses that the subject matter of statistics is concerned with inferring conclusions about a population on the basis of a sample drawn from that population. So far those identified properties of a sample necessary for establishing sound inferences include parent population identification, representativeness, randomness, sampling perspective, and likeliness of resulting sample statistics. A sampling design must be devised that ensures that these properties are satisfied; otherwise, the inferential basis for concluding comments becomes compromised.

8.1. Identifying the population to be sampled

Geographic data series involve areal units, either in aggregate surface partitioning form, surface coordinate point form, or line form. Historically, the debate about these units focuses on whether such a data series is a properly constituted sample or in fact a population. Some spatial analysts argue that "[g]reat imagination has gone into turning what appears to be a population into a sample" (Upton and Fingleton, 1985, p. 325), concluding that these imaginative arguments cannot be justified in terms of statistical theory (Summerfield, 1983, p. 148).

Recapitulating and expanding upon sections of Chapters 1 and 2, the population is the total set of items for which a parameter measurement is to be obtained. This set must be defined in such a way that one can clearly specify whether or not any item is a member of it. For example, all people who reside in the United States is a population that the U.S. Census Bureau seeks to study. A statistical population is some set of numerical values corresponding to a population. If a population has N items, and some attribute is measured, then the statistical population will include N measures, one for each item in the parent population. For instance, the income of each person who resides in the United States is a statistical population corresponding to the foregoing parent population. Once a locational context or geographic reference is attached to measures, the total set of numerical values constitutes a statistical geographic population. One popular example here is average income for each county in the United States.

Consider a planar surface that is partitioned into mutually exclusive and collectively exhaustive areal units, such as the counties of the United States. The geographic distribution of values for some phenomenon over this surface constitutes a statistical geographic population. Random samples of these counties could be drawn, although with only several thousand counties and the existence of high-speed computers, this would be a senseless exercise. Usually the number of areal units involved in geographic situations is sufficiently small that studying subsets of them saves few resources. A number of methods are commonly employed to circumvent the problem encountered with a small number of elements in a geographic population. Unfortunately these methods are often not valid. Some researchers contend that the region under study is a sample from the population of all existing regions, with the hope that findings for one region apply to all other regions. Clearly, areal units clustering in a region cannot be considered a random sample from such a parent population. Summerfield (1983, p. 145) correctly stresses that this sort of extrapolation is neither legitimatized nor justified by statistical theory. Other researchers seek to sidestep this problem by restricting attention to the region itself and treating it as a sample from a time sequence, aspiring to generalize findings to past and future moments of the region. The primary argument supporting this view contends that almost all data are imprecise and time specific. Again, though, the single-point-in-time region hardly can be viewed as a random sample in time. Summerfield (1983, p. 145) correctly stresses that without knowledge about the underlying temporal distribution for a region, statistical theory fails to furnish appropriate guidelines dictating the time interval over which such extrapolations remain valid. This context is analogous to one of extrapolating from a region to areal units immediately outside its borders.

Summerfield (1983, p. 145) also discards the viewpoint that since data are likely to contain measurement error, either because a surrogate variable has been selected or measurement precision is inadequate, then areal units embracing these data constitute a sample. If this error were random and independent, then it would allow one to view the map as a sample. Much of the controversy surrounding measurement error in U.S. census data suggests that (1) noticeable measurement error exists, and (2) this error is neither random nor independent. Standard correction techniques adopted from statistical theory have been so inadequate that considerable litigation has emerged involving the U.S. Census Bureau versus numerous municipalities. Such measurement errors are far more complicated, suggesting that for geographic data there is no possibility for repeated measurements of a population under basically identical conditions. As a result, an assessment of the distribution of this measurement error cannot be made.

Another type of population emerges from the manner in which this planar surface is partitioned into areal units. There are literally an infinite number of ways that the surface can be partitioned into n areal units (boundary configurations). In addition, the value of n for a given region can vary considerably (scale). The observed spatial partitioning can be viewed as one sampling outcome from a hypothetical population consisting of all possible spatial configurations that could be formed with the same unaggregated, microgeographic distribution of data. Summerfield (1983, p. 146) argues against this outlook, noting that most default configurations provided by administrative boundaries serve as the observed partitioning and that there tends to be little that is random about their establishment or repeated use. But as long as those phenomena upon which the network of areal units is superimposed are not the ones initially used to establish the boundaries in question, then a random partitioning of the surface seems to be a reasonable argument in support of map pattern as a sample.

A second reasonable argument posits that one can hold the configuration of areal units fixed and consider the number of different ways the set of observed numerical values can be allocated to the areal units of this configuration. This is a randomization outlook, and relates to the earlier discussions about permutations. Now the parent population consists of n! map patterns. All n! permutations over the geographic configuration are considered to be equally likely. Much of the current understanding about spatial autocorrelation rests upon this view of samples and populations. Summerfield (1983, p. 148) maintains that this argument is yet another instance in geography of misapplying the Fisherian statistical idea of "randomization" (Fisher argued that one should control factors that can be controlled, and *randomize* factors that cannot be controlled). On the other hand, a number of especially nonparametric classical statistics (see Section 4.1) view samples in precisely this fashion. With regard to spatial autocorrelation, the join counts statistics happen to be a two-dimensional extension of one of them.

A third argument supporting the existence of samples from geographic populations is similar to the preceding attitude. An observed map pattern is seen as one possible outcome of a process in which sampling error is equated with error from a probabilistic process (see Chapters 6 and 7); such processes may be called stochastic. Suppose one is driving away from an urban area into a desolate region with his or her radio tuned to some radio station. This radio receives distinct signals in the urban area, whereas these signals become increasingly weaker as distance from the urban area increases. These signals represent a discernible, understandable pattern. Meanwhile, since the radio remains capable of receiving signals, it continues to pick up sound waves. But, these sound waves are from atmospheric or other electrical sources, none of which is dominant, and when they mix together they are received as static, or noise. Analogously, social, economic, physical and/or other forces induce a given spatial distribution over a planar surface. Additional factors, however, operate independently of one another and mix with these organizing forces, introducing noise that produces a modified geographic distribution. The geographic population is that spatial distribution void of this random noise component.

Sampling designs outlined in this chapter spotlight the classical sampling approach of subset selection.

8.2. Sample size and sampling error

A prominent consideration of statistical theory is distilling information from data in the presence of sampling error. This error is attributable to computations based on a subset of the population (a statistic) rather than computations based upon the population itself (a parameter). Sampling perspectives and sample properties reviewed thus far furnish partial insights into an understanding of this error. The relationship between this sampling error and the sample size n is discussed next.

A statistic is to be interpreted within the context of its sampling distribution, which ideally is constructed from all possible samples of a given size n. In Chapter 2 the property of consistency was highlighted. This property states that as n increases, the statistic of interest should have a sampling distribution that concentrates at and around the corresponding population parameter value. This relationship helps to discriminate between meaningful and meaningless statistics. Meaningful statistics also maximally exploit the information contained in a sample (the sufficiency property). Certainly if $n = 0$, then a sample of size zero has no information; any two guesses about a population parameter value would tend to be equally respectable. In contrast, if $n = N$, then the entire population should be surveyed, and all possible information enters the computation. But what happens for the intervening values of n? Intuitively, as n approaches N, the information content of a sample should tend to increase. This contention is supported by results reported in Chapter 7 for the perspective of sampling with replacement but order unimportant. Therefore, larger values of n should be preferable to smaller values of n. But what is a large value for n? Knowing what sample size is appropriate and how it affects the expected magnitude of sampling error is a critical input into the formulation of a sampling design.

One of the complicating factors affecting any ability to attempt generalizations about appropriate sample size is the enormous number of possible population frequency distribution types that exist for a random variable X. Four extreme cases will be inspected here to illustrate this range of possibilities. Fifty random samples of various sizes have been drawn, and the arithmetic means of each calculated, producing a sample sampling distribution of $\bar{x}$. To ensure the comparability of results from these four case studies, each distribution is generated with $\mu = 1$ and $\sigma = 1$. Unfortunately, to maintain this condition, the ranges from which the values are chosen differ for each of the four distributions (see Table 8.1).

The first case is for a continuous uniform distribution, where all values of the random variable are equally likely. The probability density function for this case is

$$f(x) = 1/\text{range} = 1/3.4641;\ \mu = 1 \text{ and } \sigma = 1$$

from an interval of $-0.7321 \leq x \leq 2.7321$ (see Table 8.1).

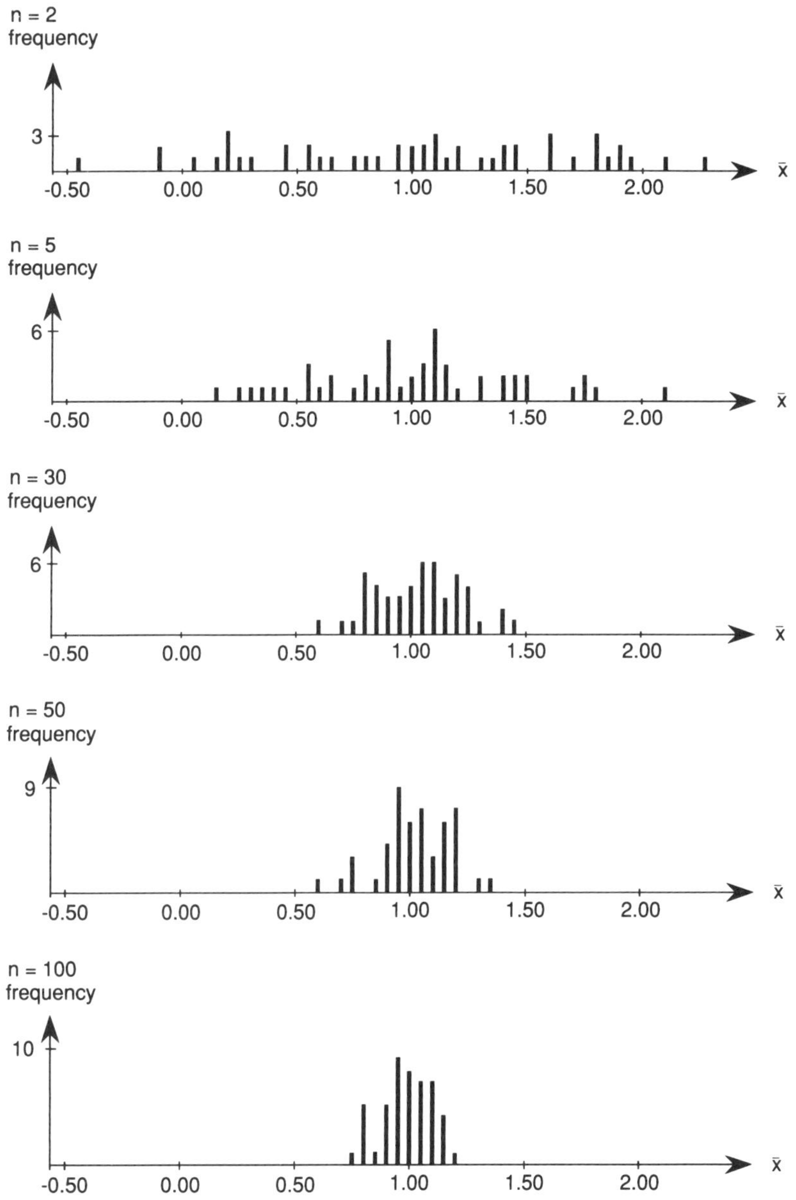

Figure 8.1: SAMPLING DISTRIBUTION OF SAMPLE MEANS FOR A UNIFORM RANDOM VARIABLE (50 SAMPLES)

TABLE 8.1

SAMPLING DISTRIBUTION STATISTICS
FOR VARIOUS SAMPLE SIZES AND FREQUENCY DISTRIBUTIONS
(from 50 samples)

Sample size	Sampling distribution							
	Uniform		Exponential		Bimodal		Normal	
n	$\bar{x}$	$s_{\bar{x}}$	$\bar{x}$	$s_{\bar{x}}$	$\bar{x}$	$s_{\bar{x}}$	$\bar{x}$	$s_{\bar{x}}$
2	1.0290	0.65433	1.0489	0.79432	1.0505	0.54210	1.0089	0.72683
5	1.0165	0.42398	1.0389	0.36526	1.0793	0.44391	1.0319	0.45984
30	1.0375	0.18676	1.0138	0.18047	1.0387	0.15478	0.9774	0.17330
50	1.0235	0.15320	0.9891	0.13716	0.9649	0.14820	1.0289	0.13678
100	0.9853	0.10927	1.0003	0.10633	1.0388	0.10331	1.0021	0.08381

Coefficient of variation[1]/range

n	Uniform	Exponential	Bimodal	Normal
2	0.636	0.422	0.632	0.404
5	0.417	0.196	0.504	0.250
30	0.180	0.099	0.183	0.099
50	0.150	0.077	0.188	0.075
100	0.111	0.059	0.122	0.047

1. Note: The coefficient of variation has been calculated with all population means and variances equal to unity in order to have standardized magnitudes and hence facilitate comparisons between distributions. Unfortunately, to achieve this result, the ranges for the four distributions had to be different; for the uniform it is 3.4641, for the exponential it is 6.2146, for the sinusoidal--which is the example of a bimodal distribution used here--it is 2.8284, and for the normal it is 6.

The simulated sampling distributions for various sample sizes (2, 5, 30, 50, 100) appear in Figure 8.1. The summary statistics for these distributions are reported in Table 8.1. One conspicuous trend that corroborates the consistency property is that as n increases, (1) the standard deviation of the sampling distribution decreases, meaning that concentration is occurring, and accordingly, (2) the frequency distribu-

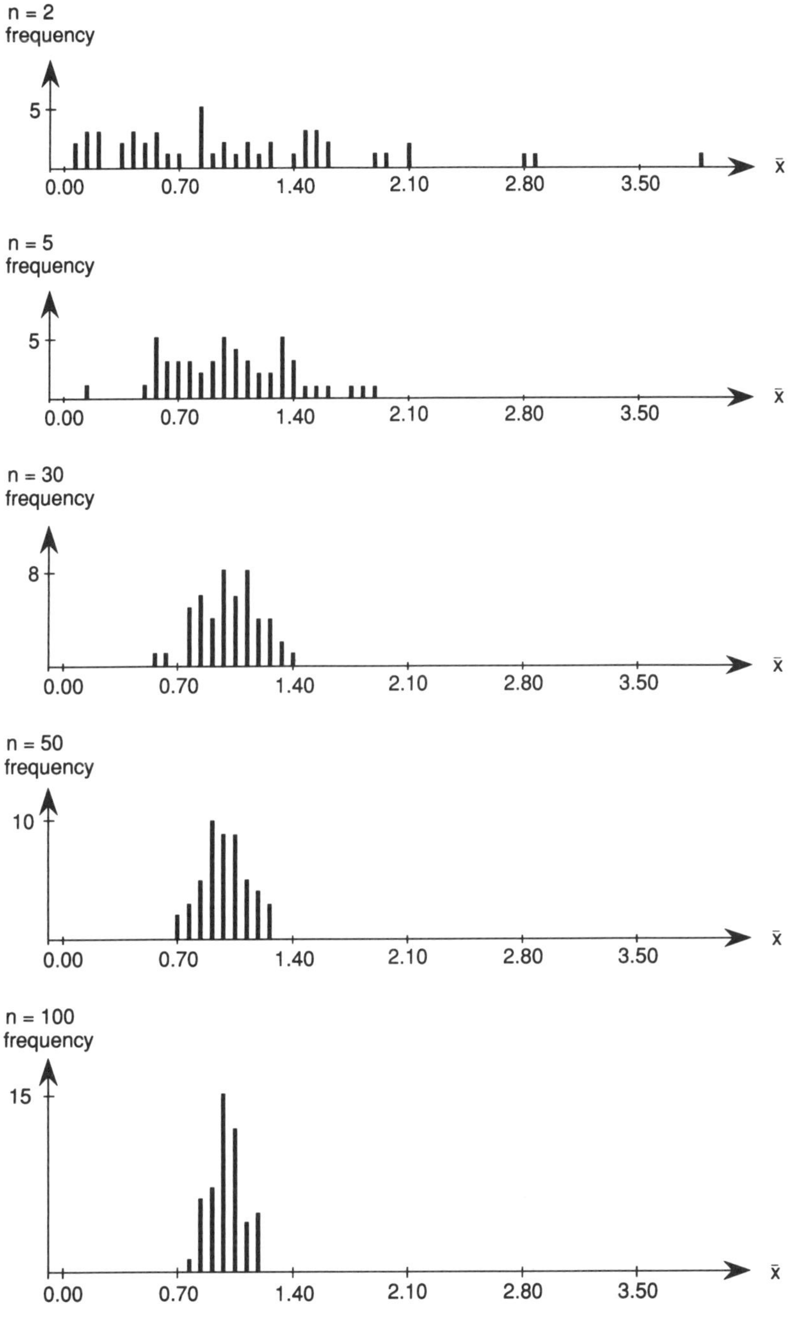

Figure 8.2: SAMPLE SAMPLING DISTRIBUTION OF SAMPLE MEANS FOR AN EXPONENTIAL RANDOM VARIABLE (50 SAMPLES)

tion displays increasing concentration. As the standard deviation sequence indicates, this concentration is occurring at a decreasing rate. The second apparent feature is that this concentration is near the population parameter value of 1.

The second example is for a continuous exponential distribution, where positive skewness is extreme; the probability density function for this case is

$$f(x) = \exp(-x); \mu = 1 \text{ and } \sigma = 1$$

from an interval of $0 < x \leq \infty$. The simulated sampling distributions for various sample sizes (2, 5, 30, 50, 100) appear in Figure 8.2. The summary statistics for these distributions also are reported in Table 8.1. The same conspicuous trends are observed here; as n increases, (1) the standard deviation of the sampling distribution decreases, meaning that concentration is occurring, and accordingly, (2) the frequency distribution displays increasing concentration about the population parameter value of 1. Apparently, extreme skewness does not obliterate the property of consistency. One noticeable difference here, however, is that the coefficient of variation signifies far more relative variability in the very small sample situations for a skewed distribution than it does for a uniform distribution; just the opposite is true for larger sample sizes. Taking into account the range of this variable (6.215 captures 99.8% of the possible values), its performance is much better than that for the uniform distribution (consider CV/range).

The third example is of a bimodal--specifically a sinusoidal--distribution, where the two modes occur at the extreme ends of the distribution; the probability density function for this case is

$$f(x) = 1/[\pi(1 + 2x - x^2)^{1/2}]; \mu = 1 \text{ and } \sigma = 1$$

from an interval of $1 - \sqrt{2} \leq x \leq 1 + \sqrt{2}$. The simulated sampling distributions for various selected sample sizes appear in Figure 8.3. The summary statistics for these distributions are reported in Table 8.1, as well. Once more, the same conspicuous trends are uncovered here; as n increases, (1) the standard deviation of the sampling distribution decreases, at a decreasing rate, meaning that concentration is occurring, and accordingly, (2) the frequency distribution displays increasing concentration about the population parameter value of 1. Apparently, extreme bimodality does not obliterate the property of consistency, either. In this case, the coefficient of variation signifies somewhat more relative variability as sample size increases, in comparison with the normal distribution, but less for small sample sizes than the skewed, peaked, or uniform distributions; it still converges upon the same trajectory for larger sample sizes. This variable's performance is poorer than it may appear, because the range is so much smaller than that for the other three variables; this restriction will decrease the variance, as has been discussed in Chapter 4.

The fourth example is of a normal distribution; the probability density function for this case is

$$f(x) = \{1/[(2\pi)^{1/2}\sigma]\}\exp[-(x - \mu)^2/(2\sigma^2)].$$

The simulated sampling distributions for various selected sample sizes appear in Figure 8.4. The summary statistics for these distributions are reported in Table 8.1, too. Once more, the same conspicuous trends are uncovered here; as n increases, (1) the standard deviation of the sampling distribution decreases, at a decreasing rate, meaning that concentration is occurring, and accordingly, (2) the frequency distribution displays increasing concentration about the population parameter value of 1. Taking into account the range of this variable (a central 6.92 captures 99.8% of the possible values), its performance is very much better than either of the other three (consider CV/range).

By the time a sample size of 30 is reached, all four sampling distributions are displaying approximately the same degree of relative variability, with the normal distribution tending to exhibit somewhat less variability than the other three distributions. By the time a sample size of 100 is reached, the uniform and bimodal distributions have nearly indistinguishable coefficients of variation. These values are very close to those for the exponential and the normal distributions, although the values for the normal distribution tend to be somewhat lower. The law of large numbers contends that all these coefficients of variation should converge onto a single value by the time n reaches infinity; presumably this value will be 0. These simulation experiments demonstrate that by the time n reaches 100, the form of the sampling distributions tends to lack any memory of the parent population frequency distribution for a random variable X. This tends to be the maximum threshold for what are classified as large sample sizes. By the time n reaches 30, little memory of the parent population frequency distribution remains, with this memory loss being less for nonnormal frequency distributions. In Chapter 7 the normal frequency distribution was put forth as an instrumental component of statistical analysis; another reason for it becomes evident here. The value of 30 tends to be a minimum threshold for what are classified as large sample sizes. Therefore, depending upon the form of the population frequency distribution for random variable X, the smallest sample size used should be a value between 30 and 100, with intermediate values being designated by the form of the underlying frequency distribution. Unfortunately, geographic data analysts rarely have the luxury of setting their sample size and often are confined to the small-sample theory domain.

Exercising control over the magnitude of sampling error through the determination of an appropriate sample size will be a subsequent topic of discussion (see Chapter 9); additional statistical tools are required before this endeavor can be undertaken. One of these tools is revealed by a cursory inspection of Figures 8.1-8.4, in which increasing sample size appears to be correlated with a convergence of each sampling distribution onto a normal distribution.

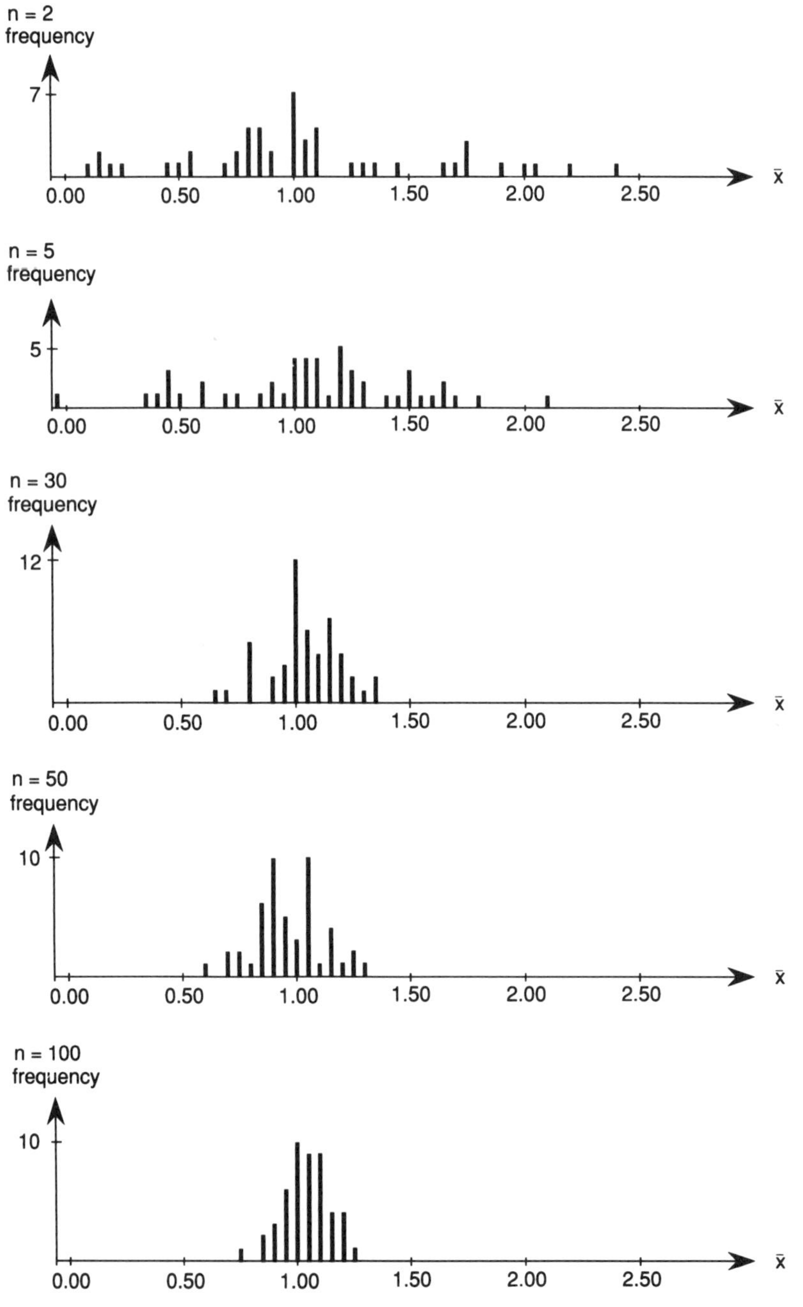

Figure 8.3: SAMPLING DISTRIBUTION OF SAMPLE MEANS FOR A BIMODAL (SINUSOIDAL) RANDOM VARIABLE (50 SAMPLES)

8.3. Drawing a random sample

The primary goal of scientific sampling is to be able to say something meaningful about the population. To collect data to achieve this end, the population must be defined very carefully, and the sample must be selected in some random fashion. Using **random numbers**--numbers in the range 0 to 9 whose selection is equiprobable, and whose resulting ordered sequence displays no pattern--classical sampling designs attempt to ensure that all elements of the population have an equal chance of being selected, and that the resulting sample is representative of a population. These two criteria of equally likely and representativeness are not always compatible and sometimes lead to design conflicts. The notion of randomness, and random numbers, is not that difficult to implement. Current techniques for generating seemingly random numbers, called pseudo-random numbers, yield quite acceptable results for very large (technically infinite) populations. In addition, these techniques are likely to improve in the future (Gleick, 1988). Similarly, for finite populations, little effort appears to be required to ensure randomness in a sample's construction (Kolata, 1990a).

The uniform probability distribution forms the basis for guaranteeing that the selection of each element in the population is equally likely. Various diagnostics can be performed on the resulting sample to decide whether or not it is sufficiently representative of its parent population; use of the binomial probability distribution to assess the reasonableness of sample composition already has been discussed as one example of such a diagnostic.

The sampling perspective becomes important here, too. In keeping with Chapter 6, the perspective of replacement, order being important, will be assumed here, even though common sense often dictates that geographers should not use replacement (a researcher wants to save time and money as well as evaluate as many elements of the population as possible). Although the sequences of probabilities change when there is no replacement (permutations and combinations), the resulting sample sets do remain equally likely; thus, some results presented here will tend to generalize to these two perspectives. Note, however, that replacement, order being unimportant, is problematic. Unfortunately, unless replacement occurs, and order is deemed important, the statistical properties of unbiased, efficient, consistent, and sufficient may well be compromised. If N is sufficiently large, then these properties will suffer little, but most geographic situations deal with relatively small population sizes.

If the population under study is finite, which often is the case in spatial data situations, then a complete list of the elements of this population can be compiled, and each element named using the consecutive integers 1, 2, . . . , N (nominal measurement scale). There are many ways to make such an assignment of numbers to population elements. Presumably the safest way to avoid error is to make the assignments utilizing as simple and easy a method as possible; the only requirement is that an assignment preserves the equally likely probabilities. A sample of size n

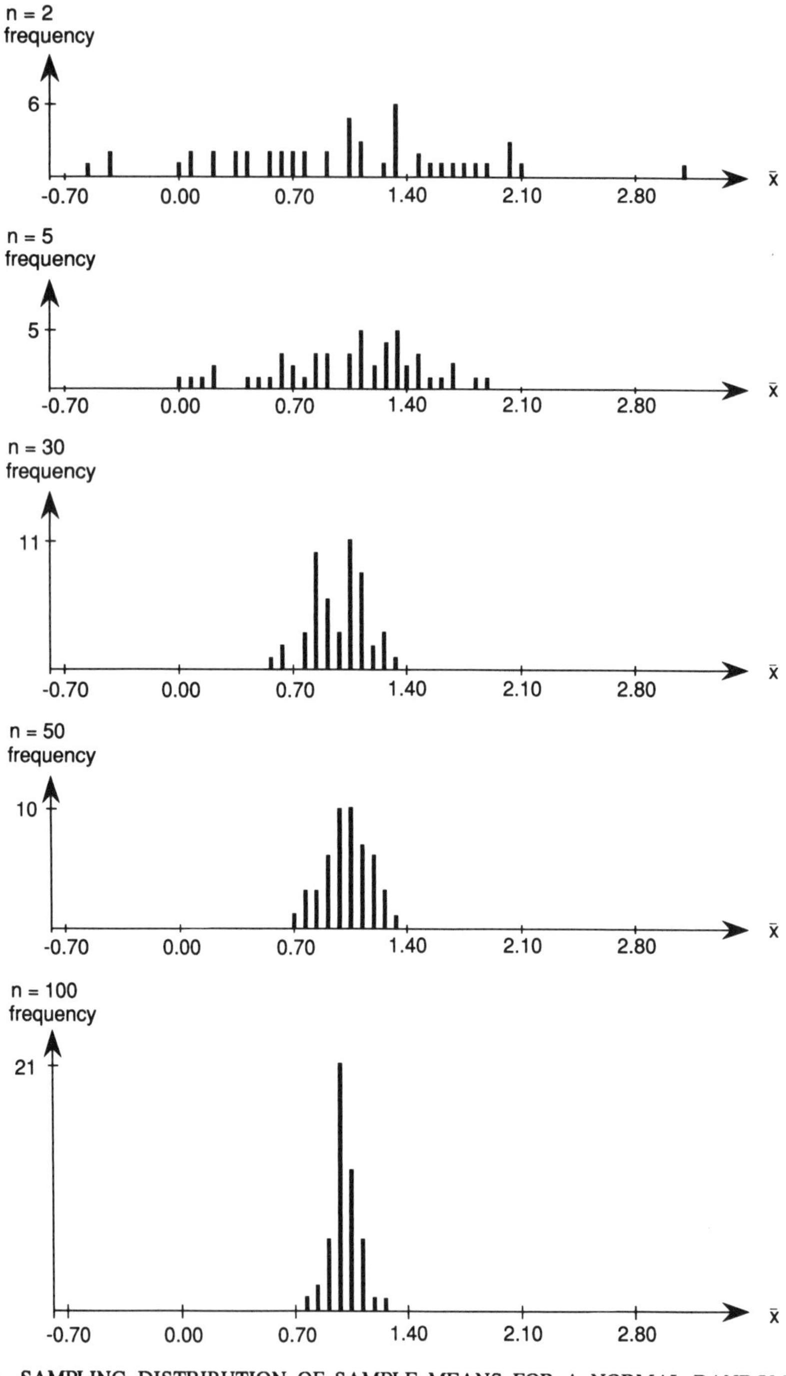

Figure 8.4: SAMPLING DISTRIBUTION OF SAMPLE MEANS FOR A NORMAL RANDOM VARIABLE (50 SAMPLES)

can be decided upon, and then a random numbers table can be consulted (see Table 7.1-A in the appendices). Recall that the random numbers table has been constructed in such a way that all single-digit integers are equally likely; the random numbers table is a collection of random digits. This table can be entered from any arbitrary point (the entry point should be selected at random), and numbers can be selected by moving through it in any direction, as long as backtracking does not occur or the appearance of the next number(s) does not influence the path being followed through this table. If $n \leq 10$, then single digits can be selected from this table. If $N \leq 10$, then any digit greater than N can be ignored. If sampling is being done without replacement, then once a digit has been selected, subsequent occurrences of it can be ignored. If order is important, then the sequence of the random numbers must be noted. A computer can be used to execute this procedure; it follows exactly the same process, except that the equation used to generate a random numbers table, rather than tabulated results, is used to determine the sequence of random numbers.

EXAMPLE PR-8.1. Selecting a sample of size 11 from the Puerto Rico data set.

The 73 areal units of Puerto Rico have been compiled into a list according to alphabetical order of their names and then numbered from 1 to 73 (see Table 1.1-A). Suppose the random numbers Table 7.1-A is entered at column 22, row 15, the sample perspective is with replacement, order being important, and movement is toward the bottom of the table. The sequence of random numbers is then 66, 67, 71, (74), 03, 30, 01, 66, 20, 73, 07, (85), and 72. Hence, the sample selected is Toa Baja, Trujillo Alto, Villalba, Aguadilla, Hatillo, Adjuntas, Toa Baja, Coamo, Yauco, Arecibo, and Yabucoa.

Note that in Example PR-8.1 two-digit numbers had to be used in order to label the areal units, the two-digit sampled numbers were arbitrarily constructed from adjacent columns of the table, and the numbers between 74 and 100 that were not used as labels were ignored (they are indicated with parentheses).

When populations do not consist of discrete elements or are infinite in size, a computer-generated sample is the only possibility for making selections. Random-number generators are based upon equations and therefore have a number of limitations. These generators are more accurately referred to as pseudo-random-number generators, producing sequences of digits that mimic truly patternless and truly unpredictable series; these computer mechanisms were used in Chapter 5 in order to sample from an infinite number of points on a surface. But every existing pseudo-random-number generator is flawed, being undercut by creeping nonrandomness in the sequence of digits, even though these flaws often are very difficult to detect. The quest for true randomness by science appears about to be at an end, however. "Mathematicians believe [that] they have found a process for making perfectly random strings of numbers." (Gleick, p. C1).

8.4. Classical sampling frameworks

One may have the impression at this point that statisticians believe the only respectable kind of sampling is random. But samples need to be representative, too. Modifications of the simple random sampling framework have been developed to ensure more representative samples. The probabilities assigned to elements of a population in these alternative frameworks are adjusted to reduce the chances of selecting an unrepresentative sample. Although statistical inference remains possible when these modifications are adopted, the number and different types of meaningful statistics they can produce are severely reduced. In addition, formulae can become considerably more complicated. If the unequal probabilities employed are known, then statistical results can be calculated in a way that incorporates these differences. If the unequal probabilities are unknown, legitimate use of statistical inference becomes impossible, because although a very representative sample is obtained, absolutely nothing is known about the sampling error associated with it.

One modification that can be introduced exploits the perspective of sampling without replacement, as it is called. This perspective is somewhat simpler than simple random sampling and involves calculating N/n, arbitrarily choosing the kth element from those numbered between 1 and N/n, and then **systematically** selecting the sequence of elements {k, k + N/n, k + 2N/n, ..., k + (n - 1)N/n} in the population. In part the validity of this method depends upon whether or not the listing of population elements represents an ordering in which some trend occurs and whether or not some periodic trend occurs every N/n elements. Ensuring that each element in the population has an equal chance of being selected becomes problematic, too, since equally likely selections occur only when the sample size n divides evenly into the population size N (N/n is an integer). Because the ordering of the elements in the population should be random, statistical methods that assume simple random sampling can be used. In other words, the ordering of the population elements needs to be scrambled between sample selections so that sample possibilities are not eliminated.

Another modification that can be introduced is to **stratify** the population, dividing it into relatively few mutually exclusive, collectively exhaustive, homogeneous categories (strata) and then randomly sampling from within each subset. Stratification ensures that each subset is represented in the selected sample, increasing its chances of being more representative than one would expect from simple random sampling. Compared with random sampling, this design tends to reduce the number of observations needed to attain a certain level of accuracy. By reducing internal variability, one effectively reduces sampling error. Each subset sample can be equal in size or proportional to the relative size of the subset. Disproportional stratified sampling may be preferable when at least one particular subset is unusually homogeneous, meaning that only a small sample from this stratum is necessary to obtain a prespecified degree of accuracy, whereas other strata are highly heterogeneous, meaning a larger sample is needed from these strata to attain the given degree of accuracy. Homogeneity and heterogeneity will tend to change as criteria vary. As mentioned earlier, frequently census data are based upon a sample and then aggregated and reported for areal units. These sorts of data may be viewed as resulting from a stratified random sample, rather than constituting a population.

Stratification is equivalent to placing constraints on sample selections, so that the sample composition is not left entirely to chance. Introducing constraints in the selection process necessarily affects upon the sampling distribution by eliminating possible random samples. For instance, Example PR-8.1 could yield any one of $73^{11} \approx 3.1373 \times 10^{20}$ samples. If the areal units of Puerto Rico are stratified by the five agricultural administrative regions, with 3 units selected from the San Juan region and 2 from each of the remaining four regions, then the total number of possible samples becomes $19^3 \times 11^2 \times 16^2 \times 14^2 \times 13^2 \approx 7.0377 \times 10^{12}$. In other words, samples such as all 11 units coming from the San Juan region have been deemed unrepresentative and have been eliminated from the sample space. In fact, most samples produced by simple random sampling have been judged unrepresentative (only 0.0000002% of the samples remain after stratification). This radical reduction in the sampling distribution basis is what scrambles understanding of the sampling error that is involved in stratified random sampling.

A third modification that can be introduced is to stratify the population, dividing it into numerous mutually exclusive, collectively exhaustive, heterogeneous **clusters** and then randomly select a subset of the clusters. Here the sample probably would include all elements contained in each of the selected clusters, although random samples could be drawn from each cluster that is selected. Not only does cluster sampling reduce the number of possible samples upon which a sampling distribution is based, but it also entails greater sampling error than does simple random sampling of the same size n. These drawbacks mean that cluster sampling, which usually is chosen because of physical convenience of data collection, often yields very poor statistical results. It tends to approximate results from a simple random sampling design only when each cluster is perfectly representative of the entire population.

Note that these primary sampling designs can be compounded into a variety of more elaborate sampling plans, analogous to combining the simple random and stratified designs into a stratified random sampling design. Space limitations prevent our doing justice to the enormous number of combinations and extensions that are feasible.

8.5. Geographic sampling: points, lines, and areas

The traditional sampling frameworks described so far are suitable for certain types of geographic problems that a spatial analyst faces. If a geographer wants to study census tract data for all major cities in the United States, then the number of census tracts is sufficiently large that simple random sampling may be worthwhile. Similarly, pixels from remotely sensed data are so numerous that simple random sampling may be justifiable. Systematic random sampling may be employed when a survey is to be administered in a region for which names and/or telephone numbers are available. As mentioned in the preceding section, analyzing census data may be viewed as stratified random sampling. Studying health risks may involve sampling those clusters of areas across a region in which the health risk occurs and then examining all inhabitants of these sampled regions (physical convenience). But usually the geographer is interested in locations on a map, meaning that often these traditional sampling frameworks become difficult to implement in their original forms. Some common spatial sampling designs are shown in Figure 8.5.

Spatially distributed samples seek to provide information about the geographic distribution under study. Any sample that is selected should render inferences about its parent surface. For instance, the point pattern analysis of Chapter 5 involves simple random sampling of points from a plane (Figure 8.5a). Clearly, then, points on a surface can be sampled. But sampling jointly in a random fashion from two independent uniform distributions leads to a Poisson distribution of sampled points. Obviously such a distribution of points may not be representative of the underlying geographic surface (for most such samples drawn, some areas will be overrepresented while others will be underrepresented). Surface representativeness can be enhanced by engaging a systematic random sample, where each of the north-south (Y) and east-west (X) axes are divided into $\sqrt{n}$ intervals, making a random selection from the first interval on each of these axes, say (x_1, y_1), and then systematically selecting the set of Cartesian coordinates $\{(x_1 + kX/\sqrt{n}, y_1 + lY/\sqrt{n}), k = 1, 2, \ldots, (\sqrt{n} - 1)$, and $l = 1, 2, \ldots, (\sqrt{n} - 1)\}$. This design is exact only if the sample size n is a perfect square (Figure 8.5b). In addition, there is a real danger that the geographically distributed phenomenon in question will display trends or periodicities over the surface. The points also could be selected from a stratified random sampling design (Figure 8.5c). The region in question could be partitioned into quadrats (or alternatively, strata), and then points could be randomly sampled from a pair of joint, independent uniform distributions (Figure 8.5d) in each quadrat (or stratum). Although the distribution of points within each quadrat would be Poisson distributed, the entire surface would assuredly be sampled.

The notion of cluster sampling may be adapted to geographic situations in one of two ways. First, as was seen in point pattern analysis, quadrats can be used where the cluster now is an area. This category of cluster sampling would be desirable mostly from a convenience point of view. If the clusters constitute a mutually exclusive and collectively exhaustive partitioning of a geographic surface, then sampling occurs either in terms of (a) the different sizes that a quadrat can take on or (b) the arbitrary position of the lines for boundaries of these quadrats. These two viewpoints often are compromised by the use of an optimal quadrat size and quadrat grid superimpositions that attempt to minimize the number of quadrats that extend

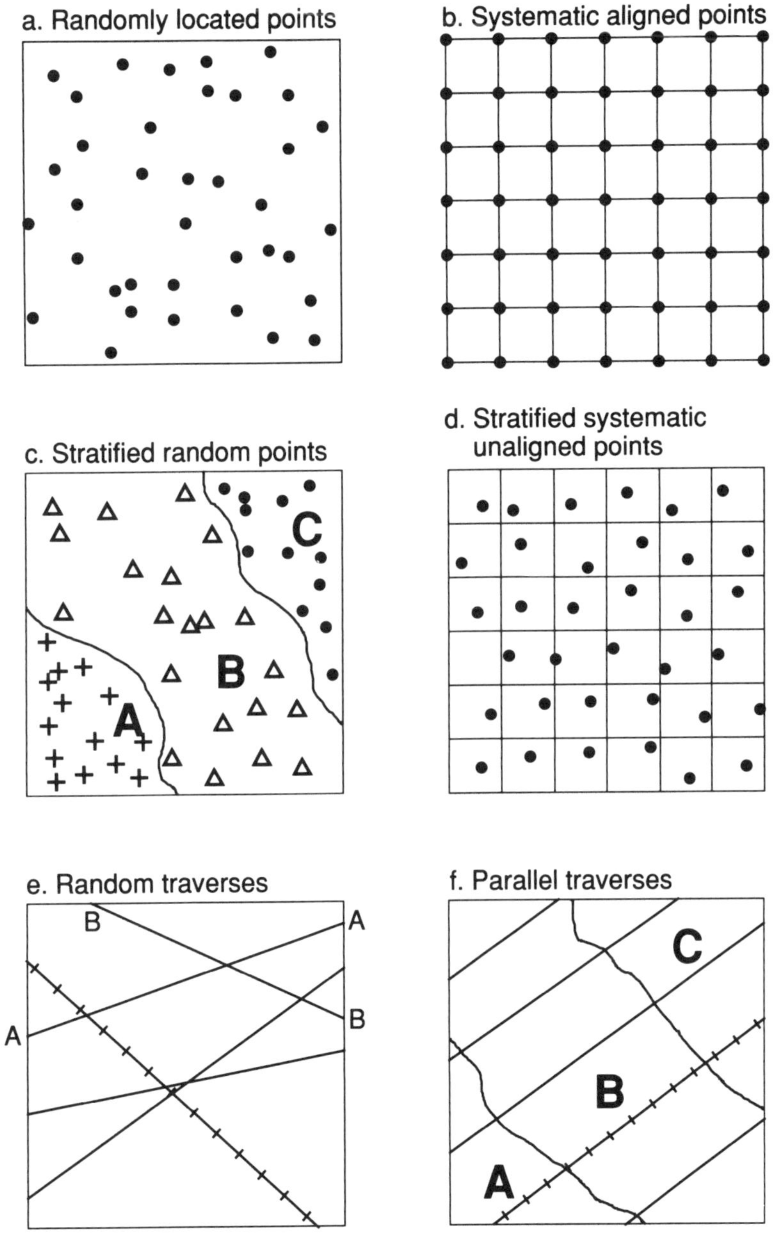

Figure 8.5: SIX COMMON GEOGRAPHIC SAMPLING DESIGNS

beyond the borders of a region. If the quadrat clusters are selected at random, and hence allowed to overlap, then each may be identified on a surface by (1) randomly selecting its centroid from a joint uniform distribution (which may well further reduce representativeness), and possibly randomly selecting an angle of orientation for it, (2) systematically selecting the set of centroids for the quadrats (as described before), and possibly randomly selecting an angle of orientation for each, or (3) partitioning the entire region into strata and randomly selecting quadrats from each subregion. Once quadrats have been selected, the resulting sample includes all relevant geographic phenomena embraced by these quadrats.

The third approach to geographic sampling involves lines, transects, or traverses, which also may be viewed as clusters of points. This category of cluster sampling is desirable for many reasons. First, truly random sampling can be expensive, because locating randomized points on real surfaces in the field is time consuming, often subject to inaccuracies, and possibly hazardous. This same constraint may not be true for samples extracted from scale models or maps. Second, collecting measurements along transects allows attributes in the area of interest to be mapped in more detail, providing additional survey information regarding the spatial pattern. For example, this method of sampling is particularly appropriate when mapping the distribution of vegetation or bedrock lithology in unsurveyed areas, where systematic traverses are the only logical means of locating sample points. Third, the researcher may not know *a priori* the nature of any existing patterns in the environmental attribute in question. A subjective choice in orienting and locating the transect based upon only a few known features may yield a set of measurements with properties analogous to those of a random sample and thus remains viable for statistical analyses. Finally, the ability to introduce some control over the sampling design (that is, orientation, placement, and sample spacing) may actually improve the representativeness of the sample drawn from an area that exhibits a high degree of ordering (e. g., spatial autocorrelation) in the attribute of interest.

Orientation, length, and number of traverse lines, as well as the frequency and spacing of measurements along each line, are common concerns in transect or grid sampling designs. Lines can be drawn randomly through a region (Figure 8.5e); however, locating them can be difficult, and randomly spaced lines tend to have a Poisson distribution. Lines that extend throughout the area of interest are more advantageous to have than are ones that cross only a small section of the region (transect AA versus transect BB of Figure 8.5e). Haggett and Board (1964) studied the effects of orientation and the number of traverse lines on the representativeness of a sample used to estimate the areal coverage of different forest cover types in a study area. They concluded that with decreased spacing, parallel traverse lines provided the highest degree of accuracy in estimating the population proportions of forest cover types (Figure 8.5f). It may not always be practical to sample a series of closely spaced transects, so a compromise between time (cost) and accuracy must be made. Such assessments can be made only if previous measurements are available or a sample pilot study is undertaken (this is true for all sampling designs). The region also can be stratified and then traverses established in each of the regional strata.

EXAMPLE CA-8.1. Sampling from the Canadian Automotive data set.

The automotive data in Chapter 1 represent a sampling process through which 53 firms were selected from an original population of 207 automotive parts firms in Ontario. To recapitulate Chapter 1, first firms with missing data were dropped, and then every third remaining firm was selected from a list of firms organized in alphabetical order. This strategy represents a systematic process of randomly choosing the first entry from the first three listed firms (observation one), and then systematically choosing every third subsequent entry. As can be seen from Figure 3.5, this selection produced a sample with 2 firms in Eastern Ontario, 12 in Central Ontario, 13 in Metropolitan Toronto, and 26 in Southwestern Ontario. Unfortunately, because the tabulated figures are for this sample, the regional distribution of the original 207 firms is not known, meaning that one cannot tell whether or not the sample used in the text represents the spatial distribution of firms in Ontario contained in the population.

EXAMPLE CA-8.2. Sampling from the Canadian Automotive data set (Example CA-8.1 continued).

To alleviate the uncertainty concerning the representativeness of spatial cover, an alternative sampling procedure would be to stratify the population by region, determine the proportion of firms in each region, select a sample size, and then choose (either randomly or systematically) the same proportion of firms from each region as in the population. The proportion of firms selected from each region represents the sampling fraction (i. e., proportion) for each region. The following table shows how many firms from each region are in a sample constructed with this spatially stratified random or stratified systematic (depending on how the firms are drawn from each region) procedure. A sample size of 53 is selected so that the membership in the existing sample can be compared to the spatially stratified sample.

	Population		Sample
Region	Number of firms	Sampling fraction	Number of firms
Eastern (K)	6	2.9	1.5 (1 or 2)
Central (L)	71	34.3	18.2 (18)
Toronto (M)	45	21.7	11.5 (11 or 12)
Southwestern (N)	83	40.1	21.3 (21)
Northern (P)	2	1.0	0.5 (0 or 1)

The existing sample has the following population K = 2, L = 12, M = 13, N = 26. A comparison of the two samples suggests that the systematic sample achieved reasonable spatial coverage without explicitly controlling for it. The notable exception is the absence of Northern Ontario firms. However, given the small population in region P, zero firms is not surprising. In addition, the spatially stratified sample selects more firms from region L, and fewer from region N, than does the systematic sample produced for Chapter 1.

Note that although a spatial stratification was chosen in the preceding example, the data also could be stratified by any categorical variable in the data set; for example, by primary product (variable PPC) to ensure proportional coverage of products.

8.6. Practical problems encountered in geographic sampling

In Chapter 5 the notion of spatial autocorrelation was presented, and it was emphasized that its presence (when positive) means that the information content of geographically referenced data is less than what is anticipated. This result suggests that the gathering of data from nearby points on a surface is a waste of resources. Accordingly, systematic point sampling (a regular grid pattern) is preferable, followed by either stratified random point sampling or systematically sampled traverses (another form of stratified sampling), whereas simple random sampling of geographic surfaces offers little advantage over either of these two forms of planar sampling, and quadrat sampling should be avoided most of the time. Ascertaining the magnitude of the sampling error associated with systematic spatial sampling is even more difficult than for aspatial systematic sampling.

Systematic sampling of the plane is hindered by the inability to alter the ordering of the surface. Consequently, systematization constrains the number of possible samples and hence reduces the basis for the sampling distribution. It tends to give satisfactory results only if the geographic ordering is random, at least in terms of characteristics that may influence the findings of the study. But the presence of spatial autocorrelation implies that the existence of this condition is unlikely. This drawback was avoided in the classical situation by allowing the elements of the population to be reordered. Otherwise, results become conditional on the ordering involved.

Transect sampling can be problematic depending upon the placement of the line. Although there are an infinite number of possible transect lines, not all the lines provide very good coverage of the area. Imagine a square region. The transect that would provide the greatest coverage would be a diagonal line connecting opposite corners of the region. Minimal coverage would be provided by a line separating one corner from the remainder of the region. For this reason, placement and orientation often are done nonrandomly. In addition to the placement of the transect, different procedures for selecting points from the transect also affect coverage. Ultimately, the number of points selected--and whether the selection process is random or systematic--will depend upon the nature of the research question and the spatial pattern of the attribute under study. All the issues raised in this chapter must then be addressed.

8.7. Summary

With the conclusion of this chapter, all the necessary pieces are in place to begin an examination of inferential procedures. Of all the basic pieces presented so far, none is more critical than the process of constructing a sample. There are many factors that constrain one's ability to sample, including cost, time available, and accessibility to the population. The quality of the sampling process provides a lower bound to the quality of any statistical analysis that might follow. In turn, the extent to which a research project can overcome the outside constraints imposed on sampling provides a lower bound on the quality of the sampling process.

Two themes discussed in this chapter are thus worth repeating. First, given the lower bound on information quality provided by the sample, a small amount of time and effort spent improving the quality of a sample at the beginning of an analysis pays dramatic dividends, not only in terms of the useful information contained in the sample, but also in the ease with which statistical results can be interpreted. Second, certainly under repeated sampling the overall "best" statistical results are obtained using random samples. But since most analysts can afford only a few samples, efforts taken to improve representativeness (such as stratification) greatly improve the useful information contained in individual samples.

8.8. Problems

1. Using the random numbers table (7.1), draw five samples of size 10 from the Puerto Rico data set, treating the 73 municipios as a population, such that

   (a) the first sample is drawn with a simple random sampling design,
   (b) the second sample is drawn with a stratified random sampling design, where the strata are the agricultural administrative regions, selecting two areal units from each strata,
   (c) the third sample is drawn with a proportional stratified random sampling design, where the strata are the coastal lowlands/interior highlands,
   (d) the fourth sample is drawn with a systematic random sampling design, and
   (e) the fifth sample is drawn with a cluster sampling design.

2. Calculate the total number of possible samples for each of the five designs used in Problem #1.

3. Draw two samples of size 10 from the Puerto Rico data set, one using a traverse and the other using a systematic spatial sampling design. Compare these results with those obtained in Problem #1.

4. It is desirable to know the median grain size for sediments deposited on a river gravel bar. Extraction of a bulk sample taken from one or several sites is considered the most representative sampling strategy but is also a time-consuming and laborious task. An alternative method to define grain-size distribution is to lay out a line transect in a study region, systematically space points along the transect, and sample one grain at each point (for example, as in the transect line shown in Figure 8.5f). The results from both sampling schemes are as follows, reported in terms of cumulative frequency distributions:

Bulk sample		Line sample	
Cumulative frequency	Size (mm)	Cumulative frequency	Size (mm)
0.10	15	0.10	20
0.25	20	0.25	48
0.40	40	0.40	80
0.60	105	0.60	120
0.75	140	0.75	160
0.90	160	0.90	220
1.00	190	1.00	310

Compute the median grain size (50th percentile--D_{50}) for each sample, and compare the results. What might account for any differences in these two estimates of the median grain size?

5. The distribution of the original 207 firms in Ontario by primary product code (variable PPC) is as follows:

PPC	0	1	2	3	4	5	6	7
Frequency	26	24	18	65	3	8	11	52

Calculate the sampling fraction for each product category and determine how many firms would be in a sample of 53 firms. Compare this stratified sample to the systematically chosen sample in Chapter 1.

8.9. References

Gleick, J. (1988) "The Quest for True Randomness Finally Appears Successful." *The New York Times*, 138 (April 19): C1 and C8.

Haggett, P., and C. Board. (1964) "Rotational and Parallel Transverses in the Rapid Integration of Geographic areas." *Annals*, Association of American Geographers, 54: 406-410.

Kolata, G. (1990) "In Shuffling Cards, 7 is Winning Number." *The New York Times*, 140 (January 9): C1 and C12.

Summerfield, M. (1983) "Population, Samples, and Statistical Inference in Geography." *The Professional Geographer*, 35: 143-49.

8.10. Additional readings

In addition to the texts listed in Chapter 1,

Berry, B., and A. Baker. (1968) "Geographic Sampling," in B. Berry and D. Marble (eds.) *Spatial Analysis: A Reader in Statistical Geography*, pp. 91-100. Englewood Cliffs, N.J.: Prentice Hall.

Dixon, C., and B. Leach. *Sampling Methods for Geographical Research.* CATMOG 17.

Kirby, R. *Choice in Field Surveying*. CATMOG 41.

Haggett, P., and C. Board. (1964) "Rotational and Parallel Transverses in the Rapid Integration of Geographic areas." *Annals*, Association of American Geographers, 54: 406-410.

Haggett, P., A. Cliff, and A. Frey. (1977) *Locational Methods*, Vol. 2 of 2nd ed. of *Location Analysis in Human Geography*. New York: Wiley, Halstead Press.

CHAPTER 9

ESSENTIAL CONCEPTS FOR INFERENTIAL STATISTICS

One of the objectives of statistics emphasized earlier in this text is to infer statements about a parent population from one of its subsets, the sample data; more specifically, the goal is to state something numerically about population parameters once sample statistics have been calculated. This objective has two closely related dimensions. First, two features of such statements are of interest, namely, (1) the numerical value itself and (2) some declaration of the degree of confidence one has in this computed numerical value. Second, one may be interested in evaluating a claim about the true value of a parameter; based upon sample information, one wants to ascertain whether or not the claim is correct, given the presence of sampling error. Estimation and hypothesis testing are two types of **statistical inference**, which refers to the technique of interpreting numerical values resulting from sampling and then making decisions based upon these results. The focus of this chapter is on parametric statistics, although most of the concepts directly extend to nonparametric statistics (see Section 4.1) as well. Accordingly, for selected inferential problems, appropriate nonparametric alternatives will be outlined.

9.1. The concept of estimation: points and intervals

Estimation involves attempting to answer a question soliciting a good guess for the unknown but true value of a population parameter. If one reports a sample statistic, such as the sample mean $\bar{x}$, the outcome will be a **point estimate**, which is the best single value that can be used to estimate a parameter. In other words, the population parameter is estimated by one point on the real number line. The point estimate for a population parameter is the value of its corresponding sample statistic. A **point estimator** is the mathematical function of n observations used to calculate a point estimate. For most problems there will be more than one reasonable point estimator. Again, then, one should realize that the properties of sufficiency, consistency, efficiency, and unbiasedness provide criteria that help distinguish between good and poor point estimators. For example, the sample mean $\bar{x}$ is sufficient because it reduces the data contained in a sample without losing information pertinent to μ. This sample statistic is consistent because, as seen in Chapter 8, as n increases, the variance of the sampling distribution decreases and $\bar{x}_n$ converges upon μ. This sample statistic has a smaller sampling distribution variance than do other measures of central tendency (it is efficient). This notion, however, is a relative one, since the smallest sampling distribution variance possible is zero, which corresponds to an absence of sampling error and hence use of the population rather than a sample. The mean of the sampling distribution of $\bar{x}$ exactly equals μ (it is unbiased). Moreover, the point estimate $\bar{x}$ tends to be the preferred measure of central tendency (as compared with a median or a mode).

Many of the conventional statistics discussed in earlier chapters furnish good point estimates--for instance, s for σ (when the denominator for s is n - 1), and r for ρ. As the figures in Chapter 8 demonstrate, however, estimators may well be sensitive to the underlying population frequency distribution being sampled. Trade-offs among these four statistical properties may be required when choosing a best estimator. In fact, the principle often invoked states that from among all unbiased estimators of a population parameter one should choose the estimator that has minimum variance.

Although a point estimate might be the best single guess for a parameter value, the presence of sampling error means that an entire interval of possible values exists. Even when a point estimator is a minimum-variance unbiased estimator, nothing is known about the estimate that is rendered by a particular sample. The presence of a continuum of possibilities suggests that point estimates may be close to parameter values but more than likely never will actually equal them. In addition to reporting a single value, or point estimate of a parameter, some knowledge of its precision is useful to have. For example, one might like to know the degree of confidence that can be attached to a guess of this sort. Believing that $\bar{x}$ is close to μ is relative, probably is arbitrarily defined, and suggests that any numerical value within a certain distance of μ would be considered an equally good guess; accordingly, estimation can be thought of in terms of an interval rather than a point. The usual measuring unit of closeness here is based upon the standard deviation of the sampling distribution of that parameter in question (see Chapter 4 for a review of this concept); this measure is known as the **standard error** of the estimate, especially since it provides a measure of deviation from some population parameter attributable to sampling error. A **confidence interval** estimate employs this concept, establishing two values for the interval boundaries and assigning a probability measure to the interval's reliability in estimating the parameter in question. Use of a confidence interval amounts to replacing a point estimate, which is a single number, by an entire interval of plausible values, with the degree of plausibility being specified by a probability measure, or **confidence level**. Conventional probabilities used to construct confidence intervals are 90%, 95%, and 99%. In other words, interval estimation means two points on the real number line are employed to establish a range within which, with a certain probability, the population parameter is thought to lie.

Good interval estimators, like good point estimators, need to be distinguished from poor ones. The two principal requirements of a good interval estimator are that (1) its accompanying width be as narrow as possible, for a given sample size, and (2) it have a very high probability of containing the unknown parameter value. This goal is achieved in part by use of standard errors to calculate interval boundaries, since, as is mentioned above, the distance separating a sample statistic from its parent population parameter value should be measured with this yardstick. Note that these two requirements usually conflict, with the probability increasing only when the interval width increases (the sample size n being held constant).

9.1.1. The meaning of a confidence interval

The construction of a confidence interval begins with the selection of a probability level. This task requires knowledge of an appropriate probability model for describing the associated sampling distribution; this issue will be addressed in the next section. In the presence of pure ignorance (the sample size is $n = 0$), or when one wants to establish a probability of 1, then the confidence interval is $(-\infty, \infty)$. This interval means that the population parameter value lies somewhere along the real number line, and hence all guesses for its value are equally reasonable (this range can be restricted by eliminating values that are physically impossible). Clearly such a confidence interval is meaningless! The classic position of statistical inference outlined in the preceding section stresses that when a sample is drawn, information contained in the sample suggests that a narrower interval than $(-\infty, \infty)$ is reasonable to posit. One does not know for certain, however, whether or not the resulting narrower confidence interval constructed from this sample contains the value of the parameter (sampling error is present), unless the bounds are maintained at infinity. Using a probability of, say, 0.90 when constructing a confidence interval translates into an expectation that in the long run (the selection of many, many samples) 90% of the confidence intervals constructed from sample data will contain the population parameter value. Moreover, the interval $(-\infty, \infty)$ must shrink to have 10% of the possible intervals failing to capture the population parameter value. Extending this notion, as the interval width shrinks to a single value (a, a), the probability that intervals will embrace the population parameter value will tend to 0.

Consider the set of confidence intervals presented in Figure 9.1. For illustrative purposes these intervals have been constructed with a knowledge of the population mean and variance, and the standard error of the sample mean; certainly, if one already knew the population parameter value of interest, one would not be trying to draw an inference about it. Fifty samples of size 49 were drawn, and 90% were confidence intervals constructed based upon $\bar{x}$ and $\sigma = 1$. Now 10% of the intervals, or 5 of them, should not include the population parameter value $\mu = 0$. In fact, exactly 5 of the intervals span a set of real numbers all of which are either strictly greater than zero (2 instances), or strictly less than zero (3 instances). The other 45 cases span a

$\bar{x}$	s	Interval boundaries
-0.212	1.063	(-0.447, 0.023)
-0.050	0.985	(-0.285, 0.185)
-0.073	1.022	(-0.308, 0.162)
-0.215	0.922	(-0.450, 0.021)
-0.028	1.070	(-0.263, 0.208)
0.076	0.974	(-0.159, 0.311)
0.179	1.142	(-0.057, 0.414)
0.101	0.835	(-0.134, 0.336)
-0.043	0.986	(-0.278, 0.193)
-0.122	1.002	(-0.357, 0.113)
-0.221	0.953	(-0.456, 0.015)
-0.170	1.082	(-0.405, 0.065)
-0.068	0.851	(-0.303, 0.167)
-0.046	0.998	(-0.281, 0.190)
-0.050	0.952	(-0.285, 0.185)
0.254	1.115	(0.018, 0.489)
-0.196	0.848	(-0.432, 0.039)
-0.268	1.057	(-0.503,-0.032)
-0.121	1.039	(-0.356, 0.115)
-0.054	1.244	(-0.290, 0.181)
-0.214	1.079	(-0.449, 0.021)
0.093	0.911	(-0.142, 0.329)
0.072	1.141	(-0.163, 0.308)
0.169	1.115	(-0.067, 0.404)
-0.013	1.152	(-0.248, 0.222)
0.048	1.111	(-0.188, 0.283)
0.142	1.013	(-0.093, 0.378)
0.108	1.119	(-0.128, 0.343)
0.126	1.063	(-0.109, 0.362)
-0.140	0.941	(-0.375, 0.095)
0.039	0.843	(-0.196, 0.274)
0.332	1.135	(0.097, 0.567)
-0.196	0.819	(-0.432, 0.039)
0.134	1.042	(-0.101, 0.370)
-0.169	0.913	(-0.404, 0.066)
-0.174	0.950	(-0.409, 0.061)
-0.020	0.817	(-0.255, 0.215)
-0.109	1.088	(-0.344, 0.126)
-0.234	1.034	(-0.469, 0.002)
0.217	0.963	(-0.018, 0.452)
0.170	0.956	(-0.066, 0.405)
0.140	1.002	(-0.096, 0.375)
-0.083	1.054	(-0.318, 0.152)
-0.068	1.072	(-0.303, 0.167)
-0.243	1.042	(-0.478,-0.007)
-0.296	1.132	(-0.531,-0.060)
0.083	1.147	(-0.152, 0.318)
-0.160	1.007	(-0.396, 0.075)
-0.203	0.931	(-0.438, 0.032)
-0.178	0.963	(-0.413, 0.057)

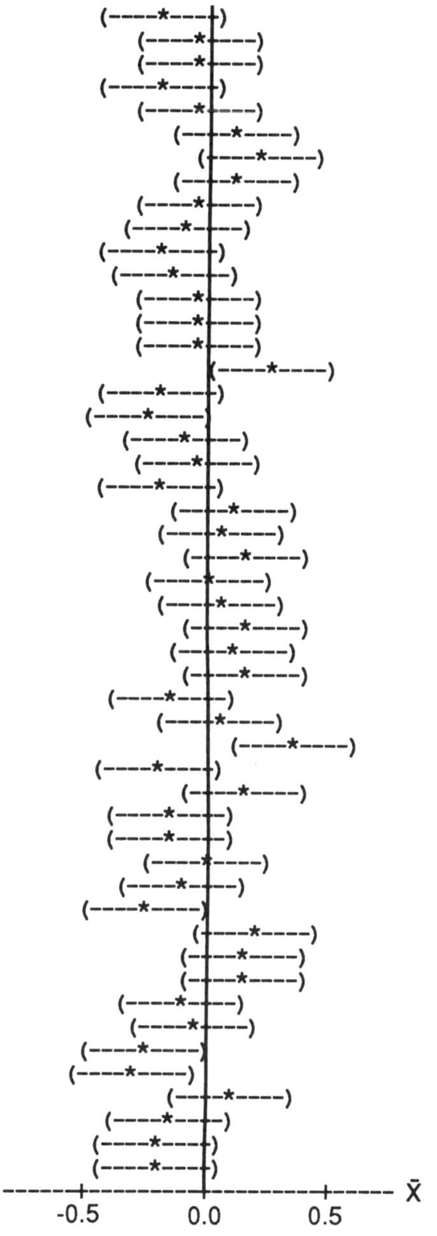

Figure 9.1: 90% CONFIDENCE INTERVALS FOR $\mu = 0$, $\sigma = 1$, $n = 49$

set of real numbers that includes zero. All these intervals are of width 0.47 (because $\sigma_{\bar{x}}$, the standard error of the sampling distribution of $\bar{x}$, is known).

The confidence interval, therefore, is not a probability statement involving some fixed interval whose bounds have been computed from a given sample; it is inappropriate to say that $P[\mu \in (a, b)] = 0.95$, where a and b are those bounds calculated for a single sample. Rather, it is a statement about what would happen if a very large number of similar intervals were to be constructed. The population parameter value is fixed, whereas the intervals vary from sample to sample. As a result, the probability that the population parameter value will fall into a given interval is either 1 or 0; either it is there, or it is not there, with the presence of sampling error causing a researcher to be uncertain about which is the case. In Figure 9.1, a vertical line constructed perpendicular to the $\bar{x}$-axis at the true value of $\mu = 0$ will pass through 90% of the comparable confidence intervals that have been constructed from the different sample data. Accordingly, one can state with 90% confidence that he or she thinks the population parameter value lies somewhere within a particular confidence interval with a 10% chance that this statement is incorrect. Confidence thus refers to the procedure for constructing intervals, not to any particular sample result.

9.2. The role of the sampling distribution in statistical estimation

Suppose a sample is representative and a good estimator of some population parameter has been selected. How does one determine those confidence interval bounds corresponding to a given probability level? As was mentioned earlier, the analysis here is cast in terms of the underlying sampling distribution. As such, a characterization of the affiliated sampling distribution for the statistic in question needs to be obtained, without resorting to complete enumeration of all possible samples. If one had to draw all possible samples of size n (which is impossible if the population is infinite in size) and then determine the standard error of the affiliated sampling distribution, the amount of work involved would more than defeat the purpose of drawing samples in order to conserve resources. Mathematical statistical theory has shown that certain probability models (as mentioned earlier, many times the normal distribution supplies an appropriate probability model for such inferential situations) accurately describe sampling distributions. These theoretical models supply information about the distributions, such as a standard error, and thus enhance the usefulness of substituting samples for populations. The law of large numbers (presented in Chapter 6) maintains that as n increases, these probability models provide increasingly better descriptions. Not only do these models supply necessary shorthand information for constructing interval estimates, but they also provide information necessary for evaluation of the quality of point estimators.

The relationship between a sampling distribution and a probability model is articulated by theorems from mathematical statistics. One of the most famous of these is the central limit theorem, which applies to the sampling distribution of the sample mean, $\bar{x}$. This theorem will be used to illustrate how the probability model for a selected sampling distribution is determined, as well as to identify pertinent information about the sampling distribution that is used to construct confidence intervals.

> Central Limit Theorem: Consider the drawing of all possible samples of size n, with replacement and order important, from a given population having finite mean μ and variance σ^2, and *any arbitrary* population frequency distribution. The mean of the sampling distribution of $\bar{x}$ for this situation is $\mu_{\bar{x}} = \mu$, the variance of this sampling distribution is $\sigma^2_{\bar{x}} = \sigma^2/n$, and as n becomes large, the sampling distribution of sample means approaches a normal distribution. The larger the value of n, the better the sampling distribution is approximated by a normal distribution.

Note that conditions exist that allow this theorem to apply when observations in a sample are being drawn from different populations, and also that circumstances exist under which the assumption of independence can be relaxed (this may be relevant to spatial autocorrelation problems); these two issues are beyond the scope of this book and hence will not be addressed here. Again one can see why the normal probability model has been linked with inferential problems here. Theorems like the central limit theorem yield theoretical generalizations for the sampling distributions of various statistics, allowing inferences to be drawn without having to totally enumerate all possible samples.

> EXAMPLE PR-9.1. Using the density of tobacco production in the Mayaguez region of the Puerto Rico data set, show that all possible samples of size 2 have a mean and a standard error equal to those values given by the central limit theorem.
>
> Assume, for this example, that the 16 observations in the Mayaguez region constitute a population. Then with replacement and order important, there are $n^r = 16^2 = 256$ samples. The "population" mean is $\mu = 20.4/16 = 1.275$, and the "population" variance is $\sigma^2 = 90.75/16 = 5.6719$. The 256 samples, with the mean for each, are listed in the following table.
>
> The sum of the 256 sample means is 326.40, so the mean of this sampling distribution is $\mu_{\bar{x}} = 326.40/256 = 1.275$, which is equivalent to the population mean of 1.275.
>
> Similarly, the variance of these 256 sample means is 726, so the variance of this sampling distribution is $\sigma^2_{\bar{x}} = 726/256 = 2.8359$, which is equivalent to the population variance divided by the sample size, or 5.6719/2.

$\bar{x}$	No. of samples	$\bar{x}$	No. of samples	$\bar{x}$	No. of samples
0.00	64	1.05	2	3.30	2
0.05	32	1.30	2	3.80	2
0.10	4	1.60	16	4.15	16
0.15	16	1.65	4	4.20	4
0.20	4	1.75	2	4.30	2
0.30	1	1.80	1	4.50	2
0.40	16	2.00	2	4.55	2
0.45	4	2.50	2	5.05	2
0.55	2	2.90	16	5.75	2
0.80	1	2.95	4	5.80	1
0.90	16	3.05	2	7.05	2
0.95	4	3.20	1	8.30	1

The central limit theorem can be used to further inspect two sets of numerical results already discussed in this text. First, Table 8.1 reports results for sample sampling distributions. In all cases the means for any sample size n should be 1 (the differences are attributable to sampling error), and the theoretical standard errors should have been as follows:

Sample size n:	2	5	30	50	100
Standard error $\sigma_{\bar{x}}$:	0.7071	0.4472	0.1826	0.1414	0.1000

Those standard errors reported in Table 8.1 differ from these theoretical ones for two reasons, namely, sampling error and the form of the population frequency distribution. Once more one can see that the normal frequency distribution performs the best over the entire range of sample sizes. Furthermore, an inspection of Figures 8.1-8.4 corroborates the contention that a sampling distribution tends to converge upon the normal distribution as n increases, regardless of the form of the population frequency distribution.

Second, those confidence intervals presented in Figure 9.1 are based upon results of this theorem. In each case the probability of 0.90 means that reference to the normal curve, which is symmetric about the population mean, results in standard normal deviates of ±1.645 that capture 90% of the area under the curve (see Table 7.2-A, which is located in the Appendices). The standard error of $\bar{x}$ theoretically is given as $1/7 \approx 0.1429$. Consequently, each of the 50 confidence intervals is constructed as follows:

$$\bar{x} + z_{\alpha/2} \cdot \sigma/\sqrt{n} < \mu < \bar{x} + z_{1-\alpha/2} \cdot \sigma/\sqrt{n}$$

where $\alpha/2$ refers to the area in one of the tails of the distribution and z is a standard normal deviate. More specifically, for this example the confidence intervals are given by

$$\bar{x} - 1.645/7 < \mu < \bar{x} + 1.645/7$$

which for the very first sample entry in Figure 9.1 gives a confidence interval of (-0.212 - 0.235, -0.212 + 0.235) = (-0.447, 0.023). Because the normal distribution is symmetric, $z_{\alpha/2} = -z_{1-\alpha/2}$, often no distinction is made between these two values; in other words, the confidence interval is presented as

$$\bar{x} - z_{1-\alpha/2} \cdot \sigma/\sqrt{n} < \mu < \bar{x} + z_{1-\alpha/2} \cdot \sigma/\sqrt{n}.$$

Clearly one can see why the width of each confidence interval in Figure 9.1 is 0.47. Also, the central limit theorem indicates that one way to shrink the width of a confidence interval, while holding the probability level constant, is to increase sample size.

EXAMPLE PR-9.2. Using the random numbers table (consult Table 7.1-A), draw 10 samples of size 4, using density of tobacco production in the Mayaguez region of the Puerto Rico data set, and construct 90%, 95%, and 99% confidence intervals for each of these samples.

Beginning in row 21 and column 21 of Table 7.1-A, moving from top to bottom through the table (shifting a single column to the right each time), and selecting horizontally adjacent pairs of digits, the following 10 samples were drawn: {1, 7, 4, 8; 15, 9, 6, 12; 8, 1, 11, 7; 15, 13, 9, 1; 15, 14, 9, 10; 6, 3, 13, 11; 13, 15, 4, 15; 6, 3, 6, 2; 10, 1, 13, 9; 7, 8, 14, 8}. Here N = 16, n = 4, $\mu = 1.275$, $\sigma = 2.3816$, and $\sigma/2 = 1.1908$. Accordingly, the observations in each of the 10 samples are as follows:

Sample municipio names	Sample mean	Sample standard deviation
Aguada, Isabela, Cabo Rojo, Lajas	2.100	4.1336
San German, Las Marias, Hormigueros, Moca	2.275	2.7801
Lajas, Aguada, Mayaguez, Isabela	2.175	4.0852
San German, Rincon, Las Marias, Aguada	2.250	2.8065
San German, Sabana Grande, Las Marias, Maricao	2.250	2.8065
Hormigueros, Anasco, Rincon, Mayaguez	0.075	0.1500
Rincon, San German, Cabo Rojo, San German	2.900	3.3486
Hormigueros, Anasco, Hormigueros, Aguadilla	0.200	0.4000
Maricao, Aguada, Rincon, Las Marias	0.800	1.6000
Isabela, Lajas, Sabana Grande, Lajas	2.125	4.1169

The confidence intervals constructed from these samples are

(a) for 90%, $\bar{x} \pm 1.645 \cdot 1.1908 = \bar{x} \pm 1.9589$,
(b) for 95%, $\bar{x} \pm 1.960 \cdot 1.1908 = \bar{x} \pm 2.3340$, and
(c) for 99%, $\bar{x} \pm 2.575 \cdot 1.1908 = \bar{x} \pm 3.0663$,

and are as follows:

Sample	90%	95%	99%
1	0.141 to 4.059	-0.234 to 4.434	-0.966 to 5.166
2	0.316 to 4.234	-0.059 to 4.609	-0.791 to 5.341
3	0.216 to 4.134	-0.159 to 4.509	-0.891 to 5.241
4	0.291 to 4.209	-0.084 to 4.584	-0.816 to 5.316
5	0.291 to 4.209	-0.084 to 4.584	-0.816 to 5.316
6	-1.884 to 2.034	-2.259 to 2.409	-2.991 to 3.141
7	0.941 to 4.859	0.566 to 5.234	-0.166 to 5.966
8	-1.759 to 2.159	-2.134 to 2.534	-2.866 to 3.266
9	-1.159 to 2.759	-1.534 to 3.134	-2.266 to 3.866
10	0.166 to 4.084	-0.209 to 4.459	0.941 to 5.191

Note that all these intervals contain the true population mean of 1.275. Note also that, as discussed, increasing confidence that an interval contains the population mean is obtained by constructing wider intervals. In addition, although the PR data set can reasonably be examined using population parameter values, the other three data sets represent samples of some population. As a result, it is not possible to construct examples parallel to this one.

In general, confidence intervals can be constructed from a sample statistic by extracting critical values (those values beyond which sample statistics are considered to be unlikely; the tail probabilities affiliated with these values are conventionally 10%, 5%, or 1%) for the appropriate probability distribution and then combining these values with the associated standard error of the sampling distribution.

9.3. Large- versus small-sample theory

Often one is faced with situations where a sample has been drawn from a population for which the variance is not known or, with reference to the sample-size results outlined in Chapter 8, where n is less than 100 or even 30. In such circumstances results tend to be far more volatile from sample to sample, and the normal approximation posited by the central limit theorem usually is poor (specification error). The question of concern here asks how one goes about making inferences when only the sample information is available, and n is relatively small (asymptotic convergence fails to be helpful).

To compensate for the increased uncertainty that accompanies either the use of one estimator in calculating the value of another estimator, or analyses employing small sample sizes (often $n < 30$), the concept of **degrees of freedom** (df) must be introduced. Several heuristics can be used to explain this concept for univariate statistics. First, degrees of freedom can be viewed in terms of constraints. Consider a set of three sample values $\{x_1, x_2, x_3\}$ whose sample mean is 5; then, if $x_1 = 2$ and $x_2 = 7$, x_3 must equal 6. In other words, one can let any two of the three values vary freely, but the constraint placed on the set of three values by $\bar{x}$ prevents the third value from being free to vary; therefore, the number of degrees of freedom refers to the number of sample values that are unconstrained. Closely related to this perspective is the one concerning substitution of one estimator in the calculation of another. In Chapter 1 the calculus of expectations was used to show that s^2 is unbiased when calculated with μ, but that s^2 must be divided by n - 1 rather than n to remain unbiased when $\bar{x}$ is substituted for μ. Moreover, $\bar{x}$ is constraining the analysis in such a way that the results pertain to a situation that is conditional on $\bar{x}$ (this estimator contains sampling error, too). This conditioning means that only n - 1 of the x_i values are free to vary; any n - 1 elements can be selected from the population, followed by selection of an nth element that ensures the correct value of $\bar{x}$. Here the number of degrees of freedom is defined as being the sample size minus the number of estimates already made in the calculation. The third viewpoint is more general; it treats degrees of freedom as an index number that identifies the correct frequency distribution to be used in an analysis. This generalization has two important implications, namely, (1) degrees of freedom can be fractional (this occurs both in multivariate analysis and spatial statistics), which fails to translate into the heuristic contexts of constraints or conditioning, and (2) degrees of freedom refer to a family of frequency distributions that allows adjustment for sample volatility.

The central limit theorem applies when the population variance is known. But what should be done when only s^2 is known? If n is large enough (certainly > 100, regardless of the form of the population frequency distribution), s^2 tends to be a good estimate of σ^2 and may be substituted for it in the procedures first discussed. If the population frequency distribution is approximately normal (this assumption is critical), and n is relatively small (< 30), then that specification error affiliated with an application of the central limit theorem is not negligible. In order to compensate for this error, the normal distribution can be replaced with Student's t distribution, where df = n - 1 (a degree of freedom is lost because the use of $\bar{x}$ is constraining the analysis). Student's t distribution is actually a family of distributions in which the exact shape is determined by the degrees of freedom present in the analysis. This distribution is presented in Table 9.1-A. Each t distribution is bell-shaped and centered at zero. Note that as n increases to ∞, the spread of Student's t distribution decreases, converging upon the normal distribution. This is consistent with what has been presented in the previous section regarding the central limit theorem. In comparison with the normal distribution for small sample sizes, probabilities associated with the tails of the Student's t distribution are much larger, whereas those associated with the central portion are much smaller (the sum of all probabilities still must be 1). This increased variability in the sampling distribution occurs when n is small because s no longer is likely to be close to σ, causing random variation to be present both in the sample mean and the estimate of the standard error; more uncertainty is introduced into the sampling distribution by replacing σ^2 with s^2.

One way of quantifying the convergence of the Student's t distribution onto the standard normal distribution is to inspect a sequence of critical values that corresponds to a fixed tail probability. Figures 8.1-8.4 present results for selected sample sizes. Their equivalents here are as follows, for a 5% upper-tail region:

df:	1	4	29	49	99	7000
t value:	6.3138	2.1318	1.6991	1.6766	1.6604	1.6450

The normal distribution value here is 1.6449. Moreover, by the time df = 30, there is a reasonably close correspondence between the Student's t and normal distributions.

What do these findings mean in terms of the confidence intervals that have been reported in Figure 9.1? Now this interval is defined as

$$\bar{x} + t_{\alpha/2,\nu} \bullet s/\sqrt{n} < \mu < \bar{x} + t_{1-\alpha/2,\nu} \bullet s/\sqrt{n}$$

where ν denotes the number of degrees of freedom. Again, symmetry means that $t_{\alpha/2,\nu} = -t_{1-\alpha/2,\nu}$, so that confidence intervals often are presented as

$$\bar{x} - t_{1-\alpha/2,\nu} \bullet s/\sqrt{n} < \mu < \bar{x} + t_{1-\alpha/2,\nu} \bullet s/\sqrt{n}.$$

What is the impact of this modification on the confidence interval? Consider a population consisting of the set of numbers {0, 1, 2}. If n = 2, then all possible samples yield

Sample	$\bar{x}$	s^2	90% confidence interval: σ is unknown		90% confidence interval: σ is known
0,0	0.0	0.0	0.0 ± 6.3138 • 0.0	versus	0.0 ± 1.6449 • 0.5774
0,1; 1,0	0.5	0.5	0.5 ± 6.3138 • 0.5	versus	0.5 ± 1.6449 • 0.5774
0,2; 2,0	1.0	2.0	1.0 ± 6.3138 • 1.0	versus	1.0 ± 1.6449 • 0.5774
1,1	1.0	0.0	1.0 ± 6.3138 • 0.0	versus	1.0 ± 1.6449 • 0.5774
1,2; 2,1	1.5	0.5	1.5 ± 6.3138 • 0.5	versus	1.5 ± 1.6449 • 0.5774
2,2	2.0	0.0	2.0 ± 6.3138 • 0.0	versus	2.0 ± 1.6449 • 0.5774

One can see from this example that none of the s^2 values is close to $\sigma^2 = 0.6667$ ($\sigma = .8165$). This particular illustration shows that the t distribution attempts to compensate for additional variability by shrinking the width of some confidence intervals (s = 0), and expanding the width of other confidence intervals (s = 0.7071 or 1.4142). The confidence interval based upon a t distribution for the first entry in Figure 9.1 is (-0.212 - 1.6772 • 1.063/7, -0.212 + 1.6772 • 1.063/7) = (-0.467, 0.043), which as expected is wider than the corresponding interval based upon a normal distribution.

EXAMPLE PR-9.3. Using the 10 samples of size 4 drawn in Example PR-9.2 for density of tobacco production (DTB) in the Mayaguez region of the Puerto Rico data set, construct 90%, 95%, and 99% confidence intervals for each of these samples treating σ^2 as unknown.

For a two-tail critical region, the appropriate t statistic for 3 degrees of freedom is $t_{3,0.95} = 2.3534$, $t_{3,0.975} = 3.1825$, and $t_{3,0.995} = 5.8410$. The sample standard deviations appear in Example PR-9.2, and each must be divided by 2, the square root of the sample size.

Sample	90%	95%	99%
1	-2.764 to 6.964	-4.478 to 8.578	-9.972 to 14.172
2	-0.996 to 5.546	-2.149 to 6.699	-5.844 to 10.394
3	-2.632 to 6.982	-4.326 to 8.676	-9.756 to 14.106
4	-1.053 to 5.553	-2.216 to 6.716	-5.946 to 10.446
5	-1.053 to 5.553	-2.216 to 6.716	-5.946 to 10.446
6	-0.102 to 0.252*	-0.164 to 0.314*	-0.363 to 0.513*
7	-1.040 to 6.840	-2.428 to 8.228	-6.880 to 12.680
8	-0.271 to 0.671*	-0.437 to 0.837*	-0.968 to 1.368
9	-1.083 to 2.683	-1.746 to 3.346	-3.873 to 5.473
10	-2.719 to 6.969	-4.426 to 8.676	-9.898 to 14.148

where * denotes confidence intervals that do not contain the population mean.

Note that, unlike the results from Example PR-9.2, not all the intervals contain the population mean of 1.275. However, as before, increasing confidence requires wider intervals.

9.4. Determining a minimum sample size

One useful extension of the confidence interval concept is to the problem of determining an appropriate sample size n once the magnitude of tolerable sampling error has been specified. The solution to this problem is similar to beginning with a confidence interval and working backward from it. Given the definition of a likely statistic value, which corresponds to the probability level assigned to a confidence interval (or an unlikely statistic, which corresponds to the probability level assigned to the tails), then in terms of a normal distribution

$$z_\alpha \bullet \sigma/\sqrt{n} = D$$

where D is the maximum absolute deviation one is willing to tolerate with a sample result. D must be preset, and refers to the centering of the interval on the population parameter value (just as for a confidence interval). Solving this equation for n yields

$$n = (z_\alpha \bullet \sigma/D)^2.$$

This result emphasizes the foregoing conjecture that in the absence of information, or the presence of total ignorance, one is willing to tolerate an infinite amount of error ($D = \infty$, and $n = 0$). In contrast, if one is willing to tolerate no error ($D = 0$), then one must take an infinite sample (or the entire population, if it is finite).

EXAMPLE 9.1. Determine the necessary sample size for the problem presented in Figure 9.1 when $|D| \leq 0.1$, $|D| \leq 1$, and $|D| \leq 2$.

Since $\mu = 0$ and $\sigma = 1$, one is seeking, respectively, sample statistics that capture the intervals (-0.1, 0.1), (-1, 1), and (-2, 2). The results are as follows:

$\lvert D\rvert \leq$	$z_{0.95}$ (90% interval)	$z_{0.975}$ (95% interval)	$z_{0.995}$ (1% interval)
0.1	$n \approx 271$	$n \approx 385$	$n \approx 664$
1.0	$n \approx 3$	$n \approx 4$	$n \approx 7$
2.0	$n \approx 1$	$n \approx 1$	$n \approx 2$

This example clearly demonstrates that a small degree of tolerance requires much larger sample sizes (one seeks considerable certainty) than does a great degree of tolerance, and that as the size of the associated confidence interval increases, the required sample size must increase (more of the sampling distribution must be captured).

9.5. Hypotheses and hypothesis testing

The four preceding sections dealt with point and interval estimation. Also mentioned in the introduction of this chapter was the inferential problem of evaluating a claim about a true value of a parameter; this claim is stated as a hypothesis. A single-value expectation is posited by the **null hypothesis** (denoted by H_0), which is the hypothesis on which a researcher wishes to focus attention. The term null may be defined as "of no value, effect, or consequence," which insinuates that H_0 should be identified with the hypothesis of no difference (a precise value), no improvement (in quality control), no change (from current opinion), no departure (from prevailing beliefs), no relationship, and the such. A second hypothesis about the parameter value, called the **alternate hypothesis** (denoted here by H_a), must accompany this null hypothesis and is constructed in such a way that the two hypotheses are mutually exclusive and collectively exhaustive of all parameter-value possibilities. Most often, then, the alternate hypothesis is a statement about the same population parameter differing in value from that specified in the null hypothesis; a few exceptions include certain quality control situations, where only two values are believed to exist. The decision as to which claim becomes H_0 and which becomes H_a is important, since treatment of these two hypotheses is not symmetrical. This formulation of pairs of hypotheses allows rejection of the null to imply acceptance of the alternative.

The computed sample statistic is evaluated to see whether it is a likely value from the sampling distribution that is associated with the hypothesized (under H_0) parameter; the null hypothesis allows a sampling distribution to be constructed. The sample statistic is evaluated under the assumption that the null hypothesis is a true statement. The sampling distribution that can be constructed under H_0 is partitioned into two sections (see Figure 9.2). The first, which corresponds to a confidence interval, identifies statistics that are consistent with the null hypothesis, and has probability $1 - (\alpha_1 + \alpha_2)$. The second, called the **critical region**, refers to those values that are unlikely under the null hypothesis, and references the tail regions (extremes) of the sampling distribution. The tails, therefore, contain the set of statistics con-

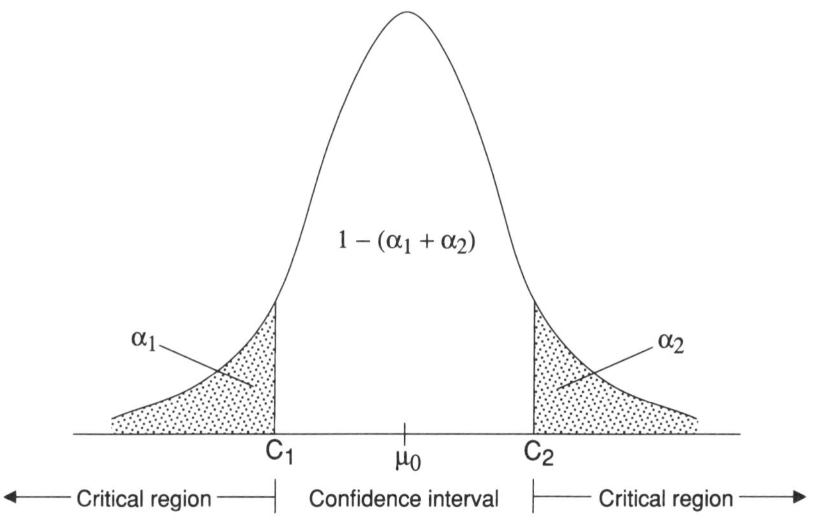

Figure 9.2: SAMPLING DISTRIBUTION OF SAMPLE STATISTIC, $\bar{x}$, IF H_0: $\mu = \mu_0$ IS TRUE, WHERE C_j ARE CRITICAL VALUES AND $(\alpha_1 + \alpha_2)$ IS THE SIGNIFICANCE LEVEL

sistent with H_a. The probability associated with the critical region is called the **level of significance**, denoted by α (10%, 5%, or 1%). For a left-hand one-tail situation the probability is $\alpha_1 = \alpha$ ($\alpha_2 = 0$). For a right-hand one-tail situation the probability is $\alpha_2 = \alpha$ ($\alpha_1 = 0$). For a two-tail situation the probability is $\alpha_2 = \alpha_1 = \alpha/2$. If the null hypothesis is true, then the probability that the sample statistic falls into the critical region is called the level of significance. **Critical values** (C_1 and C_2 in Figure 9.2) identify the boundaries of critical regions.

As was seen with confidence intervals, traditional levels of significance are 10%, 5%, and 1%. If one does not care whether the deviation from the population parameter value is above it or below it, then the alternate hypothesis involves the inequality $\neq$, and the level of significance must be divided between both the right-hand and left-hand tails of the sampling distribution; this is exactly the approach used in confidence interval construction. In other words, there are two critical regions, each of 5% for a 10% level of significance, each of 2.5% for a 5% level of significance, and each of 0.5% for a 1% level of significance. Sometimes, however, deviations are restricted to being only above or below the hypothesized value; physical restrictions may be involved, or in terms of quality control one may be interested only in whether or not improvements have occurred. If the alternate hypothesis is concerned only with values below the hypothesized number, then the alternate hypothesis involves the inequality $<$, and the level of significance and critical region are equivalent. If the alternate hypothesis is concerned only with values above the hypothesized number, then the alternate hypothesis involves the inequality $>$, and again the level of significance and critical region are equivalent.

9.5.1. Errors in hypothesis testing

A sample is drawn from somewhere, and the pair of hypotheses partitions all possible sample outcomes into two distinct groups. Thus, the sample comes from either the population described by the null hypothesis or a population described by the alternate hypothesis. Because the sampling distribution under H_0 is divided into a critical region and a confidence region, sampling error that leads to an unlikely statistic under H_0 can lead to an error when deciding from which population the sample was drawn. In this context the null hypothesis is the favored claim, with the burden of proof resting with H_a. There is no reason to disbelieve H_0 unless the sample evidence strongly contradicts it. This format stresses why one should not simply posit a "strawman" null hypothesis; H_0 should state that previously accepted conceptual thinking remains valid.

Errors in the inferential process are of two types. If a sample is drawn from that population described by H_0, and the conclusion reached is that in fact it was drawn from this population, then a correct decision has been made. But if an unlikely statistic is obtained, then an incorrect decision has been made. This category of error is called a **Type I error**; the researcher does not believe that the null hypothesis is true when in fact it is true. The probability of a Type I error is given by the size of the critical region (level of significance). In contrast, if a sample is drawn from a population described by H_a and the conclusion reached is that in fact it was not drawn from the population described by H_0, then again a correct decision has been made. But, if an unlikely statistic is obtained under H_a that is likely under H_0, then an incorrect decision has been made. This category of error is called a **Type II error**; the researcher believes that the null hypothesis is true when in fact it is false. The probability of a Type II error, often denoted by β, unfortunately is no simple matter to compute and refers to the notion of power [i. e., 1 - P(Type II error)], a concept that will be briefly discussed in a subsequent section of this chapter.

The interpretation that accompanies these statistical decisions has a subtly important phraseology. Formal logic behind this required phrasing of statistical interpretations involves the existence of a null hypothesis and an alternate hypothesis population. First, it is possible that both these populations exist, meaning both H_0 and $\sim H_0$ (read not H_0) are true. In this instance the two sampling distributions could overlap (see Figure 9.3), and one would be unable to know for certain which population yielded a selected sample. Second, it is possible that only the population under $\sim H_0$ exists (see Figure 9.3) and that the population under H_0 is a figment of one's imagination. In other words, when H_a is true, H_0 can be either true or false. Third, it is possible that only the population under H_0 exists (see Figure 9.3) and that a population under H_a is a figment of one's imagination. But since a sample has been drawn, then both populations cannot be a figment of one's imagination, since they partition the universe into two mutually exclusive and collectively exhaustive sets. In other words, when H_a is false, then H_0 must be true. Consequently, when an unlikely statistic is obtained under the null hypothesis, the researcher concludes that H_0 is to be rejected. However, if a likely statistic is obtained under the null hypothesis, then because both H_0 and $\sim H_0$ are possible, a researcher's conclusion must be to fail to reject H_0; the null hypothesis cannot be accepted. Moreover, the question being addressed asks whether a statistic is significantly different from the expected value posited by H_0, or is the difference from this expected value more than likely due to sampling error. (Is the probability that the difference observed is due to sampling error a reasonable one?)

Statistical decision possibilities employing these expressions may be summarized in tabular form as follows:

	H_0 is true	H_0 is false
Fail to reject H_0	correct decision	Type II error (β)
Reject H_0	Type I error (α)	correct decision

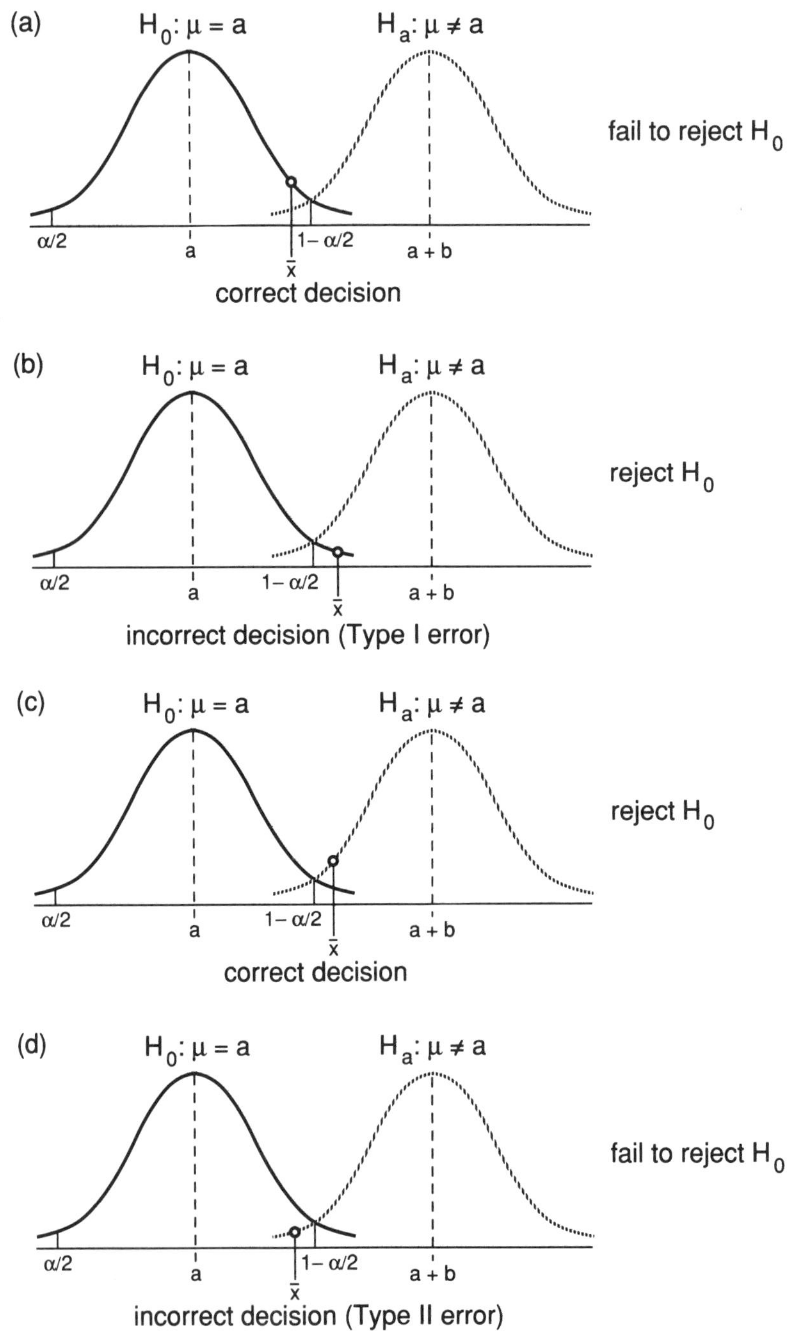

Figure 9.3: THE SAMPLE WAS DRAWN FROM (a) H_0, AND CORRECTLY IDENTIFIED AS SO; (b) H_0, BUT INCORRECTLY IDENTIFIED AS BEING FROM H_A; (c) H_A, AND CORRECTLY IDENTIFIED AS SO; (d) H_A, BUT INCORRECTLY IDENTIFIED AS BEING FROM H_0

A researcher tries to control these errors by assigning a small probability to each of them; ideally this assignment should be made before the research is undertaken. The best situation is to select a decision-making strategy that simultaneously minimizes α and β. Usually this is not possible, because these two errors are inversely related. In practice, first one fixes the probability of a Type I error, and then attempts to minimize of the conditional risk of a Type II error (again this refers to the notion of statistical power). Hence, the probability assigned to each error type will depend upon the seriousness of that error; the more serious an error type is, the less often one will be willing to allow it to occur (the smaller the probability assigned to it should be). Because most spatial data do not refer to life-threatening or financially disastrous situations, attention almost always is restricted to the Type I error risk (the cost of making a Type II error often is thought to be negligible).

Spatial scientists rarely seem to be concerned about Type II errors, but perhaps not justifiably so. Often it is the costs associated with committing a Type I error that are important. For example, suppose a researcher hypothesizes that there is no relationship between the distance children are from sources of automobile exhaust and the lead content in their bloodstream (causing lead poisoning). If a sample suggests that the null hypothesis should be rejected, when it is true (Type I error), then the government may unnecessarily require automobiles to install pollution abatement equipment. If a sample suggests that the null hypothesis should not be rejected, when it is false (Type II error), then no action will be recommended and children will continue to suffer from lead poisoning. In addition, consider the problem faced by a national health-control agency when confronted by the geographic diffusion of some virus. The null hypothesis insinuates that this disease will not diffuse into a given region. If a sample suggests that the null hypothesis fails to be rejected, when it is true (Type I error), then some people who are vaccinated will contract the virus from the vaccine. If a sample suggests that the null hypothesis fails to reject, when it is false (Type II error), then no vaccine will be administered, and some people will contract the disease from its natural source. Normally these are not the types of Type II error consequences faced in spatial analysis, and so most emphases are placed on controlling Type I error.

9.5.2. Critical values and test statistics

Once a statistic has been computed for some sample, a statistical decision may be made by comparing it to the reference sampling distribution probability model associated with the null hypothesis. This is done by converting the observed statistic to a test statistic, which quite often means subtracting the proposed population parameter value under H_0 from the observed statistic, and then dividing this difference by the standard error of the sampling distribution (while this has been the format presented thus far in this book, other test statistic forms can be constructed, and will be presented subsequently). The critical value has been defined as the first value in the critical region. If a test statistic lies beyond the critical value (it falls into the critical region), then one decides in favor of H_a; if a test statistic lies closer to the mean of the sampling distribution than does the critical value, then one fails to reject the null hypothesis.

EXAMPLE PR-9.4. Returning to Example PR-9.2, let the probability of a Type I error be set at 5%. Use the first sample drawn to evaluate the null hypothesis H_0: $\mu = 1.275$, versus the alternate hypothesis H_a: $\mu \neq 1.275$.

The sample drawn is Aguada ($x_1 = 0.0$), Isabela ($x_2 = 8.3$), Cabo Rojo ($x_3 = 0.0$), and Lajas ($x_4 = 0.1$). For this sample, $\bar{x} = 2.1$, and $s = 4.1336$. If one knows that $\sigma = 2.3816$, then the test statistic is

$$(2.1 - 1.275)/(2.3816/2) = 0.693$$

which is distributed as a standard normal score, except that $n = 4 < 30$, which means that a conservative researcher still would refer to the t distribution in order to select a critical value against which to compare this test statistic. The critical values for this two-tail situation are ± 1.96 for a normal distribution and ± 3.1825 (using 3 df) for a t distribution. If σ is unknown, then the test statistic becomes

$$(2.1 - 1.275)/(4.1336/2) = 0.399$$

which is distributed as a t statistic with 3 df. Consequently, one would fail to reject H_0, since 0.399 is between 0 and the critical value of 3.1825.

A complete interpretation of this statistical decision involves two additional statements. The second statement refers to whether or not evidence has been found in the sample that pertains to H_a; accordingly, there is no evidence to indicate that the mean density of tobacco production in the population is greater than 1.275. The third statement refers to the probability of a Type I error; as such, 95% of the time one would expect to get a z (t) test statistic that falls within the appropriate confidence interval when in fact the mean density of tobacco production in the population is 1.275.

EXAMPLE PR-9.5. Redo Example PR-9.3 using the sample of Isabela ($x_1 = 8.3$), Las Marias ($x_2 = 3.2$), San German ($x_3 = 5.8$), and San Sebastian ($x_4 = 1.8$), and the alternate hypothesis H_a: $\mu > 1.275$.

For this sample, $\bar{x} = 4.775$, and $s = 2.8756$. If one knows that $\sigma = 2.3816$, then the test statistic is

$$(4.775 - 1.275)/(2.3816/2) = 2.939$$

which if considered to be distributed as a standard normal score would have a critical value for this one-tail situation of 1.6449, whereas if considered to be distributed as t would have a critical value of 2.3534. If σ is unknown, then the test statistic becomes

$$(4.775 - 1.275)/(2.8756/2) = 2.434$$

which is distributed as a t statistic with 3 df. Consequently, one would reject H_0 in the first instance and fail to reject it in the second instance.

A complete interpretation of this statistical decision involves two additional statements, too. Suppose one has rejected H_0. The second statement would be that there is evidence to indicate that the mean density of tobacco production in the population is other than 1.275. The third statement would be that only 5% of the time would one expect to get a z (t) test statistic that falls within the appropriate confidence interval when in fact the mean density of tobacco production in the population is 1.275.

9.5.3. Power of a test

Recall that the power of a test is the probability that the null hypothesis is rejected when it should have been rejected. Assuming a fixed sample size n, the greater the ability of a test statistic to eliminate false hypotheses, the greater is its power. If two test statistics have the same Type I error probability, α, then the test statistic with the greater power is the one that tends to be preferable. But power requires a knowledge of both the exact form of the population and the degree to which the hypothesized parameter value differs from the true value. Moreover, power of a test statistic depends upon which alternative hypothesis actually is correct.

A power function can be defined that combines the probability of both a Type I and a Type II error. For some parameter θ this function takes on the form

$$\pi(\theta) = \begin{cases} \alpha(\theta) \text{ if } H_0 \text{ is true} \\ \beta(\theta) \text{ if } H_0 \text{ is false.} \end{cases}$$

This function should be designed in such a way that the power will be small when the null hypothesis is true. Graphically it appears as a logistic-type curve for a right-hand-tail critical region and an inverted normal curve for a two-tail critical region (see Figure 9.4).

For example, the power function $\pi(\mu)$ is defined for each possible value of μ as

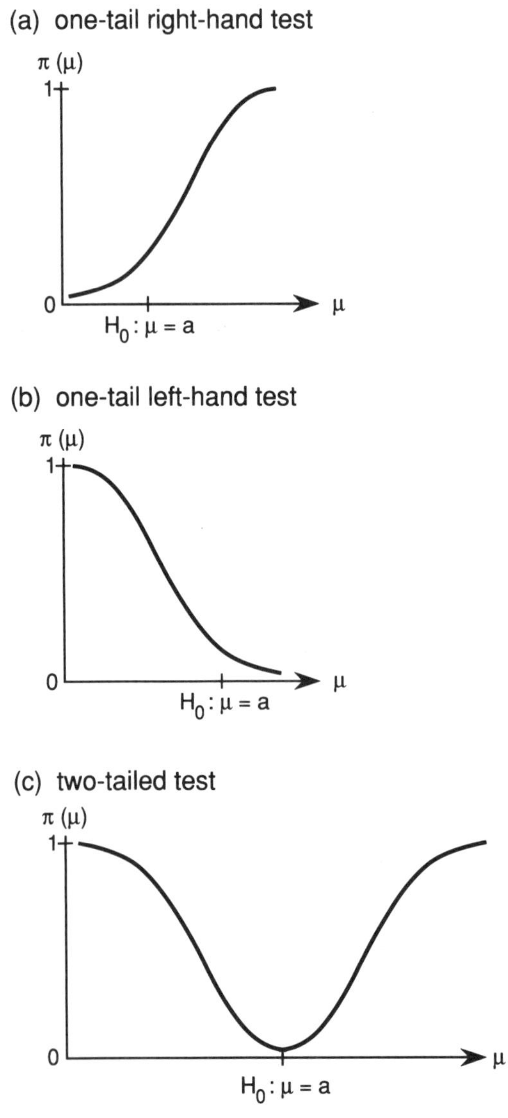

Figure 9.4: GRAPHICAL PORTRAYAL OF A POWER FUNCTION

$\pi(\mu)$ = P(rejecting H_0 when μ = the true population parameter value).

For any alternative value of μ, $\beta(\mu) = 1 - \pi(\mu)$. Considering each possible alternative hypothesis in turn, where μ_0 denotes the hypothesized population mean value and $\Phi(z)$ denotes the cumulative probability distribution function of the standard normal distribution, the power function for each of the three statistical test situations is as follows:

H_a: $\mu > \mu_0$, $\pi(\mu) = 1 - \Phi[z_\alpha + (\mu_0 - \mu)/(\sigma/\sqrt{n})]$

H_a: $\mu < \mu_0$, $\pi(\mu) = \Phi[-z_\alpha + (\mu_0 - \mu)/(\sigma/\sqrt{n})]$

H_a: $\mu \neq \mu_0$, $\pi(\mu) = 1 - \Phi[z_{\alpha/2} + (\mu_0 - \mu)/(\sigma/\sqrt{n})] + \Phi[-z_{\alpha/2} + (\mu_0 - \mu)/(\sigma/\sqrt{n})]$

The algebraic manipulations required to determine these equations are similar to those used earlier to construct confidence intervals and minimum sample size.

EXAMPLE PR-9.6. Returning to Example PR-9.3, compute the power function value for each of the three possibilities $\mu_0 = 1$, $\mu_0 = 1.275$, and $\mu_0 = 2$, where $(1 - 1.275)/1.1908 = -0.23$, $(1.275 - 1.275)/1.1908 = 0$, and $(2 - 1.275)/1.1908 = 0.61$.

$\mu_0 = 1$: $1 - \Phi(1.96 - 0.23) + \Phi(-1.96 - 0.23) = 1 - 0.9582 + 0.0143 = 0.0561$
$\mu_0 = 1.275$: $1 - \Phi(1.96 - 0) + \Phi(-1.96 - 0) = 1 - 0.975 + 0.025 = 0.05$
$\mu_0 = 2$: $1 - \Phi(1.96 - 0.61) + \Phi(-1.96 - 0.61) = 1 - 0.9115 + 0.0051 = 0.0936$

Example PR-9.6 illustrates some of the complexities associated with calculating power. A more detailed treatment of this topic is beyond the scope of this book.

9.6. Summary

The basic concepts in this chapter (estimates, sampling versus sample versus population distributions, types of errors, and others) comprise a necessary set of "tools" with which inferential analyses are built. It is essential to master the nature of these tools, including when and how to use them, before proceeding. All too often published statistical analyses are not flawed in terms of the data but rather illustrate the incorrect or incomplete application of statistical tools. As seen in Chapter 8, the process of collecting data is constrained in a number of ways. It is the job of the statistician to work with the data as they are collected. Blaming poor analyses on the data is similar to blaming the wood for a poor carpentry job; the wood may be flawed, but more likely the carpenter is the source of the problem.

9.7. *Problems*

1. Verify the parameter values and number of samples calculated in Example PR-9.1

2. Verify the sample means, sample standard deviations, and confidence intervals calculations appearing in Example PR-9.2.

3. Verify the confidence intervals in Figure 9.1. In addition, construct 95% and 99% confidence intervals for the first five sample results appearing in this figure.

4. Verify the confidence intervals in Example PR-9.3.

5. Using the random numbers table (Table 7.1-A), draw 10 samples of size 5 from the Puerto Rico data set, for the density of milk production variable (DMLK), and construct confidence intervals for them (a) assuming that σ is known (see Chapter 4) and (b) assuming that σ is unknown. Determine which intervals contain the population mean, and which do not. Compare the results from (a) and (b).

6. Draw a sample of size 10 from the density of tobacco production variable (DTB) for all of Puerto Rico. Test the null hypothesis H_0: $\mu = 6.334$, with $\sigma = 20.46$, assuming that (a) the normal distribution is a reasonable approximation to the sampling distribution, and (b) the t distribution must be used here because the sample size is too small.

7. Verify the sample sizes determined in Example 9.1.

8. In Problem #3, if one wanted to have an estimate of the mean that deviated from the true mean by no more than 0.5 unit, with a probability of 0.95, what is the minimum sample size that would be appropriate? What happens when this probability is increased to 0.99? What happens when this deviation is increased to 1.5?

9. Verify the calculations appearing in Examples PR-9.4 and PR-9.5.

10. Returning to Example PR-9.6, sketch the power function for this problem. Then sketch the comparable power function for a one-tail critical region.

9.8. Additional readings

In addition to the texts listed in Chapter 1,

Davidson, N. *Causal Inferences from Dichotomous Variables.* CATMOG 9.

CHAPTER 10

SIMPLE HYPOTHESIS TESTING

Hypotheses and their relationship to sampling distributions were introduced in Chapter 9. A basic appeal of hypothesis testing is that some sort of central limit theorem in mathematical statistics produces the desired mean and the standard error without one having the need to enumerate all possible samples of size n. Exploiting this feature, we outline simple hypothesis tests in this chapter.

10.1. The sequence of steps in hypothesis testing

Hypothesis testing involves a structured sequence of steps. Step 1 is to formulate the null hypothesis; this task was described in detail in Chapter 9. It involves identification of the parameter of interest as well as the corresponding statistic measuring it. Step 2 is to formulate an alternate hypothesis; this task also was described in Chapter 9. Often these two steps are the most critical ones, as they define the nature and scope of the research problem of interest. They are formulated by inspecting the statement that is to be investigated and then dividing all possible outcomes into two distinct groups.

Step 3 is to determine the test criteria. This step involves specifying a level of significance (deciding upon the size of the critical region) and determining a test statistic. As was emphasized in earlier chapters, traditional levels of significance employed are 10%, 5%, and 1% (these values are split in half for each tail of a two-tail H_a); recall that the choice made here must reflect Type I and Type II error concerns. Selection of the test statistic is a function of (1) sample size and degrees of freedom, (2) assumptions regarding the underlying variable and sampling distribution probability models, (3) the measurement scale involved, and (4) the estimator being used. Invalid assumptions can result in specification errors, some of which can be quite sizeable. Conclusions drawn pertain only to the posited set of assumptions. Important assumptions include (1) independence of observations (spatial autocorrelation latent in spatial data usually violates this assumption), (2) random sampling, with replacement and order important, and (3) normally distributed random variables.

Frequently a test statistic is from either the t distribution or a standard normal distribution; other relevant test statistics will be presented in this chapter and when possible related to these two distributions. The test statistic is a random variable whose value is used to make the statistical decision whether to reject H_0 or fail to reject H_0. The probability model describing the sampling distribution for this test statistic is determined as a result of assuming that the null hypothesis is true. In this context, then, the critical region is the set of values for a test statistic that will result in a decision to reject H_0 (these are unlikely statistics under H_0). As was stated before, the critical value is the first value in the critical region.

Step 4 is to calculate the value of the test statistic, which usually is a statistic having no inherent descriptive qualities but which is used in testing statistical hypotheses. Typically the value of the test statistic is based upon sample information. In many cases this computation involves dividing the difference between the estimator and the hypothesized value by the standard error. Again one must appreciate the usefulness of a central limit theorem that renders theoretical formulae for the sampling distribution mean and standard error. In other words, a quantity is computed from sample data that varies in a known way in accordance with probability theory. Unfortunately only a few test statistics have known sampling distributions. The need to substitute sample information into such formulae again focuses attention on the notion of degrees of freedom (see Chapter 9).

Step 5 is to make of a statistical decision and then to interpret it. A decision is made by comparing the value of the computed test statistic from Step 4 with the test criterion established in Step 3. If the test statistic falls within the critical region, then H_0 is rejected; if it does not fall into the critical region, then H_0 must fail to be rejected. Interpretation involves the three statements outlined in Chapter 9, namely, (a) whether or not the null hypothesis is rejected, (b) what evidence is provided by the sample data, and (c) a probability statement relating to either a confidence interval or a level of significance. Because statistical decisions depend upon what is done in Steps 1-4, ideally all choices that will affect decision-making should be made prior to tabulating results.

10.2. Goodness-of-fit tests

In Chapter 3 the usefulness of histograms was established. Selected theoretical frequency distributions, together with their respective importance, were discussed in Chapter 7. The form of the underlying frequency distributions has an important impact on statistical properties such as consistency. This impact was illustrated in Chapter 8. Because of these and other issues, a researcher may be interested in the question of whether or not the parent population frequency distribution conforms to some theoretical frequency distribution. In other words, does the theoretical frequency distribution furnish a good fit (hence the term **goodness-of-fit**) to the empirical frequency distribution? If it does not, then the accompanying specification error may render invalid any statistical decision that is reached.

The chi square statistic, which is a sum of squares of independent standard normal variables, was first presented in Chapter 4. The approximate chi square formula is

$$\sum_{i=1}^{k} (e_i - o_i)^2/e_i = \sum_{i=1}^{k} o_i^2/e_i - n$$

where e_i is the expected frequency for a category i, o_i is the frequency for category i observed in the sample, and k is the number of categories. This formula approximates the exact chi square, which is given as

$$2 \cdot \sum_{i=1}^{k} o_i \ln(o_i/e_i)$$

where ln refers to the natural logarithm of a number. The restriction on this second formula is that $|o_i - e_i| < e_i$. The approximate chi square formula has been popular mostly because the exact chi square was very difficult to compute before the advent of scientific hand calculators. Realize, though, that any numerical differences between these two calculations tend to be attributable to specification error associated with the use of the approximate chi square. As a result, even though the approximation is good when $e_i \geq 5$, if a calculated chi square statistic is very close to the critical value, then the exact chi square value should be calculated.

In addition to possible specification error, sample size affects a chi square test statistic; however, the law of large numbers maintains that this impact will diminish as n increases. Again, sample size and degrees of freedom need to be inspected. When a histogram is constructed, k categories or intervals are defined. Now a sample of size n is to be distributed over this set of k intervals, yielding a sample frequency distribution. Except for the restriction on frequencies to be nonnegative, any number of observations can be assigned to the first interval (o_1), any number of the remaining observations can be assigned to the second interval (o_2), and so on. But since the intervals are mutually exclusive and collectively exhaustive, meaning $\sum_{i=1}^{k} o_i = n$, then o_k must equal $n - \sum_{i=1}^{k-1} o_i$. In other words, sample size places a constraint upon the analysis; the conditional probabilities for the categories must sum to unity. Accordingly, there are k frequency values free to vary, and a constraint attributable to sample size, resulting in $\upsilon = k - 1$ degrees of freedom. Table 10.1-A (where the extension -A denotes that the table is located in the appendices) presents the theoretical chi square distribution for selected probability levels and degrees of freedom. A contrived example is presented now in order to help illustrate these three features.

EXAMPLE 10.1. The sampling distribution of the chi square statistic for k = 2, n = 10, and a uniform parent population frequency distribution.

The uniform distribution in the population means that $e_1 = e_2 = 5$, which satisfies the restrictions for both the approximate and the exact chi square test statistics. The sampling distribution is constructed from the following information:

Sample composition		Number of	Chi square	
k_1	k_2	samples	Approximate	Exact
10	0	2 • 1	10.0	∞
9	1	2 • 10	6.4	7.361
8	2	2 • 45	3.6	3.855
7	3	2 • 120	1.6	1.646
6	4	2 • 210	0.4	0.403
5	5	252	0.0	0.000

Note: Frequencies for some compositions are multiplied by 2 because the order of assignment of frequency pairs to categories does not alter the chi square result; this set of six values correctly sums to 2^{10} (see Chapter 6).

In Example 10.1 there is 1 degree of freedom. The exact chi square deviates more noticeably from the approximate chi square calculation as sample representativeness decreases; the representative compositions of {5, 5} and {6, 4} produce essentially the same test statistic value. Furthermore, in terms of tail probabilities,

P(approximate chi square ≥ 3.75) = 112/1024 ≈ 0.109
P(exact chi square ≥ 3.75) = 22/1024 ≈ 0.021
P(theoretical chi square with $\upsilon = 1 \geq 3.75$) = 0.0528

This comparison demonstrates the potential severity of specification error, either due to the use of the approximate rather than the exact chi square or assuming that the theoretical chi square distribution is applicable to small sample situations.

In the context of Chapter 4 chi square is used to measure relationships between categorical variables. The problem being addressed here may be viewed as asking whether or not a relation exists between a sample frequency distribution, whose histogram intervals are equivalent to categories, and some corresponding theoretical frequency distribution histogram. In Chapter 7 the procedure for calculating a theoretical frequency distribution histogram is presented. The task that remains now is to outline a hypothesis-testing setup for this problem. This type of comparison is called goodness-of-fit, for it tests how good the correspondence is between a sample and a theoretical frequency distribution.

Assumptions:

(a) there must be at least two mutually exclusive and collectively exhaustive categories in the histogram;

(b) the observations are independent; and,
(c) the expected frequency e_i must satisfy applicable restrictions, or corrective action must be taken in order to ensure that specification error remains negligible (categories where the restriction does not hold can be amalgamated with adjacent ones).

Test criteria:

(a) for chi square goodness-of-fit tests only a one-tailed critical region is to be used; and,
(b) the test statistic is approximately distributed as χ^2 with k - r - p - 1 degrees of freedom, where r is the number of categories lost in consolidation in order to satisfy the appropriate restriction, and p is the number of parameters that have to be estimated.

The following are examples of chi square goodness-of-fit tests for the four theoretical frequency distributions discussed in this text. First, consider a uniform frequency distribution.

EXAMPLE 10.2. A goodness-of-fit test for the random numbers appearing in Table 7.1-A; Example 7.1 revisited.

H_0: the frequency distribution of digits in the population conforms to a uniform distribution.

If the observed values equal the expected values, the numerator of the chi square formula becomes 0 for each term in the summation. Therefore, this is equivalent to saying that H_0: $\chi^2 = 0$.

H_a: the frequency distribution of digits in the population does not conform to a uniform distribution

which is equivalent to saying that H_a: $\chi^2 > 0$.

Since under the null hypothesis of a uniform frequency distribution in the population all digits are equally likely, the expected frequency is constant, equalling 1800 • (1/10). As such, no frequency distribution parameters need to be estimated from the sample, and so the number of degrees of freedom is υ = 10 - 1 (1 degree of freedom is lost because of the sample size, n = 1800). The corresponding critical value for α = 0.05 is (from Table 10.1-A) 16.9190. Therefore, for both the approximate and the exact test statistics, one would conclude that (1) H_0 fails to be rejected, (2) there is no evidence to indicate that the frequency distribution of the 10 digits does not conform to a uniform distribution in the population, and (3) 95% of the time one would expect to obtain a chi square value that falls within the confidence interval [0, 16.9190) when in fact the frequency distribution for the 10 digits does conform to a uniform distribution in the population (hence, all that is being observed is sampling error).

Digit	o_i	e_i	$(o_i - e_i)^2/e_i$	$o_i\ln(o_i/e_i)$
1	190	180	0.55556	10.2728
2	186	180	0.20000	6.0989
3	182	180	0.02222	2.0111
4	169	180	0.67222	-10.6568
5	168	180	0.80000	-11.5908
6	178	180	0.02222	- 1.9888
7	190	180	0.55556	10.2728
8	155	180	3.47222	-23.1774
9	193	180	0.93889	13.4585
(1)0	189	180	0.45000	9.2213
Column sum	1800	1800	7.68889	7.8432/2

The difference between the approximate and the exact chi square values, namely, 0.1543 (the approximate statistic slightly underestimates the true statistic here), is attributable to specification error. Clearly this amount is irrelevant unless the observed approximate chi square test statistic is nearly equal to the relevant critical value. Furthermore, the conclusion reached for this example is consistent with the argument put forth earlier that those numbers appearing in Table 7.1-A furnish an appropriate tool for drawing random samples.

Next, consider a binomial frequency distribution.

EXAMPLE PR-10.1. A goodness-of-fit test for the binomial frequency distribution presented in Example PR-7.2.

H_0: the sampling frequency distribution for coastal lowlands locations conforms to a binomial distribution in the population,

which is equivalent to saying that H_0: $\chi^2 = 0$.

H_a: the sampling frequency distribution for coastal lowlands locations does not conform to a binomial distribution in the population,

which is equivalent to saying that H_a: $\chi^2 > 0$.

Under the null hypothesis of a binomial frequency distribution in the population, $(10!)/[x!(10 - x)!] \bullet p^x \bullet (1 - p)^x$ are the probabilities of all outcomes, and the expected values are 100 times these probabilities. When the probability p is known, no frequency distribution parameters need to be estimated from the sample, and so the number of degrees of freedom begins as $\upsilon = 11 - 1$ (1 degree of freedom is lost because of the sample size, n = 100); when p must be estimated from the sample data, the number of degrees of freedom begins as $\upsilon = 11 - 2$. Invoking the restriction of $e_i \geq 5$ results in a loss of 5 tail categories and hence an additional loss of 5 degrees of freedom, respectively yielding $\upsilon = 5$ when p is known and $\upsilon = 4$ when p is unknown. The corresponding critical values for $\alpha = 0.05$ are (from Table 10.1-A) 9.4877 and 11.0705. Therefore, for both situations one would conclude that (1) H_0 fails to be rejected, (2) there is no evidence to indicate that the sampling frequency distribution does not conform to a binomial distribution in the population, and (3) 95% of the time one would expect to obtain a chi square value that falls within the confidence interval [0, 9.4877), for the approximate statistic, and [0, 11.0705), for the exact statistic, when in fact the sampling frequency distribution does conform to a binomial distribution in the population.

Count of outcomes	o_i	Probability known e_i	Probability known $(o_i - e_i)^2/e_i$	Probability unknown e_i	Probability unknown $(o_i - e_i)^2/e_i$
0 - 3	8	7.5444	0.02752	6.6768	0.262230
4	10	13.4945	0.90494	12.5641	0.523285
5	22	21.9395	0.00017	21.2530	0.026256
6	26	24.7704	0.06104	24.9659	0.042833
7	18	19.1771	0.07225	20.1102	0.221427
8 - 10	16	13.0741	0.65481	14.4299	0.170814
Column sum	100	100.0000	1.72072	99.9999	1.246852

The approximate chi square value is sufficiently less than the critical value that computing its exact counterpart is not merited (this value in fact equals $2 \cdot 1.52867$ when p is unknown, with the accompanying restriction requiring Classes 0-3 and 9-10 to be consolidated, and $2 \cdot 1.11372$ when p is known, with the accompanying restriction requiring Classes 0-3 to be consolidated). Note here that the restrictions not only result in a loss of degrees of freedom but also cause the chi square statistic to be insensitive to what happens in its tails (this is where the consolidation occurs in order for the relevant restriction to be satisfied). Also, the increased variability suggested by the difference $|2.47543 - 1.246882|$, due to estimating p, is sizeable.

The third type of theoretical frequency distribution discussed in Chapter 7 was the Poisson.

EXAMPLE 10.3. A goodness-of-fit test for the Poisson frequency distribution; Example 7.2 revisited.

H_0: the frequency distribution of points by quadrat in the population conforms to a Poisson distribution with $\mu = 1.00$, which is equivalent to saying that H_0: $\chi^2 = 0$.

H_a: the frequency distribution of points by quadrat in the population does not conform to a Poisson distribution with $\mu = 1.00$,

which is equivalent to saying that H_a: $\chi^2 > 0$.

No. of points/quadrat	o_i	e_i	$(o_i - e_i)^2/e_i$	$o_i \ln(o_i/e_i)$
0	10	9.19699	0.070113	0.83709
1	8	9.19699	0.155789	-1.11548
2	4	4.59849	} 0.023497	-0.55773
3+	3	2.00753		1.20512
Column sum	25	25.00000	0.249399	0.73800/2

Under the null hypothesis of a Poisson frequency distribution in the population, the number of points per quadrat have probabilities equal to $\mu^x \cdot \exp(-\mu)/(x!)$, and the expected values are 25 times these probabilities. When μ must be estimated from the sample data, the number of degrees of freedom begins as $\upsilon = 4 - 2$. To compute the approximate chi square test statistic, the restriction of $e_i \geq 5$ results in a loss of 1 tail category--an additional loss of 1 degree of freedom--yielding $\upsilon = 1$. To compute the exact chi square test statistic, the restriction of $|o_i - e_i| < e_i$ results in no loss of tail categories. The corresponding critical values for $\alpha = 0.05$ are, respectively, (from Table 10.1-A) 3.8415 and 5.9915. Therefore, for both situations one would conclude that (1) H_0 fails to be rejected, (2) there is no evidence to indicate that the frequency distribution of points by quadrat does not conform to a Poisson distribution in the population, and (3) 95% of the time one would expect to obtain a chi square value that falls within the confidence interval [0, 3.8415), for the approximate statistics, and [0, 5.9915), for the exact statistic, when in fact the frequency distribution of points by quadrat does conform to a Poisson distribution in the population.

The finding in this example that more degrees of freedom are lost in category consolidation with the approximate chi square than with the exact chi square is more characteristic than what was seen in Example 10.2; usually that restriction associated with the exact chi square is less binding than is its counterpart for the approximate chi square. In addition, the conclusion reached for this example is consistent with the argument put forth earlier that drawing a random sample from a bivariate uniform distribution generates a Poisson distribution of points in geographic space.

Finally, consider a normal frequency distribution.

EXAMPLE PR-10.2. A goodness-of-fit test for the normal frequency distribution; Example PR-7.3 revisited.

H_0: the frequency distribution of DCD in the population conforms to a normal distribution,

which is equivalent to saying that H_0: $\chi^2 = 0$.

H_a: the frequency distribution of DCD in the population does not conform to a normal distribution,

which is equivalent to saying that H_a: $\chi^2 > 0$.

Interval boundaries			o_i	e_i	$(o_i - e_i)^2/e_i$	$o_i\ln(o_i/e_i)$
60.09	-	∞	2	1.6644		0.36737
48.13	-	60.09	9	9.9207}	0.02955	-0.87659
36.17	-	48.13	19	24.9149	1.40422	-5.14951
24.21	-	36.17	31	24.9149	1.48620	6.77416
12.25	-	24.21	10	9.9207}	0.01486	0.07962
-∞	-	12.25	2	1.6644		0.36737
Column sum			73	73.0000	2.93483	3.12484/2

Under the null hypothesis of a normal frequency distribution in the population, intervals have probabilities equal to those presented in Table 7.2-A, and the expected values are 73 times these probabilities. When μ and σ must be estimated from the sample data, the number of degrees of freedom begins as $\upsilon = 6 - 3$. To compute the approximate chi square test statistic, the restriction of $e_i \geq 5$ results in a loss of 2 tail categories (one on each side), and hence an additional loss of 2 degrees of freedom, yielding $\upsilon = 1$. To compute the exact chi square test statistic, the restriction of $|o_i - e_i| < e_i$ results in no loss of tail categories. The corresponding critical values for $\alpha = 0.05$ are, respectively, (from Table 10.1-A) 3.8415 and 7.8147. Therefore, for both situations one would conclude that (1) H_0 fails to be rejected, (2) there is no evidence to indicate that the frequency distribution of DCD does not conform to a normal distribution in the population, and (3) 95% of the time one would expect to obtain a chi square value that falls within the confidence interval [0, 3.8415), for the approximate statistic, and [0, 7.8147), for the exact statistic, when in fact the frequency distribution of DCD does conform to a normal distribution in the population.

EXAMPLE OR-10.1. A goodness-of-fit test for the normal frequency distribution; Example OR-7.1 revisited.

H_0: the frequency distribution of SNOW in the population conforms to a normal distribution,

which is equivalent to saying that H_0: $\chi^2 = 0$.

H_a: the frequency distribution of SNOW in the population does not conform to a normal distribution,

which is equivalent to saying that H_a: $\chi^2 > 0$.

Interval boundaries			o_i	e_i	$(o_i - e_i)^2/e_i$	$o_i\ln(o_i/e_i)$
123.81	-	∞	1	0.6384		0.4488
105.03	-	123.81	4	3.8052}		0.1997
86.25	-	105.03	11	9.5564	0.2857	1.5475
67.47	-	86.25	8	9.5564	0.2857	-1.4222
48.69	-	67.47	3	3.8052}		-0.7133
-∞	-	48.69	1	0.6384		0.4488
Column sum			28	28.0000	0.5714	1.0186/2

Under the null hypothesis of a normal frequency distribution in the population, intervals have probabilities equal to those presented in Table 7.2-A, and the expected values are 28 times these probabilities. When μ and σ must be estimated from the sample data, the number of degrees of freedom begins as $\upsilon = 6 - 3$. To compute an approximate chi square test statistic the restriction of $e_i \geq 5$ results in the definition of only two categories, hence a loss of an additional four degrees of freedom. An approximate chi square therefore cannot be tested. To compute the exact chi square test statistic, the restriction of $|o_i - e_i| < e_i$ results in no loss of tail categories. The corresponding critical value for $\alpha = 0.05$ is (from Table 10.1-A) 7.8147. Therefore, one would conclude that (1) H_0 fails to be rejected, (2) there is no evidence to indicate that the frequency distribution of SNOW does not conform to a normal distribution in the population, and (3) 95% of the time one would expect to obtain a chi square value that falls within the confidence interval [0, 7.8147) when in fact the frequency distribution of SNOW does conform to a normal distribution in the population.

As is seen especially in Examples PR-10.1 and 10.3, the chi square test statistic can be insensitive to variation in the tails of a distribution, with this variation disappearing during aggregation. An alternative, and often more powerful, test statistic is one due to Kolmogorov-Smirnov (denoted here by K-S). Its appeal lies in an absence of restrictions pertaining to class interval expected values. This statistic is defined as follows:

$$\text{K-S} = \max|CFD_o - CFD_e|$$

where CFD_o is the cumulative relative frequency distribution for the observed histogram, CFD_e is the cumulative relative frequency distribution for the expected histogram as specified by H_0, max indicates that the maximum difference is to be determined, and $|\bullet|$ denotes absolute value. Thus the K-S statistic is the maximum absolute difference between the two relative cumulative frequency distributions. Values for these relative distributions must always lie between 0 and 1 (they are probabilities), and the last pair of intervals always will have values of unity.

Assumptions:

(a) there must be at least two mutually exclusive and collectively exhaustive categories in the histogram;
(b) the population frequency distribution for a variable being studied is continuous;
(c) the variable is measured at the ordinal, interval, or ratio scale; and
(d) observations are independent.

Test criteria:

(a) for K-S tests only a two-tail critical region can be used;
(b) sample size, rather than the number of intervals or degrees of freedom, becomes the relevant index; and
(c) the cumulative relative frequency distribution is constructed by ranking observations in ascending order, and then determining the proportions of observations in a sample that are less than or equal to the upper limit of each interval.

For discrete variables the K-S statistic gives only approximate results, and tend to suffer only a slight increase in the risk of a Type II error. Table 10.2-A presents the theoretical K-S distribution for selected probability levels and sample sizes.

The following are examples of K-S goodness-of-fit tests for the four theoretical frequency distributions discussed in this text. First, consider a uniform frequency distribution analyzed in Example 10.2.

EXAMPLE 10.4. Example 10.2 revisited.

Digit	CFD_o	CFD_e	$\|CFD_o-CFD_e\|$	Digit	CFD_o	CFD_e	$\|CFD_o-CFD_e\|$
1	0.10556	0.1	0.00556	6	0.59611	0.6	0.00389
2	0.20889	0.2	0.00889	7	0.70167	0.7	0.00167
3	0.31000	0.3	0.01000	8	0.78778	0.8	0.01222*
4	0.40389	0.4	0.00389	9	0.89500	0.9	0.00500
5	0.49722	0.5	0.00278	(1)0	1.00000	1.0	0.00000

(Note: * identifies the K-S statistic for this problem.) Since n = 1800, for α = 0.05 the critical value of the K-S statistic is $1.36/\sqrt{1800} = 0.0321$. Consequently, the statistical conclusions remain the same as those stated in Example 10.2.

Next, consider the binomial frequency distribution analyzed in Example 10.1.

EXAMPLE PR-10.3. Example PR-10.1 revisited.

Count of outcomes	CFD_o	Probability known CFD_e	$\lvert CFD_o - CFD_e \rvert$	Probability unknown CFD_e	$\lvert CFD_o - CFD_e \rvert$
0	0.01	0.00019	0.00981	0.00015	0.00985
1	0.01	0.00277	0.00723	0.00229	0.00771
2	0.04	0.01853	0.02147	0.01584	0.02416*
3	0.08	0.07544	0.00456	0.06677	0.01323
4	0.18	0.21039	0.03039*	0.19241	0.01241
5	0.40	0.42978	0.02978	0.40494	0.00494
6	0.66	0.67749	0.01749	0.65460	0.00540
7	0.84	0.86926	0.02926	0.85570	0.01570
8	0.94	0.96669	0.02669	0.96201	0.02201
9	0.99	0.99603	0.00603	0.99531	0.00531
10	1.00	1.00000	0.00000	1.00000	0.00000

(Note: * identifies the K-S statistic for this problem.) Since n = 100, for α = 0.05 the critical value of the K-S statistic is $1.36/\sqrt{100} = 0.136$. Consequently, the statistical conclusions remain the same as those stated in Example 10.3.

Now consider the Poisson frequency distribution analyzed in Example 10.4.

EXAMPLE 10.7. Example 10.4 revisited.

No. of points/ quadrat	CFD_o	CFD_e	$\lvert CFD_o - CFD_e \rvert$	No. of points/ quadrat	CFD_o	CFD_e	$\lvert CFD_o - CFD_e \rvert$
0	0.40	0.36788	0.03212	6	1.00	0.99992	0.00008
1	0.72	0.73576	0.01576	7	1.00	0.99999	0.00001
2	0.88	0.91970	0.03970	8	1.00	1.00000	0.00000
3	1.00	0.98101	0.01899	9	1.00	1.00000	0.00000
4	1.00	0.99634	0.00366	10	1.00	1.00000	0.00000
5	1.00	0.99941	0.00059				

(Note: * identifies the K-S statistic for this problem.) Since n = 25, for α = 0.05 the critical value of the K-S statistic is $1.36/\sqrt{25} = 0.272$. Consequently the statistical conclusions remain the same as those stated in Example 10.3.

This example emphatically stresses the degree of detail present in tails that K-S can handle but that chi square misses through consolidation.

Finally, consider the normal frequency distribution analyzed in Example PR-10.2.

EXAMPLE PR-10.4. Example PR-10.2 revisited.

Interval boundaries	CFD_o	CFD_e	$\lvert CFD_o - CFD_e \rvert$
60.09 - ∞	0.02740	0.02280	0.00460
48.13 - 60.09	0.15068	0.15870	0.00802
36.17 - 48.13	0.41096	0.50000	0.08904*
24.21 - 36.17	0.83562	0.84130	0.00568
12.25 - 24.21	0.97260	0.97720	0.00460
-∞ - 12.25	1.00000	1.00000	0.00000

(Note: * identifies the K-S statistic for this problem.) Since $n = 73$, for $\alpha = 0.05$ the critical value of the K-S statistic is $1.36/\sqrt{73} = 0.159$. Consequently the statistical conclusions remain the same as those stated in Example PR-10.2.

EXAMPLE OR-10.2. Example OR-10.1 revisited.

Interval boundaries	CFD_o	CFD_e	$\lvert CFD_o - CFD_e \rvert$
123.81 - ∞	0.0357	0.0228	0.0129
105.03 - 123.81	0.1786	0.1587	0.0199
86.25 - 105.03	0.5715	0.5000	0.0715*
67.47 - 86.25	0.8572	0.8413	0.0159
48.69 - 67.47	0.9643	0.9772	0.0129
-∞ - 48.69	1.0000	1.0000	0.0000

(Note: * identifies the K-S statistic for this problem.) Since $n = 28$, for $\alpha = 0.05$ the critical value of the K-S statistic is $1.36/\sqrt{28} = 0.257$. Consequently the statistical conclusions remain the same as those stated in Example PR-10.2.

These results illustrate two interesting features of statistical analysis. First, each statistic is tuned to nuances and idiosyncracies in sample data. In part these subtleties include aggregate versus disaggregate tail variation, as is apparent from a comparison of the individual class values of the K-S and chi square statistics in the foregoing examples. Second, the chi square and the K-S test statistics for any of these examples are not independent, since the same sample data are used to compute both of them (separate samples selected from the same parent population would need to be drawn and a chi square statistic calculated for one of them and a K-S statistic calculated for the remaining sample, to have independence). Therefore, joint discussion of the chi square and K-S results does not involve a 5% level of significance, even though each result separately did involve this 5% level.

10.3. Contingency table analysis

The notion of contingency table analysis was presented in Chapter 4, followed by a discussion of its relationship to conditional probabilities in Chapter 6 (see Example PR-6.2). A **contingency table** is an arrangement of data into a cross-classification scheme using two categorical variables. Usually the data are frequency counts. The inferential question to be addressed asks whether or not deviations between cell expectations based upon the null hypothesis statement of independent classification variables is due to sampling error. This is an applied context in which the chi square test statistic is most commonly found.

Assumptions:

(a) there must be at least two mutually exclusive and collectively exhaustive categories for each of the two classification variables;
(b) the observations are independent; and,
(c) the expected frequency e_i must satisfy applicable restrictions (see Section 10.2), or corrective action must be taken in order to ensure that specification error remains negligible (standard correction adjustments have been developed, and should be resorted to).

Test criteria:

(a) for chi square analysis tests only a one-tailed critical region is to be used; and,
(b) the test statistic is approximately distributed as χ^2 with (r - 1)(c - 1) = (rc - r - c + 1) degrees of freedom, where r is the number of row categories and c is the number of column categories (there are rc cross-classifications free to vary; once the marginal totals have been listed, each row and each column has a constraint placed upon it; since both the row and column totals equal the sample size, this constraint is counted twice, and so one degree of freedom must be added back).

The following is an example of chi square contingency table analysis, where a sample has been drawn from the Puerto Rico data set.

EXAMPLE PR-10.5. Test the independence of the coastal lowlands/-interior highlands versus the rural/urban binary classification variables.

H_0: the classification of a municipio as coastal lowland or interior highland is independent of its classification as either rural or urban,

which is equivalent to saying that H_0: $\chi^2 = 0$.

H_a: the classification of a municipio as coastal lowland or interior highland is not independent of its classification as either rural or urban,

which is equivalent to saying that H_a: $\chi^2 > 0$.

A random sample of 30 municipios was drawn from the Puerto Rico data set. This sample produced the following contingency table:

	Rural	Urban	Total
Coastal lowlands	6 (6.5)	7 (6.5)	13
Interior highlands	9 (8.5)	8 (8.5)	17
Total	15	15	30

Recall that under the null hypothesis of independence, the joint probability of occurrence in a cross-classification cell equals the product of the corresponding row and column marginal probabilities [e. g., P(coastal lowlands and rural) = P(coastal lowlands)•P(rural) = (13/30)(15/30)]. The expected value of each cell (reported in the parentheses in each cell) then is computed by multiplying these joint probabilities (which in fact are relative frequencies) by the sample size n [e. g., coastal lowlands and rural = 30(13/30)(15/30) = 6.5]. The number of degrees of freedom for this example equals (2 - 1)(2 - 1) = 1. The approximate chi square test statistic value is (the restriction of $e_i > 5$ is met in all cells)

$$(6 - 6.5)^2/6.5 + (7 - 6.5)^2/6.5 + (9 - 8.5)^2/8.5 + (8 - 8.5)^2/8.5 = 0.136.$$

The critical value here for a 5% level of significance (from Table 10.1-A) is 3.8415. Therefore, one would conclude that (1) H_0 fails to be rejected, (2) there is no evidence to indicate that the two classification variables are not independent in the population, and (3) 95% of the time one would expect to obtain a chi square value that falls within the confidence interval [0, 3.8415) when in fact the two classification variables are independent in the population.

The sample chi square is sufficiently less than the critical value that calculation of an exact chi square test statistic does not seem worthwhile. This example appears to illustrate a Type II error, however, since the population parameter value of χ^2 = 4.360 for the complete set of 73 municipios (remember that the presence of spatial autocorrelation would mean that the observations are not independent). In a classical context, then, the difference between 4.360 and 0.136 would be attributable to sampling error. With regard to degrees of freedom, once the following arbitrary assignment is made, all other cell entries are determined as follows. With 6 as the first entry in the coastal lowlands row, a row sum of 13 requires that the second entry be 7. With 6 as the first entry in the rural column, a column of 15 requires that the second entry be 9. Finally, using either the second row or column, the difference between the first element and the marginal total requires that the second element be 9.

	Rural	Urban	Total
Coastal lowlands	6	13 - 6 = 7	13
Interior highlands	15 - 6 = 9	17 - 9 = 15 - 7 = 8	17
Total	15	15	30

So, as determined previously, there can be only 1 degree of freedom associated with this problem.

EXAMPLE CA-10.1. Test the independence of the location (PC1) and the primary product code (PPC) classification variables from the Canadian Automotive data set. See Example CA-6.1.

H_0: the classification of an auto parts firm by location is independent of its classification by primary product, which is equivalent to saying that H_0: $\chi^2 = 0$.

H_a: the classification of an auto parts firm by location is not independent of its classification by primary product,

which is equivalent to saying that H_a: $\chi^2 > 0$.

The 53 firms in the CA data set produced the following contingency table from Example CA-6.1.

Primary product (PPC)

Location (PC1)	X	Y	Z	Total
K	0 (0.7)	0 (0.6)	2 (0.7)	2
L	6 (4.1)	3 (3.6)	3 (4.3)	12
M	5 (4.4)	5 (3.9)	3 (4.7)	13
N	7 (8.8)	8 (7.9)	11 (9.3)	26
Total	18	16	19	53

Once again, recall that under the null hypothesis of independence, the joint probability of occurrence in a cross-classification cell equals the product of the corresponding row and column marginal probabilities. The expected value of each cell (reported in the parentheses in each cell) then is computed by multiplying these joint probabilities by the sample size n = 53. The approximate chi square test statistic value is not calculated, since the restriction of $e_i > 5$ is not met in all cells. Instead, only the exact chi square value is calculated, since the relevant, less restrictive, assumption ($|o_i - e_i| < e_i$) is violated in only two cells of the first row. Therefore, the first row combined with the second row to produce the following table:

There are (r - 1)(c - 1) = (3 - 1)(3 - 1) = 4 degrees of freedom in this problem. Note that combining the first two rows cost two degrees of freedom

Primary product (PPC)

Location (PC1)	X	Y	Z	Total
K & L	6 (4.8)	3 (4.2)	5 (5.0)	14
M	5 (4.4)	5 (3.9)	3 (4.7)	13
N	7 (8.8)	8 (7.9)	11 (9.3)	26
Total	18	16	19	53

and that every cell still satisfies the assumption that ($|o_i - e_i| < e_i$). The critical value here for a 5% level of significance (from Table 10.1-A) is 9.4877.

$$\begin{aligned}\chi^2 = {} & 2[6 \ln(6/4.8) + 3 \ln(3/4.2) + 5 \ln(5/5.0) \\ & + 5 \ln(5/4.4) + 5 \ln(5/3.9) + 3 \ln(3/4.7) \\ & + 7 \ln(7/8.8) + 8 \ln(8/7.9) + 11 \ln(11/9.3)] \\ = {} & 2.4190\end{aligned}$$

Since 2.4190 < 9.4877 one would conclude that (1) H_0 fails to be rejected, (2) there is no evidence to indicate that the two classification variables are not independent in the population, and (3) 95% of the time one would expect to obtain a chi square value that falls within the confidence interval [0, 9.4877) when in fact the two classification variables are independent in the population.

10.4. Inference about a population mean

Various descriptive statistics were surveyed in Chapters 4, 5, and 6 for which simple hypothesis testing can be performed. The first to be considered here pertains to measures of central tendency, namely, the population arithmetic mean. The form of the central limit theorem presented in Chapter 9 is for this situation, specifying (1) the relevant standard error and (2) the need for a normality assumption. Both these features will be exploited here. The pair of hypotheses most often employed is

H_0: $\mu = a$
H_a: $\mu \neq a$,

where a is the anticipated value of the population arithmetic mean.

Assumptions:

(a) for small samples (n < 30) the population should be normally distributed [this assumption is not required for large samples (n > 100)];
(b) the population standard deviation is known;
(c) the sample is drawn randomly from the population; and,
(d) the observations are independent.

Test criteria:

(a) the standard normal distribution is used, where $z = (\bar{x} - \mu)/(\sigma/\sqrt{n})$; or
(b) although the standard normal distribution is appropriate, for small n (perhaps < 30), some researchers revert to the t distribution in accommodate the increase in variability due to small sample size.

A random sample of size 5 was drawn from the Puerto Rico data set (with replacement, order important) for illustrative purposes here and in subsequent sections of this chapter. This sample is as follows:

Municipio	R/U	DMLK	DSGR	Municipio	R/U	DMLK	DSGR
Guanica	1	4.4	9.9	Guanica	1	4.4	9.9
Las Piedras	0	60.0	0.9	Carolina	1	43.0	4.0
Coamo	0	10.9	0.0				

EXAMPLE PR-10.6 Inference about the population mean for the density of milk production (DMLK) in the Puerto Rico data set, where $\sigma = 49.4879$ (calculated from Table 1.1-A).

From the sample of 5 municipios listed, $\bar{x}_{DMLK} = 24.54$. Using the population mean calculated from Table 1.1-A, if one hypothesizes

H_0: $\mu = 27.0863$ (calculated from Table 1.1-A)
H_a: $\mu \neq 27.0863$,

then $z = (24.5400 - 27.0863)/(49.4879/\sqrt{5}) = -0.1151$. For a 5% level of significance, the critical values are $\pm$ 1.96 (see Table 7.2-A). Therefore, one would conclude that (1) H_0 fails to be rejected, (2) there is no evidence to indicate that the mean does not equal 27.0863 in the population, and (3) 95% of the time one would expect to obtain a z-score that falls within the confidence interval (-1.96, 1.96) when in fact the population mean equals 27.0863.

A two-tail test was used here because the researcher often does not care in which direction the sample deviation occurs; if for theoretical or conceptual reasons this direction becomes important, then a one-tail test should be utilized.

In most situations knowledge of the population variation but not the population mean is unlikely. In this case assumption (b) from the preceding situation is dropped, and the following are used.

Test criteria:

(a) the Student's t distribution is used, where $t = (\bar{x} - \mu)/(s/\sqrt{n})$; and
(b) σ is replaced with s in the standard error formula.

Because an estimate is used to compute the standard error, it places a constraint on the analysis, and so degrees of freedom become relevant again. Since only one constraint has been invoked, the number of degrees of freedom becomes $\upsilon = n - 1$.

EXAMPLE PR-10.7. Inference about the population mean for the density of sugar cane production in the Puerto Rico data set, where σ is unknown.

From the sample constructed in the previous example, $\bar{x}_{DSGR} = 4.9400$ and $s = 4.7648$. If a researcher does not believe that sugar cane is produced in Puerto Rico and hence hypothesizes

$$H_0: \mu = 0$$
$$H_a: \mu > 0,$$

then $t = (4.9400 - 0)/(4.7648/\sqrt{5}) = 2.3183$. For a 5% level of significance (see Table 9.1-A), the critical value with 4 degrees of freedom is 2.1318. Therefore, the conclusion is that (1) reject H_0, (2) there is evidence to indicate that the mean does not equal 0 in the population, and (3) only 5% of the time would one expect to obtain a t score that falls within the critical region $(2.1318, \infty)$ when in fact the population mean equals 0.

A one-tail test was employed here because it is physically impossible to have negative agricultural productivity. Note that if a 1% level of significance was used in this example, the appropriate critical value would be 3.747, and hence a Type II error would be committed (from Table 1.1-A, μ is calculated to be 6.0452).

As a brief but important digression, one can ask what happens when the assumption of independent observations is violated. Consider the geographic distribution of coffee production density (DCF) in the Mayaguez Agricultural Administrative Region. Here $\bar{x} = 9745.5$, and because the population variance is assumed to be unknown for problems treated in this section, the test statistic is $t = 2.93$, which is significantly greater than zero. This particular geographic distribution exhibits a moderate level of positive spatial autocorrelation; presentation of procedures for calculating this statistic are beyond the scope of this book. Taking into account this prevailing level of spatial autocorrelation reduces the test statistic to t = 1.37, which is not significantly greater than zero, implying that the presence of nonzero spatial autocorrelation has resulted in an incorrect statistical decision. In other words, the violation of assumption (d) in spatial data series at least tends to affect the probability of Type I errors. This source of specification error means that the standard table critical values are incorrect if the data are spatially autocorrelated.

10.5. Inference about a population variance

The second descriptive statistic considered here pertains to measures of dispersion, namely, the population variance (the standard deviation could be treated, but it is customary to talk about the variance instead). Many times problems arise that require an inference about such variability. For instance, an environmental engineer building a dam needs to consider the variance in water runoff. If the variance is too large and ignored, then the dam will (1) tend to overflow or be ineffective as a flood control device, or (2) without proper planning be underfilled and ineffective as a recreational resource. A central limit type of theorem for this situation specifies that the sampling distribution conforms to a chi square distribution.

Assumptions:

(a) the random variable in the population conforms to a normal distribution;
(b) the population has a finite variance;
(c) the sample is drawn randomly from the population; and,
(d) the observations are independent.

Test criteria:

(a) the one- or two-tail chi square distribution is used;
(b) the sampling distribution for the quantity $(n - 1)s^2/\sigma^2$ conforms to a chi square distribution with (n - 1) degrees of freedom.

Since the chi square distribution is not symmetrical, individual tail critical values must be separately determined. The mean of this sampling distribution is (n - 1), whereas the standard error is $[2(n - 1)]^{1/2}$; these properties may be exploited for untabulated degrees-of-freedom situations.

EXAMPLE PR-10.8. Inference about the population variance for density of milk production in the Puerto Rico data set.

Given the sample listed in Section 10.4, $s_{DMLK}^2 = (25.4729)^2$. If one hypothesizes that

$$H_0: \sigma = 49.4879 \text{ (calculated from Table 1.1-A)}$$
$$H_a: \sigma \neq 49.4879,$$

then $\chi^2 = (5 - 1)(25.4729)^2/(49.4879)^2 = 1.0598$. For a 5% level of significance (see Table 10.1-A), the critical values with 4 degrees of freedom are 0.4844 and 11.1433. Therefore, one would conclude that (1) H_0 fails to be rejected, (2) there is no evidence to indicate that the variance does not equal 49.4879 in the population, and (3) 95% of the time one would expect to obtain a chi square value that falls within the confidence interval (0.4844, 11.1433) when in fact the population variance equals 49.4879.

A two-tail test would be appropriate here if one wanted to implement an agricultural production policy for sugarcane in which target production levels control for variability; the mean amount being on target would not ensure that the policy is working correctly throughout the island. The two-tail situation also leads to the construction of confidence intervals for the population variance, whose formula is given by

$$s^2[(n - 1)/\chi_{\upsilon,1-\alpha/2}^2] \leq \sigma^2 \leq s^2[(n - 1)/\chi_{\upsilon,\alpha/2}^2].$$

EXAMPLE PR-10.8, continued. Construct a confidence interval for the sample variance for density of milk production.

$$(25.4729)^2[(5 - 1)/11.1433] \leq \sigma^2 \leq (25.4729)^2[(5 - 1)/0.4844]$$

$$232.9180 \leq \sigma^2 \leq 5358.1225 \text{ with probability } 0.95.$$

Indeed, 2449.0522 (which is 49.4879^2) lies within this interval. If a confidence interval for the standard deviation is wanted, then since all quantities involved must be nonnegative, one needs only to take the square root of each of the three quantities in this inequality expression.

10.6. Inference about a population correlation coefficient

The third descriptive statistic pertains to measures of association, namely, the population linear correlation coefficient. In most studies, once the value of a correlation coefficient has been calculated from sample data, a necessary question to consider asks whether or not the value of this statistic indicates that there is a relationship between the two variables in question in the population from which the sample has been drawn. This is one of the most often asked questions in quantitative analyses. As was indicated in Chapter 4, regardless of the level of measurement, all correlation coefficients can be calculated with the product moment formula. The phi, point biserial, and Spearman's rank are special reductions of the product moment formula. But a central limit type of theorem for this situation requires knowledge of the measurement scale; in general it specifies that the approximate sampling distribution conforms to a t distribution, whereas exact sampling distributions are measurement-scale specific.

Assumptions:

(a) the two variables conform to a particular bivariate distribution;
(b) there are at least 3, preferably at least 5, and more favorably at least 20 observations;
(c) the sample pairs are drawn randomly from the population; and,
(d) the observations are independent.

The sample size n must be at least 3, since if n = 2 the correlation coefficient always will be unity (one can always draw a straight line through two points).

Test criterion:

(a) the sampling distribution for the quantity $(r - 0)/[(1 - r^2)/(n - 2)]^{1/2}$ can be approximated by a t distribution with (n - 2) degrees of freedom.

Three points are noteworthy here. First, this test statistic is based upon a variance distribution, and as such in some situations is only an approximation that may embrace noticeable specification error. Second, the standard error of r equals $[(1 - \rho^2)/(n - 2)]^{1/2}$; but if ρ were known, there would be no statistical inference problem, and so r must be substituted for ρ in the standard error formula. Third, because for situations when $\rho \neq 0$ the sampling distribution becomes highly skewed, some confidence intervals will have bounds that fall outside of the closed interval [-1, 1]; selected ad hoc solutions to this problem will be outlined shortly.

EXAMPLE PR-10.9. Inference about selected population correlation coefficients for the Puerto Rico data set.

Given the random sample of size n = 5 listed in Section 10.4, where r(DMLK, DSGR) = -0.6036, r_s(DMLK, DSGR) = -0.6842, and r_{pbi}(R/U, DMLK) = -0.3910. If one hypothesizes in each of these cases that

$$H_0: \rho = 0$$
$$H_a: \rho \neq 0,$$

then the respective test statistics are $t = -0.6036/\{[1 - (-0.6036)^2]/(5 - 2)\}^{1/2} = -1.3111$, $t = -0.6842/\{[1 - (-0.6842)^2]/(5 - 2)\}^{1/2} = -1.6250$, and finally $t = -0.3910/\{[1 - (-0.3910)^2]/(5 - 2)\}^{1/2} = -0.7358$. For a 5% level of significance (see Table 9.1-A), the critical values with 3 degrees of freedom are ± 3.1825. Therefore, the conclusion in each of these three cases is that (1) H_0 fails to be rejected, (2) there is no evidence to indicate that the two variables in question are not independent in the population (recall that a value of 0 indicates independence), and (3) 95% of the time one would expect to obtain a t statistic that falls within the confidence interval (-3,1825, 3.1825) when in fact the two variables are independent in the population.

A two-tail test would be appropriate here because one would be indifferent as to whether an existing relationship were direct or indirect. The two-tail situation also leads to the construction of confidence intervals for the population correlation coefficient, whose formula is given by

$$r - t_{n-2,\alpha/2}[(1 - r^2)/(n - 2)]^{1/2} \leq \rho \leq r + t_{n-2,\alpha/2}[(1 - r^2)/(n - 2)]^{1/2}.$$

One rule of thumb applied here when confidence interval bounds fall outside of the interval [-1, 1] is to round to the appropriate correlation limit (i. e., -1 or 1).

EXAMPLE PR-10.9, continued. Construct a confidence interval for the sample point-biserial correlation coefficient for R/U and DMLK.

$$-0.3910 - 3.1825\{[1 - (-0.3910)^2]/(5 - 2)\}^{1/2} \leq \rho \leq -0.3910 + 3.1825\{[1 - (-0.3910)^2]/(5 - 2)\}^{1/2}$$

$$-2.0821 \leq \rho \leq 1.3002 \text{ with probability 0.95, or}$$

$$-1 \leq \rho \leq 1 \text{ with probability 0.95.}$$

Indeed, 0 lies within this interval; this result is consistent with that for the significance test. But realize that the small-sample variability is so overwhelming here that a guess without sample information would be just as good (by definition a correlation coefficient must lie between -1 and 1 with probability 1!).

What can be done when specification error renders a confidence interval bound that is meaningless? The primary assumption that is being employed for the general standard error of r previously computed is that either the two random variables or the errors affiliated with these random variables in the population conform to a normal distribution. Certain supplemental assumptions can be posited in order to remove some of the undesirable features of this standard error estimator; these assumptions are correlation coefficient (or measurement-scale) specific. First, if a phi coefficient is being calculated, then

Test criterion:

(a) the sampling distribution for the quantity $n\phi^2$ can be approximated by a χ^2 distribution with 1 degree of freedom ($n\phi^2$ reduces to the approximate χ^2).

Second, if a Spearman's rank coefficient is being calculated, especially when the ranks are distinct (i. e., no ties), then

Test criteria:

(a) the standard normal distribution is used, where $z = (r_s - 0)/[1/(n - 1)^{1/2}]$; or

(b) although the standard normal distribution is appropriate, for small n (perhaps < 30), some researchers revert to the t distribution in order to accommodate the increase in variability due to small sample size.

Note that the standard error presented here does not require the estimation of any quantity. Third, if a product moment correlation coefficient is being calculated, then

Test criteria:

(a) the sampling distribution of the transformed variable $0.5\ln[(1 + r)/(1 - r)]$ displays desired statistical properties (see Chapter 2) with respect to the population parameter ρ;

(b) the standard normal distribution is used, where $z = \{0.5\ln[(1 + r)/(1 - r)] - 0\}/[1/(n - 3)^{1/2}]$; or

(c) although the standard normal distribution is appropriate, for small n (perhaps < 30), some researchers revert to the t distribution in order to accommodate the increase in variability due to small sample size.

Note that the standard error presented here also does not require the estimation of any quantity.

EXAMPLE PR-10.10. Comparing the general and specific test criteria; Examples PR-10.5 and PR-10.9 revisited.

In Example PR-10.5 an approximate chi square statistic was computed, its value being 0.1357. The corresponding phi coefficient for this problem is the negative root of $[0.1357/30]^{1/2}$, which is -0.0673. The chi square test statistic for the phi coefficient is $30(-0.0673)^2 = 0.1359$ (the difference here is due to rounding error). A confidence interval for this phi coefficient is given by

$$P(n\phi^2 \leq \chi^2_{\alpha/2,1}) = 0.95, \text{ or } P[-(\chi^2_{1-\alpha,1}/n)^{1/2} \leq \phi \leq (\chi^2_{1-\alpha,1}/n)^{1/2}] = 0.95$$

which in this case gives the 95% confidence interval of

$$-0.3578 \leq \phi \leq 0.3578.$$

The general approximation 95% confidence interval counterpart, with the critical value $\pm t_{.025,28}$, is

$$-0.0673 - 0.1890 \cdot 2.0484 \leq \phi \leq -0.0673 + 0.1890 \cdot 2.0484, \text{ or}$$
$$-0.4544 \leq \phi \leq 0.3198.$$

Meanwhile, converting the interval/ratio measures for variables DMLK and DSGR to ordinal measures results in a Spearman's rank correlation coefficient of -0.684. The approximate standard error of this coefficient equals $\{[1 - (-0.6842)^2]/(5 - 2)\}^{1/2} = 0.4211$, whereas the exact standard error equals $1/(5 - 1)^{1/2} = 0.5$ (this calculation is complicated by the presence of a tie). The respective two-tail critical values for a 5% level of significance are ±3.1825 and ±1.96 (see Tables 7.2-A and 9.1-A). Thus, the comparative confidence intervals are

$$-0.6842 - 3.1825 \cdot 0.4211 \leq \rho_s \leq -0.6842 + 3.1825 \cdot 0.4211, \text{ or}$$
$$-1 \leq \rho_s \leq 0.6560, \text{ and}$$

$$-0.6842 - 1.96 \cdot 0.5 \leq \rho_s \leq -0.6842 + 1.96 \cdot 0.5, \text{ or}$$
$$-1 \leq \rho_s \leq 0.2958.$$

Finally, the product moment correlation coefficient for variables DMLK and DSGR needs to be transformed such that $0.5 \ln[(1 - 0.6071)/(1 + 0.6071)] = -0.7043$. The standard error here is $1/(5 - 3)^{1/2} = 1/\sqrt{2}$. Accordingly, the approximate 95% confidence interval is

$$-0.6071 - 3.1825 \cdot 0.4588 \leq \rho \leq -0.6071 + 3.1824 \cdot 0.4588, \text{ or}$$
$$-1 \leq \rho \leq 0.8530.$$

In contrast, the exact 95% confidence interval is

$$-0.7043 - 1.96 \cdot 0.7071 \leq 0.5 \ln[(1 + \rho)/(1 - \rho)]$$
$$\leq -0.7043 + 1.96 \cdot 0.7071, \text{ or}$$
$$-4.1804 \leq \ln[(1 + \rho)/(1 - \rho)] \leq 1.3632.$$

This interval then can be algebraically manipulated to yield a confidence interval for ρ itself, such that

$$[\exp(-4.1804) - 1]/[\exp(-4.1804) + 1] \leq \rho \leq [\exp(1.3632) - 1]/[\exp(1.3632) + 1], \text{ or}$$
$$-0.9699 \leq \rho \leq 0.5926.$$

It becomes apparent with this example that the exact sampling distributions tend to have smaller standard errors and hence tighter confidence intervals. The respective population values for these correlation coefficients are $\phi = -0.244$, $\rho_{pbi} = 0.329$, $\rho_s = -0.134$, and $\rho = -0.206$; in all four cases both the approximate and the exact confidence intervals captured these population values.

Besides defining an exact sampling distribution, the product moment correlation coefficient transformation allows a more general hypothesis to be tested, namely,

H_0: $\rho = a$, $|a| \leq 1$
H_a: $\rho \neq a$.

The other correlation coefficient test statistics do not generalize to a null hypothesis of nonzero correlation in the population; moreover, this is one of the few situations in classical univariate statistics where a null hypothesis of nonrandomness or dependence can be evaluated. The single change to the test criterion becomes

Test criterion:

(b) the standard normal distribution is used, where $z = \{0.5\ln[(1 + r)/(1 - r)] - 0.5\ln[(1 + a)/(1 - a)]\}/[1/(n - 3)^{1/2}]$.

In other words, rather than deviations from zero, one is studying deviations from $0.5 \ln[(1 + a)/(1 - a)]$; the standard error remains the same.

EXAMPLE PR-10.11. Inference about the population product moment correlation coefficient for DMLK and DSGR in the Puerto Rico data set.

H_0: $\rho = -0.2038$
H_a: $\rho \neq -0.2038$

Returning to the sample data set for this chapter (see Section 10.4), we have $z = \{0.5 \ln[(1 - 0.6036)/(1 + 0.6036)] - 0.5 \ln[(1 - 0.2038)/(1 + 0.2038)]\}/[1/(5 - 3)^{1/2}] = (-0.6988 + 0.2067)/(0.7071) = -0.6959$. For a 5% level of significance (see Table 7.2-A), the critical values are ± 1.96. Therefore, the conclusion is that (1) H_0 fails to be rejected, (2) there is no evidence to indicate that the correlation between the two variables in question is not equal to -0.2038 in the population, and (3) 95% of the time one would expect to obtain a z-score value that falls within the confidence interval (-1.96, 1.96) when in fact the correlation between the two variables is -0.206 in the population.

10.7. Inference about spatial statistics

10.7.1. Inference about two-dimensional point patterns

Under the null hypothesis a Poisson process is assumed to generate point patterns. In this situation the sampling distribution of the nearest neighbor statistic has a mean $\mu_R = 1$ and a standard error $\sigma_R = [(4 - \pi)/(n\pi)]^{1/2}$; note that this standard error does not require the estimation of any quantity.

Assumptions:

(a) the first-order nearest neighbor distance random variable in the population conforms to one generated by a Poisson distribution of points that is homogeneous over the planar surface;
(b) the sample is drawn randomly from the population; and,
(c) the observations are independent.

Test criteria:

(a) the standard normal distribution is used, where $z = (R - 1)/[(4 - \pi)/(n\pi)]^{1/2}$; or
(b) although the standard normal distribution is appropriate, for small n (perhaps < 30), some researchers revert to the t distribution in order to accommodate the increase in variability due to small sample size.

While the statistic R has been shown to be highly sensitive to the definition of the area of a region under study (see Chapter 5), one can see here that the variance is independent of area.

EXAMPLE PR-10.12. Example PR-5.7 revisited.

H_0: the geographic pattern of municipio centroids is random (i. e., generated by an independent Poisson process),

which is equivalent to saying that H_0: $\mu_R = 1$.

H_a: the geographic pattern of municipio centroids is nonrandom (i. e., generated by something other than an independent Poisson process),

which is equivalent to saying that H_a: $\mu_R \neq 1$.

The nearest neighbor statistic for the geographic distribution of municipio centroids in the Mayaguez Agricultural Administrative Region was found to be 1.7305 (see Chapter 5). With n = 16, its standard error is $[(4 - \pi)/(16\pi)]^{1/2} = 0.1307$. Hence, z = (1.7305 - 1)/0.1307 = 5.5891. For a 5% level of significance, the critical values are ±1.96 (see Table 7.2-A). Therefore, the conclusion is that (1) reject H_0, (2) there is evidence to indicate that the geographic distribution of municipio centroids in the Mayaguez Agricultural Administrative Region is not generated by an independent Poisson process, and (3) only 5% of the time would one expect to obtain a z-score that falls within the critical region {(-∞, -1.96), (1.96, ∞)} when in fact the geographic distribution of municipio centroids in the Mayaguez Agricultural Administrative Region is generated by an independent Poisson process.

A confidence interval for this nearest neighbor statistic may be constructed as follows:

$$R - z_{1-\alpha/2}[(4 - \pi)/(16\pi)]^{1/2} \leq \mu_R \leq R + z_{1-\alpha/2}[(4 - \pi)/(16\pi)]^{1/2}$$

which for Example PR-10.12 yields $1.5272 \leq \mu_R \leq 1.9338$. Again realize that physically meaningless bounds can be rendered by the confidence interval formula (i. e., values less than 0 and greater than 2.15). Note that the formula for the variance of the sampling distribution of the nearest neighbor statistic R often is stated incorrectly in statistical geography texts. The formula presented here is correct!

In terms of quadrat analysis, the null hypothesis that a Poisson process generates a point pattern means a goodness-of-fit test may be conducted on the observed frequency distribution, as discussed in Section 10.2. An alternative approach to the testing of this hypothesis is reminiscent of the coefficient of variation. Because a Poisson distribution has $\mu = \sigma^2$, the ratio of these two parameters will be unity (i. e., $\sigma^2/\mu = 1$). A central limit type of theorem for this situation specifies that the appropriate sampling distribution conforms to a chi square distribution. The following assumption needs to be appended to the preceding three:

(d) $n > 6$ and $\bar{x} > 1$.

<u>Test criterion</u>:

(a) $(k - 1)s^2/\bar{x}$ follows a chi square distribution with (k - 1) degrees of freedom, where k is the number of quadrats.

EXAMPLE 10.6. Example 5.8 revisited, where $\mu = 1$.

H_0: the geographic pattern of the computer-generated set of points is random (i. e., generated by an independent Poisson process),

which is equivalent to saying that H_0: $\sigma^2/\mu = 1$.

H_a: the geographic pattern of the computer-generated set of points is nonrandom (i. e., generated by something other than an independent Poisson process),

which is equivalent to saying that H_a: $\sigma^2/\mu \neq 1$.

The variance/mean ratio statistic for the geographic distribution of simulated points, where k = 25 (the number of quadrats used), is 24(26/24)/(25/25) = 26. For a 5% level of significance, with k - 1 = 24 degrees of freedom, the critical chi square value is 36.4150 (see Table 10.1-A). Therefore, one would conclude that (1) H_0 fails to be rejected, (2) there is no evidence to indicate that the geographic distribution of the computer-generated set of points is not the outcome of an independent Poisson process, and (3) 95% of the time one would expect to obtain a chi square statistic that falls within the confidence [0, 36.4150) when in fact the geographic distribution of computer-generated set of points is the outcome of an independent Poisson process.

10.7.2. Inference about spatial dependency

The last spatial statistic to be discussed here is for spatial autocorrelation. Before tackling the problem of inference about spatial dependency, we pursue two useful digressions. First, recall that the Spearman's rank correlation coefficient has a sampling error that may be derived by considering all possible permutations of distinct ranks. For instance, given that n = 3, then using the sampling without replacement and order important perspective, there are (3!)(3!) = 36 possibilities. The mean of the 36 affiliated r_s values is 0, and the standard error is $1/(3 - 1)^{1/2}$. Second, an example closer to the issue of geographic configuration is provided by the "runs" statistic.

Suppose that observations can display only one of two properties (hence, they can be dichotomized). Let a run be defined as an uninterrupted sequence of data in which consecutive observations possess one of these properties, with one run ending and another beginning when an observation in the complete sample sequence switches to displaying the other property in question. Denote the number of runs in a complete sample sequence by V. Given this context, statistic V measures the degree of randomness displayed by the data sequence; it is a nonparametric statistic. This classical statistic is a one-dimensional predecessor to the two-dimensional join count statistics and may be interpreted in a way analogous to that for spatial autocorrelation. If all observations in a sequence that display one of the two properties are clustered together, then V = 2, and similar values cluster together (positive autocorrelation). If the sequence is such that the two properties alternate, then V approaches n (depending upon how many observations display each property), and dissimilar values cluster together (negative autocorrelation). Moreover, the total number of runs in a sequence is a good indicator of the degree of randomness that prevails, for it indexes the predictability or amount of systematic variation exhibited by a series.

Let n_1 be the number of observations displaying one of the two properties, and let n_2 be the number of observations displaying the other property so that $n_1 + n_2 = n$.

Assumptions:

(a) the underlying continuous random variable is measured at least at the ordinal scale;
(b) the two categories have identical, although perhaps unknown, statistical distributions in their respective populations;
(c) the sample is drawn randomly from the population; and,
(d) the observations are independent.

Test criteria:

(a) if both n_1 and n_2 are less than or equal to 20, then the exact sampling distribution critical values may be retrieved from Table 10.3-A; and,
(b) for larger n the sampling distribution of V is approximately normal, with $\mu_V = 2n_1n_2/n + 1$ and $\sigma_V = \{2n_1n_2(2n_1n_2 - n_1 - n_2)/[n^2(n - 1)]\}^{1/2}$, meaning that $z = (V - \mu_V)/\sigma_V$.

Note that although the variable in question is not assumed to be normally distributed in its parent population, the sampling distribution of V will be approximately normal even when n is relatively small. Also, note that the standard error presented here does not involve the estimation of any quantity. The dichotomizing criterion may well affect a result, mostly because of the crudity of measuring instruments.

The theory upon which the sampling distribution for V is based employs the sampling perspective of without replacement, order important. Accordingly, the total number of possibilities is $n!/(n_1!n_2!)$; the mean of these V values is $\mu_V = 2n_1n_2/n + 1$ and the standard error of these V values is, given that $n_1 + n_2 = n$, $\sigma_V = \{2n_1n_2(2n_1n_2 - n)/[n^2(n - 1)]\}^{1/2}$. These formulae are derived from the randomization approach; they do not apply to the situation where each observation could take on either of the two properties (yielding 2^n possible samples).

EXAMPLE 10.7. The runs statistic for the first row of uniform random numbers in Table 7.1-A (remember that $0 \equiv 10$).

H_0: the linear sequence of digits is random in the population,

which is equivalent to saying that H_0: $\mu_V = 2n_1n_2/n + 1$.

H_a: the linear sequence of digits is not random in the population,

which is equivalent to saying that H_a: $\mu_V \neq 2n_1n_2/n + 1$.

Suppose that the criterion is "below the mean of 5.5 is -" and "above the mean of 5.5 is +." Then the sequence is

++-+---+++--+---+---+++++-++-++-++----+---++----+-++-+----++

and $V = 29$. Here $n_1 = 32$ (number of -) and $n_2 = 28$ (number of +), $\mu_V = 2 \cdot 32 \cdot 28/60 + 1 = 30.8667$, and $\sigma_V = [2 \cdot 32 \cdot 28 \cdot (2 \cdot 32 \cdot 28 - 60)/(60^2 \cdot 59)]^{1/2} = 3.8227$. Thus, $z = (29 - 30.8667)/3.8227 = -0.4883$. For a 5% level of significance (see Table 7.2-A), the critical values are ± 1.96. Therefore, the conclusion is that (1) H_0 fails to be rejected, (2) there is no evidence to indicate that the sequence of 60 single digits is not random in the population, and (3) 95% of the time one would expect to obtain a z-score value that falls within the confidence interval (-1.96, 1.96) when in fact the sequence of 60 single digits is random in the population.

Next, suppose that the criterion is "even is -" and "odd is +." Then the sequence is

++++-++-+---+-+---++-+-+++-+-++--++-++++++--+-+++++----+++--

and $V = 30$. Here $n_1 = 25$ (number of -) and $n_2 = 35$ (number of +), $\mu_V = 2 \cdot 25 \cdot 35/60 + 1 = 30.1667$, and $\sigma_V = [2 \cdot 25 \cdot 35 \cdot (2 \cdot 25 \cdot 35 - 60)/(60^2 \cdot 59)]^{1/2} = 3.7315$. Thus, $z = (30 - 30.1667)/3.7315 = -0.0447$. Therefore, with this second classification criterion one would draw the same conclusions as before.

These two example criteria demonstrate the possibility of even subtly different computational results.

Some insight into the potential small-sample specification error level is obtained by comparing results from the normal approximation with those of the tabulated critical values (see Table 10.3-A). Suppose that $n_1 = 3$ and $n_2 = 6$. Then $\mu_V = 5$ and $\sigma_V = (3/2)^{1/2}$. Along the lines of confidence intervals, the critical values can be computed (since both the mean and the standard error are known simply from the subsample sizes) such that

$$\pm 1.96 = (V - 5)/1.2247 \Rightarrow 2.6 \leq V \leq 7.4$$

which suggests that the pair of critical values should be 3 and 7; the table values are 2 and 8. In other words, if the normal approximation is used for this sample of size 9, both critical values are off by an additive factor of 1.

The two preceding illustrations of sampling distribution construction via randomization show that this method is respected and employed in classical statistics and counters to some degree the arguments put forth by Summerfield against its use in spatial data analysis (see Chapter 8). It forms one of the bases for sampling distribution construction for the spatial autocorrelation indices presented in Chapter 5 (the join count statistics, Moran Coefficient, and Geary Ratio). The expected values, given a randomization perspective, of these statistics are given in that chapter; as review,

$$\begin{aligned}
E(WW) &= (\textstyle\sum_{i=1}^{n} \sum_{j=1}^{n} c_{ij})n_1(n_1 - 1)/[2n(n - 1)] \\
E(BB) &= (\textstyle\sum_{i=1}^{n} \sum_{j=1}^{n} c_{ij})n_2(n_2 - 1)/[2n(n - 1)] \\
E(BW) &= (\textstyle\sum_{i=1}^{n} \sum_{j=1}^{n} c_{ij})n_2 n_1/[n(n - 1)] \\
E(MC) &= -1/(n - 1), \text{ and} \\
E(GR) &= 1.
\end{aligned}$$

Using a central limit type of theorem for the join count statistics WW and BB specifies that the appropriate sampling distribution conforms to a normal distribution, with the following standard errors:

$$\begin{aligned}
\sigma_{WW} = \Big[&\{n_1 n_2(n_1 - 1)/[4n(n - 1)(n - 2)(n - 3)]\}\{2(n_2 - 1)(\textstyle\sum_{i=1}^{n} \sum_{j=1}^{n} c_{ij}) \\
&+ 4(n_1 - 2)[\textstyle\sum_{i=1}^{n} (\sum_{j=1}^{n} c_{ij})^2] \\
&+ 2(3n + 3n_1 - 2nn_1 - 3)(\textstyle\sum_{i=1}^{n} \sum_{j=1}^{n} c_{ij})^2/[n(n - 1)]\}\Big]^{1/2}, \text{ and}
\end{aligned}$$

$$\begin{aligned}
\sigma_{BB} = \Big[&\{n_2 n_1(n_2 - 1)/[4n(n - 1)(n - 2)(n - 3)]\}\{2(n_1 - 1)(\textstyle\sum_{i=1}^{n} \sum_{j=1}^{n} c_{ij}) \\
&+ 4(n_2 - 2)[\textstyle\sum_{i=1}^{n} (\sum_{j=1}^{n} c_{ij})^2] \\
&+ 2(3n + 3n_2 - 2nn_2 - 3)(\textstyle\sum_{i=1}^{n} \sum_{j=1}^{n} c_{ij})^2/[n(n - 1)]\}\Big]^{1/2}
\end{aligned}$$

In addition, for the BW statistic

$$\sigma_{BW} = \Big[\{n_1 n_2/[n(n - 1)(n - 2)(n - 3)]\}\{2(n_1 - 1)(n_2 - 1)(\textstyle\sum_{i=1}^{n} \sum_{j=1}^{n} c_{ij})$$

$$+ [n^2 - n + 2 - 4n_1n_2][\sum_{i=1}^{n} (\sum_{j=1}^{n} c_{ij})^2] + [2n_1n_2(2n - 3) - n(n - 1)^2](\sum_{i=1}^{n} \sum_{j=1}^{n} c_{ij})^2/[n(n - 1)]\}]^{1/2}.$$

Meanwhile, the Moran Coefficient has a standard error, based upon this randomization perspective, of

$$\sigma_{MC} = \{[2n(n - 1)\{n^2 - 3n + 3 - n(n - 1)[\sum_{i=1}^{n} (x_i - \bar{x})^4]/[\sum_{i=1}^{n} (x_i - \bar{x})^2]^2\}(\sum_{i=1}^{n} \sum_{j=1}^{n} c_{ij}) - 4n^2(n - 1)\{1 - 2[\sum_{i=1}^{n} (x_i - \bar{x})^4]/[\sum_{i=1}^{n} (x_i - \bar{x})^2]^2\}[\sum_{i=1}^{n} (\sum_{j=1}^{n} c_{ij})^2] + 2\{n^2 + n - 3 - 3n(n - 1)[\sum_{i=1}^{n} (x_i - \bar{x})^4]/[\sum_{i=1}^{n} (x_i - \bar{x})^2]^2\}(\sum_{i=1}^{n} \sum_{j=1}^{n} c_{ij})^2]/[(n - 1)^2(n - 2) \times(n - 3)(\sum_{i=1}^{n} \sum_{j=1}^{n} c_{ij})^2]\}^{1/2}$$

whereas the Geary Ratio has a standard error, based upon this perspective, of

$$\sigma_{GR} = \{[2(n - 1)\{n^2 - 3n + 3 - n(n - 1)[\sum_{i=1}^{n} (x_i - \bar{x})^4]/[\sum_{i=1}^{n} (x_i - \bar{x})^2]^2\}(\sum_{i=1}^{n} \sum_{j=1}^{n} c_{ij}) - (n - 1)\{n^2 + 3n - 6 - n(n^2 - n + 2)[\sum_{i=1}^{n} (x_i - \bar{x})^4]/[\sum_{i=1}^{n} (x_i - \bar{x})^2]^2\} \times[\sum_{i=1}^{n} (\sum_{j=1}^{n} c_{ij})^2] + \{n^2 - 3 - n(n - 1)^2[\sum_{i=1}^{n} (x_i - \bar{x})^4]/[\sum_{i=1}^{n} (x_i - \bar{x})^2]^2\} \times(\sum_{i=1}^{n} \sum_{j=1}^{n} c_{ij})^2]/[n(n - 2)(n - 3)(\sum_{i=1}^{n} \sum_{j=1}^{n} c_{ij})^2]\}^{1/2}.$$

As can be seen, all these standard errors are far more complicated in their computations than are those for traditional statistics, and none of them involves the estimation of any quantity. Simpler versions exist, but they assume normality; these will not be covered in this text.

Assumptions:

(a) the underlying random variable is measured at the nominal scale for the join count statistic, and at the interval/ratio scale for MC and GR;
(b) the statistical distribution of the variate is unknown;
(c) the sample is drawn randomly from the population, and is of at least size 4 (with n_1 and $n_2 \geq 2$); and,
(d) the observations are independent.

Test criteria:

(a) the standard normal distribution is used, where $z = (WW - \mu_{WW})/\sigma_{WW}$, $z = (BB - \mu_{BB})/\sigma_{BB}$, $z = (BW - \mu_{BW})/\sigma_{BW}$, $z = (MC - \mu_{MC})/\sigma_{MC}$, and $z = (GR - \mu_{GR})/\sigma_{GR}$; or
(b) although the standard normal distribution is appropriate, for small n (perhaps < 30), some researchers revert to the t distribution in order to accommodate the increase in variability due to small sample size.

EXAMPLE PR-10.13. Example PR-5.8 revisited.

H_0: the geographic distribution of urban propensity in the Mayaguez Agricultural Administrative Region is random,

which is equivalent to saying

$$H_0\colon \mu_{WW} = (\textstyle\sum_{i=1}^{n} \sum_{j=1}^{n} c_{ij})n_1(n_1 - 1)/[2n(n - 1)], \text{ or}$$
$$H_0\colon \mu_{BB} = (\textstyle\sum_{i=1}^{n} \sum_{j=1}^{n} c_{ij})n_2(n_2 - 1)/[2n(n - 1)], \text{ or}$$
$$H_0\colon \mu_{BW} = (\textstyle\sum_{i=1}^{n} \sum_{j=1}^{n} c_{ij})n_1n_2/[n(n - 1)].$$

H_a: the geographic distribution of urban propensity in the Mayaguez Agricultural Administrative Region is not random,

which is equivalent to saying

$$H_a\colon \mu_{WW} \neq (\textstyle\sum_{i=1}^{n} \sum_{j=1}^{n} c_{ij})n_1(n_1 - 1)/[2n(n - 1)], \text{ or}$$
$$H_a\colon \mu_{BB} \neq (\textstyle\sum_{i=1}^{n} \sum_{j=1}^{n} c_{ij})n_2(n_2 - 1)/[2n(n - 1)], \text{ or}$$
$$H_a\colon \mu_{BW} \neq (\textstyle\sum_{i=1}^{n} \sum_{j=1}^{n} c_{ij})n_1n_2/[n(n - 1)].$$

For this example $n = 16$, $\sum_{i=1}^{n} \sum_{j=1}^{n} c_{ij} = 64$, and $\sum_{i=1}^{n} (\sum_{j=1}^{n} c_{ij})^2 = 280$. For the join count statistics, $n_1 = 4$, $n_2 = 12$, $E(WW) = 1.6$, $E(BB) = 12.8$, and $E(BW) = 17.6$. Now,

$$\sigma_{WW} = [\{4 \bullet 12 \bullet (4 - 1)/[4 \bullet 16 \bullet (16 - 1)(16 - 2)(16 - 3)]\}\{2 \bullet (12 - 1) \bullet 64 + 4 \bullet (4 - 2) \bullet 280 + 2 \bullet (3 \bullet 16 + 3 \bullet 4 - 2 \bullet 16 \bullet 4 - 3) \bullet 64^2/[16 \bullet (16 - 1)]\}]^{1/2} = 1.0046$$

$$z = (1 - 1.6)/1.0046 = -0.5973$$

$$\sigma_{BB} = [\{12 \bullet 4 \bullet (12 - 1)/[4 \bullet 16 \bullet (16 - 1)(16 - 2)(16 - 3)]\}\{[2 \bullet (4 - 1) \bullet 64 + 4 \bullet (12 - 2) \bullet 280 + 2 \bullet (3 \bullet 16 + 3 \bullet 12 - 2 \bullet 16 \bullet 12 - 3) \bullet 64^2/[16 \bullet (16 - 1)]\}]^{1/2} = 1.9370$$

$$z = (19 - 17.6)/1.9370 = 0.7228$$

$$\sigma_{BW} = [\{4 \bullet 12/[16 \bullet (16 - 1)(16 - 2)(16 - 3)]\}\{2(4 - 1)(12 - 1) \bullet 64 + (16^2 - 16 + 2 - 4 \bullet 4 \bullet 12) \bullet 280 + [2 \bullet 4 \bullet 12(2 \bullet 16 - 3) - 16 \bullet (16 - 1)^2] \bullet 64^2/[16 \bullet (16 - 1)]\}]^{1/2} = 2.1732$$

$$z = (12 - 12.8)/2.1732 = -0.3681.$$

For a 5% level of significance (see Table 7.2-A), the critical values are

± 1.96. Therefore, the conclusion from each of these three statistics is that (1) H_0 fails to be rejected, (2) there is no evidence to indicate that the geographic distribution of urban propensity in the Mayaguez Agricultural Administrative Region is not random in the population, and (3) 95% of the time one would expect to obtain a z-score value that falls within the confidence interval (-1.96, 1.96) when in fact the geographic distribution of urban propensity in the Mayaguez Agricultural Administrative Region is random in the population.

Confidence intervals can be constructed for these join count statistics, too. Such constructions take on the same form as before, namely,

$$WW - z_{1-\alpha/2}\sigma_{WW} \le \mu_{WW} \le WW + z_{1-\alpha/2}\sigma_{WW}$$
$$BB - z_{1-\alpha/2}\sigma_{BB} \le \mu_{BB} \le BB + z_{1-\alpha/2}\sigma_{BB}, \text{ and}$$
$$BW - z_{1-\alpha/2}\sigma_{BW} \le \mu_{BW} \le BW + z_{1-\alpha/2}\sigma_{BW}.$$

One should recognize the possibility that these three statistics will not lead to consistent inferences.

EXAMPLE PR-10.14. Example PR-5.11 revisited.

H_0: the geographic distribution of density of coffee production (DCF) in the Mayaguez Agricultural Administrative Region is random,

which is equivalent to saying

$$H_0: \mu_{MC} = -1/(n - 1), \text{ or}$$
$$H_0: \mu_{GR} = 1.$$

H_a: the geographic distribution of density of coffee production (DCF) in the Mayaguez Agricultural Administrative Region is not random,

which is equivalent to saying

$$H_a: \mu_{MC} \ne -1/(n - 1), \text{ or}$$
$$H_a: \mu_{GR} \ne 1.$$

For this example n = 16, the sum of the 1s in the Mayaguez Region areal unit configuration table is $\sum_{i=1}^{n} \sum_{j=1}^{n} c_{ij} = 64$, the squared row totals sum for this table is $\sum_{i=1}^{n} (\sum_{j=1}^{n} c_{ij})^2 = 280$, $\sum_{i=1}^{n} (x_i - \bar{x})^4 = 2.580944\times10^{18}$, and $[\sum_{i=1}^{n} (x_i - \bar{x})^2]^2 = 7.074792\times10^{18}$. For the Moran Coefficient E(MC) = -1/15. Now,

$$\sigma_{MC} = [\{2 \cdot 16 \cdot 15 \cdot [16^2 - 3 \cdot 16 + 3$$

$- 16 \bullet 15 \bullet 2.580944{\times}10^{18}/(7.074792{\times}10^{18})] \bullet 64 - 4 \bullet 16^2 \bullet 15[1$
$- 2 \bullet 2.580944{\times}10^{18}/(7.074792{\times}10^{18})] \bullet 280 + 2 \bullet [16^2 + 16 - 3$
$- 3 \bullet 16 \bullet 15 \bullet 2.580944{\times}10^{18}/(7.074792{\times}10^{18})] \bullet 64^2\}/(15^2 \bullet 14 \bullet 13$
$\times 64^2)]^{1/2} = 0.1264$

$z \;= [0.2438 - (-1/15)]/0.1264 = 2.4562$

$\sigma_{GR} = \; [\{2 \bullet 15 \bullet [16^2 - 3 \bullet 16 + 3$
$- 16 \bullet 15 \bullet 2.580944{\times}10^{18}/(7.074792{\times}10^{18})] \bullet 64 - 15 \bullet [16^2$
$+ 3 \bullet 16 - 6 - 16 \bullet (16^2 - 16 + 2) \bullet 2.580944{\times}10^{18}/(7.074792{\times}10^{18})]$
$\times 280 + [16^2 - 3 - 16 \bullet 15^2 \bullet 2.580944{\times}10^{18}/(7.074792{\times}10^{18})]$
$\times 64^2\}/(16 \bullet 14 \bullet 13 \bullet 64^2)]^{1/2} = 0.2196$

$z \;= (0.6671 - 1)/0.2196 = -1.5159$

For a 5% level of significance the critical values are ±1.96 (see Table 7.2-A). Therefore, based upon the MC, the conclusion is that (1) reject H_0, (2) there is evidence to indicate that the geographic distribution of DCF in the Mayaguez Agricultural Administrative Region is not random in the population, and (3) only 5% of the time would one expect to obtain a z-score value that falls within the critical region $\{(-\infty, -1.96), (1.96, \infty)\}$ when in fact the geographic distribution of DCF in the Mayaguez Agricultural Administrative Region is random in the population. The Geary Ratio implies an opposite conclusion here.

Confidence intervals can be constructed for these spatial autocorrelation indices, too. Such constructions take on the same form as before, namely,

$MC - z_{1-\alpha/2}\sigma_{MC} \leq \mu_{MC} \leq MC + z_{1-\alpha/2}\sigma_{MC}$ and

$GR - z_{1-\alpha/2}\sigma_{GR} \leq \mu_{GR} \leq GR + z_{1-\alpha/2}\sigma_{GR}.$

Once more, note that physically meaningless bounds could be obtained here (e. g., a lower bound for GR that is less than zero).

10.8. Summary

Only the very simple null hypothesis situations have been addressed in this chapter. The first extensions to more complicated circumstances are presented in the next chapter.

Several issues appear in this chapter that are repeated throughout the remaining chapters. The importance of sample size in inferential tests is critical. Sample size restricts the assumptions that can safely be made about the sampling distribution of a statistic. Furthermore, as sample size decreases, the variability arising in the sample statistics reduces the amount of information contained. Ultimately, a sample size that is too small may render inferential statements impossible.

The impact of using a computationally easy approximate formula, as opposed to an exact formula, has been demonstrated. When the inferential decision is clear, the effect of this substitution (which introduces specification error) is negligible. But when a sample statistic lies close to the critical value, the result can be a reversal of the inferential decision that is made. Clearly the approximate version of a statistic should be used with caution.

Finally, complexities encountered when the effect of space is introduced into a problem increase with the sophistication of the research question being addressed. This complexity is the unfortunate, but inescapable, burden imposed upon spatial analysis by the inherent importance of locational information. This fact will be all too obvious in chapters to come.

10.9. Problems

1. Verify the chi square calculations presented in Examples 10.1, 10.3, PR-10.1, PR-10.2, and PR-10.5. Then replicate Example PR-10.2 for DMLK.

2. Verify the K-S calculations in Examples 10.4, 10.5, PR-10.3, and PR-10.4.

3. State and test the appropriate null hypothesis for Problem #5 of Chapter 3, which also has been treated as Problem # 10 of Chapter 7; first employ the chi square statistic and then employ the K-S statistic.

4. Verify the calculations pertaining to inference about a population mean in Examples 10.2, PR-10.6, and PR-10.7.

5. State and test the appropriate null hypotheses for Example PR-4.1.

6. Construct appropriate 90%, 95%, and 99% confidence intervals for Example PR-4.1.

7. Verify the calculations pertaining to inference about a population variance in Example PR-10.8.

8. State and test the appropriate null hypotheses for Example PR-4.6.

9. Construct appropriate 90%, 95%, and 99% confidence intervals for Example PR-4.6.

10. Verify the calculations pertaining to inference about a population correlation coefficient in Examples PR-10.9, PR-10.10, and PR-10.11.

11. State and test the appropriate null hypotheses for Examples CA-4.1, PR-4.9, PR-4.10, PR-4.11, and PR-4.12.

12. Construct appropriate 90%, 95%, and 99% confidence intervals for Problem #11.

13. Show that if n = 2, then the sample correlation coefficient r always equals unity.

14. Using a randomization sampling perspective, show that if n = 3 and X and Y are ranked, with these rankings being distinct, then $E(r_s) = \rho_s$, and $VAR(r_s) = (1/\sqrt{2})^2$. Does this result hold when sampling with replacement occurs?

15. Verify the calculations pertaining to inference about a point pattern in Example PR-10.13.

16. State and test the appropriate null hypothesis for Example PR-5.7.

17. Construct appropriate 90%, 95%, and 99% confidence intervals for Problem #16.

18. Simulate a Poisson point process on a unit square, doing 5 replications and using 9 quadrats and 18 points, to illustrate that the standard error is $\sigma_R = [(4 - \pi)/(n\pi)]^{1/2}$.

19. Replicate Example 10.6.

20. Verify the calculations pertaining to inference about a random sequence in Example 10.7.

21. Construct the sampling distribution of the runs statistic V for the set of classifications {A, A, A, B, B}. Compute μ_V and σ_V from the results of this sampling distribution, verifying your calculations with the formulae for these two quantities.

22. Verify the calculations pertaining to inference about spatial autocorrelation in Example PR-10.13.

23. State and test the appropriate null hypothesis for the geographic distribution of the coastal lowlands/interior highlands classification variable in the Mayaguez

Agricultural Administrative Region.

24. Construct appropriate 90%, 95%, and 99% confidence intervals for Problem #23.

25. Verify the calculations pertaining to inference about spatial autocorrelation in Example PR-10.14.

26. State and test the appropriate null hypothesis for inference about spatial autocorrelation for the variable DMLK in Example PR-5.9.

27. Construct appropriate 90%, 95%, and 99% confidence intervals for Problem #26.

28. Consider the set of numbers {0, 1, 1, 2} and the 2-by-2 regular square grid configuration of areal units

A	B
C	D

Construct the sampling distribution of MC and of GR for this situation. Then compute μ_{MC} and σ_{MC} and μ_{GR} and σ_{GR} from the results of this sampling distribution, verifying your calculations with the formulae for these two pairs of quantities.

29. Construct an appropriate contingency table for PC1, as shown in Example CA-10.1, and variable SPC found in the Table 1.2-A. Test the two categorical variables for independence using the chi square test.

10.10. Additional readings

In addition to the texts listed in Chapter 1,

Davidson, N. *Causal Inferences from Dichotomous Variables.* CATMOG 9.

Pringle, D. *Causal Modelling: The Simon-Blalock Approach.* CATMOG 27.

Henkel, R. *Tests of Significance.* SAGE 4.

CHAPTER 11

COMPLEX HYPOTHESIS TESTING

Simple hypotheses and their relationship to sampling distributions were treated in Chapter 10. The nature and scope of hypothesis testing was introduced in that chapter, together with inferential problems involving single parameters. Assumptions and test criteria were spelled out in some detail for the most common problems encountered in, or as precursors to, spatial data analysis. Attention now is turned to hypotheses involving more than one parameter. Many of the topics that appeared in Chapter 10 for single-parameter inferences will be found in this chapter for multiple simultaneous parameter inferences.

11.1. Differences between population variances

The order of presentation in this chapter deviates somewhat from that followed in Chapter 10 because, as seen with the inference about a population mean problem, test statistics tend to be very sensitive to knowledge about the population variance.

One theoretical assumption frequently invoked in geographic analysis is that a homogeneous or stationary process is operating over a planar surface. The analysis of point processes described in Chapter 10 explicitly emphasized this issue. This homogeneity notion implies that the statistical distribution of some variable in two different regions is the same. Accordingly, in order to conduct a comparative statistical analysis of two regions, it is necessary to determine whether the sample variance for these two regions differs because of (1) sampling error (the null hypothesis) or (2) the failure of the process in question to be homogeneous over space (the alternate hypothesis). Of course, the resulting statistical decision may involve a Type I or a Type II error! Many hypothesis-testing situations are very sensitive to this variance homogeneity assumption. Confidence intervals will not be dealt with in this section, since the focus is usually on evaluating the homogeneity assumption. However, if constructing confidence intervals is important, the format of Chapter 10 could be used here as well.

11.1.1. The difference between two independent population variances

In Chapter 10 a hypothesis about a single population variance was treated, and its associated sampling distribution was chi square in form. The simplest extension of this hypothesis is a situation involving two population variances with a null hypothesis H_0: $\sigma_1^2 = \sigma_2^2$. Attempting to compare a chi square sampling distribution for s_1^2 with a chi square sampling distribution for s_2^2 makes this particular form of the null hypothesis difficult to test. One way to streamline this hypothesis testing situation is to divide both sides of the equation by one of the variances; in practice, this is done by dividing the smaller sample variance into the larger sample variance. Accordingly, the null hypothesis is transformed into H_0: $\sigma_1^2/\sigma_2^2 = 1$, where the population with the larger variance is labelled "1" and the other population is labelled "2." Clearly, since the larger sample variance is always in the numerator, $\sigma_1^2/\sigma_2^2 \in [1, \infty)$, and the only acceptable alternative hypothesis is H_a: $\sigma_1^2/\sigma_2^2 > 1$. Moreover, by construction, a two-tail test situation would not be appropriate here. Now, the null hypothesis has a sampling distribution that is obtainable from results reported in Chapter 10 for inference about a single population variance, namely,

$$\chi_1^2/\chi_2^2 = [(n_1 - 1)s_1^2/\sigma_1^2]/[(n_2 - 1)s_2^2/\sigma_2^2] = (\sigma_2^2/\sigma_1^2)[(n_1 - 1)s_1^2]/[(n_2 - 1)s_2^2].$$

But under the null hypothesis $\sigma_1^2/\sigma_2^2 = 1 = \sigma_2^2/\sigma_1^2$, meaning the ratio becomes $[\chi_1^2/(n_1 - 1)]/[\chi_2^2/(n_2 - 1)] = 1(s_1^2/s_2^2)$. This ratio of two sample variances constitutes a test statistic for the difference between two population variances, under the null hypothesis. This test statistic is distributed as the ratio of two chi square distributions divided by their respective degrees of freedom, a ratio whose sampling distribution conforms to the F distribution. Noteworthy properties of the F distribution include (1) F is nonnegative in value (each variance must be at least 0), (2) F is asymmetrical (it is skewed to the right) but with the advantage that the numerator and denominator chi squares are arbitrary (thus, $F_{\alpha,\upsilon_1,\upsilon_2} = 1/F_{1-\alpha,\upsilon_2,\upsilon_1}$), (3) as with the t distribution, there is a different F distribution for each pair of degree-of-freedom values, υ_1 and υ_2, and (4) $t_{1-\alpha,\upsilon_2} = \sqrt{F_{1-\alpha,1,\upsilon_2}}$. As shown in Chapter 10, because s_j^2 has $n_j - 1$ degrees of freedom, for an F ratio, then, $\upsilon_1 = n_1 - 1$ and $\upsilon_2 = n_2 - 1$. The last property of the F statistic suggests that a t statistic can be viewed as a variance ratio test statistic in which there is only one degree of freedom associated with the numerator variance.

The next example illustrates the difference of variances test.

Assumptions:

(a) the random variable in each of the two populations conforms to a normal distribution;

(b) s_1^2 and s_2^2 are independent estimates;

(c) independent samples are drawn randomly from each population; and,

(d) the observations are independent.

Note that no assumptions are made about either the two population means or the two sample sizes involved (although $n_j \geq 2$ for any sample variance to be detected and for positive degrees of freedom to be available).

Test criteria:

(a) the one-tail (upper) F distribution is used, having $n_1 - 1$ and $n_2 - 1$ degrees of freedom; and,

(b) customarily the larger sample variance is placed in the numerator of this F ratio.

Because of the enormous number of possible F distributions, only a few critical values tend to be tabulated (for example, see Table 11.1-A found in the appendices). If a researcher chooses to use the lower tail, then he or she should remember that the critical values become $1/F_{1-\alpha,\upsilon_2,\upsilon_1}$. The following example uses a one-tail test.

EXAMPLE PR-11.1. The difference between selected regional variances of DCD (density of farm size) in Puerto Rico.

The findings reported in Example PR-10.2 indicate that the variable DCD should be normally distributed in its parent population; additional test statistics (not reported in this book) imply that this variable should be normally distributed in each of its five regional parent populations as well. Its unbiased sample variances for the Agricultural Administrative Regions of Puerto Rico are as follows:

San Juan	Arecibo	Mayaguez	Ponce	Caguas
$(13.378)^2$	$(8.820)^2$	$(10.611)^2$	$(9.168)^2$	$(6.726)^2$

The hypothesis

H_0: the variance of DCD is the same in both the San Juan and Mayaguez regions,

is equivalent to saying H_0: $\sigma_{SJ}^2/\sigma_M^2 = 1$, or H_0: $F = 1$, then $F = 13.378^2/10.611^2 = 1.590$. For a 5% level of significance (see Table 11.1-A), the critical value with 19 - 1 = 18 degrees of freedom in the numerator and 16 - 1 = 15 degrees of freedom in the denominator is 2.3399. Therefore, the conclusion is that (1) H_0 fails to be rejected, (2) there is no evidence to indicate that the variances in the San Juan and Mayaguez regions are not equal in the population, and (3) 95% of the time one would expect to obtain an F ratio value that falls within the confidence interval (1, 2.3399) when in fact the two population variances are equal. This result suggests that a homogeneous geographic process is operating.

Next, if the hypothesis is

H_0: the variance of DCD is the same in both the San Juan and Caguas regions,

which is equivalent to saying H_0: $\sigma_{SJ}^2/\sigma_C^2 = 1$, or H_0: $F = 1$,

then $F = 13.378^2/6.726^2 = 3.956$. For a 5% level of significance (see Table 11.1-A), the critical value with 19 - 1 = 18 degrees of freedom in the numerator and 13 - 1 = 12 degrees of freedom in the denominator is 2.5684. Therefore, the conclusion is that (1) H_0 is rejected, (2) there is evidence to indicate that the variances in the San Juan and Caguas regions are not equal in the population, and (3) only 5% of the time would one expect to obtain an F ratio value that falls within the critical region (2.5684, ∞) when in fact the two population variances are equal. This result implies that a nonhomogeneous geographic process is operating.

But how can the pair of contradictory results obtained in this example be combined? This question will be answered in Section 11.1.3.

11.1.2. The difference between two correlated population variances

A slightly more complicated extension of the hypothesis presented in Chapter 10 involves two population variances for each of two different variables and the same set of areal units. In other words, compare two pairs of variances, or four values in total. Quite often the spatial problem of interest asks whether or not the variance of a variable is changing through time; once more the notion of a homogeneous process comes into play. In contrast with Section 11.1.1, where two distinct sets of areal units were used to obtain data, correlation enters into the analysis here because the same set of areal units is used to obtain data (these same areal units are the source of communality). The null hypothesis remains H_0: $\sigma_1^2 = \sigma_2^2$, while assumption (b) is discarded. Now

Test criterion:

(a) a one-tail t distribution is used, with n - 2 degrees of freedom, having a mean of 0 (the null hypothesis becomes H_0: $\sigma_1^2 - \sigma_2^2 = 0$), and a standard error of $2s_1s_2[(1 - r^2)/(n - 2)]^{1/2}$.

Because σ_1 and σ_2 are unknown (if they were known, there would be no statistical problem to solve), these two quantities respectively have been replaced in the standard error formula by their estimates s_1 and s_2, so that the t distribution will involve two constraints and hence have n - 2 degrees of freedom. Also, a one-tail test situation is promoted here in keeping with the test format of two independent variances, with the smaller of the two variances always being subtracted from the larger of the two (H_a: $\sigma_1^2 - \sigma_2^2 > 0$). Recall the relationship between t and F that was mentioned in Section 11.1.1.

EXAMPLE PR-11.2. The difference between the variances of DCD in 1969 and 1974 for selected regions in Puerto Rico.

The supplemental data for this problem are as follows (recall that the 1974 DCD data appear in Table 1.1-A):

San Juan Region			Caguas Region	
Municipio	DCD_{1969}	$DCD_{1969}-DCD_{1974}$	Municipio	DCD_{1969}
Barranquitas	40.2	- 2.2	Aguas Buenas	33.9
Bayamon/Catano	10.9	- 1.5	Aibonito	41.6
Carolina	21.3	1.4	Caguas	38.7
Ceiba	10.7	2.5	Cayey	46.2
Comerio	41.3	11.3	Cidra	38.5
Corozal	48.5	- 7.8	Gurabo	47.6
Dorado	60.3	34.1	Humacao	17.5
Fajardo	35.8	11.3	Juncos	48.6
Guaynabo	18.0	- 3.0	Las Piedras	43.0
Loiza/Canovanas	35.4	11.7	Maunabo	30.0
Luquillo	37.9	14.6	Naguabo	48.4
Naranjito	32.2	2.0	San Lorenzo	33.9
Orocovis	40.5	13.1	Yabucao	53.5
Rio Grande	14.1	0.2		
San Juan/ Rio Piedras	3.5	- 0.6		
Toa Alta	34.7	- 7.0		
Toa Baja	36.4	- 8.2		
Trujillo Alto	18.4	2.0		
Vega Alta	22.6	9.1		

Additional test statistics (not reported in this book) imply that for 1969 as well as 1974 the DCD variable should be normally distributed in both the San Juan and Caguas regional parent populations. If the hypothesis is

H_0: the 1969 and 1974 variances of DCD are equal in the San Juan region,

which is equivalent to saying H_0: $\sigma_{1969}^2 = \sigma_{1974}^2$, or H_0: $\sigma_{1969}^2 - \sigma_{1974}^2 = 0$,

then

$$t = (14.708^2 - 13.378^2)/[2 \bullet 13.378 \bullet 14.708 \bullet \{[1 - 0.716^2]/(19 - 2)\}^{1/2}]$$
$$= 0.561.$$

For a one-tail 5% level of significance (see Table 9.1-A), the critical value with 17 degrees of freedom is +1.7396. Therefore, the conclusion is that (1) H_0 fails to be rejected, (2) there is no evidence to indicate that the variance of DCD in the San Juan region is changing through time in the population, and (3) 95% of the time one would expect to obtain a t statistic that falls within the confidence interval (0.00, 1.7396) when in fact the two population variances are equal.

Next, if the hypothesis is

H_0: the 1969 and 1974 variances of DCD are equal in the Caguas region,

then $t = (9.657^2 - 6.726^2)/[2 \bullet 6.726 \bullet 9.657 \bullet \{[1 - 0.699^2]/(13 - 2)\}^{1/2}] = 1.714$. Choose a one-tail 10% level of significance (see Table 9.1-A). A 10% rather than the usual 5% level was chosen because the massive urbanization that took place between 1969 and 1974 could be expected to reduce this variance. The corresponding critical value with 11 degrees of freedom is -1.3634. Therefore, the conclusion is that (1) H_0 is rejected, (2) there is evidence to indicate that the 1969 and 1974 variances of DCD in the Caguas region are not equal in the population, and (3) only 10% of the time would one expect to obtain a t statistic of the observed magnitude when in fact the two population variances are equal. The choice of a 10% critical region implies less of a willingness to attribute a given difference in variances (a larger absolute value in the test statistic) to sampling error than does a 5% critical region.

The particular test statistic computation presented in this section re-emphasizes the need to study correlation coefficients.

11.1.3. The difference between k independent population variances

A considerably more complicated extension of the hypothesis presented in Chapter 10 involves k population variances. Here the null hypothesis becomes H_0: $\sigma_1^2 = \sigma_2^2 = \ldots = \sigma_k^2$, and the alternate hypothesis becomes H_a: there exists (denoted $\exists$) a pair of variance values for groups g and h such that $\sigma_g^2 \neq \sigma_h^2$, $g \neq h$. In other words, which implication from Example PR-11.1 is most reasonable? Whereas assumption (b) was discarded in Section 11.1.2, here it is retained, with both it and assumption (a) being modified as follows:

<u>Assumptions</u>:

(a) the random variable in each of the k populations conforms to a normal distribution; and,

(b) $s_1^2, s_2^2, \ldots, s_k^2$ are independent estimates.

Now

<u>Test criterion</u>:

(a) call the test statistic in this problem B; the sampling distribution for the quantity

$$B = 3(k - 1)\{(n - k)\ln[\textstyle\sum_{j=1}^{k} \sum_{i=1}^{n_j} (x_{ij} - \bar{x}_j)^2/(n - k)] - \sum_{j=1}^{k} (n_j - 1)$$
$$\times\ln[\textstyle\sum_{j=1}^{k} \sum_{i=1}^{n_j} (x_{ij} - \bar{x}_j)^2/(n_j - 1)]\}/[3(k - 1) - 1/(n - k) + \sum_{j=1}^{k} 1/(n_j - 1)],$$

which conforms to a chi square distribution with (k - 1) degrees of freedom.

For a true null hypothesis, the population value of this quantity becomes

$$B = 3(k - 1)\{(n - k)\ln(\sigma^2) - \textstyle\sum_{j=1}^{k} (n_j - 1)\ln(\sigma^2\}/[3(k - 1) - 1/(n - k) + \sum_{j=1}^{k} 1/(n_j - 1)] = 0$$

as expected, since the null hypothesis is equivalent to H_0: $\chi^2 = 0$.

EXAMPLE PR-11.3. The difference between the variance of DCD among the five agricultural administrative regions in Puerto Rico.

If the hypothesis is

H_0: the variance of DCD in the San Juan, Arecibo, Mayaguez, Ponce, and Caguas regions is equal,

which is equivalent to saying H_0: $\sigma_{SJ}^2 = \sigma_A^2 = \sigma_M^2 = \sigma_P^2 = \sigma_C^2$,

then

$$B = 3 \bullet (5 - 1)\{(73 - 5)\ln(10.378^2) - [(19 - 1)\ln(13.378^2) + (11 - 1)\ln(8.820^2) + (16 - 1)\ln(10.611^2) + (14 - 1)\ln(9.168^2) + (13 - 1)\ln(6.726^2)]\}/\{3 \times(5 - 1) - 1/(73 - 5) + [1/(19 - 1) + 1/(11 - 1) + 1/(16 - 1) + 1/(14 - 1) + 1/(13 - 1)]\} = 6.868.$$

For a 5% level of significance (see Table 10.1-A), the chi square critical value with 4 degrees of freedom is 9.4877. Therefore, the conclusion is that (1) H_0 fails to be rejected, (2) there is no evidence to indicate that the five regional variances of DCD are not equal in the population, and (3) 95% of the time one would expect to obtain a chi square statistic value that falls within the confidence interval (0, 9.4877) when in fact the five population variances are equal.

The finding reported in Example PR-11.3 suggests that the geographic process in question is homogeneous, and that the non-homogeneity conclusion reached in Example PR-11.1 is probably a Type II error.

11.2. The difference between two population means

In Chapter 10 the problem of drawing an inference about a population mean, when σ is either known or, more commonly, unknown, was outlined. Whereas this might be an interesting question to answer in many research projects, spatial analysts usually are concerned with the question of whether or not either two regional means are equal or the regional means of two variables are equal. In other words, the null hypothesis of concern becomes H_0: $\mu_1 = \mu_2$. This particular null hypothesis can be rewritten as H_0: $\mu_1 - \mu_2 = 0$, a form that facilitates the writing of an appropriate test statistic (remember that this convenient modification appears with the difference-of-variance F ratio, too) and leads to the problem name "difference of means." Since a researcher usually does not care which of the two populations means is larger (otherwise a one-tail test would be adopted), the accompanying alternate hypothesis becomes H_a: $\mu_1 - \mu_2 \neq 0$. As in the case of the single population mean, the sampling distribution for the difference of means follows a t distribution. But, not surprisingly, the form of this test statistic is a function of the existing knowledge about the population variances.

11.2.1. The difference between two independent population means

First of all, assume that the two population variances, σ_1^2 and σ_2^2, are known. In this instance the standard error of the sampling distribution of the difference of means, $\bar{x}_1 - \bar{x}_2$, is given by $[\sigma_1^2/n_1 + \sigma_2^2/n_2]^{1/2}$. This sampling distribution result looks like a direct extension of the single mean standard error discussed in Chapter 10.

Assumptions:

(a) for small samples (n_1, $n_2 < 30$) the random variable in each of the two populations conforms to a normal distribution [this part of the assumption is not required for large samples (n_1, $n_2 > 100$)], respectively having means μ_1 and μ_2, and variances σ_1^2 and σ_2^2;

(b) σ_1^2 and σ_2^2 are known;

(c) independent samples of size n_1 and n_2, respectively, are drawn randomly from each population; and,

(d) the observations are independent.

Test criteria:

(a) the standard normal distribution provides the test statistic, where

$$z = [(\bar{x}_1 - \bar{x}_2) - (\mu_1 - \mu_2)]/[\sigma_1^2/n_1 + \sigma_2^2/n_2]^{1/2} = (\bar{x}_1 - \bar{x}_2)/[\sigma_1^2/n_1 + \sigma_2^2/n_2]^{1/2}$$

under H_0; or,

(b) although the standard normal distribution is appropriate, for small n_j (perhaps < 30), some researchers revert to the t distribution (where degrees of freedom become an index of information context) to accommodate the increase in variability due to small sample size.

Given the test criteria outlined here, appropriate confidence intervals can be constructed as follows:

$$(\bar{x}_1 - \bar{x}_2) - z_{1-\alpha/2}[\sigma_1^2/n_1 + \sigma_2^2/n_2]^{1/2} \leq (\mu_1 - \mu_2) \leq (\bar{x}_1 - \bar{x}_2) + z_{1-\alpha/2}[\sigma_1^2/n_1 + \sigma_2^2/n_2]^{1/2}.$$

Of course, these look very similar to those found in Chapter 10.

EXAMPLE PR-11.4. The difference between selected regional means of DCD in Puerto Rico.

From previous analysis it is known that in Puerto Rico the variable DCD is normally distributed with variance 11.8765^2. The five agricultural administrative regional means for this variable are as follows:

San Juan	Arecibo	Mayaguez	Ponce	Caguas
26.4368	39.6364	38.8250	44.2857	35.4769

If the hypothesis is

H_0: the mean of DCD is the same in both the Arecibo and Mayaguez regions,

which is equivalent to saying H_0: $\mu_A - \mu_M = 0$,

then $z = (39.6364 - 38.8259)/[11.8765^2/11 + 11.8765^2/16]^{1/2} = 0.1744$. For a two-tail 5% level of significance (see Table 7.2-A), the critical values for the z-score are ±1.96. Therefore, the conclusion is that (1) H_0 fails to be rejected, (2) there is no evidence to indicate that the means of DCD in the Arecibo and Mayaguez regions are not equal in the population, and (3) 95% of the time one would expect to obtain a z-score value that falls within the confidence interval (-1.96, 1.96) when in fact the two population means are equal. This result implies that a homogeneous geographic process is operating.

Next, if the hypothesis is

H_0: the mean of DCD is the same in both the San Juan and Ponce regions,

which is equivalent to saying H_0: $\mu_{SJ} - \mu_P = 0$,

then $z = (26.4368 - 44.2857)/[11.8765^2/19 + 11.8765^2/14]^{1/2} = -4.2668$. Therefore, the conclusion is that (1) H_0 is rejected, (2) there is evidence to indicate that the means of DCD in the San Juan and Ponce regions are not equal in the population, and (3) only 5% of the time would one expect to obtain a z-score that falls within the critical region $\{(-\infty, -1.96), (1.96, \infty)\}$ when in fact the two population means are equal. This result implies that a nonhomogeneous geographic process is operating.

But, again, how can the pair of contradictory results obtained in this example be combined? This question will be answered in Section 11.3.

Assume assumption (b) is violated. This is the far more common case of population variances being unknown. Knowledge about whether or not the regional variances are equal is needed in order to determine the appropriate sampling distribution on which to base an inference about a difference of means; test procedures like those outlined in the preceding sections of this chapter can be used to acquire this knowledge. But to retain the proper levels of significance, variance tests should be performed on a sample other than that used to evaluate difference of means. In practice this separation of test results is not done, meaning that unadjusted significance levels for the sequential testing procedure are technically incorrect.

Suppose the F ratio test statistic implies that the variances for the two regional populations are equal. Accordingly, assumption (b) becomes

(b) σ_1^2 and σ_2^2 are unknown but equal.

Accompanying this assumption change is an estimated standard error based upon the pooled unbiased regional variance estimates, such that

<u>Test criteria</u>:

(a) the t distribution furnishes the correct test statistic here, where

$$t = [(\bar{x}_1 - \bar{x}_2) - (\mu_1 - \mu_2)]/[s_p(1/n_1 + 1/n_2)^{1/2}] = (\bar{x}_1 - \bar{x}_2)/[s_p(1/n_1 + 1/n_2)^{1/2}]$$

under H_0, with the pooled variance estimate given by

$$s_p^2 = [(n_1 - 1)s_1^2 + (n_2 - 1)s_2^2]/(n_1 + n_2 - 2);\text{ and,}$$

(b) since two parameter estimates are required when computing the standard error, two constraints are placed on an analysis, with the result that $(n_1 + n_2 - 2)$ degrees of freedom are available.

As before, a two-tail test situation is usually of interest.

EXAMPLE PR-11.5. The difference between selected regional means of DCD in Puerto Rico.

From Example PR-11.1 it is known that in Puerto Rico the variable DCD is normally distributed with equal population variance for the San Juan and Mayaguez agricultural administrative regions.

If the hypothesis is

H_0: the mean of DCD is the same in both the San Juan and Mayaguez regions,

which is equivalent to saying H_0: $\mu_{SJ} - \mu_M = 0$,

then $t = (26.4368 - 38.8250/\{[(18 \bullet 13.378^2 + 15 \bullet 10.611^2)/(19 + 16 - 2)] \bullet (1/19 + 1/16)\}^{1/2} = -2.9931$. For a two-tail 5% level of significance (see Table 9.1-A), the critical values for the t statistic are ±2.0360 (using 33 degrees of freedom and linear interpolation). Therefore, the conclusion is that (1) H_0 is rejected, (2) there is evidence to indicate that the means of DCD in the San Juan and Mayaguez regions are not equal in the population, and (3) only 5% of the time would one expect to obtain a t statistic that falls within the critical region $\{(-\infty, -2.0360), (2.0360, \infty)\}$ when in fact the two population means are equal.

Next, if the hypothesis is

H_0: the mean of DCD is the same in both the Arecibo and Mayaguez regions,

which is equivalent to saying H_0: $\mu_A - \mu_M = 0$ (the accompanying F ratio here is not significant),

then

$$t = (39.6364 - 38.8250)/\{[(10 \bullet 8.820^2 + 15 \bullet 10.611^2)/(11 + 16 - 2)] \times (1/11 + 1/16)\}^{1/2} = 0.2085.$$

For a two-tail 5% level of significance (see Table 9.1-A), the critical values for the t statistic are ±2.0595 (using 25 degrees of freedom). Therefore, the conclusion is that (1) H_0 fails to be rejected, (2) there is no evidence to indicate that the means of DCD in the Arecibo and Mayaguez regions are not equal in the population, and (3) 95% of the time would one expect to obtain a t statistic that falls within the confidence interval (-2.0595, 2.0595) when in fact the two population means are equal.

Next, suppose the F ratio test statistic implies that variances for the two regional populations are unequal. Accordingly, assumption (b) becomes

(b) σ_1^2 and σ_2^2 are unknown and unequal.

The sampling distribution for this problem is unknown, but an approximate sampling distribution has been identified. Thus, accompanying this assumption change is an estimated standard error based upon the individual unbiased regional variance estimates, such that

Test criteria:

(a) the t distribution furnishes the correct test statistic here, where

$t = [(\bar{x}_1 - \bar{x}_2) - (\mu_1 - \mu_2)]/(s_1^2/n_1 + s_2^2/n_2)^{1/2} = (\bar{x}_1 - \bar{x}_2)/(s_1^2/n_1 + s_2^2/n_2)^{1/2}$ under H_0; and,

(b) the approximation requires that the degrees of freedom serve as an index here (see Section 9.3), rather than referring to constraints, with the appropriate number being given by the minimum of $\{(n_1 - 1),(n_2 - 1)\}$.

Recognize that the estimated approximate standard error for this case is equivalent to that for the case of known variance, with σ_j being replaced with s_j.

EXAMPLE PR-11.6. The difference between selected regional means of DCD in Puerto Rico.

From Example PR-11.1 it is known that in Puerto Rico the variable DCD is normally distributed with unequal population variance for the San Juan and Caguas agricultural administrative regions.

If the hypothesis is

H_0: the mean of DCD is the same in both the San Juan and Caguas regions,

which is equivalent to saying H_0: $\mu_{SJ} - \mu_C = 0$,

then $t = (26.4368 - 35.4769)/(13.378^2/19 + 6.726^2/13)^{1/2} = -2.5170$. For a two-tail 5% level of significance (see Table 9.1-A), the critical values for the t statistic are ± 2.1604 (using 13 degrees of freedom). Therefore, the conclusion is that (1) H_0 is rejected, (2) there is evidence to indicate that the means of DCD in the San Juan and Caguas regions are not equal in the population, and (3) only 5% of the time would one expect to obtain a t statistic that falls within the critical region $\{(-\infty, -2.1604), (2.1604, \infty)\}$ when in fact the two population means are equal.

Next, if the hypothesis is

H_0: the mean of DCD is the same in both the Mayaguez and Caguas regions,

which is equivalent to saying H_0: $\mu_A - \mu_M = 0$ (the accompanying F ratio here is significant at the 10% level),

then $t = (38.8250 - 35.4769)/(10.611^2/16 + 6.726^2/13)^{1/2} = 1.0324$. For a two-tail 5% level of significance, the critical values for the t statistic still are ± 2.1604. Therefore, the conclusion is that (1) H_0 fails to be rejected, (2) there is no evidence to indicate that the means of DCD in the Mayaguez and Caguas regions are not equal in the population, and (3) 95% of the time would one expect to obtain a t statistic that falls within the confidence interval (-2.1604, 2.1604) when in fact the two population means are equal.

Because the difference of means test for unknown but assumed unequal variances is an approximation, this test statistic does not easily generalize. This result is in contrast to the case for equal variances, which generalizes to ANOVA (see Section 11.3.1).

11.2.2. The difference between two correlated population means

A situation may be encountered in which two variables for a single set of areal units are being studied. In other words, assumption (c) of Section 11.2.1 is being relaxed. For example, the data may be agricultural productivity for two different time periods. Accordingly the data are composed of pairs of measurements that more than likely are correlated. These circumstances lead to a test for significant difference between means distinct from that for independent samples. This situation is encountered in geographical analyses in which interventions or impacts are to be assessed; the concept of using paired data for this purpose has a built-in ability to remove numerous otherwise uncontrollable factors, such as changes in consumer demands or microclimatic variations.

The simplest way to approach this problem is to address the correlation that is present through differencing the two variables in question. This differencing is consistent with rewriting the null hypothesis in the difference of means test to correct for the correlated variances. The correction adjusts for the presence of nonzero correlation (see Section 11.1.2) by calculating the population variance as $\sigma_D^2 = \sigma_1^2 + \sigma_2^2 - 2\rho\sigma_1\sigma_2$, where ρ is the correlation between the two variables in question. A positive correlation between the paired measurements will reduce the size of this variance, whereas a negative correlation will increase it. Accordingly, the appropriate null hypothesis is H_0: $\mu_D = \mu_1 - \mu_2 = 0$. Note that by computing differences, this case reduces to the simple null hypothesis of drawing an inference about a population mean where the population variance is unknown.

Assumptions:

(a) σ_1^2 and σ_2^2 are unknown;

(b) a sample of size n is drawn randomly from the population; and,

(c) the observations are independent.

Test criteria:

(a) the t distribution is used for the test statistic, in which the two variables for each observation are combined into a differenced variate, $D_i = X_{1,i} - X_{2,i}$, and where s_D^2 is the unbiased variance estimate and under H_0

$$t = (\bar{x}_D - 0) /(s_D/\sqrt{n}) = \bar{x}_D/(s_D/\sqrt{n}); \text{ and,}$$

(b) since there is only one constraint placed upon the sampling distribution involved (estimating σ_D), then (n - 1) degrees of freedom are available.

Given the test criteria outlined here, appropriate confidence intervals can be constructed as follows:

$$\bar{x}_D - t_{n-1,1-\alpha/2}s_D/\sqrt{n} \leq \mu_1 - \mu_2 \leq \bar{x}_D + t_{n-1,1-\alpha/2}s_D/\sqrt{n}.$$

EXAMPLE PR-11.7. Example PR-11.2 revisited.

The results reported in Example PR-11.2 imply that the correlated population variances of DCD in 1969 and 1974 for the San Juan region should be equal in the population. A more appropriate question to ask is whether or not the average DCD is the same for these two time periods. The values for D_i (= $DCD_{1974,i} - DCD_{1969,i}$) are presented in Example PR-11.2. Here $\bar{x}_D = -3.179$, $s_D = 10.647$, and so $t = -3.179/(10.647/\sqrt{19}) = -1.301$. For a two-tail 5% level of significance (see Table 9.1-A), the critical values for the t statistic are ±2.1009 (using 18 degrees of freedom). Therefore, the conclusion is that (1) H_0 fails to be rejected, (2) there is no evidence to indicate that the mean of DCD in the San Juan region is not equal in both the 1969 and 1974 populations, and (3) 95% of the time one would expect to obtain a t statistic that falls within the confidence interval (-2.1009, 2.1009) when in fact the two population means are equal.

Had the correlation between the two sets of measures in the foregoing example been ignored, then the test of equality of variance for independent samples would have yielded F = 1.207 (nonsignificant), and the difference of means test for independent samples with equal but unknown variances would have yielded t = -0.165, having 36 degrees of freedom. Clearly the probability of the observed difference, when related to the t distribution, would have been dramatically changed by the specification error that would be present. This is the same sort of problem that arises when nonzero spatial autocorrelation is present but is ignored.

This section illustrates that correlated sample results can be dealt with effectively. Most statistical procedures for independent samples have correlated sample counterparts. In order to illustrate various impacts attributable to specification error, only a select few of these procedures will be discussed in this text. An advanced book should be consulted for others.

11.2.3. The difference between two population means for nonnormal variables

Assumption (a) of Section 11.2.1 stressed the importance of underlying normal distributions when the sample sizes are small. But what happens when this assumption does not hold? How can the specification error that results from its violation be managed? In order to address these types of questions, the spatial analyst once again must consult nonparametric statistics (see Chapter 10). It is reasonable to ask whether or not nonparametric statistics always should be used. However, even though some scientists would argue that this is the case, when a data set satisfies all the assumptions of a given parametric test, the comparable nonparametric alternatives are almost always markedly less powerful [as mentioned before, the power of a test is defined as 1 - P(Type II error)]. Furthermore, a number of parametric test statistics tend to be robust (that is, the sampling distribution of the test statistic is insensitive to assumption departures) against certain assumption violations (specification error). In terms of differences of means, the central limit theorem indicates that as the sample size increases, the test statistic becomes increasingly less sensitive to a violation of the normality assumption. In contrast, difference of means tests tend to be extremely sensitive to the equality of variances assumption, whereas tests of variance equality are very sensitive to violations of the normality assumption.

Consider the discarding of assumption (a) in Section 11.2.1 when the two sample sizes, n_1 and n_2, are small. The Mann-Whitney U test statistic is a nonparametric alternative to the t test statistic for the difference between means for two independent samples. Here assumption (a) is replaced with

(a) the random variable under study is continuous.

The procedure for computing U is as follows:

Step 1: Merge the two groups of data, and rank their values in ascending order (recalling from Chapter 4, ties are equated to the mean of their positional ranks).

Step 2: Sum the ranks for each of the two samples, denoting these sums as R_1 and R_2 [note that $R_1 + R_2 = n(n + 1)/2$].

Step 3: Let $U_a = n_1n_2 + n_2(n_2 + 1)/2 - R_2$ and $U_b = n_1n_2 + n_1(n_1 + 1)/2 - R_1$.

Step 4: Set U, the test statistic, equal to the smaller of U_a and U_b (note that $U_a + U_b = n_1n_2$).

Since R_1 and R_2 always sum to a fixed value, or equivalently U_a and U_b always sum to a fixed value, only one of these two computations needs to be considered. This suggests that exactly the same sample information is summarized in both pairs of these statistics. Further, converting an interval/ratio scale to an ordinal scale essentially transforms any frequency distribution into a uniform distribution; this is always the conversion and is exact if there are no ties.

What is the range of numerical possibilities here? If the two samples are identical (they consist of exactly the same set of distinct values), then the difference between their respective sample means is zero, and $R_1 = R_2 = [n(n + 1)/2]/2$. Suppose that $n_1 = n_2 = n/2$. Then

$$U_a = U_b = (n/2)(n/2) + (n/2)(n/2 + 1)/2 - [n(n+1)/2]/2 = n^2/8 = U.$$

In contrast, suppose all the small values appear in one sample, and all the large values appear in the other sample. Then

$$\begin{aligned}
R_1 &= (n/2)(n/2 + 1)/2 \\
R_2 &= n(n + 1)/2 - (n/2)(n/2 + 1)/2 \\
U_a &= (n/2)(n/2) + (n/2)(n/2 + 1)/2 - [n(n + 1)/2 - (n/2)(n/2 + 1)/2] = 0 \\
U_b &= (n/2)(n/2) + (n/2)(n/2 + 1)/2 - (n/2)(n/2 + 1)/2 = n^2/4 \\
U &= 0.
\end{aligned}$$

Therefore, the test statistic U ranges between 0 and $n^2/8$.

<u>Test criteria</u>:

(a) for values of $n_j \le 10$, the critical values of U are given in Table 11.2-A; or,

(b) for $n_j > 10$, the standard normal (Z) distribution is used for the test statistic, where $z = (U - n_1n_2/2)/[n_1n_2(n_1 + n_2 + 1)/12]^{1/2}$, $\mu_U = n_1n_2/2$, and $\sigma_U = [n_1n_2(n_1 + n_2 + 1)/12]^{1/2}$.

The Z distribution is referred to here in part because the standard error of the sampling distribution of U depends only upon the individual sample sizes and does not require the estimation of any parameters.

EXAMPLE PR-11.8. The difference between selected regional means of DMLK in Puerto Rico.

A goodness-of-fit test implies that DMLK is not normally distributed in its parent population; additional test statistics (not reported in this book) imply that this variable should not be normally distributed within the regions of Arecibo (A) and Mayaguez (M), either. Merging and ranking these two samples yields the following data:

DMLK:	0.0	0.0	0.0	0.0	0.0	0.1	0.1	0.5	2.1
Region:	A	M	M	M	M	M	M	M	M
Rank:	3	3	3	3	3	6.5	6.5	8	9

DMLK:	2.7	2.9	3.5	4.4	4.9	5.7	6.1	6.9	22.5
Region:	A	M	M	M	A	M	M	M	M
Rank:	10	11	12	13	14	15	16	17	18

DMLK:	32.3	39.1	39.6	40.9	46.7	50.4	68.1	89.2	368.6
Region:	A	A	M	A	A	A	A	A	A
Rank:	19	20	21	22	23	24	25	26	27

If the hypothesis is

H_0: the mean of DMLK is the same in both the Arecibo and Mayaguez regions,

which is equivalent to saying H_0: $\mu_A = \mu_M$,

then $R_A = 213$, $R_M = 165$, $U_a = 11 \cdot 16 + 16 \cdot (16 + 1)/2 - 165 = 147$, $U_b = 11 \cdot 16 + 11 \cdot (11 + 1)/2 - 213 = 29$, and so $U = 29$, and hence

$$z = (29 - 11 \cdot 16/2)/[11 \cdot 16 \cdot (11 + 16 + 1)/12]^{1/2} = -2.911.$$

For a two-tail 5% level of significance (see Table 7.2-A), the critical values for the z-score are ±1.96. Therefore, the conclusion is that (1) H_0 is rejected, (2) there is evidence to indicate that the means of DMLK in the Arecibo and Mayaguez regions are not equal in the population, and (3) only 5% of the time one would expect to obtain a z-score that falls within the critical region {(-∞, -1.96), (1.96, ∞)} when in fact the two population means are equal.

If the apparent normality assumption violation was ignored, then the conclusion would be that these two samples came from populations having different variances (F = 95.920), and the corresponding difference of means t statistic would be 2.360, which is quite close to the 5% critical value of 2.2281 (10 degrees of freedom); moreover, by decreasing the critical region to 1%, this assumption violation would cause the statistical decision to be reversed. In addition, adjusting the DMLK values (i. e., treating the square root of this variable) so that they more closely conform to a normal distribution would cause the difference between variances F ratio to be reduced to 7.789 and the difference of means t statistic to be increased to 3.116. Consequently, overlooking the normality assumption violation would introduce specification error into this analysis that both would inflate the F ratio about 12-fold and potentially could reverse the statistical decision reached.

EXAMPLE OR-11.1. The difference between the PRECIP values of the northern and southern regions in Ontario.

A goodness-of-fit test implies that PRECIP is not normally distributed in its parent population (see Figure 4.2). Additional tests on samples from the population imply that this variable should not be normally distributed within the northern (N) and southern (S) regions of Ontario, either. Merging and ranking these two samples yields the following data:

PRECIP:	22.0	25.0	26.5	26.5	28.0	28.5	28.5
Region:	N	N	N	N	S	N	N
Rank:	1	2	3.5	3.5	5	7	7

PRECIP:	28.5	29.5	30.0	30.5	31.0	31.0	31.5
Region:	N	S	S	N	S	N	N
Rank:	7	9	10	11	12.5	12.5	14

PRECIP:	32.0	32.0	32.5	34.0	34.0	34.5	35.0
Region:	S	N	N	S	S	S	S
Rank:	15.5	15.5	17	18.5	18.5	20	21.5

PRECIP:	35.0	35.5	36.0	36.0	36.0	37.0	37.0
Region:	S	S	S	S	S	N	S
Rank:	21.5	23	25	25	25	27.5	27.5

If the hypothesis is

H_0: the mean of PRECIP is the same in both the northern and southern Ontario regions,

which is equivalent to saying H_0: $\mu_N = \mu_S$,

then $R_N = 128.5$, $R_S = 277.5$, $U_a = 13 \cdot 15 + 15 \cdot (15 + 1)/2 - 277.5 = 37.5$, $U_b = 13 \cdot 15 + 13 \cdot (13 + 1)/2 - 128.5 = 157.5$, and so $U = 157.5$, and hence $z = (157.5 - 13 \cdot 15/2)/[13 \cdot 15 \cdot (13 + 15 + 1)/12]^{1/2} = 2.764$. For a two-tail 5% level of significance (see Table 7.2-A), the critical values for the z-score are ±1.96. Therefore, the conclusion is that (1) H_0 is rejected, (2) there is evidence to indicate that the means of PRECIP in northern and southern Ontario regions are not equal in the population, and (3) only 5% of the time one would expect to obtain a z-score that falls within the critical region $\{(-\infty, -1.96), (1.96, \infty)\}$ when in fact the two population means are equal.

If the apparent normality assumption violation is ignored, then the conclusion might be that these two samples came from populations having different variances (F = 1.96), and the corresponding difference of means t statistic would be 3.43. Consequently, overlooking the normality assumption violation would introduce specification error into this analysis, which would raise the test statistic value from 2.764 to 3.43.

11.3. The difference between k independent population means

A considerably more complicated extension of the hypothesis presented in Chapter 10 involves k population means. Here the null hypothesis becomes H_0: $\mu_1 = \mu_2 = \ldots = \mu_k$, and the alternate hypothesis becomes H_a: $\exists$ g and h such that $\mu_g \neq \mu_h$, $g \neq h$. In other words, which implication from Example PR-11.4 is most reasonable?

11.3.1. Classical analysis of variance

To generalize the difference between means problem, those assumptions spelled out in Section 11.2.1 need to be modified here.

Assumptions:

(a) the variable under study is measured at an interval or ratio scale;

(b) the variable is normally distributed within each of the k regions, having mean μ_j in region j;

(c) the regional variances σ_j^2 are unknown but equal to σ^2 (i. e., constant, stationary, or homogeneous variance);

(d) independent samples of size n_j are drawn randomly from each population (with $\sum_{j=1}^{k} n_j = n$); and,

(e) the observations are independent.

Now each datum will need two subscripts; subscript i will still denote the observation number, and subscript j will denote the region number, yielding the notation x_{ij} for observation i in region j. Accordingly, $\sum_{j=1}^{k} \sum_{i=1}^{n_j} x_{ij}/n = \bar{x}_T$, the overall sample mean for all regions pooled together, and $\sum_{i=1}^{n_j} x_{ij}/n_j = \bar{x}_j$ for the sample mean of region j. The analytical goal in testing H_0 becomes somewhat complicated, since the presence of more than two means does not allow a single difference of means equalling zero statement to be posited. To circumvent this complication, the null hypothesis is tested in a rather indirect fashion by partitioning the total variation displayed by the variable under study into two components, namely, variation within regions, and variation between regions. Since variation is being analyzed, the statistical procedure employed here has been named analysis of variance and goes by the acronym ANOVA.

Test criteria:

(a) the total variation $\sum_{j=1}^{k} \sum_{i=1}^{n_j} (x_{ij} - \bar{x}_T)^2$ is partitioned into the sum of the within-regions variation $\sum_{j=1}^{k} \sum_{i=1}^{n_j} (x_{ij} - \bar{x}_j)^2$ and the between-regions variation $\sum_{j=1}^{k} n_j(\bar{x}_j - \bar{x}_T)^2$;

(b) unbiased estimates of the variance terms here are obtained by computing

$$s_T^2 = \sum_{j=1}^{k} \sum_{i=1}^{n_j} (x_{ij} - \bar{x}_T)^2/(n - 1),$$

where n - 1 is the number of degrees of freedom for the overall variation (see Section 9.3),

$$s_w^2 = \sum_{j=1}^{k} \sum_{i=1}^{n_j} (x_{ij} - \bar{x}_j)^2/(n_j - 1),$$

with an aggregate of $\sum_{j=1}^{k} (n_j - 1) = n - k$ degrees of freedom (there are k mean estimates, $\bar{x}_j$, and hence k constraints here), and

$$s_b^2 = \sum_{j=1}^{k} n_j(\bar{x}_j - \bar{x}_T)^2/(k - 1),$$

with k - 1 degrees of freedom (there are k means free to vary, restricted by the single constraint of the overall mean, $\bar{x}_T$); and,

(c) because variances are being evaluated, the test statistic is

$$F = s_b^2/s_w^2 = [\sum_{j=1}^{k} n_j(\bar{x}_j - \bar{x}_T)^2/(k - 1)]/ [\sum_{j=1}^{k} \sum_{i=1}^{n_j} (x_{ij} - \bar{x}_j)^2/(n - k)],$$

which is distributed with k - 1 degrees of freedom in the numerator and n - k degrees of freedom in the denominator.

Note that (1) the degrees of freedom are additive (i. e., n - 1 = n - k + k - 1), and (2) in contrast to the difference between two independent variances F ratio, in ANOVA the between-regions variance estimate *always* is in the numerator, regardless of whether it is larger or smaller than the within-regions variance estimate.

For a true null hypothesis, the population value of this F ratio becomes

$$F = [\sum_{j=1}^{k} n_j(\mu_j - \mu_T)^2/(k - 1)]/[\sum_{j=1}^{k} \sum_{i=1}^{n_j} (x_{ij} - \mu_j)^2/(n - k)]$$

$$= [\sum_{j=1}^{k} n_j(\mu - \mu)^2/(k - 1)]/[\sum_{j=1}^{k} \sum_{i=1}^{n_j} (x_{ij} - \mu)^2/(n - k)] = 0$$

since all the regional means are equal. Furthermore, $E(s_b^2) = E(s_w^2)$, which causes the expected value of the F ratio to be 1. Therefore, the pair of hypotheses concerning equality of k regional means are translated into

H_0: $F = \sigma_b^2/\sigma_w^2 = 1$
H_a: $F > 1$.

The one-tail test is relevant here because when the null hypothesis is true, perfectly representative samples will yield an F ratio value of 0 (see the previous discussion).

ANOVA

Source of variation	Sum of squares	Degrees of freedom	Mean squares	F ratio
Between regions	$\sum_{j=1}^{k} n_j(\bar{x}_j - \bar{x}_T)^2$	n - k	$s_w^2 = \sum_{j=1}^{k} n_j(\bar{x}_j - \bar{x}_T)^2/(n - k)$	s_b^2/s_w^2
Within regions	$\sum_{j=1}^{k} \sum_{i=1}^{n_j} (x_{ij} - \bar{x}_j)^2$	k - 1	$s_b^2 = \sum_{j=1}^{k} \sum_{i=1}^{n_j} (x_{ij} - \bar{x}_j)^2/(k - 1)$	
Total	$\sum_{j=1}^{k} \sum_{i=1}^{n_j} (x_{ij} - \bar{x}_T)^2$	n - 1		

Almost without exception, the output of an ANOVA is summarized in tabular form. Its format is presented here in the table entitled ANOVA. The "mean squares" terms are the unbiased variance estimates.

EXAMPLE PR-11.9. The difference between the five regional means of DCD in Puerto Rico.

In Example PR-11.3 the inference drawn was that the five regional variances of DCD are equal in the population. The conclusion reached in Example 10.5 was that DCD is normally distributed in the parent populations. DCD is measured at the interval/ratio scale (see Chapter 1), and the five regional samples are independent in that they are different sets of municipios. If the assumption of zero spatial autocorrelation is overlooked, then the hypothesis is

H_0: the means of DCD in the San Juan, Arecibo, Mayaguez, Ponce, and Caguas regions are equal,

which is equivalent to saying H_0: $\mu_{SJ} = \mu_A = \mu_M = \mu_P = \mu_C$.

The detailed information for this example is, with $\bar{x}_T = 2640.7/73 = 36.174$,

Region	$n_j - 1$	$\bar{x}_j$	$\sum_{i=1}^{n_j}(x_{ij} - \bar{x}_j)^2$	$n_j(\bar{x}_j - \bar{x}_T)^2$
SJ	18	502.3/19 = 26.437	3221.264	$19 \cdot (26.437 - 36.174)^2$ = 1801.374
A	10	436.0/11 = 39.636	777.965	$11 \cdot (39.636 - 36.174)^2$ = 131.840
P	15	621.2/16 = 38.825	1688.770	$16 \cdot (38.825 - 36.174)^2$ = 112.445
M	13	624.0/14 = 44.193	1092.617	$14 \cdot (44.286 - 36.174)^2$ = 921.264
C	12	461.2/13 = 35.477	542.823	$13 \cdot (35.477 - 36.174)^2$ = 6.316
Column sum	68		7323.439	2973.239

Note that the column labelled $\sum_{i=1}^{n_j}(x_{ij} - \bar{x}_j)^2$ can be obtained directly from the regional variances estimates of DCD presented in Example PR-11.1. These details tabulate into the standard ANOVA table (except for rounding error) as follows:

ANOVA of DCD

Source of variation	Sum of squares	Degrees of freedom	Mean squares	F ratio
Between regions	2973.239	4	2973.239/4 = 743.310	743.310/107.698 = 6.902
Within regions	7323.439	68	7323.439/68 = 107.698	
Total	10296.678	72		

For a 5% level of significance (see Table 11.1-A), the F ratio critical value with 4 and 68 degrees of freedom is 2.5231 (using linear interpolation). Therefore, the conclusion is that (1) H_0 is rejected, (2) there is evidence to indicate that at least one of the five regional means of DCD is not equal to the others, or equivalently, not equal to the population mean, and (3) only 5% of the time one would expect to obtain an F ratio value that falls within the critical region (2.5231, ∞) when in fact the five population means are equal.

The finding reported here suggests that the geographic process in question is nonhomogeneous and that the homogeneity conclusion reached in Example PR-11.4 is a Type II error.

Two problems associated with analysis of variance merit attention at this time. The first deals with the question of which pairs of means out of the k(k - 1)/2 possibilities (this is a combinatorial count, since order is not important) differ. In other words, what is the best alternate hypothesis? The solution to this problem requires concepts from multivariate analysis, but it is mentioned here so that the reader may be aware that a solution, albeit rather complex, does exist. The second problem deals with cross-classifications. The regional classification constitutes a multinomial nominal scale variable (see Chapter 2). A researcher may be interested in the simultaneous effects of several such variables, thus exploring the resulting cross-classification groupings. Rarely are more than three classification variables treated at a time, although the technical expertise and statistical theory exist for handling any number of these classification variables. Most introductory statistical textbooks address the case of at least two classification variables, noting that such an analysis must be balanced (the number of observations in each of the cross-classification groups must be the same; a solution also exists for proportional numbers of observations). Unfortunately, in spatial analysis this technical restriction of a balanced design infrequently holds, and such cross-classifications are almost always unbalanced. Again, the general problem of unequal numbers of cross-classification observations has a complex solution requiring concepts from multivariate analysis and so will not be discussed here (see Chapter 13).

11.3.2. Analysis of variance for nonnormal or heterogeneous variance variables

Section 11.2.3 presents the solution for the difference between two population means when the frequency distribution of a variable is nonnormal. Generalizing this solution to k population means is equivalent to discarding assumption (b) of Section 11.3.1; assumption (c) can also be discarded. The Kruskal-Wallis H statistic is a nonparametric alternative to the F ratio for classical analysis of variance. It is a test statistic much like the Mann-Whitney U statistic, based upon an ascending ranking of data values (again, any underlying frequency distribution essentially is converted to a uniform distribution). It is an analysis of variance of ranks and actually is a difference between medians test. As before, assumption (b) of Section 11.3.1 is replaced with

(b) the random variable under study is continuous.

The preliminary calculations for computing H are as follows:

Step 1: Merge the k groups of data, and rank their values in ascending order (recalling from Chapter 4, ties are equated to the mean of their positional ranks).

Step 2: Sum the ranks r_{ij} for each of the k samples, denoting the sum for region j as $R_j = \sum_{i=1}^{n_k} r_{ij}$ [note that $\sum_{j=1}^{k} R_j = n(n + 1)/2$].

$E(r_{ij}) = (n + 1)/2$, and $E(R_j/n_j) = (n + 1)/2$. Test statistic H measures the extent to which the k individual R_j/n_j values deviate from their common expected value of $(n + 1)/2$.

Test criteria:

(a) $H = \{12\sum_{j=1}^{k} (R_j^2/n_j)/[n(n + 1)] - 3(n + 1)\}/[1 - \sum_{j=1}^{p} (t_i^3 - t_i)/(n^3 - n)]$,

where t_i is the number of observations tied for a given rank, of which there are p; and,

(b) if $n_j > 5$ for each region j, then the sampling distribution of H follows a chi square distribution with k - 1 degrees of freedom.

For $n_j \le 5$, the exact sampling distribution of H is determined by the randomization method discussed earlier; clearly, construction of this sampling distribution is computationally tedious.

What is the range of numerical possibilities here? If the k samples are identical (they consist of exactly the same set of distinct values), then the difference between their respective sample means is zero, and $R_j = [n(n + 1)/2]/k$. Suppose that $n_j = n/k$ (a balanced setup). Then

$$H = \{(12k \bullet [n(n + 1)/2/k]^2/(n/k)/[n(n + 1)] - 3(n + 1)\}/[1 - (n/k)(k^3 - k)/(n^3 - n)] = 0$$

which is the desired result for χ^2 under H_0. In contrast, if no values are the same, then $t_i^3 - t_i = 0$ in every case, and the denominator reduces to 1.

EXAMPLE PR-11.10. The difference between the five regional means of DMLK in Puerto Rico.

Noted in Example PR-11.8 is the failure of the frequency distribution of DMLK to conform to a normal distribution; this assumption also is violated in the Arecibo and Mayaguez regions. A test of variance homogeneity yields B = 110.698 (see Section 11.2.3 for the formula for B), which is highly significant. But whether or not this large value is due to a false null hypothesis, specification error attributable to the nonnormality of the variable, or both is not known. As mentioned before, analyzing the square root of DMLK results in a frequency distribution that does conform to a normal distribution; this finding holds for the variable DMLK as well as for this variable in each of the five regions. Computing B for this transformed data yields a value of 26.309, which is significant and whose magnitude should lack any serious specification error. Classical ANOVA results for this transformed variable yield F = 8.41, which is highly significant. Again, though, is this large value due to a false null hypothesis, or specification error attributable to the nonhomogeneity of variance? These parametric results suggest that perhaps a nonparametric analysis of DMLK is in order; thus, extending Example PR-11.8 to all five regions yields the following ranked data:

DMLK:	0.0	0.0	0.0	0.0	0.0	0.0	0.0	0.0	0.0
Region:	P	M	P	M	SJ	M	P	M	C
Rank:	6	6	6	6	6	6	6	6	6

DMLK:	0.0	0.0	0.1	0.1	0.5	1.7	1.7	2.0	2.1
Region:	A	P	M	M	M	P	SJ	SJ	P
Rank:	6	6	12.5	12.5	14	15.5	15.5	17	18.5

DMLK:	2.1	2.2	2.3	2.7	3.1	3.5	3.8	4.4	4.6
Region:	M	SJ	P	A	SJ	M	M	M	P
Rank:	18.5	20	21	22	23	24	25	26	27

DMLK:	4.9	5.0	5.3	5.3	5.7	6.1	6.9	6.9	7.8
Region:	A	P	P	P	M	M	M	SJ	P
Rank:	28	29	30.5	30.5	32	33	34	35	36

DMLK:	8.6	9.6	10.7	10.8	10.9	12.0	15.0	15.2	16.6
Region:	P	SJ	C	SJ	P	SJ	C	C	C
Rank:	37	38	39	40	41	42	43	44	45

DMLK:	18.0	19.2	20.8	22.5	23.7	27.4	28.6	32.2	32.5
Region:	C	SJ	SJ	M	C	SJ	SJ	A	C
Rank:	46	47	48	49	50	51	52	53	54

DMLK:	39.1	39.6	40.9	43.0	45.1	46.7	49.1	50.4	60.0
Region:	A	M	A	SJ	SJ	A	SJ	A	C
Rank:	55	56	57	58	59	60	61	62	63

DMLK:	66.6	68.1	68.2	75.2	82.8	89.2	91.1	94.0	134.4
Region:	SJ	A	C	C	C	A	SJ	SJ	C
Rank:	64	65	66	67	68	69	70	71	72

DMLK:	368.6
Region:	A
Rank:	73

The hypothesis is

H_0: the means of DMLK in the San Juan, Arecibo, Mayaguez, Ponce, and Caguas regions are equal,

which is equivalent to saying H_0: $\mu_{SJ} = \mu_A = \mu_M = \mu_P = \mu_C$.

Calculating the Kruskal-Wallis test statistic for these data renders

$$H = \{12 \bullet (817^2/19 + 550^2/11 + 361^2/16 + 310^2/14 + 663^2/13)/[73(73 + 1)] - 3(73 + 1)\}/\{1 - [(11^3 - 11) + 5(2^3 - 2)]/(73^3 - 73)\} = 25.672.$$

For a 5% level of significance (see Table 10.1-A), the critical value with 5 - 1 = 4 degrees of freedom is 9.4877. Therefore, the conclusion is that (1) H_0 is rejected, (2) there is evidence to indicate that the mean of DMLK for at least one of the five agricultural administrative regions is not equal to the others or, equivalently, to the population, and (3) only 5% of the time one would expect to obtain a chi square statistic that falls within the critical region (9.4877, ∞) when in fact the five population means are equal. Note that the detected difference is not strictly due to specification error!

EXAMPLE OR-11.2. The difference between three means of runoff (RUN) for drainage basins in Ontario.

The variable RUN for Ontario drainage basins is standardized such that runoff values are amounts per unit area, allowing for direct comparisons between basins and regions. The mechanisms by which water is yielded from these basins are probably influenced by size. In other words, a working hypothesis is that, per unit area, smaller basins yield more water than larger basins. To investigate this notion, the mean runoff for small-sized basins (S: < 1000 km²), medium-sized basins (M: 1000-3000 km²), and large-sized basins (L: > 3000 km²) can be compared. Although the runoff values for the entire data set approximate a normal distribution, the normality of the three subsets may be questionable due to the small and variable sample sizes (n_j = 8 or 12). Without formal goodness-of-fit tests, results from classical ANOVA become suspect here. Hence, a nonparametric analysis of RUN is in order. Ordering and ranking the data yield

RUN:	97	132	140	155	157	169	181	218	243	252
Basin:	L	L	M	L	L	L	S	M	S	S
Rank:	1	2	3	4	5	6	7	8	9	10

RUN:	271	273	275	277	278	287	308	311	341	374
Basin:	L	L	S	S	L	M	S	S	S	S
Rank:	11	12	13	14	15	16	17	18	19	20

RUN:	385	387	397	410	419	455	471	471
Basin:	M	M	M	M	S	S	M	S
Rank:	21	22	23	24	25	26	27.5	27.5

The hypothesis is

H_0: the means of unit runoff (RUN) in small, medium, and large drainage basins in Ontario are equal,

which is equivalent to saying H_0: $\mu_S = \mu_M = \mu_L$.

Calculating the Kruskal-Wallis test statistic for these data renders

$$H = \{12 \bullet (205.5^2/12 + 144.5^2/8 + 56^2/8)/[28(28 + 1)] - 3(28 + 1)\}/\{1 - (2^3 - 2)/(28^3 - 28)\} = 9.375.$$

For a 5% level of significance (see Table 10.1-A), the critical value with 3 - 1 = 2 degrees of freedom is 5.991. Therefore, the conclusion is that (1) H_0 is rejected, (2) there is evidence to indicate that the mean of RUN for at least one of the three size groups is not equal to the others or, equivalently, to the population, and (3) only 5% of the time one would expect to obtain a chi square statistic that falls within the critical region (5.991, ∞) when in fact the three population means are equal.

There is considerable confidence that the inference drawn with statistic H is reliable and free of specification error problems. Its power is approximately 95% of that of ANOVA, for large samples; hence, this increase in confidence is attained at a relatively small price.

Just as in classical ANOVA, multiple pairwise comparisons of medians can also be made. Further, nonparametric alternatives exist for correlated sample problems, too. The interested reader should consult a more comprehensive statistics text for discussion of these procedures.

11.4. The difference between two correlation coefficients

Simple hypotheses concerning correlation coefficients were treated in Chapter 10. In Section 10.6 the more thorough understanding of a product moment correlation coefficient was recognized through the ability to test a null hypothesis of the form H_0: $\rho = a$; insufficient knowledge about sampling distributions for the other correlation coefficients prevents the testing of parallel null hypotheses for them. Analogously, the ability to draw inferences about correlation coefficients may be extended to differences between pairs of coefficients *only* for the product moment case. With respect to this particular correlation coefficient, recall that inferential difficulties attributable to the skewness of the sampling distribution of r_j are resolved by transforming this statistic into $[\ln(1 + r_j) - \ln(1 - r_j)]/2$, which causes the resulting sampling distribution to conform to a normal distribution.

11.4.1. The difference between two correlation coefficients for independent samples

Again the question being addressed by a researcher asks whether or not the relationship between two interval/ratio variables is stationary over geographic regions. In other words, the null hypothesis becomes H_0: $\rho_1 = \rho_2$. Given the question being posed, the alternate hypothesis usually is stated as H_a: $\rho_1 \neq \rho_2$. As with other hypothesis testing situations, the analysis here is simplified considerably if the null hypothesis is rewritten as a difference, which in its logarithmic form becomes H_0: $\{[\ln(1 + \rho_1) - \ln(1 - \rho_1)] - [\ln(1 + \rho_2) - \ln(1 - \rho_2)]\}/2 = 0$.

Assumptions:

(a) both variables under study are measured at an interval or ratio scale;

(b) the pair of variables is jointly normally distributed within each of the two regions;

(c) independent samples of size n_j are drawn randomly from each of the two populations; and,

(d) the observations are independent.

Test criteria:

(a) the standard normal distribution provides the test statistic, where

$$z = [\{[\ln(1 + r_1) - \ln(1 - r_1)] - [\ln(1 + r_2) - \ln(1 - r_2)]\}/2 - \{[\ln(1 + \rho_1) - \ln(1 - \rho_1)] - [\ln(1 + \rho_2) - \ln(1 - \rho_2)]\}/2]/[1/(n_1 - 3) + 1/(n_2 - 3)]^{1/2}$$

$$= [\{[\ln(1 + r_1) - \ln(1 - r_1)] - [\ln(1 + r_2) - \ln(1 - r_2)]\}/2]/[1/(n_1 - 3) + 1/(n_2 - 3)]^{1/2}$$

under H_0; or,

(b) although the standard normal distribution is appropriate, for small n (perhaps < 30), some researchers revert to the t distribution in order to accommodate the increase in variability due to small sample size.

Given the test criteria outlined here, appropriate confidence intervals can be constructed as follows:

$$\{[\ln(1 + r_1) - \ln(1 - r_1)] - [\ln(1 + r_2) - \ln(1 - r_2)]\}/2 - z_{1-\alpha/2}[1/(n_1 - 3) + 1/(n_2 - 3)]^{1/2}$$

$$\leq \{[\ln(1 + \rho_1) - \ln(1 - \rho_1)] - [\ln(1 + \rho_2) - \ln(1 - \rho_2)]\}/2 \leq$$

$$\{[\ln(1 + r_1) - \ln(1 - r_1)] - [\ln(1 + r_2) - \ln(1 - r_2)]\}/2 + z_{1-\alpha/2}[1/(n_1 - 3) + 1/(n_2 - 3)]^{1/2}.$$

Of course this interval result looks very similar to those found in Chapter 10. But, exact algebraic manipulations cannot be applied here (as illustrated in Chapter 10) to simplify this confidence interval.

EXAMPLE PR-11.11. The difference between the correlation coefficient for DCD and DMLK, and DCD and DSGR for selected regions in Puerto Rico.

Testing for conformity to a joint normal distribution requires concepts from multivariate analysis; accordingly, for illustrative purposes here, joint normality will be assumed. For the two variables DCD and DMLK, the sample correlation coefficients in the Ponce and the Caguas regions are $r_P = 0.185$ and $r_C = 0.151$. If the hypothesis is

H_0: the correlation between DCD and DMLK is the same in both the Ponce and the Caguas regions,

which is equivalent to saying H_0: $\rho_P = \rho_C$,

then

$$z = [\{[\ln(1 + 0.186) - \ln(1 - 0.186)] - [\ln(1 + 0.150) - \ln(1 - 0.150)]\}/2]/[1/(14 - 3) + 1/(13 - 3)]^{1/2} = 0.085.$$

For a two-tail 5% level of significance (see Table 7.2-A), the critical values for the z-score are ±1.96. Therefore, the conclusion is that (1) H_0 fails to be rejected, (2) there is no evidence to indicate that the correlation between DCD and DMLK in the Ponce and Caguas regions is not equal in the population, and (3) 95% of the time one would expect to obtain a z-score value that falls within the confidence interval (-1.96, 1.96) when in fact the two population correlation coefficients are equal. This result implies that a homogeneous geographic process is operating.

Next, for the two variables DMLK and DSGR, the sample correlation coefficients in the San Juan and the Caguas regions are $r_{SJ} = 0.413$ and $r_C = -0.376$. If the hypothesis is

H_0: the correlation between DMLK and DSGR is the same in both the San Juan and Caguas regions,

which is equivalent to saying H_0: $\rho_{SJ} = \rho_C$,

then

$$z = [\{[\ln(1 + 0.414) - \ln(1 - 0.414)] - [\ln(1 - 0.374) - \ln(1 + 0.374)]\}/2]/[1/(19 - 3) + 1/(13 - 3)]^{1/2} = 2.068.$$

For a two-tail 5% level of significance (see Table 7.2-A), the critical values for the z-score are ±1.96. Therefore, the conclusion is that (1) H_0 is rejected, (2) there is evidence to indicate that the correlation coefficients for DMLK and DSGR in the San Juan and Caguas regions are not equal in the population, and (3) only 5% of the time would one expect to obtain a z-score that falls within the critical region $\{(-\infty, -1.96), (1.96, \infty)\}$ when in fact the two population correlation coefficients are equal. This result implies that a nonhomogeneous geographic process is operating.

Applying a test of significance in a situation of this kind is simple. Interpreting the substantive meaning of such a difference may be difficult.

11.4.2. The difference between two correlation coefficients for correlated samples

Engaging in analyses that approach being multivariate in nature will necessitate that a researcher work with correlation coefficients computed for the same set of areal units (a source of communality in statistical results). Comparing correlation coefficients frequently is required in this context. But the method described in the preceding section does not apply here. Rather, if three correlation coefficients are calculated, a comparison of any pair (there are three possibilities--a combinatorial count, since order is not important) requires that the effects summarized in the third coefficient be partialed out of the resulting comparison. For example, if the correlations between variables 1 and 3, and 2 and 3 are to be compared, then the correlation between variables 2 and 3 has to be removed first (if this correlation is zero, then there is nothing to remove).

Assumptions:

(a) both variables under study are measured at an interval or ratio scale;

(b) the pair of variables is jointly normally distributed within the region being studied;

(c) a sample of size n is drawn randomly from the population; and,

(d) the observations are independent.

Test criteria:

(a) the t distribution furnishes the test statistic, where

$$t = [(r_{12} - r_{13}) - (\rho_{12} - \rho_{13})]/\{2(1 - r_{12}^2 - r_{13}^2 - r_{23}^2 + 2r_{12}r_{13}r_{23})/[(n - 3) \times(1 + r_{23})]\}^{1/2}$$

$$= (r_{12} - r_{13})/\{2(1 - r_{12}^2 - r_{13}^2 - r_{23}^2 + 2r_{12}r_{13}r_{23})/[(n - 3)(1 + r_{23})]\}^{1/2}$$

under H_0, with n - 3 degrees of freedom.

The standard error involves three estimates, namely, r_{12} (the correlation between variables 1 and 2), r_{13} (the correlation between variables 1 and 3), and r_{23} (the correlation between variables 2 and 3), and therefore has three constraints attached to it. This test statistic is a generalization of the generic one presented in Chapter 10; this generalization is recognizable by analogy but requires multivariate statistical concepts to be formally established.

EXAMPLE PR-11.12. The difference between the correlation coefficient for DCD and DMLK, and DCD and DSGR for the San Juan region in Puerto Rico.

The sample correlation coefficients in this region are $r_{DCD,DMLK} = 0.038$, $r_{DCD,DSGR} = -0.006$, and $r_{DMLK,DSGR} = 0.414$ (see Chapter 4 for sample calculations). If the hypothesis is

H_0: the correlation between DCD and DMLK is the same as that between DCD and DSGR in the San Juan region,

which is equivalent to saying H_0: $\rho_{DCD,DMLK} = \rho_{DCD,DSGR}$,

then

$$t = [0.037 - (-0.006)]/\{2[1 - 0.037^2 - (-0.006)^2 - 0.413^2 + 2(0.037)(-0.006)(0.413)]/[(19 - 3)(1 + 0.413)]\}^{1/2} = 0.159.$$

For a two-tail 5% level of significance (see Table 9.1-A), the critical values for the t statistic are ±2.1199 (using 16 degrees of freedom). Therefore, the conclusion is that (1) H_0 fails to be rejected, (2) there is no evidence to indicate that the correlation between DCD and DMLK and DCD and DSGR are not equal in the San Juan population, and (3) 95% of the time one would expect to obtain a t statistic that falls within the confidence interval (-2.1199, 2.1199) when in fact the two population correlation coefficients are equal.

With a result like that obtained in Example PR-11.12, there is a reasonable probability that the difference between the two correlation coefficients $r_{DCD,DMLK}$ and $r_{DCD,DSGR}$ can be explained in terms of sampling error.

11.5. Summary

This chapter has explored common techniques in complex hypothesis testing. Although these tests are often ends in themselves, they are equally often used as intermediate steps in more sophisticated data analyses. The ability to compare sample means, variances, and correlations with a population, or with other samples, says little about the relationship between sampled populations. The ability to mathematically describe this relationship requires identification of the rules of correspondence involved (regression analysis), which is the focus of Chapter 12.

11.6. Problems

1. Verify the calculations in Examples PR-11.1 through PR-11.11.

2. Reevaluate the results of Example PR-11.4 assuming that n_1 and n_2 are too small for negligible specification error to be attributable to the approximation.

3. Test the null hypothesis H_0: the variance of DCD is the same in both the Arecibo and Ponce regions.

4. Test the null hypothesis H_0: the variances for DMLK and DSGR in the Caguas region are equal.

5. First using DMLK, and then using LOG(DMLK + 1), test the null hypothesis H_0: the variance of this variable does not differ across the five agricultural administrative regions of Puerto Rico.

6. For the case of $n_1 = n_2 = 2$, and populations $P_1 = \{1, 2\}$ and $P_2 = \{1, 2\}$, construct the sampling distribution for the difference of means. Show that this distribution is approximately normally distributed, has a mean of zero, and has a variance of 0.25. Then, construct the confidence interval for each of the 16 possible samples, identifying which contain and which do not contain the population difference of means.

7. Test the null hypothesis H_0: the mean of DCD is the same in both the Arecibo and Caguas regions, assuming that (1) $\sigma_A^2 = \sigma_C^2 = 11.8765$, and (2) the regional variances are unknown.

8. Consulting Example PR-11.2, test the null hypothesis H_0: the mean of DCD during 1969 equals the mean of DCD during 1974 in the Caguas region.

9. First, show that DTB does not conform to a normal distribution. Then, test the null hypothesis H_0: the mean of DTB is the same in both the San Juan and Ponce regions.

10. Redo Example PR-11.5 as an ANOVA problem, and show that the F ratio value obtained is equal to the t^2 value of Example PR-11.5.

11. Draw a random sample of size $n_j = 2$ for DCD, j = 1, 2, 3, 4, 5. How many possible samples are there for this situation? Perform an ANOVA for these data. How do these results compare with those reported in Example PR-11.9?

12. Compare ANOVA results for LOG(DMLK + 1) reported in Examples PR-11.9 and PR-11.10.

13. Conduct an ANOVA test for RUN, as in example OR-11.2, and compare the ANOVA results with the results of the Kruskal-Wallis analysis given in OR-11.2.

14. Test the null hypothesis H_0: the five regional means for DTB are equal.

15. Test the null hypothesis H_0: the correlation between DCD and DMLK is the same in both the San Juan and the Caguas regions.

16. Test the null hypothesis H_0: the correlation between DCD and DMLK is the same as that between DCD and DSGR in the Ponce region.

11.7. Additional readings

In addition to the texts listed in Chapter 1,

Silk, J. *The Analysis of Variance.* CATMOG 30

Iversen, G. and N. Norporth. *Analysis of Variance.* SAGE 1.

Siegel, S. (1956) *Nonparametric Statistics for the Behavioral Sciences.* New York: McGraw-Hill.

CHAPTER 12

SIMPLE LINEAR REGRESSION

The notion that a descriptive equation can be written for each observation in a sample or a population was outlined in Chapter 4 (see Section 4.3). This idea has been closely linked to that of relationships between variables. An entire set of problems revolves around the theme of determining whether or not two variables are related; hypothesis testing for this situation was summarized in Chapters 10 and 11. Going one step further, another set of problems concerns the prediction of one variable from another. This class of problems is important within two different contexts:

(1) one variable may have a causal impact upon another variable; this is not often the case in geographical analyses; and,

(2) data may be much easier to collect on one of two related variables, and a researcher wishes to exploit this redundancy of information in order to predict elusive variable values, given available variable values.

In the case of prediction, then, the labelling of variables as X and Y no longer is arbitrary. The variable to be predicted is viewed as being dependent upon the variable from which a prediction is to be made, and as such, conventionally Y is designated the dependent variable and X is designated the independent variable. In other words, a change in the value of variable X is related, in some manner, to a corresponding (and, it is hoped, predictable) change in the value of variable Y. However, rarely can the exact value of an occurrence be predicted. Statistically speaking, it is sufficient if the predictions are, on the average, correct (i. e., unbiased). This problem type differs from correlation analysis in that the goal is to establish the statistical rule of correspondence (an equation) that connects two variables. While in most circumstances no causal relationship is asserted, it is possible to construct causal models using this perspective. The statistical technique applicable for estimating this rule is **simple linear regression**, and the resulting line is called a **regression line**. In general there are a number of valid reasons for seeking to uncover this rule. Those found most relevant by spatial analysts are addressed in this chapter. Again, this sort of analysis falls under the heading of complex hypothesis testing, since simultaneous inferences about multiple parameters are to be drawn.

Before beginning a discussion of specific applications, it is important to distinguish between the steps in the proper application of a statistical procedure and the reason why the analysis is being undertaken. The focus of this book is necessarily on the former. Often, however, the most difficult part of an analysis is encountered in the latter. In other words, once theory regarding some spatial process is brought to bear on a problem, the appropriate statistical procedure is chosen. Identifying and using a suitable body of theory to formulate and articulate meaningful questions is one of the most difficult aspects of an analysis. Never forget that statistical techniques can be properly used to analyze ridiculous as well as meaningful questions. "Good" spatial analysis requires the application of relevant analytical techniques to useful questions. The correct application of techniques is necessary, but not sufficient. Even less acceptable is the use of statistical techniques to "explore" data in an effort to find questions for which the answers are known; such "fishing expeditions," sometimes called data mining, often lead to very curious hypotheses.

12.1. The concept of linear regression

As has been the case throughout this book, linear relationships are assumed to underlie all statistical relationships. In other words, if a scatter diagram is constructed for two variables, its scatter of points will be well approximated by a straight line. The equation of this line would be of the form

$$Y_i = \alpha + \beta X_i + \xi_i$$

where α is the population intercept, β is the population slope of the equation and represents the change in the value of Y_i for a unit change in the value of X_i, and ξ_i is the random error associated with observation i; the corresponding statistics are denoted as a and b. An example of such a line is shown in Figure 12.1. Because α is a constant over the set of observations, it is equivalent to a mean (it is a conditional mean). Accordingly, the mean of the error terms ξ_i must be zero, otherwise the value of the mean would be added to the intercept term α. For the same reason, the sample estimates of ξ_i, denoted e_i, will have a mean of zero. The simplest perspective shedding light on the meaning of this error again refers to sampling error. The variable X is assumed to be fixed, and a random sample of Ys is selected. This viewpoint yields a conception of error like that illustrated in Figure 12.2, in which all the error is attributed to the Y variable; the X variable is assumed to be error-free.

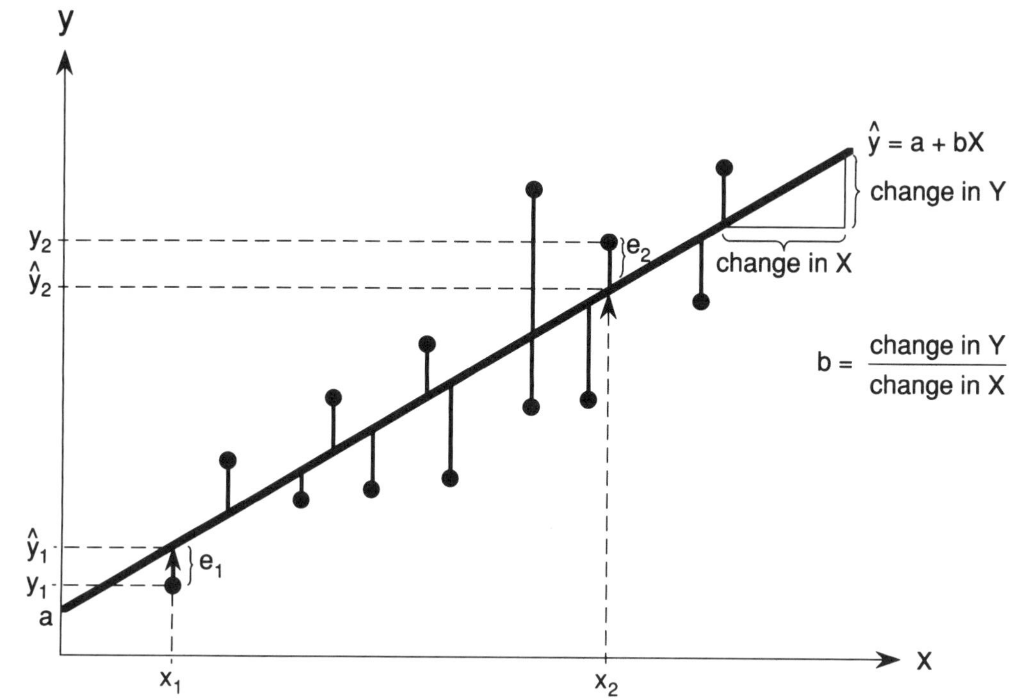

Figure 12.1: AN EXAMPLE REGRESSION LINE

Intuitively speaking, statistical estimation of the parameters α and β (denoted by the letters a and b) should minimize error. The problem, then, is how to minimize error. If a linear relationship seems appropriate, the best-fitting straight line is found by using the method of **ordinary least squares** (OLS), which is one way of minimizing the sample error captured by the $e_i = y_i - a - bx_i$ terms. The calculated errors e_i are estimates of the ξ_i values and are called residuals, since they represent the remaining, or residual, portion of variance not explained by the regression line. Figure 12.3 shows the residuals for two values of Y, given the two values x_1 and x_2, as e_1 and e_2, respectively. This criterion results in the minimization (MIN) problem of

$$\underset{a,b}{\text{MIN}}: \ \sum_{i=1}^{n} e_i^2 = \sum_{i=1}^{n} (y_i - a - bx_i)^2$$

where the sample error values are squared in order for the minimum to occur at zero. Minimizing this expression locates the regression line in a position such that the sum of squares of distances from the observed points (x_i, y_i) to the line taken parallel to the Y-axis is a minimum. It is important to note that the objective of OLS is not to minimize the residual for each value of Y (for example e_1 and e_2 in Figure 12.3), but

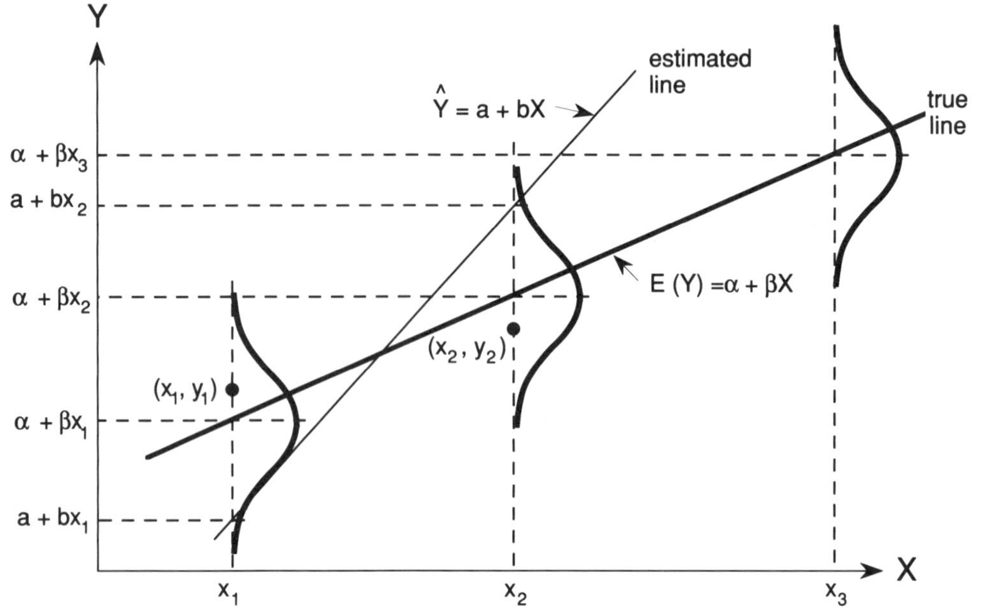

Figure 12.2: DISTRIBUTION OF REGRESSION ERROR

rather to minimize the sum of all squared errors. Since a relatively larger value of e_i, once squared, contributes more to the sum of squared errors than does a smaller squared value, OLS tries to avoid large residuals as much as possible.

Other criteria for positioning the regression line include measuring distance parallel to the X-axis (this is equivalent to regressing X on Y), shown as e_i' in Figure 12.4, or perpendicular to the regression line (this is equivalent to what is done in some more advanced multivariate statistical analyses), shown as e_i'' in Figure 12.4. In fact, there are infinitely many possibilities between e_i and e_i''.

To be correct on average, ordinary least squares establishes the following equation

$$\mu_{y|x} = \alpha + \beta X$$

which states that the expected value of variable Y is conditional and hence non-stationary, changing as a function of X. For instance, in Figure 12.1 the value of $\mu_{y|x}$ occurs where the population regression line cuts through the sampling distribution of error around a given X value. Using this perspective, then, ξ is linked to sampling error.

The solution to the foregoing minimization problem yields the estimates

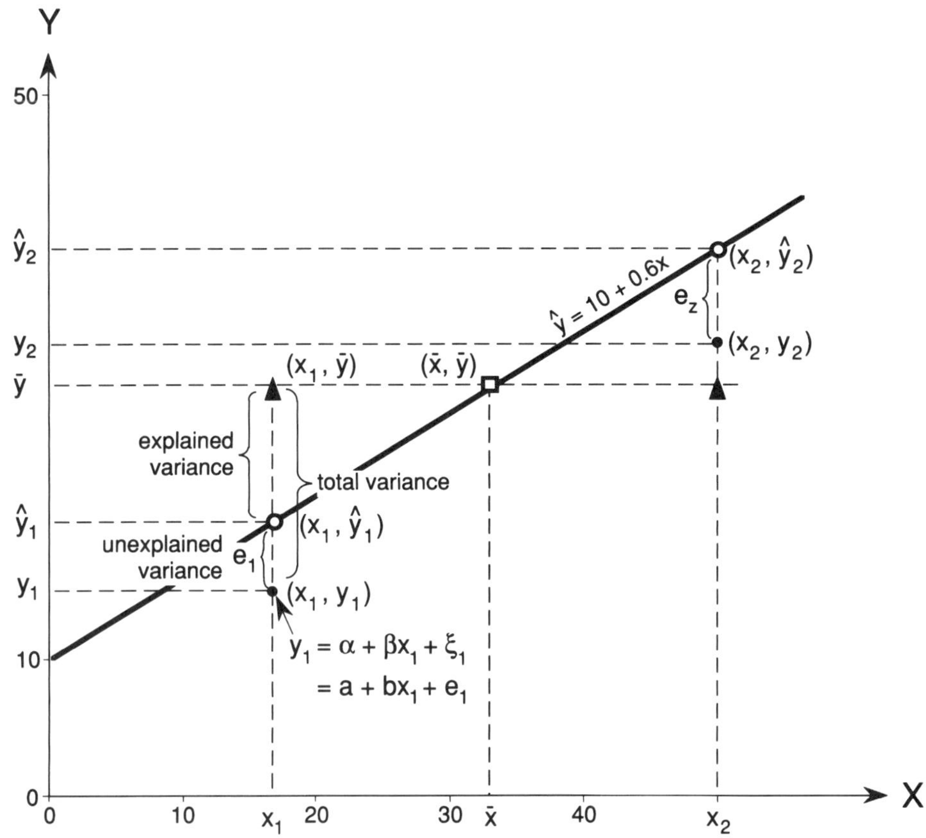

Figure 12.3: THE RESIDUAL AND VARIANCE COMPONENTS OF A REGRESSION LINE

$$a = [(\Sigma_{i=1}^{n} y_i)(\Sigma_{i=1}^{n} x_i^2) - (\Sigma_{i=1}^{n} x_i)(\Sigma_{i=1}^{n} x_i y_i)]/[n\Sigma_{i=1}^{n} x_i^2 - (\Sigma_{i=1}^{n} x_i)^2] = \bar{y} - b\bar{x} \text{ and}$$

$$b = [n\Sigma_{i=1}^{n} y_i x_i - (\Sigma_{i=1}^{n} y_i)(\Sigma_{i=1}^{n} x_i)]/[n\Sigma_{i=1}^{n} x_i^2 - (\Sigma_{i=1}^{n} x_i)^2].$$

Two features of these estimates should be familiar. First, the denominator of each relates to the sample variance of variable X. Second, the numerator for b is quite similar to what is found in the product moment correlation coefficient formula. In fact, the intrinsic relationship between correlation and regression is further demonstrated by rewriting the estimate of b as

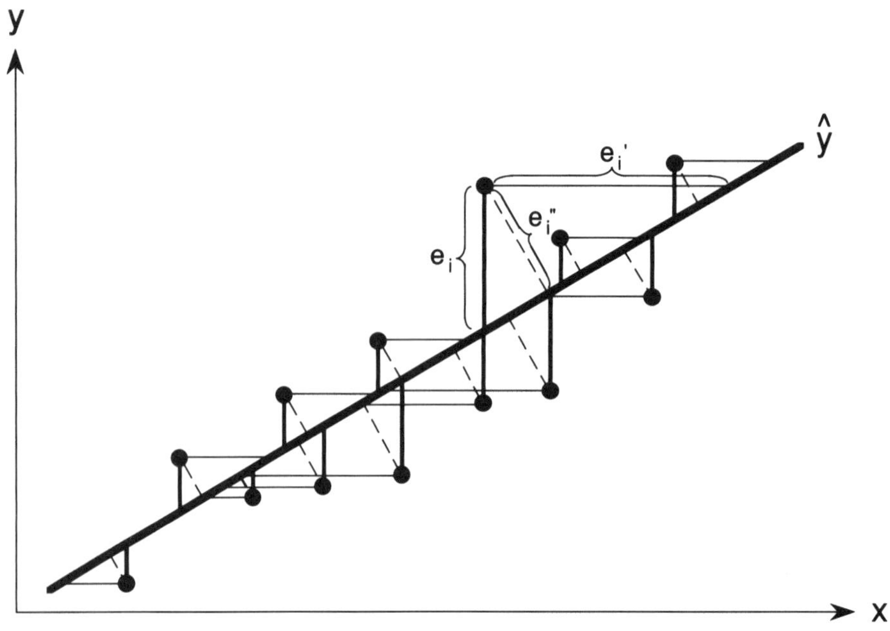

Figure 12.4: VARIOUS ERROR DISTANCES (HORIZONTAL, PERPENDICULAR, VERTICAL)

$b = r_{xy}s_y/s_x$.

Algebraic manipulations producing this version of b also indicate that if both variables X and Y are converted to z-scores, then a = 0 and $b = r_{xy}$.[1] This finding further reinforces the notion that parameter α is a conditional mean. This conversion to z-scores does nothing more than relabel the axes. Subtraction of the means, $\bar{x}$ and $\bar{y}$, shifts the two axes in such a way that the origin is relocated in the middle of the scatter of points. Division of the resulting differences by the sample standard deviations, s_x and s_y, alters the magnitude of numbers along the two axes in such a way that the sample variance becomes 1, causing b to become equal to r_{xy}. Consequently, the scatter of points does not change, and the angle between the regression

[1]The algebraic manipulations are as follows:

$$\begin{aligned} Y &= a + bX = (\bar{y} - b\bar{x}) + bX = \bar{y} + b(X - \bar{x}) \\ &= \bar{y} + (r_{xy}s_y/s_x)(X - \bar{x}) \\ Y - \bar{y} &= (r_{xy}s_y/s_x)(X - \bar{x}) \\ (Y - \bar{y})/s_y &= r_{xy}(X - \bar{x})/s_x \\ z_y &= 0 + r_{xy}z_x. \end{aligned}$$

line and the horizontal axis does not change; rather, only the amplitude of numbers changes.

In Chapter 4 the relation between correlation and the regression line was described in such a way that as $|r_{xy}| \to 1$, the scatter of points more closely approximated the regression line, with $|r_{xy}| = 1$ giving a perfect alignment. This idea can be used to evaluate the goodness-of-fit of the regression line to a scatter of points. To do so, r_{xy} needs to be rewritten in terms of the regression equation, with

$$\hat{Y} = a + bX$$

giving $|r_{xy}| = r_{Y,\hat{Y}}$. Because $\hat{Y}$ is a prediction of Y, this correlation coefficient must be nonnegative. In addition, $r_{Y,\hat{Y}}^2$, called the **coefficient of determination** and sometimes denoted by R^2, provides a measure of the proportion of variance in variable Y accounted for by variable X. If the proportion reaches 100%, then the error variable e is equal to zero, indicating that Y can be predicted perfectly from X.

Three variance terms can be defined for each value of Y predicted from a value of X. In Figure 12.3, these three terms are:

(1) **Total variance**, the difference between the mean value of Y and each y_i;

(2) **Explained variance**, the difference between $\bar{y}$ and $\hat{y}_i$, the estimated value of y_i; and,

(3) **Unexplained variance**, the difference between y_i and $\hat{y}_i$, that is, the residual.

The explained variance plus the unexplained variance equals total variance for each observation, and the sum of squares of the explained variances plus the sum of squares of the unexplained variances equals the sum of squares of the total variances. The coefficient of determination also can be viewed as a ratio of the sum of the squared explained variances divided by the sum of the squared total variances. This is the context within which statistical explanation is cast; the variation in Y is explained through the variation in X. The coefficient of determination is a very popular goodness-of-fit measure, but many better measures exist.

EXAMPLE PR-12.1. Estimating the regression equation in which Y = DBN and X = DFRM for the Arecibo Region of Puerto Rico.

In this example the hypothesis poses the question of whether or not density of banana/plantain production can be predicted from density of farms in a municipio; farm density is a much easier variable to measure than is banana/plantain production, and theoretically no cause-and-effect relationship is expected to hold here. If there is a relationship between these two variables, and the government desires to monitor banana/plantain cultivation, then surveying the density of farms is one simple way of attaining a glimpse of banana/plantation production.

Municipio	DBN	DBN^2	DBN • DFRM	DFRM	$DFRM^2$
Arecibo	0.3	0.09	1.35	4.5	20.25
Barceloneta/ Florida	0.6	0.36	3.36	5.6	31.36
Camuy	0.4	0.16	4.84	12.1	146.41
Ciales	9.4	88.36	79.90	8.5	72.25
Hatillo	0.9	0.81	8.01	8.9	79.21
Lares	25.6	655.36	458.24	17.9	320.41
Manati	0.1	0.01	0.30	3.0	9.00
Morovis	3.4	11.56	43.86	12.9	166.41
Quebradillas	0.4	0.16	4.28	10.7	114.49
Utuado	9.0	81.00	164.70	18.3	334.89
Vega Baja	0.3	0.09	1.44	4.8	23.04
Column sum	50.4	837.96	770.28	107.20	1317.72

Based upon these calculations, the regression equation parameter estimates are as follows:

$$a = (50.4 \bullet 1317.72 - 107.20 \bullet 770.28)/(11 \bullet 1317.72 - 107.20^2)$$
$$= -5.3815 \text{ and}$$

$$b = (11 \bullet 770.28 - 50.4 \bullet 107.20)/(11 \bullet 1317.72 - 107.20^2)$$
$$= 1.0224.$$

The correlation between DBN and DFRM is 0.6856, the means are $\bar{x}_{DBN}$ = 4.5818 and $\bar{x}_{DFRM}$ = 9.7455, and the standard deviations are s_{DBN} = 7.7913 and s_{DFRM} = 5.2250. Accordingly,

$$a = \bar{y} - b\bar{x} = 4.5818 - 1.0224 \bullet 9.7455 = -5.3820 \text{ and}$$

$$b = r_{xy}s_y/s_x = 0.6856 \bullet (7.7913/5.2250) = 1.0223$$

which agree with the preceding computations, except for rounding error. The correlation coefficient $r_{DBN,D\hat{B}N}$ = 0.6856 = $r_{DBN,DFRM}$, meaning that the coefficient of determination equals 0.6856^2 = 0.4700; in other words, 47% of the statistical variation in DBN is accounted for by DFRM. $\sum_{i=1}^{11} e_i/11$ = -0.0004, which deviates from 0 only by rounding error. The ordinary least squares minimum occurs here at $\sum_{i=1}^{11} e_i^2$ = 321.69. The estimated values of DBN together with the estimated error terms are as follows:

Municipio	a + b • DFRM	= $\hat{DBN}$	e = (DBN - $\hat{DBN}$)
Arecibo	-5.3815 + 1.0224 • 4.5	= -0.7807	1.0807
Barceloneta/ Florida	-5.3815 + 1.0224 • 5.6	= 0.3439	0.2561
Camuy	-5.3815 + 1.0224 • 12.1	= 6.9895	-6.5895
Ciales	-5.3815 + 1.0224 • 8.5	= 3.3089	6.0911
Hatillo	-5.3815 + 1.0224 • 8.9	= 3.7179	-2.8179
Lares	-5.3815 + 1.0224 • 17.9	= 12.9195	12.6805
Manati	-5.3815 + 1.0224 • 3.0	= -2.3143	2.4143
Morovis	-5.3815 + 1.0224 • 12.9	= 7.8075	-4.4075
Quebradillas	-5.3815 + 1.0224 • 10.7	= 5.5582	-5.1582
Utuado	-5.3815 + 1.0224 • 18.3	= 13.3284	-4.3284
Vega Baja	-5.3815 + 1.0224 • 4.8	= -0.4740	0.7740

The four desirable statistical properties (see Section 2.5) can be established for estimates a and b only when certain major assumptions hold. These assumptions allow inferences similar to those outlined in Chapters 10 and 11.

Assumptions:

(a) variable Y is measured at the interval or ratio scale;
(b) variable X is measured without error, whereas any measurement error in variable Y is random;
(c) the relationship between Y and X is linear;
(d) the error variable ξ has zero mean;
(e) the error variable ξ has constant variance σ^2;
(f) all pairs of errors are uncorrelated; and,
(g) the error variable ξ is normally distributed.

Assumption (f) in part suggests that a random sample has been drawn and is at odds with nonzero spatial autocorrelation latent in geographical data sets. Assumptions (f) and (g) imply that individual observation errors are independent random variables. Assumption (g) allows hypothesis testing and confidence interval estimation. Gross violations of these assumptions introduce serious specification error into a regression analysis. As seen previously, this specification error may cause statistical results to be sample specific, possibly leading to completely opposite statistical decisions with different samples (the Type I error probability is wrong). With this in mind, researchers always should conduct diagnostic analyses on data to determine whether or not these assumptions seem reasonable.

The scatter diagram permits a visual inspection for a linear relationship between the X and Y variables. A goodness-of-fit chi square or Kolmogorov-Smirnov statistic (see Chapter 10) tests whether or not the residuals appear to come from a normally distributed population. The Moran Coefficient (see Chapter 5) tests for the presence of spatial autocorrelation in the residuals, in accordance with assumption (f). Note that this is not the same problem as spatial autocorrelation latent in either the Y or the X variables, whose discussion is beyond the scope of this book. The sampling distribution of this statistic is complicated here by the additional constraints placed upon it through the estimation of the regression parameters a and b (hence, two additional degrees of freedom are lost). The new expected value and standard error are

$$E(MC) = -\{1 + n\sum_{i=1}^{n} z_i \sum_{j=1}^{n} c_{ij}z_j/[(n - 1)\sum_{i=1}^{n} \sum_{j=1}^{n} c_{ij}]\}/(n - 2) \text{ and}$$

$$\sigma_{MC} = (2/\sum_{i=1}^{n} \sum_{j=1}^{n} c_{ij})^{1/2}$$

where the z_i are the z-score values of the X variate, and σ_{MC} is the asymptotic (i. e., limiting large-sample) standard error. A more precise standard error formula is very complex, and requires concepts from multivariate analysis; accordingly it is not presented here. In addition, constant variance may be visually assessed by searching for unusual patterns in the scatter diagram of e_i plotted against $\hat{y}_i$. The residuals in this plot should be contained in a horizontal band centered at $e_i = 0$; funnel or bow scatterings suggest nonconstant variance.

EXAMPLE PR-12.2. Example PR-12.1 continued.

DBN is measured at the ratio scale (a meaningful interval and an absolute zero exist for density of bananas/plantains measures), and $\bar{e} = -0.0004$. The relationship between DBN and DFRM appears to be only roughly linear, as is indicated by the scatter diagram presented in Figure 12.5. The plot for evaluating constant variance also appears in this figure, and its funnel shape suggests that the variance is not constant across the set of errors, a violation of assumption (e). Next, spatial autocorrelation among the residuals is measured with MC. First, consider the connectivity matrix for municipios in the Arecibo Region (see Figure 4.1):

Municipio	A	B/F	Ca	Ci	H	L	Ma	Mo	Q	U	VB
Arecibo (A)	0	1	0	1	1	0	0	0	0	1	0
Barceloneta/ Florida (B/F)	1	0	0	1	0	0	1	0	0	0	0
Camuy (Ca)	0	0	0	0	1	1	0	0	1	0	0
Ciales (Ci)	1	1	0	0	0	0	1	1	0	1	0
Hatillo (H)	1	0	1	0	0	1	0	0	0	1	0
Lares (L)	0	0	1	0	1	0	0	0	0	1	0
Manati (Ma)	0	1	0	1	0	0	0	1	0	0	1
Morovis (Mo)	0	0	0	1	0	0	1	0	0	0	1
Quebradillas (Q)	0	0	1	0	0	0	0	0	0	0	0
Utuado (U)	1	0	0	1	1	1	0	0	0	0	0
Vega Baja (VB)	0	0	0	0	0	0	1	1	0	0	0

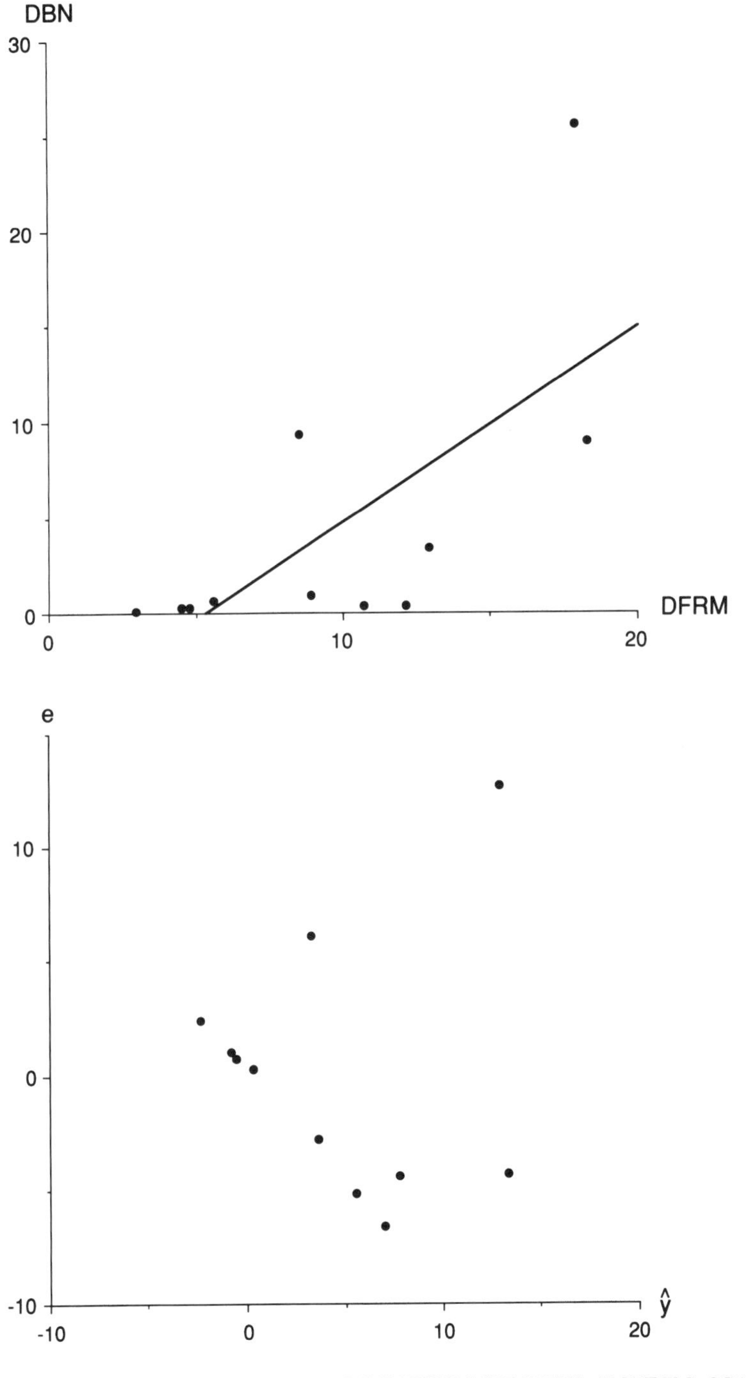

Figure 12.5: (TOP) SCATTER DIAGRAM FOR EVALUATING LINEARITY; (BOTTOM) SCATTER DIAGRAM FOR EVALUATING CONSTANT VARIANCE

Converting the DFRM values to z-scores and utilizing the regression residuals yields

Municipio	z_i	$\sum_{j=1}^{n} c_{ij}z_j$	$z_i \sum_{j=1}^{n} c_{ij}z_j$	e_i^2	$\sum_{j=1}^{n} c_{ij}e_j$	$e_i \sum_{j=1}^{n} c_{ij}e_j$
A	-1.0039	0.4436	-0.4454	1.168	-0.7973	-0.862
B/F	-0.7934	-2.5333	2.0099	0.066	9.5867	2.457
Ca	0.4506	1.5816	0.7127	43.415	4.7062	-31.009
Ci	-0.2384	-0.8473	0.2020	37.106	-4.9829	-30.353
H	-0.1618	2.6446	-0.4279	7.938	2.8457	-8.018
L	1.5607	1.9261	3.0059	160.817	-13.7340	-174.166
Ma	-1.2910	-1.3745	1.7745	5.829	2.7151	6.555
Mo	0.6037	-2.4759	-1.4948	19.420	9.2800	-40.896
Q	0.1827	0.4506	0.0823	26.602	-6.5890	33.984
U	1.6372	0.1566	0.2564	18.728	17.0363	-73.725
VB	-0.9465	-0.6873	0.6505	0.599	-1.9925	-1.543
Column sum	-0.0001	-0.7155	6.3261	321.688	18.0743	-317.576

MC $= (11/36)(-317.576/321.688) = -0.3017.$

E(MC) $= -(1/9)[1 + (11/10)(6.3261/36)] = -0.1326.$

$\sigma_{MC} = 1/\sqrt{18} = 0.2357$ (the exact small-sample value is 0.1734).

Thus, z = [-0.3017 - (-0.1326)]/0.2357 = -0.7174, implying that these residuals are not spatially autocorrelated in the population. Supplemental statistical analysis (not reported in this book) indicates that the error terms were drawn from a population whose frequency distribution conforms to a normal distribution. In summary, of the seven assumptions listed at the beginning of this chapter, only assumption (e), regarding variance homogeneity, seems to be violated. The effect of this violation will be examined later in the chapter.

The diagnostics presented in this section comprise the minimum set that should be reviewed. Additional diagnostics can be performed on regression results, and a specialized reference book of such diagnostics should be consulted.

12.2. Inference in regression analysis

A number of inferential problems are affiliated with the regression line. The most obvious is testing a null hypothesis about the intercept and slope terms. The two most popular pairs of hypotheses here are

H_0: $\alpha = 0$ H_0: $\beta = 0$
H_a: $\alpha \neq 0$ H_a: $\beta \neq 0$

Test criteria:

(a) the t distribution with (n - 2) degrees of freedom is used as the test statistic for the estimated intercept a, where

$$t = (a - \alpha)/\{s_b^2[(n - 1)s_x^2/n + \bar{x}^2]\}^{1/2} = a/\{s_b^2[(n - 1)s_x^2/n + \bar{x}^2]\}^{1/2}$$

under the null hypothesis (note that the denominator of this test statistic is the standard error of a), and

(b) the t distribution with (n - 2) degrees of freedom is used as the test statistic for the estimated slope b, where

$$t = (b - \beta)/\{(s_y^2 - b^2s_x^2)/[s_x^2(n - 2)]\}^{1/2} = b/\{(s_y^2 - b^2s_x^2)/[s_x^2(n - 2)]\}^{1/2}$$

under the null hypothesis (note that the denominator of this test statistic is the standard error of b).

A t distribution is needed here because both standard errors s_a and s_b include terms that are estimated with the sample data; specifically they require estimates for σ_x and σ_y (recall that the same constraint is associated with s_x and $\bar{x}$, and that b is calculated from s_x and s_y), placing two constraints on the analysis. The two criteria are presented together because their two t statistics are for simultaneously testing the pair of null hypotheses; in other words, H_0: $\alpha = 0$ is tested given that b already has been calculated, and vice versa. Furthermore, although the foregoing null hypotheses are the most common ones posited, regression analysis is not restricted to these special cases of zero; any real number values can be put into these null hypotheses, with the same standard errors applying (deviations from zero tend to deteriorate the power of these tests, though).

EXAMPLE PR-12.3. Example PR-12.1 continued.

If the hypothesis is

H_0: the regression intercept for the equation DBN = $\alpha + \beta \cdot$ DFRM + ξ equals zero in the Arecibo region,

which is equivalent to saying H_0: $\alpha = 0$,

then $t = -5.3815/\{0.3621^2[(10/11) \cdot 5.2250^2 + 9.7455^2]\}^{1/2} = -1.3579$. For a 5% level of significance (see Table 9.1-A, found in the appendices), the critical values with $11 - 2 = 9$ degrees of freedom are ± 2.2622. Therefore, the conclusion is that (1) H_0 fails to be rejected, (2) there is no evidence to indicate that the intercept of the regression equation in the Arecibo region is not equal to zero in the population, and (3) 95% of the time one would expect to obtain a t statistic that falls within the confidence interval defined as (-2.2622, 2.2622) when in fact the regression equation intercept is zero in the population.

Next, if the hypothesis is

H_0: the regression slope for the equation $DBN = \alpha + \beta \cdot DFRM + \xi$ equals zero in the Arecibo region,

which is equivalent to saying H_0: $\beta = 0$,

then $t = 1.0224/\{(7.7913^2 - 1.0224^2 \cdot 5.2250^2)/[5.2250^2 \cdot (11 - 2)]\}^{1/2} = 2.8257$. The critical values remain as before. Therefore, the conclusion is that (1) H_0 is rejected, (2) there is evidence to indicate that the slope of the regression equation in the Arecibo region is not equal to zero in the population, and (3) only 5% of the time one would expect to obtain a t statistic that falls within the critical region $\{(-\infty,-2.2622), (2.2622,\infty)\}$ when in fact the slope of the regression line equals zero in the population.

Of course, values other than zero could be used. However, as stated before, substantive conceptual or theoretical considerations must be consulted in order to establish meaningful nonzero values for these two hypotheses.

Confidence intervals can be constructed for these two situations in exactly the same manner as used in Chapters 10 and 11;

$$a - t_{1-\alpha/2,n-2}\{s_b^2[(n - 1)s_x^2/n + \bar{x}^2]\}^{1/2} \le \alpha \le a + t_{1-\alpha/2,n-2}\{s_b^2[(n - 1)s_x^2/n + \bar{x}^2]\}^{1/2}$$

$$b - t_{1-\alpha/2,n-2}\{(s_y^2 - b^2 s_x^2)/[s_x^2(n - 2)]\}^{1/2} \le \beta \le b + t_{1-\alpha/2,n-2}\{(s_y^2 - b^2 s_x^2)/[s_x^2(n - 2)]\}^{1/2}.$$

One confusing aspect of these confidence intervals is that α is used to denote both the population intercept term and the critical region; this is a bewildering notational usage commonly found in elementary statistic textbooks and is dealt with effectively in Chapter 13. The student is cautioned to think carefully about notational issues in these sorts of perplexing circumstances. A confidence interval for the regression line takes on a slightly different flavor. Assume that x_0 is any value of the regressor variable contained within the range of the sample data. An unbiased point estimator is $E(y|x_0) = a + b \bullet x_0$. By assumption (g), $E(y|x_0)$ is normally distributed and so will have variance

$$VAR[E(y|x_0)] = \sigma_{y|x}^2[1/n + (x_0 - \mu_x)^2/(n\sigma_x^2)].$$

Recall that the population regression line being estimated is $\mu_{y|x} = \alpha + \beta X$, where $\alpha = \mu_y - \beta\mu_x$ when $X = \mu_x$ and $Y = \mu_y$, so both the population and sample regression lines always pass through the pair of variable means. Substituting the mean of X into the formula for $VAR[E(y|x_0)]$ produces the expression σ^2/n, which is the variance of the sampling distribution of the sample mean presented in Section 9.2. As x_0 moves away from this mean value, though, $VAR[E(y|x_0)]$ must increase, since the term $(x_0 - \mu_x)^2/(n\sigma_x^2)$ always will be nonnegative. Consequently, the confidence interval for the regression line will curve away from this line rather than parallel it (older quantitative geography texts incorrectly display parallel lines, often plus or minus one standard deviation about the line), as is shown in Figure 12.6. In order to estimate this confidence interval, terms in the variance formula need to be estimated from the sample data. As shown before, μ_x is replaced by $\bar{x}$ and $n\sigma_x^2$ is replaced by $(n - 1)s_x^2$. The additional required parameter estimate is $s_{y|x}^2 = \sum_{i=1}^{n} e_i^2/(n - 2)$. Therefore, the confidence interval is constructed using the following formula:

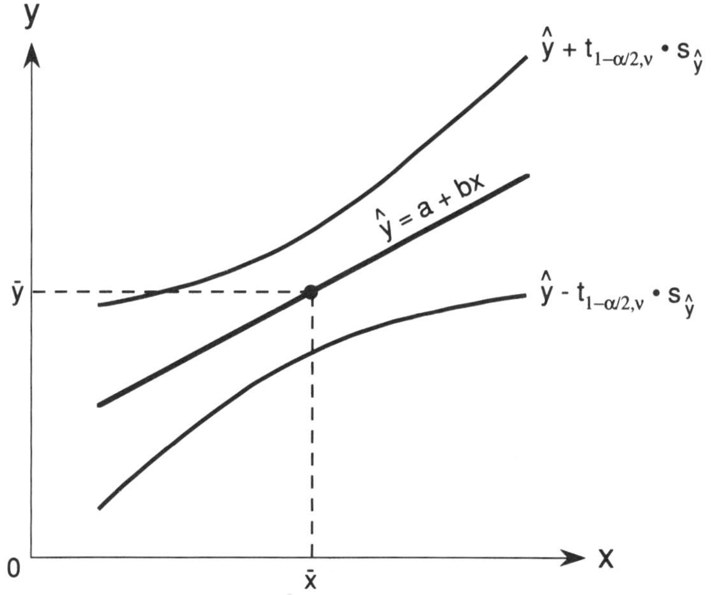

Figure 12.6: CHARACTERISTIC CONFIDENCE INTERVAL

$$y|x_0 - t_{1-\alpha/2,n-2}s_{y|x}\{1/n + n(x_0 - \bar{x})^2/[n \sum_{i=1}^{n} x_i^2 - (\sum_{i=1}^{n} x_i)^2]\}^{1/2}$$

$$\leq E(y|x_0) \leq y|x_0 + t_{1-\alpha/2,n-2}s_{y|x}\{1/n + n(x_0 - \bar{x})^2/[n \sum_{i=1}^{n} x_i^2 - (\sum_{i=1}^{n} x_i)^2]\}^{1/2}.$$

It is important to emphasize here that this interval pertains only to values of the X variate that fall into the interval $[x_{min}, x_{max}]$.

EXAMPLE PR-12.4. Example PR-12.1 continued.

Both the regression line and the 95% regression line confidence intervals may be constructed as follows:

Municipio	DFRM; DFRM$_0$	D$\hat{B}$N	CI boundaries Upper	Lower
Arecibo	4.5	- 0.7807	5.1408	-6.7022
Barceloneta/ Florida	5.6	0.3439	5.6489	-4.9611
Camuy	12.1	6.9895	11.4998	2.4792
Ciales	8.5	3.3089	7.5123	-0.8945
Hatillo	8.9	3.7179	7.8541	-0.4183
Lares	17.9	12.9195	20.7414	5.0976
Manati	3.0	- 2.3143	4.5498	-9.1784
Morovis	12.9	7.8075	12.6341	2.9809
Quebradillas	10.7	5.5582	9.7102	1.4062
Utuado	18.3	13.3284	21.4315	5.2253
Vega Baja	4.8	- 0.4740	5.2719	-6.2199

From Example PR-12.2, $s_{y|x} = 35.7432^{1/2}$, $\bar{x} = 9.7455$, $\sum_{i=1}^{n} x_i^2 = 1317.72$, and $(\sum_{i=1}^{n} x_i)^2 = (107.20)^2$. From Table 9.1-A, $t_{0.975,9} = 2.2622$. For the first entry in this table, $x_0 = -0.7807$. Thus, the upper and lower bounds of the 95% confidence interval for this first entry become

upper bound: $-0.7807 + 2.2622 \bullet (35.7432)^{1/2}\{1/11 + 11 \bullet (4.5 - 9.7455)^2/(11 \bullet 1317.72 - 107.20^2)\}^{1/2} = 5.1408$, and

lower bound: $-0.7807 - 2.2622 \bullet (35.7432)^{1/2}\{1/11 + 11 \bullet (4.5 - 9.7455)^2/(11 \bullet 1317.72 - 107.20^2)\}^{1/2} = -6.7022$.

The hyperbolic curvature of the confidence intervals is apparent in this example, as can be seen in Figure 12.7.

For simplicity, in Example PR-12.4 the observed x_i and the conditional x_0 are the same; this need not be the case for graphical purposes.

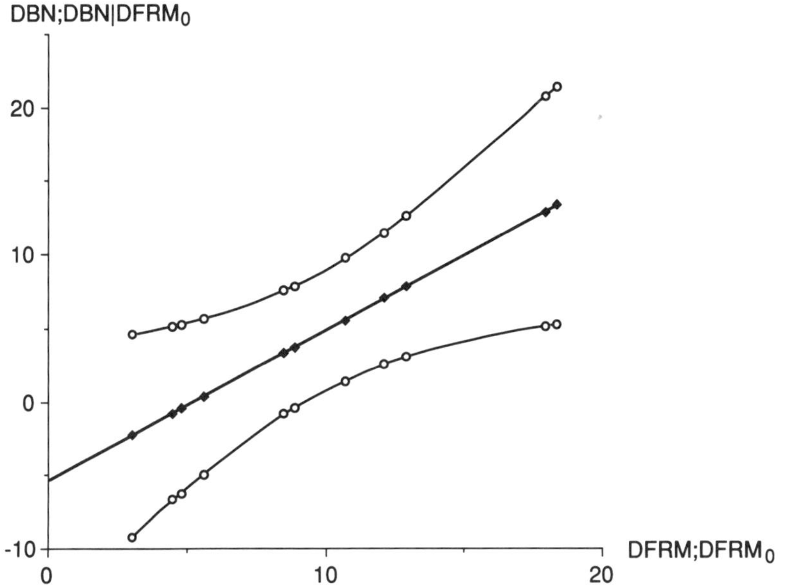

Figure 12.7: CONFIDENCE INTERVAL FOR D$\hat{B}$N OF THE ARECIBO REGION

The last four examples have been used to develop the theory behind OLS regression. To this end, these examples have illustrated the theory in separate sections. Following is a set of integrated examples that outlines the steps an analyst might follow in a comprehensive regression analysis.

When confronted with several variables relating to a particular topic, a researcher must select dependent and independent variables. Sometimes this selection is easy, for it is based upon a thorough understanding of existing theory behind a set of physical, economic, or social processes. At other times, if a goal of the analysis is to establish some cause-and-effect relationships among a group of variables, then it may be difficult to know in advance how to choose the dependent and the independent variables. As discussed at the beginning of the chapter, the dependent and independent variables in a regression analysis not only must conform to all the statistical assumptions outlined previously, but also must make sense on theoretical or conceptual grounds.

EXAMPLE SG-12.1. Constructing simple linear regression models for the Sentinel Glacier data set.

The set of all possible pairwise correlations is a useful set of summary calculations used in exploratory data analysis. As an aside, the sort of carefully constructed exploratory data analysis used in the following example is *not* the same as a "fishing expedition," as discussed at the end of this chapter. This set of correlation coefficients may be tabulated, producing a table commonly called a correlation matrix (see Chapter 13). The correlation matrix for the Sentinel Glacier data is as follows:

Variable	ELA	BN	PRECIP	TEMP
ELA	1.000	-0.957	-0.786	0.259
BN	-0.957	1.000	0.668	-0.251
PRECIP	-0.786	0.668	1.000	-0.014
TEMP	0.259	-0.251	-0.014	1.000

The diagonal entries in this table or matrix include the correlation value of a particular variable with itself, which by definition is always 1.000 (a variable must be perfectly related to itself; the amount of information contained in a variable and itself is perfectly redundant). The upper right and lower left areas are mirror images (that is, the first row is identical to the first column, the second row is identical to the second column, and so on). This property is known as **symmetry**. Thus, results are required to be interpreted on only half of the matrix. The high negative correlation between BN and ELA is expected, since when the net mass balance of the glacier is positive and high, the line separating the glacier's zone of accumulation and zone of ablation (ELA) is low. Both of these variables are *dependent* upon climatic conditions at and around the glacier. In this case either of the variables, BN or ELA, can be used to construct a simple linear regression against either of the climatic variables, PRECIP or TEMP.

EXAMPLE SG-12.2. Constructing and testing simple linear regression models for the Sentinel Glacier data set.

Results from the correlation analysis reported in Example SG-12.1 provide a starting point for considering simple OLS regression models. The OLS best-fit regression lines for ELA on PRECIP and ELA on TEMP are shown in Figures 12.8 and 12.9. The regression line of PRECIP accounts for 62% of the statistical variation ($R^2 = 0.62$) in ELA, whereas TEMP accounts for only 7% of the statistical variation in ELA. The graphical results presented in Figure 12.8 suggest that assumptions (a) through (c) are satisfied. As shown in example PR-12.3, inferences about both the intercept (a) and slope (b) coefficients can be made. In particular, a conclusion can be drawn about whether or not the estimates come from a population having an intercept (α) and slope (β) equal to zero. Using the computed t-statistics shown in the following table for both models, and a confidence coefficient of 95% (the **confidence coefficient** is the probability of the population parameter falling within the bounds of the confidence interval), the values for a and b for PRECIP are both highly significant. In other words, the two null hypotheses H_0: $\alpha = 0$ and H_0: $\beta = 0$ are rejected. However, neither estimate is significant for the model regressing ELA on TEMP, so neither hypothesis is rejected.

Parameter	Estimate	t value
Simple regression of ELA on PRECIP		
Intercept	2198.37	29.42
Slope	-0.35698	-5.24
Simple regression of ELA on TEMP		
Intercept	1195.92	2.12
Slope	49.75	1.11

An additional method of testing the results is to assess the overall regression model. This feat is accomplished by comparing the mean sums of squares associated with the explained variance with the mean sums of squares associated with the unexplained or error variance. This ratio should conform to the F distribution and can be tested in the same fashion as in ANOVA models (see Chapter 11). In fact, ANOVA tables are standard output from computer regression packages. The following table provides the relevant information.

ANOVA

Source	Sum of Squares	df	Mean Square	F ratio
Simple regression of ELA on PRECIP				
Model	166222	1	166222	27.41
Error	103081	17	6064	
Total	269303	18		
Simple regression of ELA on TEMP				
Model	18114	1	18114	1.23
Error	251188	17	14776	
Total	269303	18		

In the ELA regressed on PRECIP model, the computed F statistic is 27.41, well beyond the 5% critical value ($F_{0.05,1,17}$) of 4.4513 (see Table 11.1-A),

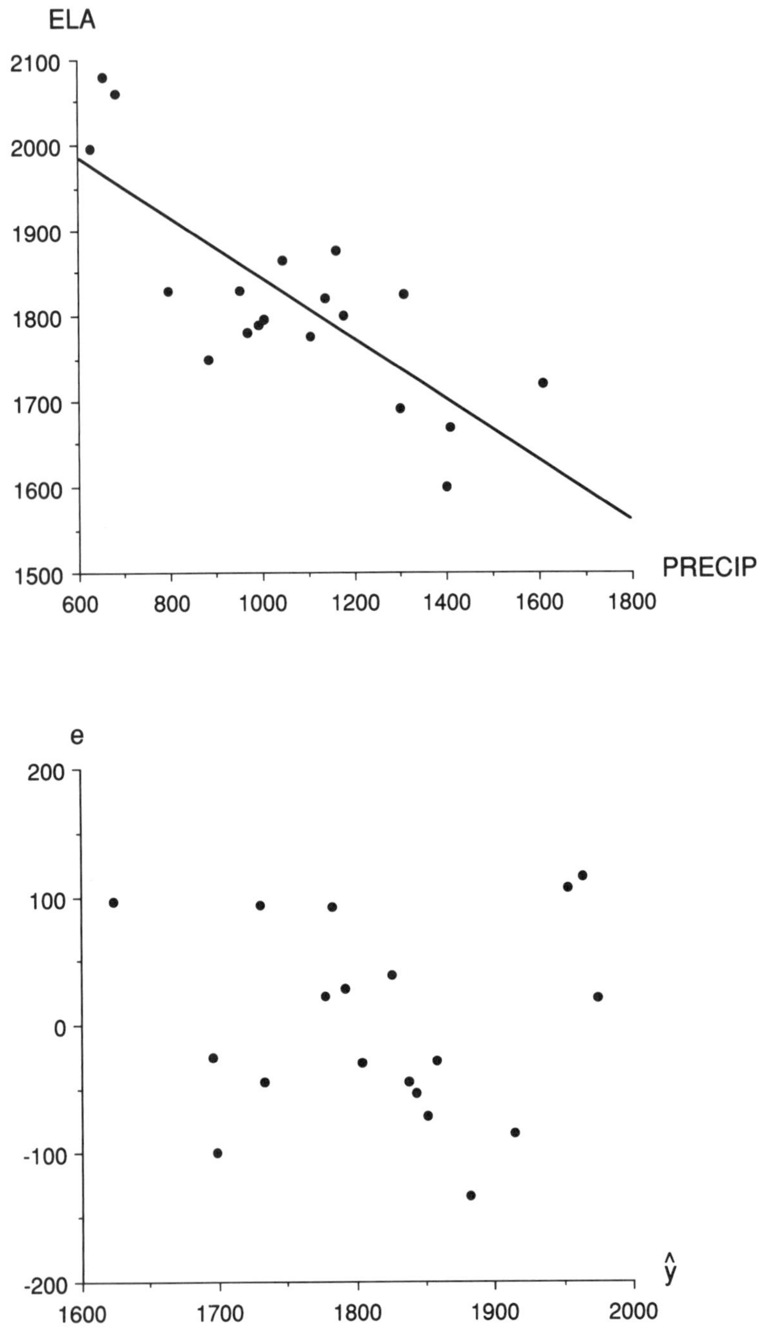

Figure 12.8: (TOP) SCATTER DIAGRAM FOR EVALUATING LINEARITY; (BOTTOM) SCATTER DIAGRAM FOR EVALUATING CONSTANT VARIANCE

implying a rejection of the null hypothesis. The F value of 1.23 in the ELA regressed on TEMP model is much closer to 1 than is 4.45, implying a failure to reject the null hypothesis. In both cases, statistical results support the conclusions derived earlier from an analysis of model parameters.

Examination of the residual plot, presented in Figure 12.9, for the regression of ELA on PRECIP suggests that there might be a slight positive trend to the residuals. However, the small sample size limits the number of observations across the independent variable X (X = PRECIP) and precludes any definitive statement about violations of assumptions (d) through (f). The conclusion, therefore, is that the PRECIP model is valid and reflects real conditions at Sentinel Glacier. The results also are consistent with theory about physical systems; this is at least as important as the statistical results. Located in the Western Coast Mountains of British Columbia in Canada, Sentinel Glacier receives a comparatively high amount of winter snowfall. Snowfall in any given year determines the winter accumulation and winter mass balance. As a result, the winter mass balance clearly dominates the net annual balance that is measured by BN; summer melting, as defined by a regional temperature index, appears to be less important than winter precipitation in the form of snow.

EXAMPLE CA-12.1. Regressing export sales (EXP$) on the total sales (SALE) of a company.

A country's ability to export manufactured goods successfully has become increasingly important to national economic health. While this is certainly true in the United States, it is of vital importance in Canada. A national government's ability to influence export sales requires an understanding of the aspects of exporting firms that affect growth in export sales. An initial hypothesis might be to determine the extent to which the size of the firm (SALE), measured in total current sales, accounts for a firm's current exporting behavior (EXP$).

The correlation between SALE and EXP$ is 0.73, suggesting that further analysis into the nature of the relationship between these two variables is warranted. The regression model to be tested is

$$\text{EXP\$} = \alpha + \beta \cdot \text{SALE} + \xi.$$

The OLS best-fit regression line for EXP$ on SALE is in Figure 12.10. This regression line accounts for 53% of the statistical variation ($R^2 = 0.53$) in EXP$. The graphical results presented in Figure 12.10 suggest that assumptions (a) through (c) are satisfied. Note, however, that these are essentially grouped data. While the measurement is ratio in scale ($0 of

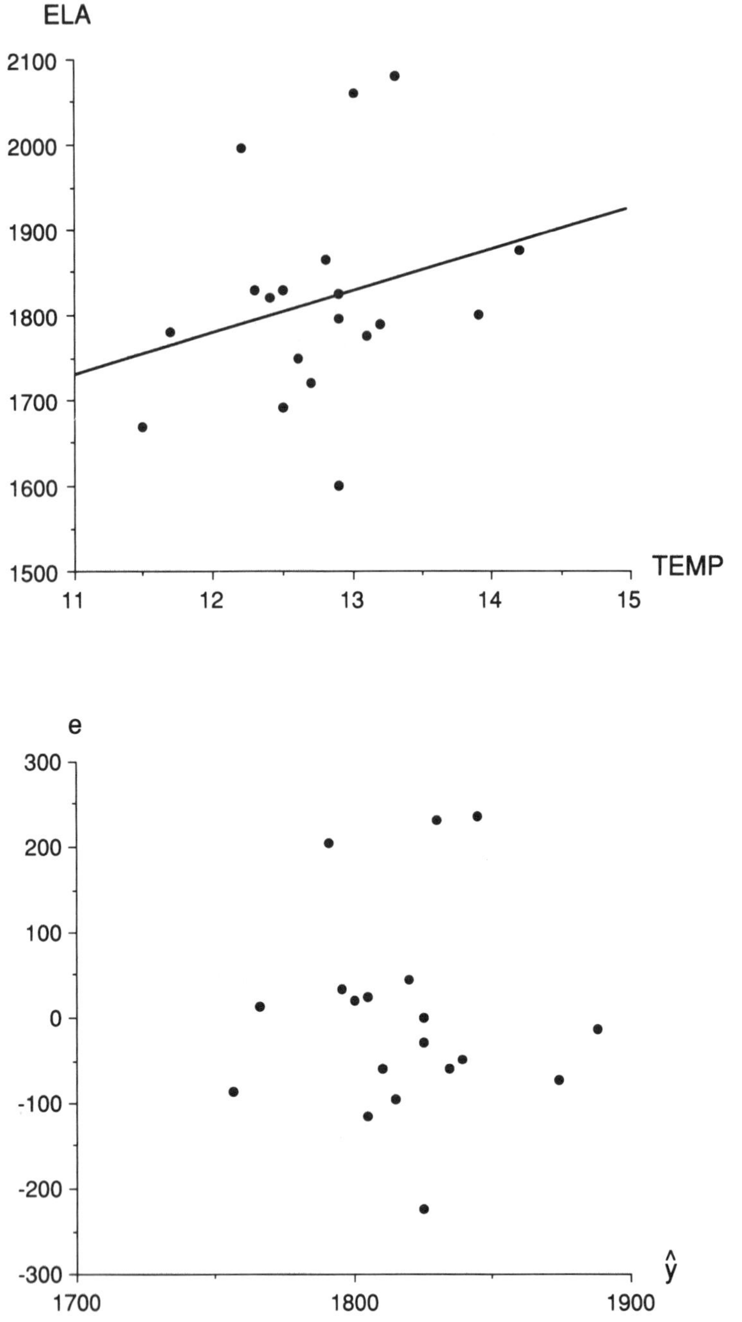

Figure 12.9: (TOP) SCATTER DIAGRAM FOR EVALUATING LINEARITY; (BOTTOM) SCATTER DIAGRAM FOR EVALUATING CONSTANT VARIANCE

sales has a precise meaning), the survey with which the data are collected report data in broad categories, with every firm in a category receiving the same measurement. This is particularly important with regard to the largest firms, in which every firm greater than \$50 million has been assigned a value of \$50,000.00 (in thousands of dollars). As shown in Example PR-12.3, inferences about both the intercept (a) and slope coefficient (b) can be made. In particular, a conclusion can be drawn concerning whether or not the estimates come from a population having an intercept (α) and a slope (β) equal to zero. Using the computed t-statistics shown in the following table, and a confidence coefficient of 95%, the value for the intercept is found not to be significant, whereas the value for the slope is found to be highly significant (see Table 9.1-A; the critical t value for a two-tail test with n - 2 = 52 - 2 = 50 degrees of freedom is ±2.009). In other words, of the two null hypotheses, H_0: $\alpha = 0$ and H_0: $\beta = 0$, the first is not rejected, whereas the second is. Substantively this finding makes sense. A y-axis intercept of 0 suggests that a firm with no sales has no export sales either. Similarly, among companies in the automotive parts industry in Southern Ontario, each dollar increase in total sales generates, on average, \$0.62 in new export sales.

Parameter	Estimate	t value
Intercept	-1271.43	-0.569
Slope	0.62	7.611

Following Example SG-12.2, the overall regression model is examined by comparing the mean sums of squares associated with the explained variance with the mean sums of squares associated with the unexplained, or error, variance. As before, this ratio should conform to the F distribution and can be tested in the same fashion as in ANOVA models. The following table provides the relevant information for this test:

ANOVA

Source	Sum of squares	df	Mean square	F ratio
Model	8093382129	1	8093382129	57.92
Error	7125948936	51	130724489	
Total	15219331065	52		

The computed F statistic is 57.92, well beyond the 5% critical value ($F_{0.05,1,52}$) of 4.0304 (using linear interpolation; see Table 11.1-A), leading to a conclusion that the null hypothesis should be rejected. These results support the conclusion derived earlier from an analysis of model parameters.

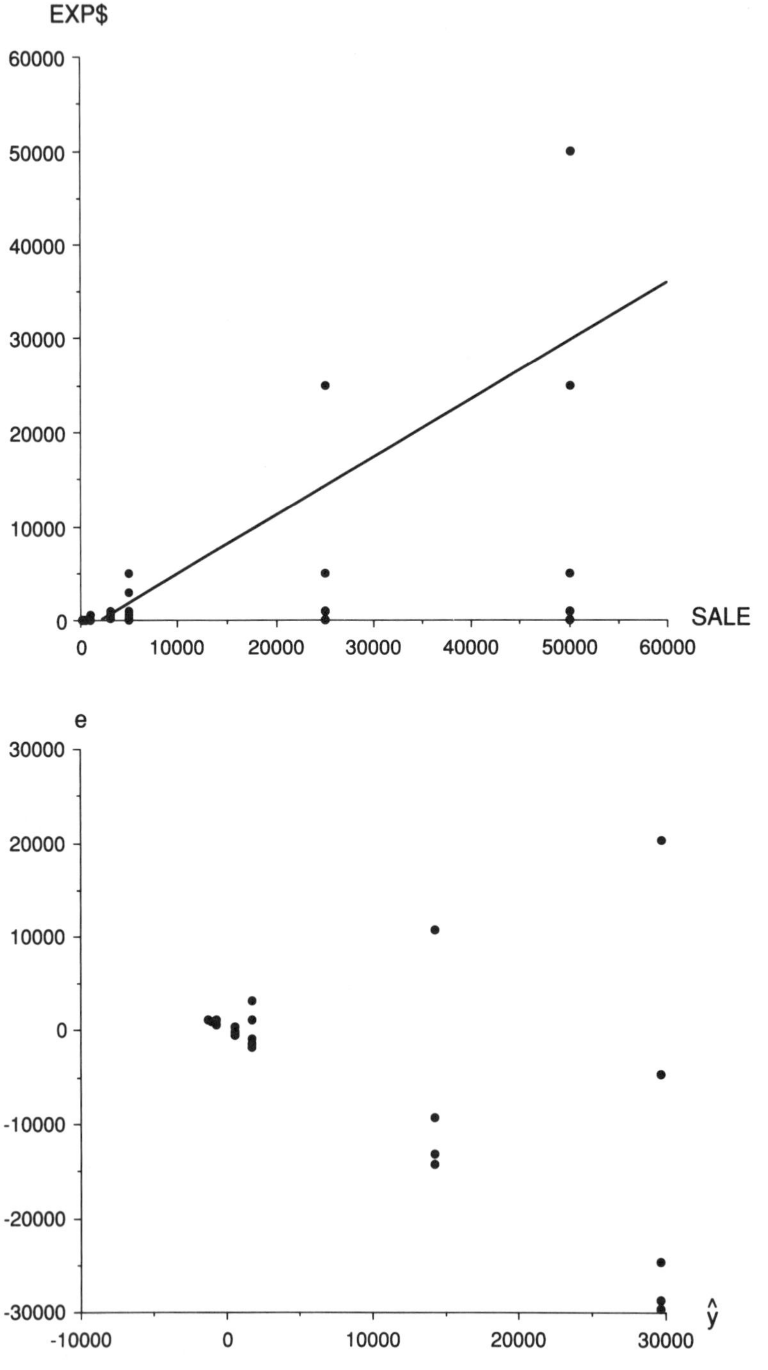

Figure 12.10: (TOP) SCATTER DIAGRAM FOR EVALUATING LINEARITY; (BOTTOM) SCATTER DIAGRAM FOR EVALUATING CONSTANT VARIANCE

Examination of the residual plots shows no obvious pattern in the residuals. The conclusion, therefore, is that the model is valid and describes actual behavior of firms; that is, larger firms export more. This finding is not very surprising. Two things to note are that a better model might be to examine the relationship between the proportion of total sales that are exported, and total sales. Such a model would remove the bias introduced by size in that just as a firm with zero sales cannot export anything, so too a firm cannot export any more than total sales. Second, remember that these are grouped data.

12.3. Common regression problems in spatial analysis

Section 12.2 has presented the conventional problem to which regression analysis is applied. This problem is concerned with prediction and the establishment of rules of correspondence between two variables. But other geographic problems can be addressed with the regression technique. This more general problem set either implicitly or explicitly involves prediction, and in some instances differs only subtly from the problem presented in Section 12.2.

12.3.1. Predicting one measure of a variable from another measure of the same variable

This problem concerns the adequacy of a measurement instrument. It is useful for checking measurements made in some context against those for the same observations made by an independent source. In this case the two specific null hypotheses to be tested are

$$H_0: \alpha = 0 \qquad H_0: \beta = 1$$
$$H_a: \alpha \neq 0 \qquad H_a: \beta \neq 1.$$

This problem can be structured so that the independent measurement source provides the Y, or dependent variable, whereas the measures made by the researcher constitute the X, or independent variable; the opposite assignment is acceptable, too. Presumably the one with the smallest measurement error should be assigned to variable X, in keeping with assumption (b) made at the beginning of the chapter. The emphasis still is on prediction.

EXAMPLE PR-12.5. Predicting the area of municipios in the Mayaguez region measured by the Commonwealth of Puerto Rico (CPR) from measures furnished by digitization (DGT) of areal unit polygon approximations; Example PR-5.1 extended.

The Commonwealth of Puerto Rico has published municipio area data. Combining these data with those obtained by applying the area estimation formula presented in Chapter 5 to digitized boundary coordinates for these same municipios has resulted in the following data set:

Municipio	$AREA_{CPR}$	$AREA_{DGT}$	Municipio	$AREA_{CPR}$	$AREA_{DGT}$
A: Aguada	30.21	26.8713	I: Las Marias	47.03	39.6011
B: Aduagilla	35.57	31.0115	J: Maricao	36.85	32.4462
C: Anasco	40.05	33.2918	K: Mayaguez	56.95	45.8654
D: Cabo Rojo	72.35	59.6684	L: Moca	50.43	43.0506
E: Guanica	36.52	34.5940	M: Rincon	14.14	11.8399
F: Hormigueros	11.16	9.3191	N: Sabana Grande	35.24	30.9182
G: Isabela	55.47	46.5532	O: San German	53.90	45.6620
H: Lajas	60.23	51.0072	P: San Sebastian	70.80	60.6780

For these data $\bar{x}_{CPR}$ = 44.1813, $\bar{x}_{DGT}$ = 37.6486, s_{CPR} = 17.6321, s_{DGT} = 14.5309, and r = 0.9962. The difference of means and difference of variances for correlated samples can be tested. Regressing $AREA_{CPR}$ on $AREA_{DGT}$ results in a = -1.3287 with s_a = 1.1354, b = 1.2088 with s_b = 0.0283, and $s_{y|x}$ = 1.5899. In the scatter diagram presented in Figure 12.11, the letters refer to the municipio as listed in the preceding table. This scatter diagram supports the notion that these two interval/ratio variables are linearly related. The only marked deviations from a straight line occur with municipios Anasco (C) and Guanica (E). A plot of e_i versus $\hat{y}_i$ supports the idea of a constant error variance (see Figure 12.11); visual inspection hints at a possible bow, however. Potential problems uncovered here are with Guanica, which already has been labelled as problematic with the linearity evaluation, and Mayaguez (K), which may well be expected to have a different error variance, since an area estimate for the island of Mona had to be subtracted from its $AREA_{CPR}$ value. The MC coefficient for the residuals from this regression is 0.1840, which has an affiliated approximate z-score of 1.4539 (with an exact z-score of 1.6918); this finding implies that the population errors should be independent. Finally, a test for normality of the residuals results in a failure to reject the null hypothesis that the population error terms conform to a normal distribution. Consequently, the assumptions spelled out in Section 12.2 appear to hold for these data.

If the hypothesis is

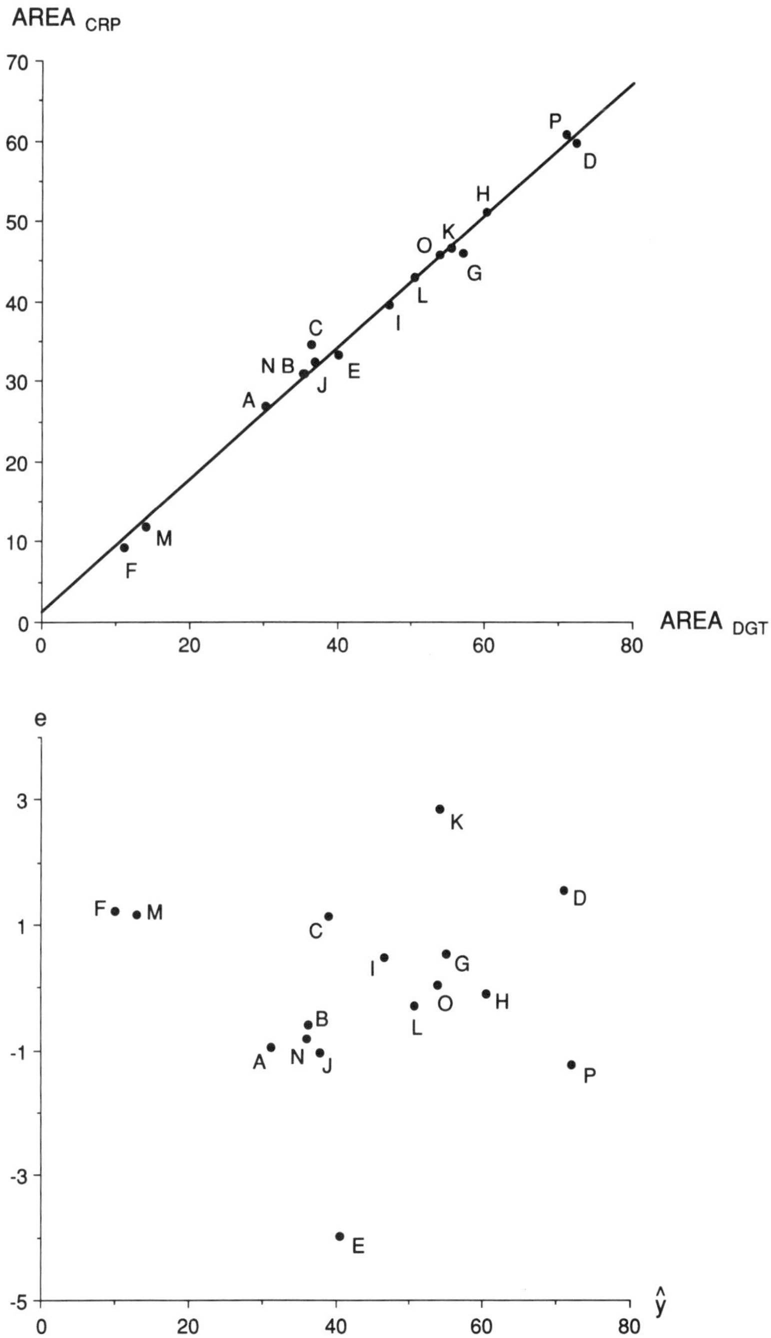

Figure 12.11: (TOP) SCATTER DIAGRAM FOR EVALUATING LINEARITY; (BOTTOM) SCATTER DIAGRAM FOR EVALUATING CONSTANT VARIANCE

H_0: the regression intercept for the equation $AREA_{CPR} = \alpha + \beta \bullet AREA_{DGT} + \xi = 0$ in the Mayaguez region,

which is equivalent to saying H_0: $\alpha = 0$,

then t = -1.3287/1.1354 = -1.1702. For a 5% level of significance (see Table 9.1-A), the critical value with 16 - 2 = 14 degrees of freedom is ±2.1448. Therefore, the conclusion is that (1) H_0 fails to be rejected, (2) there is no evidence to indicate that the intercept of the regression equation in the Mayaguez region is not equal to 0 in the population, and (3) 95% of the time one would expect to obtain a t statistic that falls within the confidence interval (-2.1448, 2.1448) when in fact the regression equation intercept is zero in the population.

Next, if the hypothesis is

H_0: the regression slope for the equation $AREA_{CPR} = \alpha + \beta \bullet AREA_{DGT} + \xi = 1$ in the Mayaguez region,

which is equivalent to saying H_0: $\beta = 1$,

then t = (1.2088 - 1)/0.0283 = 7.3781. The critical values remain the same as above. Therefore, the conclusion is that (1) H_0 is rejected, (2) there is evidence to indicate that the slope of the regression equation in the Mayaguez region is not equal to 1 in the population, and (3) only 5% of the time one would expect to obtain a t statistic that falls within the critical region $\{(-\infty, -2.1448), (2.1448, \infty)\}$ when in fact the slope of the regression line equals 1.00 in the population.

The first conclusion reached in this example suggests that there is no constant bias in the digitized area estimates, whereas the second conclusion suggests that there is a differential bias in the digitized area estimates. Two sources of deviation between these pairs of area measures are (1) a poor digitization process, and (2) the $AREA_{CPR}$ is based upon a three-dimensional surface, whereas the polygon approximation is for a two-dimensional surface. The latter source of deviation probably is more severe.

12.3.2. Forecasting a trajectory for convergence

In Section 12.2 it was emphasized that regression prediction is restricted to interpolation, limiting the x_0 values to the sample range for variable X. There always is a danger in trying to extrapolate outside of this range. But forecasting is desirable in some instances. Specifically, when an iterative procedure converges slowly, it may be desireable to project the observed trajectory. Spatial analysts find this useful for problems such as calculating spatial medians. Here either the x-coordinate or the y-coordinate of the spatial median becomes the Y regression variable, and the X regression variable is some inverse function of the iteration step number, say τ. The focus is on the constant term, since the objective is to determine the value of the regression equation parameter when $\tau = \infty$. For the spatial median problem the appropriate regression equation becomes

$$\text{x-coordinate} = \alpha + \beta/(\tau + 1) + \xi;$$

the limit of this equation is x-coordinate $= \alpha + \xi$. Exactly the same mathematical structure can be written for the y-coordinate trajectory. A value of 1 is added to the iteration counter index because the first guess is assigned an index value of 0.

EXAMPLE PR-12.6. Forecasting the geographic coordinates of the spatial median for DFRM in the Mayaguez Agricultural Administrative Region; Example PR-5.4 revisited.

The two regression equations for this problem were found to be

$$\text{x-coordinate}_i = 27.5441 - 0.0530/(i + 1) + e_i, \text{ and}$$

$$\text{y-coordinate}_i = 48.7361 - 0.1594/(i + 1) + e_i.$$

The respective correlation coefficients for these two equations are 0.926 and 0.995. The limiting situations yield a spatial median of (27.5441, 48.7361), which is slightly different from the 13th iteration result of (27.5347, 48.7180) presented in Example PR-5.4. The scatter diagrams for these two situations appear in Figure 12.12. These two graphs illustrate two features of such forecasting. First, as can be seen from the first few observations (those appearing in the far right of the graphs), the initial guess of a spatial mean was quite poor; but the algorithm does adjust for this error, quickly moving estimates toward the correct trajectory. Second, in a deterministic setting such as this, a careful application of regression analysis may well be very economical, since an infinite number of iterations cannot be completed (also, rounding error can become problematic in the iterative solution). The researcher always must view extrapolation of this type with caution, however! Notice how the trajectories are changing course as the number of iterations increases, whereas the linear forecasting regression lines intersect the Y-axis in the opposite direction.

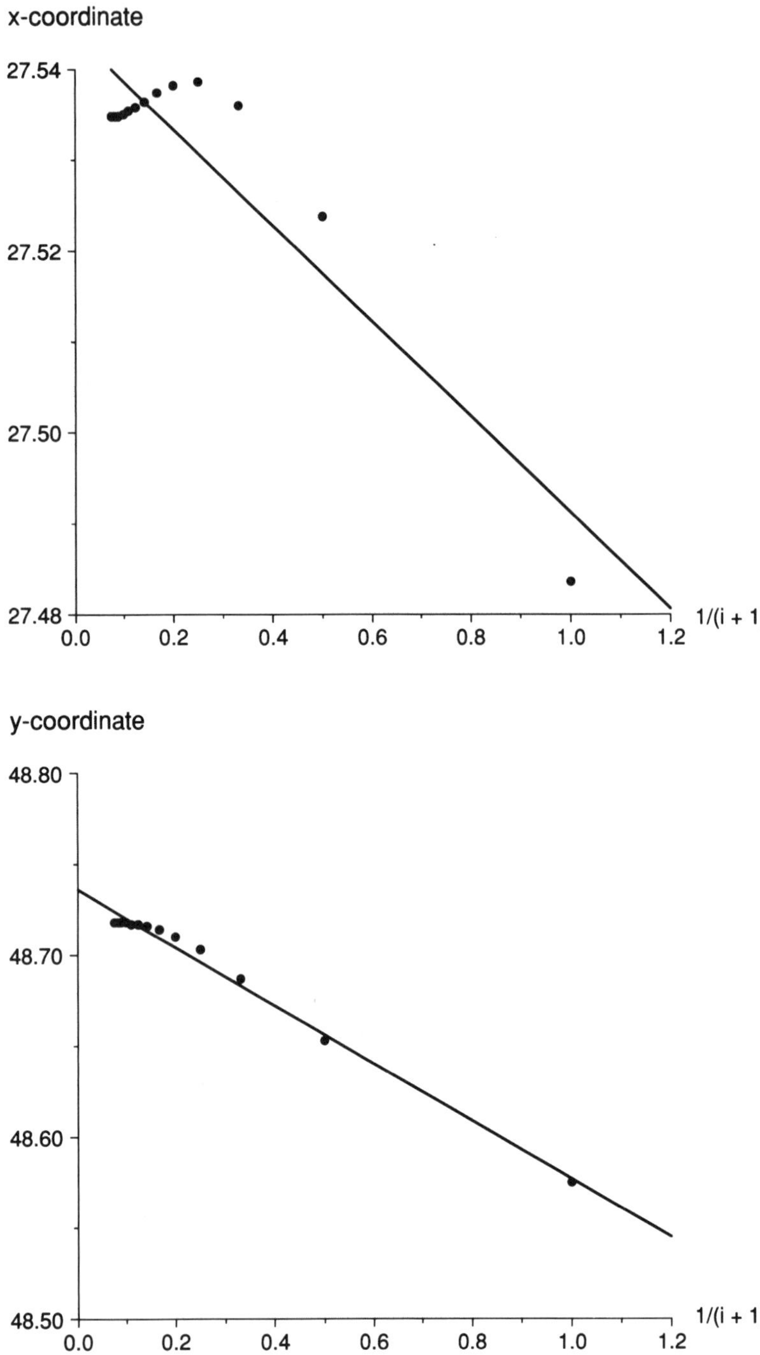

Figure 12.12: (TOP) SCATTER DIAGRAM FOR EVALUATING LINEARITY; (BOTTOM) SCATTER DIAGRAM FOR EVALUATING CONSTANT VARIANCE

12.3.3. Predicting areal unit values during one time period from those in a previous time period

Another routine problem that a spatial researcher may pursue asks whether or not some policy intervention, or other sort of event that has transpired, has made a difference in a geographic distribution of some phenomenon. Data are assembled for time period t - 1, which precedes the intervention, and for time period t, which follows the intervention. Of course many other factors may affect the geographic distribution between time periods t - 1 and t, causing any statistical decision to be speculative in nature; if the assumption is that these additional factors cause random fluctuations in the values, then the pair of relevant null hypotheses again becomes

H_0: $\alpha = 0$ $\quad$ H_0: $\beta = 1$
H_a: $\alpha \neq 0$ $\quad$ H_a: $\beta \neq 1$.

The related regression equation is

$$x_i(t) = \alpha + \beta x_i(t - 1) + \xi_i.$$

Rejection of either of these null hypotheses implies that the intervention has made a difference, either by bringing about a constant change (α is altered), a proportional change (β is altered), or both.

EXAMPLE PR-12.7. The impact of urbanization in the San Juan region of Puerto Rico on DCD between 1969 and 1974; Example PR-11.2 revisited.

The accompanying regression analysis yields a = 7.1385 and b = 0.6516, with the resulting equation

$$DCD_{1974} = 7.1385 + 0.6516 \bullet DCD_{1969} + e_i.$$

The statistics that accompany this problem include $\bar{x}_{1969} = 26.4368$, $\bar{x}_{1974} = 29.6158$, $s_{1969} = 13.3776$, $s_{1974} = 14.7082$, $r = 0.7164$, and $s_{y|x} = 9.6034$, where $Y = X_{1974}$. The corresponding t statistics for the regression parameter estimates are

$$t = 7.1385/5.0623 = 1.4101, \text{ and } t = (0.6516 - 1)/0.1539 = -2.2638.$$

The critical values for t, with 17 degrees of freedom, are ±2.1098. Employment of a one-tail test might be expected in this situation, based upon the intended outcome of the governmental policy. A more conservative test would be two-tail, since often government policies do not achieve their anticipated goals; the two-tail test would facilitate detection of this type of

> unforseen outcome. Therefore, according to the observed t statistics, the null hypothesis of $\alpha = 0$ is not rejected, and the null hypothesis that $\beta = 1$ is rejected. The scatter diagram for these two variables appearing in the top of Figure 12.13 indicates that their relationship is approximately linear, although there may be some evidence of nonlinearity [violation of assumption (c)]. The scatter diagram of e_i versus $\hat{y}_i$, presented in the bottom of Figure 12.13, supports the idea of a nonconstant error variance, with an apparent parabolic scatter of the points. The goodness-of-fit is moderately good, with the coefficient of determination indicating that 51.33% of the variance in DCD_{1974} is accounted for by DCD_{1969}. MC = 0.1102, the approximate $\sigma_{MC} = \sqrt{1/31} \approx 0.1796$, E(MC) = -0.0710, and thus z = [0.1102 - (-0.0710)]/0.1796 = 1.0090. Therefore, the population residuals are not expected to be spatially autocorrelated. Finally, the null hypothesis that the population residuals conform to a normal distribution is not rejected.

What kind of implication can be drawn from this intervention analysis? First, the inference that $\alpha = 0$ implies that there is no constant change across the municipios in density of cultivated land that is attributable to urbanization in the San Juan region. Second, the inference that $\beta \neq 1$ implies that, although those municipios having a high density of cultivated land in 1969 had a relatively high density in 1974 and those municipios having a low density of cultivated land in 1969 had a relatively low density in 1974, the pairs of municipio values for the two years are not identical. In other words, the intervention was causing a differential effect, which according to the scatter diagram tended to cause shrinkage where DCD was large, and little change where DCD was low. But the researcher needs to posit these conclusions with caution. The inferences more than likely are based upon poor estimates of σ_a and σ_b since assumption (e) is violated. Corrective action should be taken here before any understandable degree of confidence can be attached to these two inferences; these strategies are topics in more advanced texts, and will be touched upon in Section 12.4.

12.3.4. Regression as a generalized approach to statistical analysis

Researchers with a good command of parametric statistical techniques eventually realize that most statistical analyses can be cast as a regression problem. Moreover, regression serves as a unifying approach to much of what can be undertaken in statistical analysis. The advantage to recognizing this feature of the technique is that a very sizeable body of statistical theory has been developed for regression, and so any problem treated within the context of regression has this body of theory to tap. The goal of this section is simply to demonstrate this general feature of regression. The problem selected for illustrative purposes is a difference of means test for two independent samples having constant variance.

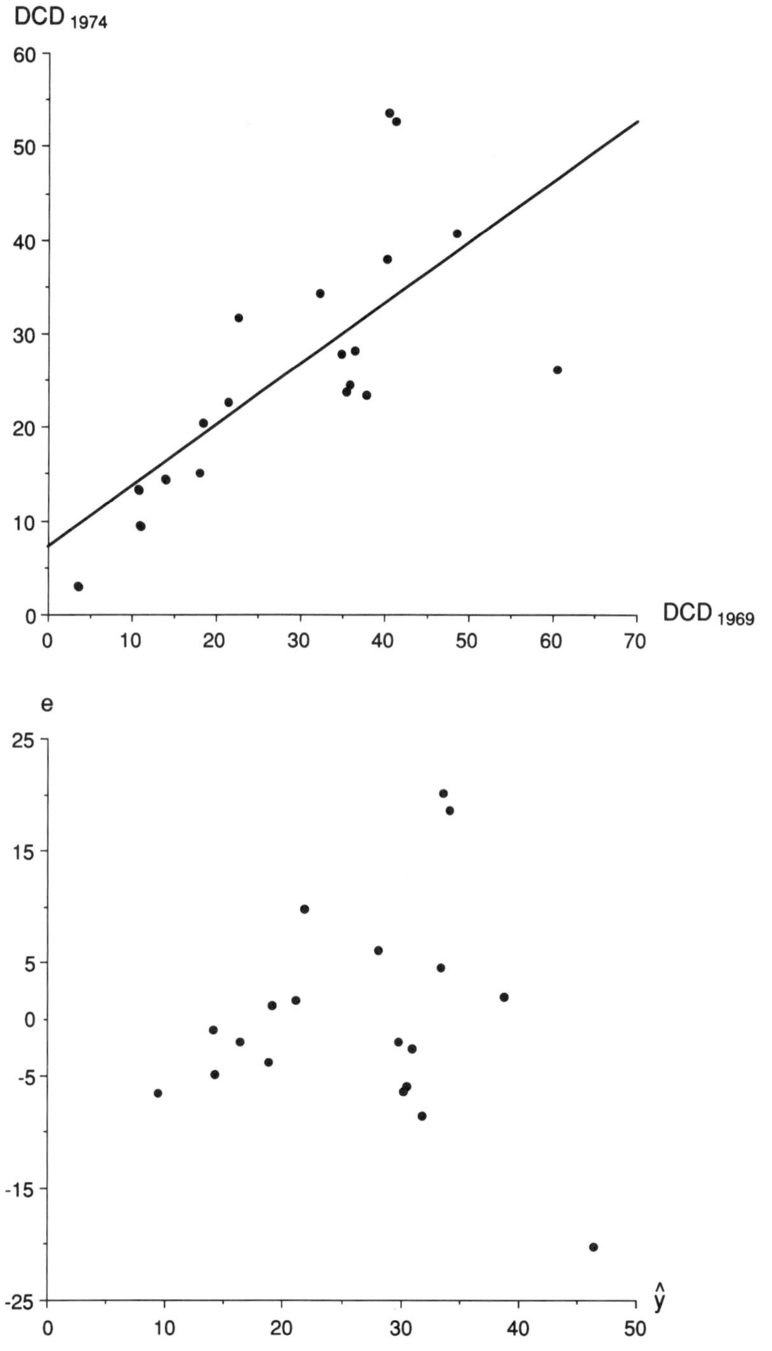

Figure 12.13: (TOP) SCATTER DIAGRAM FOR EVALUATING LINEARITY; (BOTTOM) SCATTER DIAGRAM FOR EVALUATING CONSTANT VARIANCE

A Y variable and an X variable are needed in order to undertake a regression analysis. Here Y denotes the variable whose regional means are to be tested. The X variable is a binary variable, indicating whether or not an areal unit is included in a particular region. Accordingly, the coefficient of determination will refer to a point biserial correlation coefficient (see Chapter 4). But the null hypothesis H_0: $\mu_1 = \mu_2$ has been written as a difference of two means H_0: $\mu_1 - \mu_2 = 0$ to simplify the mathematics involved. Similarly, the binary variable that is to be called X will take on the values of 1 and -1 (rather than the traditional 1 and 0) in order to simplify the mathematics involved. Given this context, the appropriate regression hypotheses are

$$H_0\!: \alpha = 0 \qquad\qquad H_0\!: \beta = 0$$
$$H_a\!: \alpha \neq 0 \qquad\qquad H_a\!: \beta \neq 0,$$

with the researcher being concerned almost exclusively with the hypothesis pertaining to β, where β is the difference of means measure (note that $\mu_1 - \mu_2 = \beta$).

EXAMPLE PR-12.8. Example PR-11.5 revisited.

Tests of variance homogeneity and normality for DCD already have been conducted. Let membership in the San Juan region be denoted by $x_i = 1$ and membership in the Mayaguez region be denoted by $x_i = -1$, which results in the regression equation

$$DCD_i = \alpha + \beta x_i + \xi_i.$$

The regression parameter estimates obtained here are b = -6.1941 and a = 32.6309, producing the regression equation

$$DCD_i = 32.6309 - 6.1941x_i + e_i$$

having the respective standard errors of $s_a = s_b = 2.0694$. The test statistic of interest is for estimate b and is t = -6.1941/2.0694 = -2.9932, which is exactly the value obtained in Example PR-11.5. The only assumption remaining to be evaluated here has to do with error independence. MC for the San Juan region is 0.3967, with an approximate z = [0.3967 - (-0.0556)]/0.1796 = 2.5180, which is significant at the 5% level. MC for the Mayaguez region is -0.0583, with an approximate z = [-0.0583 - (-0.0667)]/0.1768 = 0.0474, which is not significant at the 5% level. Thus some caution should be exercised here when drawing the difference of means inference.

The conclusion reached in Example PR-12.8 is that the variance of DCD appears to be positively spatially autocorrelated in the San Juan region. Such spatial dependence will cause the error variance estimate to be inflated, meaning that the test for equality of regional variances may well be incorrect. This in turn casts some doubt on the calculation of the difference of means test statistic; was the wrong t formula chosen? Computation of separate MC values for each region is preferred here because these two regions are not adjacent and because their means do not appear to be equal (a pooled mean should not be used in the spatial autocorrelation calculation).

Generalization of regression to other statistical techniques is a topic from multivariate analysis. The interested reader should consult some of the more applied references for this theme.

12.4. Intermediate considerations in regression analysis

The basic assumptions presented in Section 12.1, together with selected diagnostics for evaluating the appropriateness of these assumptions for a given data set. Additional considerations that are more advanced in nature merit attention before moving on. Still other quite advanced details are left to more specialized books. By acknowledging these three classes of complications, the reader certainly should recognize that proper regression analysis is an elaborately involved rather than a simple exercise.

12.4.1. Reflections on selected regression analysis features

A trait of regression analysis alluded to in the footnote of Section 12.1 concerns translating and applying a similarity transformation to a measurement scale. If the origin of a measurement scale is changed, then the regression intercept term α is affected. Since this parameter is a conditional mean, why should adding or subtracting a constant to or from either variable Y or variable X affect only this parameter? Judiciously selected values to subtract from the two variables (namely, their respective sample means) forces α to zero. Of course, adding some constant k to Y and then subtracting b•k from X, or subtracting k/b from Y and then adding k to X results in the same value of α. Meanwhile, if the X measurement scale is stretched or shrunk, then the regression slope term β is affected. Since $\hat{Y} = \alpha + \beta X$, then multiplying the measurement scale for variable X by some constant k means that $\hat{Y} = \alpha + (\beta/k)(kX)$ for the initial equation to be satisfied. Similarly, altering the measurement scale for variable Y by some constant k means that $k\hat{Y} = k\alpha + (k\beta)X$; in other words, both the intercept and the slope of the equation are altered. These ideas about measurement scale modification are useful to show, as in Section 12.1, how regression and correlation are connected as well as to deal with numerical problems that may arise when computing with numbers that are enormously large or exceptionally small. The conversion of x_i values to z-scores in Example PR-12.2 illustrates these ideas about transformations.

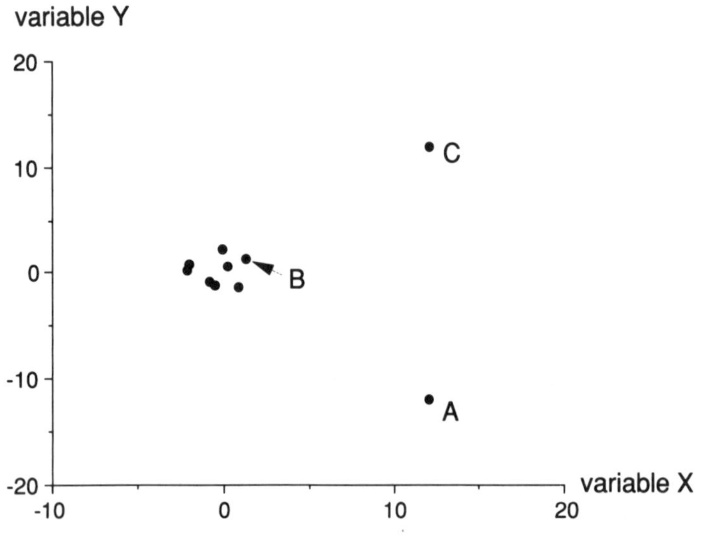

Figure 12.14: IMPACTS OF AN EXTREME POINT ON THE GOODNESS-OF-FIT OF A REGRESSION LINE

Location of extreme point	A	B	C
Correlation coefficient	-0.941	0.059	0.958

Another quality of regression analysis that often is confronted in geographic data series pertains to unusual observations. Regression equation parameter estimates a and b are very sensitive to the presence of extreme values. Such observations sometimes can be identified because they tend to have extremely large residuals; the x_i value is not atypical of the X variable, but the y_i value is an outlier with regard to the range of Y values. An interpretable magnitude for residuals is attainable by converting them from the scale upon which variable Y is measured to a z-score scale; this is achieved by computing $z_i = e_i / s_{y|x}$. Identification of an outlier suggests that the researcher needs to pay some attention to the particular data point in question. Perhaps it is due to data transcription or measurement error. Possibly the variance associated with this particular observation differs from that common to the remaining n - 1 observations. For example, in spatial interaction data a single observation may dominate the results because it is extreme, especially in its X position. In part, a degrees-of-freedom problem exists that allows a single data point to be considerably influential. If n - 1 of the data points are approximately the same, with an nth point positioned far from them, then essentially only two points--one repeatedly measured and the other singly measured--are determining the regression line; a straight line can always be fit through just two points! Figure 12.14 helps to illustrate these notions by indicating what happens in a scatter diagram when a single (extreme) point is moved about.

Consequently, in regression analysis the examination of scatter diagrams is an integral part of the analysis, with careful scrutiny and explanation of various plots affording a more thorough understanding of reality as measured by the data. Recognize, however, that statistics are available for more rigorous and exhaustive screening of data in order to objectively identify outliers and influential points.

12.4.2. Common correction techniques: variable transformations

What should be done when assumptions are violated? Why might some data points appear to be extreme when in theory they are not? (An alternative, and much more difficult, question asks why some data points might not appear to be extreme, when in theory they should.). The problem being pursued here was alluded to in Chapter 11, when the Puerto Rico data set variable DMLK was found not to conform to a normal distribution, while its modified form $DMLK^{1/2}$ did appear to conform to a normal distribution. After examining scatter diagrams and frequency distributions, to modify variable values by using transformations a researcher attempts to achieve the desired target plots and distributions. For example, in physical geography, precipitation data are often positively skewed with a high frequency of low values and a low frequency of high values. These data are often normalized using a cube-root transformation, in which each observed value is replaced by its cube-root (for example, an observed value of 8 becomes 2, and an observed value of 27 becomes 3). It must be realized that a data transformation that improves normality test statistics, unfortunately, may deteriorate linearity test statistics or variance equality statistics. In other words, many times there are trade-offs that must be faced and assessed. Since the goal of any analysis ultimately is to provide useful information with which to address questions arising from theory, a complicated transformation of the data performed to meet the requirements of the statistical procedure could easily render the results uninterpretable, and thus of little real use.

The concept of a variable transformation can be approached in one of several ways in regression analysis. First, if the scatter diagram of Y versus X indicates a nonlinear relationship, then perhaps something simple can be attempted. For example, if the scatter diagram appears to be parabolic in nature, then the equation $Y = \alpha + \beta X^2 + \xi$ is explored. In an iterative fashion this is the approach that was used to decide upon the form of the equation for Example PR-12.6; the variable transformation $X = 1/(\tau + 1)$ was not intuitively obvious from the initial scatter diagram. Second, rather than X the variable Y may need to be transformed. For instance, if $Y = (\alpha + \beta X + \xi)^2$, then $Y^{1/2} = \alpha + \beta X + \xi$. Third, the desired model might be multiplicative, rather than additive, in structure. Now $Y = \alpha X^{\beta}\xi$; here the regression line is estimated by replacing X with ln(X) and Y with ln(Y). The equation is intrinsically linear, having the form $\ln(Y) = \ln(\alpha) + \beta\ln(X) + \ln(\xi)$. Generally speaking, transformations have their most noticeable effect when the ratio of the maximum to the minimum value of a variable exceeds three.

The range of possibilities being alluded to here is characterized by transforming the regression equation into

$$Y = \alpha + \beta X^r + \xi, \text{ or } Y^r = \alpha + \beta X + \xi,$$

for some exponent r. In advanced analyses statistical techniques that allow r to be estimated are employed. In introductory analyses, recognizing what is achieved by selected values of r tends to be sufficient. First, when $r < 0$ and r then is applied to variable X, the equation becomes $Y = \alpha + \beta/X^{|r|} + \xi$. The most common value of r utilized is -1; this transformation is useful when a frequency distribution displays extreme negative skewness (see Chapter 3 for pictorial displays of skewness; in general, power transformations are applicable here), when a scatter diagram exhibits a curve that is noticeably concave toward the X-axis, or when standard deviations are found to be proportional to the square of the individual observation means. Second, when $r = 0$ and r then is applied to variable X, the equation becomes $Y = \alpha + \beta\ln(X) + \xi$. This transformation is useful when a frequency distribution displays extreme positive skewness or when variances are found to be proportional to the square of the individual observation means. Since the data are positively skewed, a transformation is needed that will reduce values in the upper tail by a greater amount than those located in the lower tail of the frequency distribution. Third, when $r > 0$ and r then is applied to variable X, the equation becomes $Y = \alpha + \beta X^r + \xi$. The most common value of r utilized in this class of values, besides 1, is 1/2 (in the example of precipitation data, r = 1/3); this transformation is useful when a frequency distribution displays moderate positive skewness or when variances are found to be proportional to the individual observation means. Negative skewness suggests that an appropriate reciprocal transformation is needed. Furthermore, many more complex transformations can be explored, too. But in the end, the use of especially concocted transformations complicates interpretation.

EXAMPLE PR-12.9. Deciding upon a variable transformation for DTB in the Puerto Rico data set when Agricultural Administrative Regions are important.

The variable DTB has the following regional sample statistics:

Region	San Juan	Arecibo	Mayaguez	Ponce	Caguas
$\bar{x}_{DTB}$	11.28	5.32	1.28	0.31	12.71
sDTB	35.50	7.60	2.46	0.46	20.16
$s_{LN(DTB + 0.033)}$	2.84	2.53	2.10	1.51	2.22

Inspection of the original data suggests that $\bar{x}_{DTB} \propto s_{DTB}$ ($\mu_{DTB} = \beta\sigma^2_{DTB}$), suggesting that a logarithmic transformation should be applied to these data (the null hypothesis of normality is rejected). This suggestion is reinforced by an inspection of the frequency distribution of DTB, which is highly positively skewed. Applying a logarithmic transformation to variable DTB and adding an optimal constant 0.033, since some density values are 0, results in both a variance across the five regions that is nearly constant, and a frequency distribution for which the null hypothesis of normality is not rejected.

Example PR-12.9 completes the set of illustrations in this book of the three commonly utilized transformations, namely, $r = -1$, $r = 0$, and $r = 1/2$.

The results of this last example return discussion to the point made at the end of the introduction to this chapter. There are many corrections, transformations, and other techniques that permit a range of problems to be analyzed in the linear regression framework. This flexibility, in part, has made the OLS technique one of the most widely used techniques in statistics. However, forcing increasingly marginal problems (marginal in terms of the decreasing degree to which the model's assumptions are easily satisfied by the data) through an OLS analysis using various "fixes" quickly renders the results extremely difficult to interpret, if not meaningless, in terms of the original underlying theory. There also is an increasing chance that sample-specific, rather than population, information will be exploited. Ultimately, massive effort is likely to be wasted, and a more suitable analytical framework should be chosen.

12.5. Summary

The last three chapters have presented an introduction to simple and complex hypothesis testing. The linear regression model has been used as a framework within which to structure a rather wide range of testing situations. The remaining four chapters consider even more advanced topics, introducing some elementary (but not necessarily simple) concepts from multivariate statistics. Thus far in this text we have looked at situations involving one variable or two variables, with one an independent and the other a dependent variable. These are univariate and bivariate situations. Multivariate situations involve more than one variable or, in a regression context, multiple independent variables. While the regression framework is still applicable, complicating extensions will be encountered.

12.6. Problems

1. Verify the calculations in Examples PR-12.1 through PR-12.9.

2. Estimate the regression equation $DBN = \alpha + \beta \cdot DFRM + \xi$ for the Mayaguez region from the Puerto Rico data set. Numerically show that using z-scores in this equation forces $a = 0$ and $b = r_{DBN,DFRM}$.

3. Evaluate the regression analysis assumptions for results obtained in Problem #2.

4. Draw inferences for α and β in Problem #2. Then, write the confidence intervals for these two parameters based upon the Ponce regional data. How do these inferences compare with those for the Arecibo region (see Example PR-12.3)?

5. Numerically justify the use of the square-root transformation for variable DMLK contained in the Puerto Rico data set when Agricultural Administrative Regions are not important.

6. Numerically justify the use of the inverse transformation in Example PR-12.6.

7. Construct confidence boundaries around the regression results in Example SG-12.2. What can be said about the probability that certain net mass balances will occur in years when snowfall (PRECIP) exceeds 200 cm?

12.7. Additional readings

In addition to the texts listed in Chapter 1,

Ferguson, R. *Linear Regression in Geography*. CATMOG 15.

Gould, P. (1970) "Is *Statistics Inferens* the Geographical Name for a Wild Goose Chase?" *Economic Geography*, 46: 439-48.

Mark, D., and M. Church. (1977) "On the Misuse of Regression in Earth Science." *Journal of the International Association of Mathematical Geology*, 9: 63-75.

Mark, D., and T. Peucker. (1978) "Regression Analysis and Geographic Models." *Canadian Geographer*, 22: 51-64.

Lewis-Beck, M. *Applied Regression*. SAGE 22.

Achen, C. *Interpreting and Using Regression*. SAGE 29.

Schroeder, L., D. Sjoquist, and P. Stephan. *Understanding Regression Analysis*. SAGE 57.

PART III

A TRANSITION TO MULTIVARIATE ANALYSIS

The first two parts of this book address the themes of descriptive and inferential statistics for geographers, almost exclusively in terms of univariate analysis. The basic exception to this focus is the bivariate treatments referred to in geostatistics, correlation, and simple regression analysis. Aims of these two earlier parts included establishing criteria for determining useful statistics, outlining sources and consequences of error, describing various features of information content in data, uncovering major statistical assumptions, reviewing common sampling distribution forms, and drawing inferences about population parameters from sample data.

Multivariate analysis extends earlier results to situations in which more than one variable is being analyzed; hence, the few bivariate topics already discussed in this book initialize a transition into the more general realm of multivariate analysis. Two features of this earlier material that become a focal point of multivariate techniques are correlated sample results and sequential/simultaneous testing. In Sections 11.1.2, 11.2.2, and 11.4.2, selected procedures for handling correlated sample data were presented; the source of correlation in the collected data measuring several variables comes from using the same set of areal units, while the impact usually is some modification of a standard error formula. Multivariate analysis is concerned with the collection of data measuring many variables for the same set of areal units; again, the communality permeating these measures in part will be their shared observational units, and in part information redundancies in measures. This compound notion of "conditional-many" generalizes correlated samples to the idea known as multicollinearity. In Chapters 11 and 12, altering Type I error probability levels by sequential testing practices frequently was noted but overlooked. The correct t statistics for simultaneous testing affiliated with parameters of simple linear regression were highlighted in Chapter 12. This oversight will be redressed here. Again, because the same areal units and data are used for several tests, sequential decision making probabilities are neither independent nor constant.

The following four chapters provide an introduction to multivariate analysis. They will initiate a stronger transition into this field of study, with the presentation being restricted to only those concepts from multivariate analysis absolutely necessary to understand this transition. Multivariate analysis is numerically intensive. Accordingly, all empirical work will be completed with the aid of a computer (the SAS and MINITAB-PC commercial packages will be employed); for the sake of everyone's sanity, manual calculations will be avoided. Reference theoretical sampling distributions almost always are either the F or the chi square.

CHAPTER 13

MULTIPLE LINEAR REGRESSION

The first complication that is confronted when moving into the more general area of multivariate analysis is that tables of numerical sums and products are encountered and must be worked with. These tables have either both row and column labels that are the variables being studied, or row labels that are the areal units and column labels that are the variables. These tables are known as **matrices** (one example of a matrix already has been seen in Example SG-12.1); if a matrix has a single row or column, then it is called a **vector**, and if it has a single row and a single column, it is called a **scalar** (the type of number encountered in Chapters 1-12). The location in a matrix identified by the intersection of a row and column is called a **cell**. For square matrices (matrices with an equal number of rows and columns), the cell addressed when the row number and the column number are the same (for example, second row and second column) occupies a position on the **diagonal** of the matrix. Matrices will be denoted here by bold letters. The basic matrices needed for an excursion into multivariate analysis include

X the independent variable matrix, which has n rows, one for each areal unit, and p + 1 columns, where p is the number of variables,

Y the dependent variable vector, which has n rows and one column,

R the correlation matrix, which contains the correlation coefficient between variables i and j as entry r_{ij} (row i and column j), with $r_{ij} = r_{ji}$ (see Chapter 4), and which has p rows and p columns (this notion has been illustrated in Example SG-12.1),

I the identity matrix, which is a square matrix with the diagonal elements equal to unity, and zeroes everywhere else, and

1 the unit vector, which has n rows, one column, and all entries of unity.

Four matrix operations are needed in this discussion. First, matrix multiplication is defined such that the product matrix entry cell (i, j) contains the product of row i from the first matrix multiplied by column j from the second matrix. Multiplication of a row and a column requires that the corresponding cell entries are multiplied, and then these products are summed. Note that the row and the column being multiplied must have exactly the same number of cells. Algebraically, in matrix multiplication

$\mathbf{AB} = \mathbf{C}$ means that $c_{ij} = \sum_{k=1}^{n} a_{ik}b_{kj}$;

for example,

$$\begin{pmatrix} 1 & 2 \\ 3 & 4 \end{pmatrix}\begin{pmatrix} 6 & 4 \\ 5 & 0 \end{pmatrix} = \begin{pmatrix} 1 \cdot 6 + 2 \cdot 5 \\ 3 \cdot 6 + 4 \cdot 5 \end{pmatrix}\begin{pmatrix} 1 \cdot 4 + 2 \cdot 0 \\ 3 \cdot 4 + 4 \cdot 0 \end{pmatrix} = \begin{pmatrix} 16 & 4 \\ 38 & 12 \end{pmatrix}$$

Note that the order of matrix multiplication is important. The matrix product **AB** usually does not equal **BA**, and in fact, since the first matrix must have the same number of rows as the second matrix has columns, the product **BA** may not be possible to compute. The student should be able to show, for the preceding example, that $\mathbf{BA} \neq \mathbf{C}$.

Second, matrix inversion, which actually is a generalization of the familiar scalar inverse, is defined such that $\mathbf{A}^{-1}\mathbf{A} = \mathbf{A}\mathbf{A}^{-1} = \mathbf{I}$ (which is directly analogous to $a^{-1}a = aa^{-1} = 1$). **A** must have the same number of rows as columns if a unique matrix inversion is to be defined; that is, **A** must be a square matrix. In scalar arithmetic, in order for a^{-1} to exist, $a \neq 0$. Analogously, any given matrix can be characterized by a polynomial whose numerical value is equal to the determinant of that matrix; in order for $\mathbf{A}^{-1}$ to exist, this determinant cannot equal zero. Although a more precise definition of a determinant will be avoided here, its relation to the correlation matrix **R** will be introduced shortly. Computing matrix inverses is numerically intensive. One way of viewing it is that matrices **A** and **I** are known, while matrix $\mathbf{A}^{-1}$ is unknown and must be solved for; as such, the problem is one from intermediate algebra of solving a system of linear equations. For instance,

$$\begin{pmatrix} 1 & 2 \\ 3 & 4 \end{pmatrix}\begin{pmatrix} a & b \\ c & d \end{pmatrix} = \begin{pmatrix} 1 & 0 \\ 0 & 1 \end{pmatrix}$$

$$\begin{aligned} 1a + 2c &= 1 \quad & 3a + 4c &= 0 \\ 1b + 2d &= 0 \quad & 3b + 4d &= 1 \end{aligned}$$

Solving these four equations in four unknowns gives

$a = -2, b = 1, c = 3/2, d = -1/2.$

Hence,

$$\begin{pmatrix} 1 & 2 \\ 3 & 4 \end{pmatrix}\begin{pmatrix} -2 & 1 \\ 3/2 & -1/2 \end{pmatrix}=\begin{pmatrix} 1 & 0 \\ 0 & 1 \end{pmatrix}.$$

As an aside, the determinant of the initial matrix in this example is $-2 \neq 0$.

Third, diag(**A**) extracts the diagonal elements located in the (i, j) cells, where i = j, from a square matrix **A**, and fills the off-diagonal cells with zeroes. For instance,

$$\text{diag}\begin{pmatrix} 1 & 2 \\ 7 & 3 \end{pmatrix}=\begin{pmatrix} 1 & 0 \\ 0 & 3 \end{pmatrix}$$

Finally, matrix transpose, denoted as $\mathbf{X}^t$, may be defined as switching the rows and columns of a matrix. For example,

$$\begin{pmatrix} 1 & 2 \\ 3 & 4 \end{pmatrix}^t=\begin{pmatrix} 1 & 3 \\ 2 & 4 \end{pmatrix}$$

Although there are many other matrix operations that are extensively employed in multivariate analysis, these rather simple ones are the only essentials for the introduction to be presented here.

13.1. The matrix formulation of linear regression

Linear regression computations can be completed within the context of matrix notation. As such, the simplest, bivariate model is of the form

$$\mathbf{Y} = \alpha\mathbf{1} + \beta\mathbf{X} + \xi$$

where **Y**, **1**, **X**, and ξ are vectors of length n, and vector **1** appears because the intercept term α is constant over all areal units. To write the parameters appearing in this matrix equation in matrix form, a change in notation for the intercept term α occurs; $\beta_0 = \alpha$. This change necessitates the adding of a subscript to the slope term, so that it becomes β_1, and also alleviates the confusion concerning the double meaning of α that was mentioned in Section 12.2. Consequently,

$$\mathbf{Y} = \mathbf{X}\beta + \xi, \text{ or } \mathbf{Y} = \mathbf{1}\beta_0 + \mathbf{X}\beta_1 + \xi.$$

In the bivariate case, matrix β now denotes a (2-by-1) matrix, or column vector with $\beta^t = (\beta_0, \beta_1)$, and matrix **X** has two columns, the first being the vector **1**, which now is denoted by X_0, and the second being the independent variable, which now is denoted by X_1. Furthermore, any number of additional variables X_j can be added to this form of the linear regression equation, together with their corresponding β_j parameters. The only binding restriction is that a positive number of degrees of freedom must remain; recall that the rule of thumb is for at least 20 degrees of freedom per parameter (see Chapter 10), implying that $n \geq 20 \bullet (p + 1)$.

The respective linear regression parameter estimates are denoted by b_j. For the bivariate case, then, they are b_0 and b_1. These estimates are calculated as follows:

$$\mathbf{b} = (\mathbf{X}^t\mathbf{X})^{-1}\mathbf{X}^t\mathbf{Y}, \text{ with } E(\mathbf{b}) = \beta \text{ and } VAR(\mathbf{b}) = \sigma_{y|x}^2\text{diag}[(\mathbf{X}^t\mathbf{X})^{-1}];$$

as Example PR-13.1 illustrates, these equations produce the same results as those reported for the bivariate case in Chapter 12. Two noteworthy points merit attention at this time. First, multicollinearity increases as correlations among the X_j variables approach ±1. Along with this increase, the determinant of matrix $(\mathbf{X}^t\mathbf{X})$ approaches zero. Thus, two perfectly correlated variables cannot appear in an estimated regression equation. Second, $\sigma_{y|x}^2 = \sum_{i=1}^n e_i^2/(n - p - 1)$, where p is the number of independent variables (excluding vector $\mathbf{X}_0 = \mathbf{1}$).

EXAMPLE PR-13.1. Example PR-12.7 revisited.

Recall that **X** has two columns; the first is **1**, whereas the second is the set of values given by DCD_{1969}. The dependent variable is $Y = DCD_{1974}$. From MINITAB-PC, the matrix form of the estimated vector **b** is

$$(X^tX): \begin{pmatrix} 19.00 & 562.70 \\ 562.70 & 20558.79 \end{pmatrix} \quad X^tY: \begin{pmatrix} 502.30 \\ 17413.42 \end{pmatrix}$$

where $\sum_{i=1}^{19} 1 = 19$,

$$\sum_{i=1}^{19} 1 \bullet x_i = 1 \bullet 40.2 + 1 \bullet 10.9 + 1 \bullet 21.3 + 1 \bullet 10.7 + 1 \bullet 41.3 + 1 \bullet 48.5$$
$$+ 1 \bullet 60.3 + 1 \bullet 35.8 + 1 \bullet 18.0 + 1 \bullet 35.4 + 1 \bullet 37.9 + 1 \bullet 32.2$$
$$+ 1 \bullet 40.5 + 1 \bullet 14.1 + 1 \bullet 3.5 + 1 \bullet 34.7 + 1 \bullet 36.4 + 1 \bullet 18.4$$
$$+ 1 \bullet 22.6 = 562.7$$

(note that switching the order of the multiplications will not change this sum),

$$\sum_{i=1}^{19} x_i^2 = 40.2 \cdot 40.2 + 10.9 \cdot 10.9 + 21.3 \cdot 21.3 + 10.7 \cdot 10.7 + 41.3 \cdot 41.3 + 48.5 \cdot 48.5 + 60.3 \cdot 60.3 + 35.8 \cdot 35.8 + 18.0 \cdot 18.0 + 35.4 \cdot 35.4 + 37.9 \cdot 37.9 + 32.2 \cdot 32.2 + 40.5 \cdot 40.5 + 14.1 \cdot 14.1 + 3.5 \cdot 3.5 + 34.7 \cdot 34.7 + 36.4 \cdot 36.4 + 18.4 \cdot 18.4 + 22.6 \cdot 22.6 = 20558.79,$$

$$\sum_{i=1}^{19} 1 \cdot y_i = 1 \cdot 38.0 + 1 \cdot 9.4 + 1 \cdot 22.7 + 1 \cdot 13.2 + 1 \cdot 52.6 + 1 \cdot 40.7 + 1 \cdot 26.2 + 1 \cdot 24.5 + 1 \cdot 15.0 + 1 \cdot 23.7 + 1 \cdot 23.3 + 1 \cdot 34.2 + 1 \cdot 53.6 + 1 \cdot 14.3 + 1 \cdot 2.9 + 1 \cdot 27.7 + 1 \cdot 28.2 + 1 \cdot 20.4 + 1 \cdot 31.7 = 502.3, \text{ and}$$

$$\sum_{i=1}^{19} x_i y_i = 40.2 \cdot 38.0 + 10.9 \cdot 9.4 + 21.3 \cdot 22.7 + 10.7 \cdot 13.2 + 41.3 \cdot 52.6 + 48.5 \cdot 40.7 + 60.3 \cdot 26.2 + 35.8 \cdot 24.5 + 18.0 \cdot 15.0 + 35.4 \cdot 23.7 + 37.9 \cdot 23.3 + 32.2 \cdot 34.2 + 40.5 \cdot 53.6 + 14.1 \cdot 14.3 + 3.5 \cdot 2.9 + 34.7 \cdot 27.7 + 36.4 \cdot 28.2 + 18.4 \cdot 20.4 + 22.6 \cdot 31.7 = 17413.42.$$

$$(X^tX)^{-1}: \begin{pmatrix} 0.277875 & -0.007606 \\ -0.007606 & 0.000257 \end{pmatrix} \quad X^tY: \begin{pmatrix} 502.30 \\ 17413.42 \end{pmatrix} \quad b: \begin{pmatrix} 7.1385 \\ 0.6516 \end{pmatrix}$$

The matrix inversion computations, except for rounding error, are

$[19.0000 \cdot 0.277875 + 562.70 \cdot (-0.007606)] = 1$,
$[19.0000 \cdot (-0.007606) + 562.70 \cdot 0.000257] = 0$,
$[562.70 \cdot 0.277875 + 20558.79 \cdot (-0.007606)] = 0$, and
$[562.70 \cdot (-0.007606) + 20558.79 \cdot 0.000257] = 1$.

In keeping with earlier matrix inverse cell location notation, a = 0.277875, b = c = -0.007606, and d = 0.000257. Next, for the two regression coefficients the matrix operations are equivalent to

$[0.277875 \cdot 502.30 + (-0.007606) \cdot 17413.42 = 7.1301$, and
$[(-0.007606) \cdot 502.30 + 0.000257 \cdot 17413.42] = 0.6548$,

which are the same as before, except for rounding error.

This example illustrates a prominent concern associated with multiple regression, namely, that matrix inversion can introduce considerable rounding error into a calculation (7.1385 is noticeably but not markedly different from 7.1301).

13.2. Introducing additional independent variables into the linear regression equation

The matrix form of the linear regression equation facilitates introduction of a second, third, fourth, and so on, independent variable into an analysis. But if rounding error is so problematic, why should more variables be considered? On the one hand, substantively most phenomena are correlated with many other variables; we live in a multivariate world. The coefficient of determination was used in Chapter 12 as an index of goodness-of-fit of a regression line. When more than one independent variable appears in a regression equation, this coefficient no longer refers to a simple correlation; rather, it is the square of a multiple correlation coefficient (a coefficient capturing the correlation between variable Y and an additive weighted combination of variables X_j) and must be viewed strictly in terms of its more general notation of $R_{Y,\hat{Y}}^2$, which reduces to $r_{Y,X}^2$ in the bivariate case. The introduction of additional variables into a multiple linear regression equation can never decrease this quantity; if $b_j = 0$, then the coefficient of determination will remain the same, whereas if $b_j \neq 0$, then this coefficient must increase. In addition, the only way a set of p bivariate regression lines based upon the same Y variable and set of areal units can be compared is by simultaneously including the p variables in a single multiple regression equation. Because these variables are measured for the same set of areal units, the correlation coefficients involved can be compared only within the context of a correlated sample for the correct standard errors to be calculated (see Section 11.4.2); the notion of partial correlation dominates this situation. But, even when the correlation matrix for the set of X_j measures equals the identity matrix (no multicollinearity exists in the set of independent variables), concurrent treatment of variables for the same set of areal units means that the Type I error probabilities must be adjusted. Again, this adjustment is achieved through the simultaneous treatment of the b_j estimates; conditional inferences are being drawn as in Chapter 12.

On the other hand, assumption violations may be a result of the omission of variables from a regression equation. When this is the case, often these assumption violations disappear once the missing variable is placed into the regression equation. One important facet of regression theory is that omitted (hence too few) variables tend to cause serious problems, whereas the presence of superfluous variables (i. e., variables having regression coefficients approximately equal to zero) tends not to cause major problems. Regardless of whether or not a variable is superfluous, though, if it is highly correlated with other variables in the regression equation, the resulting standard errors will tend to be inflated. The three assumptions most sensitive to missing variables are constant variance, zero spatial autocorrelation, and normality [assumptions (e), (f), and (g), respectively, from the beginning of Chapter 12].

EXAMPLE PR-13.2. Example PR-12.7 extended.

The regression of DCD_{1974} on DCD_{1969} resulted in a coefficient of determination equaling 0.5135 and a possible violation of the equal variance assumption [assumption (e) from Chapter 12]. Suppose that variable DCF also is introduced into this regression analysis. Now the estimated equation becomes, using MINITAB-PC (see Appendix 13-A),

$$\hat{DCD}_{1974} = 8.1360 + 0.5373 \cdot DCD_{1969} + 0.0019 \cdot DCF$$

Statistic	Coefficient	Standard error	t statistic
b_0	8.1360	4.1558	1.96
b_1	0.5373	0.1314	4.09
b_2	0.0019	0.0006	3.06

$s_{y|x} = 7.859$, coefficient of determination = 0.693.

Note that the coefficient of determination suggests that DCF adds an increment of about 18% to the variance accounted for in DCD_{1974}. The critical values for the t statistic, using 19 - 3 = 16 degrees of freedom and a 5% level of significance (two-tail situation), are ±2.1199. Therefore, these results (obtained with MINITAB-PC) indicate that the correct inferences are (1) fail to reject that the intercept is 0 and (2) reject that the slopes for the two independent variables are 0. This first inference reverses the one drawn in Example PR-12.7. Next, consider the plot for assessing the constant variance assumption, which appears in Figure 13.1. Not surprisingly, both the range of the residuals has shrunk, and their scatter more closely approximates a random point pattern (there now is even less evidence of nonconstant variance). Furthermore, these improvements in the linear regression model did not require a sacrifice of other assumptions; MC = -0.0731 is even closer to its expected value, and the frequency distribution of the residuals continues to support an inference of normality.

Again, attention must be directed to the possibility that improvement in satisfying one assumption yields a more dramatic violation of another assumption. Each assumption must be reassessed after each model enhancement.

EXAMPLE SG-13.1. Examples SG-4.1, SG-12.1, and SG-12.2 extended.

The scatter diagrams of net mass balance of Sentinel Glacier (BN) versus mean summer air temperature (TEMP) and BN versus total winter precipitation (PRECIP) already have been plotted separately. But what about the assessment of their joint effects? If both TEMP and PRECIP are entered into a regression equation, to account for the variation in BN, then the

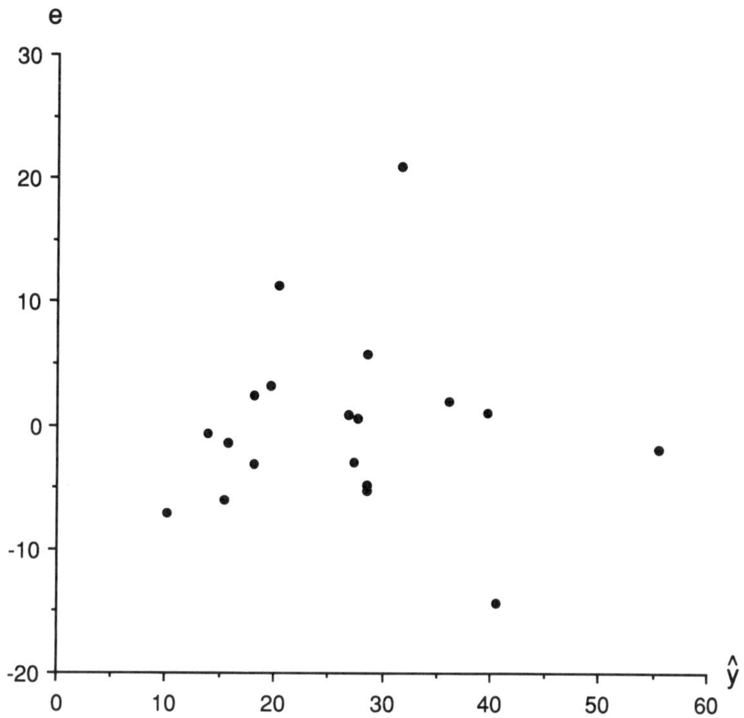

Figure 13.1: SCATTER DIAGRAM FOR EVALUATING CONSTANT VARIANCE

resulting estimated equation is

$$\hat{BN} = 2.3218 + 0.0024 \cdot PRECIP - 0.3564 \cdot TEMP.$$

Statistic	Coefficient	Standard error	t statistic
b_0	2.3218	3.394	0.68
b_1	0.0024	0.0006	3.78
b_2	-0.3564	0.2596	-1.37

$s_{y|x} = 0.7165$, coefficient of determination = 0.505.

At a 5% level of significance, the critical values for the t statistic are ±2.1199 (from Table 9.1-A, found in the appendices, using 16 degrees of freedom). Comparing the computed t statistics with these critical values suggests the following inferences: (1) fail to reject that the intercept is 0; (2) reject that the slope for the PRECIP variable is 0; and, (3) fail to reject that the slope for the TEMP variable is 0. Note that the regression coefficient for TEMP is negative, as might be expected (i. e., higher temperatures tend to be associated with lower net mass balances); this point is more fully discussed in Example SG-12.2.

This second example stresses two important considerations often emphasized in the literature. First, researchers often are concerned with whether or not the sign of a regression coefficient is correct, regardless of its statistical inference. In other words, does the numerical result make logical sense? Second, because of redundant information contained in variables, a multiple regression equation including two variables will not necessarily be consistent with or an improvement upon individual simple regression analyses.

13.3. ANOVA revisited: unbalanced two-way classifications

This section examines two characteristics of the regression technique as a generalized approach to statistical analysis (see Section 12.3.4). The first has to do with how regression analysis exploits ANOVA (most commercial statistical packages output an ANOVA table with the execution of a regression procedure). The second has to do with how regression analysis provides a solution to unbalanced multiclassificatory ANOVA problems. More specifically, in its generalized form, regression alludes to what statisticians call the general linear model (GLM).

13.3.1. The standard ANOVA table of linear regression

In Chapter 12 the coefficient of determination for regression was described as measuring the proportion of variance in variable Y explained by variable X. This notion has been generalized in this chapter for multiple linear regression. ANOVA, as presented in Chapter 11, and linked to bivariate linear regression in Chapter 12, partitions variance into that which is attributable to the classification variable (between regions) and that which is attributable to error (within regions). The counterparts to these two variance components are variance attributable to the regression equation (the proportion of variance in variable Y explained by variable X) and variance attributable to the residuals (error variance). Suitable modifications to the ANOVA table presented in Section 11.3.1, using matrix notation, result in the following one. Again, regardless of the magnitude of the estimated variance terms, the variance estimate due to regression--namely, s_R^2, is always in the numerator. In addition, note that numerical results tabulated in this manner, when the X_j are indicator variables (as mentioned in Chapters 2 and 12, only K - 1 of them need to be included), is exactly what is obtained from a difference of K independent means analysis of variance.

ANOVA

Source of variation	Sum of squares	Degrees of freedom	Mean squares	F ratio
Regression	$b^tX^tY - (1^tY)^2/n$	p	$s_R^2 = [b^tX^tY - (1^tY)^2/n]/p$	s_R^2/s_E^2
Residual error	$Y^tY - b^tX^tY$	n - p - 1	$s_E^2 = [Y^tY - b^tX^tY]/(n - p - 1)$	
Corrected total	$Y^tY - (1^tY)^2/n$	n - 1		

EXAMPLE PR-13.3. Example PR-11.9 revisited.

Binary indicator variables were created for each of the five agricultural administrative regions. Then the indicator variable for the Caguas region (an arbitrary selection) was subtracted from each of the other four regional indicator variables; this eliminates a source of perfect multicollinearity, and it allows a difference of means version of the problem to be treated. Using MINITAB-PC, the resulting regression equation is

$$\widehat{DCD} = 36.9322 - 10.4953 \bullet (I_{SJ} - I_C) + 2.7042 \bullet (I_A - I_C) + 1.8928 \bullet (I_M - I_C) + 7.3535 \bullet (I_P - I_C)$$

The accompanying ANOVA table for this equation is

ANOVA OF DCD

Source of variation	Sum of squares	Degrees of freedom	Mean squares	F ratio
Regression	2973.3	4	743.3	6.90
Residual error	7323.4	68	107.7	
Corrected total	10296.7	72		

Selection of the indicator variable to subtract is arbitrary. Although indicator variable selection for subtraction purposes will make a difference in **b** (as it should, since it will alter the specific difference of means comparisons), the ANOVA table entries will not change.

Except for rounding error, the numbers appearing in Example PR-13.3 are exactly the values computed in Chapter 11.

EXAMPLE CA-13.1. One-way ANOVA for export sales (EXP$), location (PC1), and market niche (PPC).

Continuing the analysis of export sales (EXP$) started in Example CA-12.1, two categorical variables are examined for their relationship with EXP$. PC1 is the first character in the Canadian postal code, and indicates where in Southern Ontario the plant is located (see the map in Chapter 1). PPC indicates which type of automotive parts the firm considers to be the primary product (see Chapter 1 for a discussion of this variable). The goal here is to examine whether or not location or product group is related to the level of export sales.

The distribution of observations for PC1 is $n_K = 2$, $n_L = 12$, $n_M = 13$, and $n_N = 26$. The distribution of observations for the categories of PPC is $n_0 = 7$, $n_1 = 6$, $n_2 = 5$, $n_3 = 16$, $n_4 = 1$, $n_5 = 1$, $n_6 = 3$, and $n_7 = 14$. Again from MINITAB-PC, the resulting regression equation for PC1 is

$$\text{SL}\hat{\text{E}}\text{XP} = 8830 - 8330 \bullet (I_K - I_N) + 2812 \bullet (I_L - I_N) + 5951 \bullet (I_M - I_N).$$

The accompanying ANOVA table for PC1 is

ANOVA OF PC1

Source of variation	Sum of squares	Degrees of freedom	Mean squares	F ratio
Regression	568308352	3	189436112	0.63
Residual error	14651022336	49	299000448	
Corrected total	15219331072	52		

The resulting regression equation for PPC is

$$\text{SL}\hat{\text{E}}\text{XP} = 8700 + 3586 \bullet (I_0 - I_7) - 233 \bullet (I_1 - I_7) - 2595 \bullet (I_2 - I_7)$$

$$+ 8837 \cdot (I_3 - I_7) - 3700 \cdot (I_4 - I_7) - 8600 \cdot (I_5 - I_7) + 7967 \cdot (I_6 - I_7).$$

The accompanying ANOVA table for PPC is

ANOVA OF PPC

Source of variation	Sum of squares	Degrees of freedom	Mean squares	F ratio
Regression	1886053120	7	269436160	0.91
Residual error	13333277696	45	296295072	
Corrected total	15219331072	52		

The F values for these two models indicate that in both cases the null hypothesis of equal group means is not rejected. The critical F value is 2.2403 ($F_{0.05,7,45}$ from Table 11.1-A, using linear interpolation). In addition, neither of the models explains much of the statistical variation in EXP$; R^2 for the model with PC1 = 0.04, and for the model with PPC equals 0.12.

The regression framework for ANOVA leads to two additional operational concepts. First, as mentioned in Section 11.3.1, one supplemental question asks what is the best alternate hypothesis for ANOVA. Since each regression coefficient attached to an indicator variable difference is equivalent to a particular difference of means, and the multiple regression standard errors are adjusted in order to allow simultaneous testing, then substantial light is shed upon this question by the individual regression coefficient t statistics. Significant coefficients suggest that the pair of population means are different, whereas insignificant coefficients suggest that the pair of population means are not different.

EXAMPLE PR-13.4. Example PR-13.3 continued.

MINITAB-PC output also includes the following test statistics information for vector **b**. Here the Type I error probability is the probability of exceeding the observed t statistic merely due to sampling error, under the null hypothesis, and for a two-tail situation; hence, only values less than 10%, 5%, or 1% conventionally are meaningful, depending upon the level of significance that is selected. Two of the computed Type I error probabilities (0.324 and 0.425) are greater than any of these three levels of significance. These results suggest that the best alternate hypothesis would involve $\mu_A = \mu_M = \mu_C$, but the null hypothesis of a difference in means for these three regions would not be rejected, and $\mu_{SJ} \neq \mu_C$, and $\mu_P \neq \mu_C$. However, the null hypothesis of a difference in means for these two pairs of regions is rejected, indicating that further analysis is needed before a relationship between μ_{SJ} and μ_P can be established.

Variable	Coefficient	Standard error	t	Type I error probability
1	36.9322	1.236	29.89	0.000
$(I_{SJ} - I_C$	-10.4953	2.220	-4.73	0.000
$(I_A - I_C)$	2.7042	2.720	0.99	0.324
$(I_M - I_C)$	1.8928	2.359	0.80	0.425
$(I_P - I_C)$	7.3535	2.478	2.97	0.004

This spinoff is invaluable and sets the stage for many multiple comparisons dealt with in multivariate analysis.

The regression framework for ANOVA also leads to statistical testing based upon the **extra sum of squares** principle. A contribution to the regression sum of squares made by an individual independent variable can be determined directly by comparing the regression ANOVA table including this variable in the regression analysis (called the full model) with that ANOVA table excluding this variable from the regression analysis (called the reduced model). The regression sum of squares plus the residual sum of squares always equals the corrected total sum of squares. When a given variable is included in the model, its sum of squares contribution appears in the regression sum. When this same variable is excluded from the model, its sum of squares contribution appears in the residual sum. Subtracting the residual sum of squares for the model including a given variable from the residual sum of squares for the model excluding this variable determines its contribution; hence, this difference is conditional. This principle also applies to sets of variables. Note that if each variable in a multiple regression equation is done in turn, except in very special instances, the sum of the p extra sums of squares will not necessarily equal the corrected total sum of squares due to the presence of multicollinearity in matrix **X**.

EXAMPLE PR-13.5. Example PR-13.3 extended.

Suppose the indicator variable difference $(I_P - I_C)$ is not included in the regression analysis of Example PR-13.3. Applying the extra sum of squares principle to this pair of results produces the sum of squares directly attributable to variable $(I_P - I_C)$, namely, 8271.6 - 7323.4 = 2973.3 - 2025.1 = 948.2 (except for rounding error), or

(residual error in PR-13.5) - (residual error in PR-13.3) = (regression sum of squares in PR-13.3) - (regression sum of squares in PR-13.5) = 948.2.

Similarly, the number of degrees of freedom that accompany this variation term is

4 - 3 = 69 - 68 = 1.

Then the accompanying ANOVA table for the reduced equation is

ANOVA OF DCD

Source of variation	Sum of squares	Degrees of freedom	Mean squares	F ratio
Regression	2025.1	3	675.0	5.63
Residual error	8271.6	69	119.9	
Corrected total	10296.7	72		

Dividing this extra sum of squares value for variable(s) X_j by its accompanying degrees of freedom yields a variance estimate s_j^2. Recalling that $t^2 = F$, the full-model t test statistics are computed by calculating $[s_j^2/s_E^2]^{1/2}$, where s_E^2 is from the full-model ANOVA table. This consequence demonstrates why the test statistics are conditional in a regression analysis, and hence why the accompanying Type I error probabilities are for simultaneous testing.

13.3.2. Unbalanced ANOVA

Partitioning variation in the single classification ANOVA reviewed in Chapter 11 is not affected by an unequal number of areal units in each region (although its power may be affected). Test statistic computations for this simplest ANOVA are not affected by the nature of the classification variable, either. Both of these central numerical themes become problematic when a multiple classification is unbalanced.

Suppose there are R categories in the first classification variable and C categories in the second classification variable. When a two-way ANOVA is being undertaken, the overall null hypothesis becomes

$$H_0\colon\ \mu_{11} = \mu_{12} = \ldots = \mu_{1C} = \mu_{21} = \ldots = \mu_{2C} = \ldots = \mu_{R1} = \ldots = \mu_{RC}.$$

If these two classification variables are independent, then this null hypothesis reduces to the following pair of individual ones:

$$H_0\colon\ \mu_{\cdot 1} = \mu_{\cdot 2} = \ldots = \mu_{\cdot R},\text{ and}$$

H_0: $\mu_{1\cdot} = \mu_{2\cdot} = \ldots = \mu_{C\cdot\cdot}$

The problem of proper variation partitioning to determine which null hypotheses and test statistics are appropriate is solved by employing the extra sum of squares principle and treating ANOVA in a regression context. Most introductions to two-way ANOVA in beginning texts assume balanced designs (which rarely occur with spatial data series) in which the individual sums of squares add to the corrected total sum of squares. As alluded to in Section 13.3.1, for unbalanced designs this addition equality no longer is guaranteed to hold.

ANOVA as a regression problem means there will be

$$(R - 1) + (C - 1) + (R - 1)(C - 1)$$

indicator variables entered into the full-model regression equation; remember that one arbitrarily selected classification variable is subtracted from all others in order to avoid perfect multicollinearity. The first set of indicator variables are removed to obtain the sum of squares attributable to the first classification, the second set of indicator variables are removed to obtain the sum of squares attributable to the second classification, and the third set of indicator variables are removed to obtain the sum of squares attributable to the interaction between the two classification variables. As a result there will be three reduced-model regression equation ANOVA tables. The interaction variation term *must be insignificant* for the two individual classification variables to be interpreted separately. Significant interaction implies that the impact of one classification is differential over the categories of the other classification. Testing interaction first is not a sequential procedure, since all the variation terms have been computed as conditional values (given that the other indicator variates are in the regression equation); this is why an unbalanced design does not have sums of squares that add to the total, whereas balanced designs do. In MINITAB Release 6.1 and lower, the appropriate indicator variables must be constructed, followed by execution of individual regressions. In releases MINITAB 7.0 and SAS 5.0 and higher, PROC GLM automatically will convert the classification variables into indicator variables and produce the final ANOVA table.

EXAMPLE PR-13.6. A simple two-way ANOVA for DCD, using the San Juan (SJ) and Mayaguez (M) regions as one classification and coastal lowlands (cl) and interior highlands (ih) as the other classification.

The cross-classification cells for this example produce the following sub-sample sizes:

	Coastal lowlands	Interior highlands
San Juan	9	7
Mayaguez	12	7

This tabulation indicates that an unbalanced ANOVA situation exists. The regression analysis here will involve three differenced indicator variables, namely,

$(I_{SJ} - I_M)$,
$(I_{cl} - I_{ih})$, and
$(I_{SJ} - I_M)(I_{cl} - I_{ih})$,

with its corresponding full-model

$$\hat{DCD} = b_o + b_1(I_{SJ} - I_M) + b_2(I_{cl} - I_{ih}) + b_3(I_{SJ} - I_M)(I_{cl} - I_{ih}).$$

From MINITAB-PC, the four ANOVA tables are

INDIVIDUAL ANOVAs FOR DCD

Full-model [no differenced indicator variables excluded]:

Source of variation	Sum of squares	Degrees of freedom	Mean squares	F ratio
Regression	2899.0	3	966.3	8.96
Residual error	3344.0	31	107.9	
Corrected total	6243.0	34		

Reduced-model [variable $(I_{SJ} - I_M) \cdot (I_{cl} - I_{ih})$ excluded]:

Source of variation	Sum of squares	Degrees of freedom	Mean squares	F ratio
Regression	2405.1	2	1202.6	10.03
Residual error	3837.9	32	119.9	
Corrected total	6243.0	34		

Reduced-model [variable $(I_{cl} - I_{ih})$ excluded]:

Source of variation	Sum of squares	Degrees of freedom	Mean squares	F ratio
Regression	1912.8	2	956.4	7.07
Residual error	4330.2	32	135.3	
Corrected total	6243.0	34		

Reduced-model [variable (I_{SJ} - I_M) excluded]:

Source of variation	Sum of squares	Degrees of freedom	Mean squares	F ratio
Regression	2042.1	2	1021.0	7.78
Residual error	4200.9	32	131.3	
Corrected total	6243.0	34		

The following resulting variation terms are of the form

(value from the appropriate reduced model) - (value from the full model):

sum of squares due to interaction = 3837.9 - 3344.0 = 493.9
sum of squares due to coastal lowlands/interior highlands
classification = 4330.2 - 3344.0 = 986.2
sum of squares due to agricultural administrative regions
classification = 4200.9 - 3344.0 = 856.9

The same results, except for rounding error, are obtained with SAS as TYPE III SS values.

But how are these ANOVA results commonly summarized and interpreted? Almost without exception, the output from a two-way ANOVA is summarized in tabular form. The format of this table utilizes the one-way ANOVA table as a template and is as follows, for an unbalanced design:

TWO-WAY ANOVA

Source of variation	Sum of squares	Degrees of freedom	Mean squares	F ratio
Classification variable 1	Differences between appropriate reduced models terms		Appropriate sum of squares	?
Classification variable 2	Differences between appropriate reduced models terms		Divided by the corresponding	?
Interaction	Differences between appropriate reduced models terms		Number of degrees of freedom	s_I^2/s_E^2
Within cross-classification error	Residual error from full model			
Corrected total	Corrected total from full model			

The first null hypothesis to test (always) is H_0: the two classification variables are independent, H_0: $\sigma_I^2/\sigma_E^2 = 1$; this is the test for interaction. If this null hypothesis is not rejected, then the pair of null hypotheses to be tested are H_0: the means across the first classification variable categories are equal, and H_0: the means across the second classification variable categories are equal. If this first null hypothesis is rejected, then the initial compound null hypothesis must be tested; more advanced textbooks can be consulted for handling this problem. Selected F ratio calculations are listed as a question in the preceding table and relate to the interpretation issue.

To interpret the variance decomposition of two-way ANOVA, some decisions must be made regarding the classification variables. Once more, these decisions lie dormant in the one-way ANOVA problem. Each of the two classification variables must be labelled as either fixed or random; this labelling will determine how the two F ratio values appearing as question marks in the preceding table are to be calculated. Regardless of how these two variables are labelled, the interaction F ratio always has the estimated error variance in the denominator (s_E^2) and the estimated interaction variance in the numerator (s_I^2). A classification variable is fixed if the classes into which the areal units are assigned will not change upon repeating the analysis with a different sample. For example, the agricultural administrative regions and the coastal lowlands/interior highlands are fixed; areal units cannot be moved from one region to another, or from the coast to the interior. The proper F ratio for a fixed effect places the estimated error variance in the denominator and the estimated classification variance in the numerator. In contrast, a classification variable is called random if the classes into which the areal units are assigned may vary from one sample to another. For instance, the rural/urban variable is random (see Section 1.4 for its definition), since a different detailed census sample could cause an areal unit to switch from one of these categories to the other. The proper F ratio for a random effect places the estimated interaction variance in the denominator and the estimated classification variance in the numerator.

EXAMPLE PR-13.7. Two-way ANOVA table for Example PR-13.6.

Testing for interaction first (this is a must!), the null hypothesis in question is

H_0: the agricultural administrative regions and the coastal lowlands/interior highlands classification variables are independent in the population.

For a 5% level of significance (see Table 11.1-A), the F ratio critical value with 1 and 31 degrees of freedom is 4.1919 (using linear interpolation). Therefore, the conclusion is that the two classification variables interact in the population (4.570 > 4.1919), and hence ANOVA results cannot be interpreted separately for each classification variable. However, if a 1% level of significance is selected (see Table 11.1-A), the F ratio critical value with 1 and 31 degrees of freedom is 7.6260 (using linear interpolation). Therefore, the null hypothesis that the two classification variables do not interact in the population is not rejected, and the ANOVA results can be interpreted separately for each classification variable (4.570 < 7.6260). These null hypotheses are

H_0: the mean of DCD does not differ between the agricultural administrative regions of San Juan and Mayaguez, and

H_0: the mean of DCD does not differ between the coastal lowlands and the interior highlands.

In both cases the null hypothesis is rejected, and the conclusion is that there are differences of means.

The following table was constructed using MINITAB-PC (see Appendix 13-B).

TWO-WAY ANOVA FOR DCD

Source of variation	Sum of squares	Degrees of freedom	Mean squares	F ratio
Agricultural administrative regions	856.9	1	856.9/1	856.9/107.9 = 7.942
Coastal lowlands/ interior highlands	986.2	1	986.2/1	986.2/107.9 = 9.140
Interaction	493.9	1	493.9/1	493.9/107.9 = 4.570
Within cross-classification error	3344.0	31	3344.0/31 = 107.9	
Corrected total	6243.0	34		

The degree of confidence placed in these conclusions again rests upon how well the model assumptions are satisfied. Two-way ANOVA retains the assumptions posited for one-way ANOVA, with the following slightly modified replacements:

<u>Assumptions</u>:

(b) the variable is normally distributed within each of the R • C cross-classifications, having mean μ_{ij} in cross-classification (i, j); and

(c) the cross-classification variances σ_{ij}^2 are unknown but equal to σ^2.

Testing of this battery of assumptions is somewhat more complex, albeit adhering to the same principles that have been spelled out for one-way ANOVA.

EXAMPLE CA-13.2. Two-way ANOVA table for Example CA-13.1.

A reasonable extension to Example CA-13.1 is to combine the two categorical variables into a single two-way ANOVA problem. The cross-classification cells for this example produce the following subsample sizes:

	PPC							
PC1	0	1	2	3	4	5	6	7
K	0	0	0	0	0	0	0	2
L	2	3	1	3	0	0	1	2
M	1	1	3	5	0	0	0	3
N	4	2	1	8	1	1	2	7

Because of the presence of any subsample sizes of 0, categories must be combined, as was done in Example CA-3.2. This aggregation of the data produces the following aggregated cross-classification:

	PPC		
PC1	0,1,2	3	4,5,6,7
K,L,N	13	11	16
M	5	5	3

This second tabulation indicates that an unbalanced ANOVA situation exists. The regression analysis results are

TWO-WAY ANOVA FOR EXP$

Source of variation	Sum of squares	Degrees of freedom	Mean squares	F ratio
Aggregated PC1	214838544	1	214838544	0.78
Aggregated PPC	1437489664	2	718744832	2.62
Interaction	840291264	2	420145632	1.53
Within cross-classification error	1.2904×10^{10}	47	274545792	
Corrected total	1.5219×10^{10}	52		

Testing for interaction first, the null hypothesis in question is

H_0: the postal code locational classification and the primary product code classification variables are independent in the population.

For a 5% level of significance (see Table 11.1-A), the F ratio critical value with 2 and 47 degrees of freedom is 3.2069 (using linear interpolation). Therefore, the conclusion is that the two classification variables do not interact in the population (1.53 < 3.2069), and hence ANOVA results can be interpreted separately for each classification variable. These null hypotheses are

H_0: the mean of aggregated PPC does not differ between the postal code locations, and

H_0: the mean of aggregated PC1 does not differ between the primary product code groups.

In both cases a fixed effect is appropriate to assume, and hence the null hypothesis would not be rejected (for PPC, clearly 0.78 is not significantly greater than 1; for PC1, the same critical F value is used as for the interaction term), and the conclusion is that there are no differences of means. Consequently, neither PPC nor PC1 appears to be a predictor of export sales of a firm.

Extensions to a three-way ANOVA problem entail four interaction terms, one for each pair of classification variables, and one for all three classification variables jointly. The table format follows the same layout as is presented here. Again, the interaction terms must be analyzed first.

13.4. Summary

Regression analysis was used as the vehicle to extend the discussion into the world of multivariate analysis. There is a wide range of multivariate techniques that could be examined. Many of these can be presented in the context of a "regression type" of problem, with suitable adjustments according to the types of dependent and independent variables employed in the analysis. This common structure was illustrated in this chapter for the case of unbalanced ANOVA problems. The last three chapters in the book present an introduction to three additional multivariate techniques commonly used in spatial analysis: analysis of covariance, trend surface analysis, and cluster analysis. The first two of these techniques fit nicely into the framework provided by regression analysis. By the end of Chapter 15, the general structure of this framework should be clear. The final chapter departs from the regression model to present a technique (actually a group of techniques) that, while important, do not fit easily into the discussion presented in the third part of this text.

13.5. Problems

1. Verify the calculations in Examples PR-13.1 through PR-13.7.

2. Redo Example PR-12.8 using a matrix solution, and identify the extent of rounding error that occurs.

3. Extend Example PR-12.8 by including the second variable DFAM, converting this problem to a multiple regression one, and show that this extension filters out the spatial autocorrelation detected in the San Juan region residuals without compromising the Mayaguez residuals situation (see Example PR-14.2).

4. Redo Example PR-13.3 using variable $DMLK^{1/2}$.

5. Complement Example PR-13.4 with an analysis that will help determine the relationship between μ_{SJ} and μ_M.

6. Complete a two-way ANOVA for DCD using the San Juan/Mayaguez regional classification variable together with the rural/urban classification variable.

7. Redo Example PR-13.6 using variable $DMLK^{1/2}$, completing a two-way ANOVA table with the numerical results, and interpreting the associated null hypotheses.

13.6. Additional Readings

In addition to the texts listed in Chapter 1,

Wrigley, N. *Introduction to the Use of Logit Models in Geography.* CATMOG 10.

Tinkler, K. *Introductory Matrix Algebra.* CATMOG 48.

Draper, N., and H. Smith. (1981) *Applied Regression Analysis*, 2nd ed. New York: Wiley.

Montgomery, D., and E. Peck. (1982) *Introduction to Linear Regression Analysis*. New York: Wiley.

Namboodiri, K. *Matrix Algebra*. SAGE 38.

APPENDIX 13-A

COMPUTER CODE FOR EXAMPLE PR-13.2

MINITAB-PC code

The master macro file is as follows:

```
NOECHO
READ 'DATA.132' C2-C4
REGRESS C3 2 C2 C4 C30 C31;
RESID C30;
COEF C32.
PLOT C30 C31
PRINT C32
LET K1 = 19
READ 'SANJUAN.CON' 19 19 M1
LET C1=C30
EXEC 'MORAN.CAL'
END
```

where the contents of the files called upon are as follows:

DATA.132

```
40.2  38.0   3279.7
10.9   9.4    758.2
21.3  22.7     11.0
10.7  13.2     23.7
41.3  52.6    708.0
48.5  40.7   2816.2
60.3  26.2      8.5
35.8  24.5     15.6
18.0  15.0    193.7
35.4  23.7    720.3
37.9  23.3     22.7
32.2  34.2   1591.7
40.5  53.6  13265.8
14.1  14.3     14.0
 3.5   2.9     31.1
34.7  27.7     16.0
36.4  28.2      0.0
18.4  20.4      7.2
22.6  31.7     59.0
```

MORAN.CAL

```
CENTER C1 C75;
LOCATION.
MULT M1 C75 C76
LET K2 = SUM(C75*C76)
LET K3 = SUM(C75**2)
SET C77
K1(1)
END
MULT M1 C77 C78
LET K4 = SUM(C78)
LET K5 = (K1/K4)*(K2/K3)
PRINT K5
END
```

SANJUAN.CON

```
0  0  0  0  1  1  0  0  0  0  0  1  1  0  0  0  0  0  0
0  0  0  0  1  0  0  0  1  0  0  1  0  0  0  1  1  0  0
0  0  0  0  0  0  0  0  0  1  0  0  0  0  1  0  0  1  0
0  0  0  0  0  0  0  1  0  0  1  0  0  1  0  0  0  0  0
1  1  0  0  0  0  0  0  0  0  0  1  0  0  0  0  0  0  0
1  0  0  0  0  0  0  0  0  0  0  1  1  0  0  1  0  0  1
0  0  0  0  0  0  0  0  0  0  0  0  0  0  0  1  1  0  1
0  0  0  1  0  0  0  0  0  0  1  0  0  0  0  0  0  0  0
0  1  0  0  0  0  0  0  0  0  0  0  0  0  1  0  0  0  0
0  0  1  0  0  0  0  0  0  0  0  0  0  1  0  0  0  0  0
0  0  0  1  0  0  0  1  0  0  0  0  0  1  0  0  0  0  0
1  1  0  0  1  1  0  0  0  0  0  0  0  0  0  1  0  0  0
1  0  0  0  0  1  0  0  0  0  0  0  0  0  0  0  0  0  0
0  0  0  1  0  0  0  0  0  1  1  0  0  0  0  0  0  0  0
0  0  1  0  0  0  0  0  1  0  0  0  0  0  0  0  0  1  0
0  1  0  0  0  1  1  0  0  0  0  1  0  0  0  0  1  0  1
0  1  0  0  0  0  1  0  0  0  0  0  0  0  0  1  0  0  0
0  0  1  0  0  0  0  0  0  0  0  0  0  0  1  0  0  0  0
0  0  0  0  0  1  1  0  0  0  0  0  0  0  0  1  0  0  0
```

SAS code

```
OPTIONS LINESIZE=80;
TITLE 'REGRESSION SAS CODE FOR EXAMPLE PR-13.2';
DATA STEP1;
INPUT DCD69 DCD74 DCF C1-C19;
LIST;
CARDS;
40.2 38.0  3279.7   0 0 0 0 1 1 0 0 0 0 0 1 1 0 0 0 0 0 0
10.9  9.4   758.2   0 0 0 0 1 0 0 0 1 0 0 1 0 0 0 1 1 0 0
21.3 22.7    11.0   0 0 0 0 0 0 0 0 0 1 0 0 0 0 1 0 0 1 0
10.7 13.2    23.7   0 0 0 0 0 0 0 1 0 0 1 0 0 1 0 0 0 0 0
41.3 52.6   708.0   1 1 0 0 0 0 0 0 0 0 0 1 0 0 0 0 0 0 0
48.5 40.7  2816.2   1 0 0 0 0 0 0 0 0 0 0 1 1 0 0 1 0 0 1
60.3 26.2     8.5   0 0 0 0 0 0 0 0 0 0 0 0 0 0 0 1 1 0 1
35.8 24.5    15.6   0 0 0 1 0 0 0 0 0 0 1 0 0 0 0 0 0 0 0
18.0 15.0   193.7   0 1 0 0 0 0 0 0 0 0 0 0 0 0 1 0 0 0 0
35.4 23.7   720.3   0 0 1 0 0 0 0 0 0 0 0 0 0 1 0 0 0 0 0
37.9 23.3    22.7   0 0 0 1 0 0 0 1 0 0 0 0 0 1 0 0 0 0 0
32.2 34.2  1591.7   1 1 0 0 1 1 0 0 0 0 0 0 0 0 0 1 0 0 0
40.5 53.6 13265.8   1 0 0 0 0 1 0 0 0 0 0 0 0 0 0 0 0 0 0
14.1 14.3    14.0   0 0 0 1 0 0 0 0 0 1 1 0 0 0 0 0 0 0 0
 3.5  2.9    31.1   0 0 1 0 0 0 0 0 1 0 0 0 0 0 0 0 0 1 0
34.7 27.7    16.0   0 1 0 0 0 1 1 0 0 0 0 1 0 0 0 0 1 0 1
36.4 28.2     0.0   0 1 0 0 0 0 1 0 0 0 0 0 0 0 0 1 0 0 0
18.4 20.4     7.2   0 0 1 0 0 0 0 0 0 0 0 0 0 0 1 0 0 0 0
```

```
22.6 31.7    59.0   0 0 0 0 0 1 1 0 0 0 0 0 0 0 0 1 0 0 0
;
PROC PRINT; RUN;
PROC REG;
   MODEL DCD74 = DCD69 DCF/P;
   OUTPUT OUT=SASREG
   PREDICTED = DCDHAT
   RESIDUAL = DCDRES;
RUN;
PROC PLOT DATA=SASREG;
   PLOT DCDRES*DCDHAT;
RUN;
PROC TRANSPOSE DATA=SASREG PREFIX=ET OUT=ETDATA;
   VAR DCDRES;
DATA STEP2;
SET SASREG;
ESQ = DCDRES*DCDRES;
ARRAY CONN{19} C1-C19;
ARRAY ECONN{19} EC1-EC19;
   CSUM = 0;
   DO I=1 TO 19;
   CSUM = CSUM + CONN{I};
   ECONN{I} = DCDRES*CONN{I};
   END;
RUN;
DATA STEP3;
   SET STEP2;
   IF _N_=1 THEN SET ETDATA;
RUN;
DATA STEP4;
SET STEP3;
ARRAY ETMTX{19} ET1-ET19;
ARRAY ECONN{19} EC1-EC19;
   ETCE = 0;
   DO I=1 TO 19;
   ETCE = ETCE + ETMTX{I}*ECONN{I};
   END;
RUN;
PROC MEANS DATA=STEP4 MEAN;
   VAR CSUM ESQ ETCE;
   OUTPUT OUT=REGSA MEAN=CSUM ESQ ETCE;
RUN;
DATA STEP5;
```

```
SET REGSA;
MC = ETCE/(CSUM*ESQ);
PROC PRINT; RUN;
ENDSAS;
```

APPENDIX 13-B

COMPUTER CODE FOR EXAMPLE PR-13.6

MINITAB-PC code

The master macro file is as follows:

```
NOECHO
READ 'DATA.136' C1-C3
GLM C1=C2 C3 C2*C3
END
```

where the contents of the files called upon are as follows:

DATA.132

```
41.7  3  0
28.1  3  0
37.0  3  0
38.0  1  1
 9.4  1  0
48.3  3  0
22.7  1  0
13.2  1  0
52.6  1  1
40.7  1  1
26.2  1  0
24.5  1  0
39.0  3  0
15.0  1  0
44.5  3  1
32.3  3  0
62.5  3  0
40.3  3  1
23.7  1  0
23.3  1  0
42.0  3  1
26.4  3  0
37.9  3  1
34.2  1  1
53.6  1  1
21.6  3  0
14.3  1  0
24.3  3  1
49.4  3  1
 2.9  1  0
```

```
45.9  3  1
27.7  1  1
28.2  1  0
20.4  1  1
31.7  1  0
```

SAS code

The SAS file, which contains the data, is as follows:

```
OPTION LINESIZE=80;
TITLE 'GLM SAS CODE FOR EXAMPLE PR-13.6';
DATA GRIFFITH;
INPUT DCD REG CI;
LIST;
CARDS;
  41.7  3  0
  28.1  3  0
  37.0  3  0
  38.0  1  1
   9.4  1  0
  48.3  3  0
  22.7  1  0
  13.2  1  0
  52.6  1  1
  40.7  1  1
  26.2  1  0
  24.5  1  0
  39.0  3  0
  15.0  1  0
  44.5  3  1
  32.3  3  0
  62.5  3  0
  40.3  3  1
  23.7  1  0
  23.3  1  0
  42.0  3  1
  26.4  3  0
  37.9  3  1
  34.2  1  1
  53.6  1  1
  21.6  3  0
  14.3  1  0
  24.3  3  1
  49.4  3  1
   2.9  1  0
  45.9  3  1
  27.7  1  1
```

```
  28.2  1  0
  20.4  1  1
  31.7  1  0
;
PROC PRINT; RUN;
PROC GLM;
   CLASS REG CI;
   MODEL DCD = REG CI REG*CI; RUN;
ENDSAS;
```

CHAPTER 14

AN INTRODUCTION TO ANALYSIS OF COVARIANCE

Chapter 13 presented ANOVA as a regression problem, showing how nominally measured data can be used to predict interval/ratio data. The accompanying correlation matrix **R** then consists of point biserial and phi coefficient correlation values. But what does a researcher do if the assumptions of ANOVA are not met? One option is to seek an appropriate variable transformation, following the discussion in Chapter 12. The presence of outliers in the e_i versus $\hat{y}_i$ plot (the residual plot) from an ANOVA, however, suggests that an uncontrolled for, or uncontrollable, interval/ratio variable may be missing from the regression equation. Analysis of covariance (ANCOVA) is a technique that combines features of ANOVA and regression, augmenting the ANOVA model with one or more concomitant interval/ratio variables that are related to variate Y. The goal here is to reduce the variance of the residual or error terms in the ANOVA model, making an analysis more precise (the coefficient of determination should increase) and more powerful in comparing classification variable means. The covariate X is introduced to statistically control for certain effects, thus reducing the residual variance. Conceptually this strategy is equivalent to holding variable X statistically constant and then observing the residual variation in variable Y. The spatial analyst is interested in relating some variate Y to a classification variable (represented by a set of indicator variables), controlling for or partialling out their common relationship with geographically distributed variable X. Such statistical control was treated in Section 11.4.2 with regard to partial correlation. An alternative for accomplishing this sort of statistical control would involve estimating a separate bivariate regression equation for each geographic region. Three major drawbacks of this latter approach are that (1) any interaction between the regional classification variable and variate X would be overlooked, (2) simultaneous comparisons of regional parameters would be difficult to make (Do the intercepts and/or slopes differ in magnitude by more than that which can be attributable to sampling error?), and (3) larger sample sizes would be required in order to retain a reasonable number of degrees of freedom. Analysis of covariance circumvents all three of these problems.

A simple way of identifying the effects of covariance analysis is to consider graphically the combined results of analysis of variance and regression analysis. The slope of the regression line between the response variable **Y** in the analysis of variance and the concomitant variable **X** is used to adjust the values of the K category means ($\bar{Y}_{\cdot j}$). The effect of **X** on **Y** is

$$\bar{Y}_{\cdot j} = \bar{Y}_{\cdot j} - b(\bar{X}_{\cdot j} - \bar{X}_{\cdot\cdot}).$$

The term on the left is the adjusted category mean using the slope b between **X** and **Y** as a reference. Figure 14.1 shows how the means of the three groups (L, M, N) are moved parallel to the regression line of **Y** on **X** to the common point of $\bar{X}_{\cdot\cdot}$ (the grand

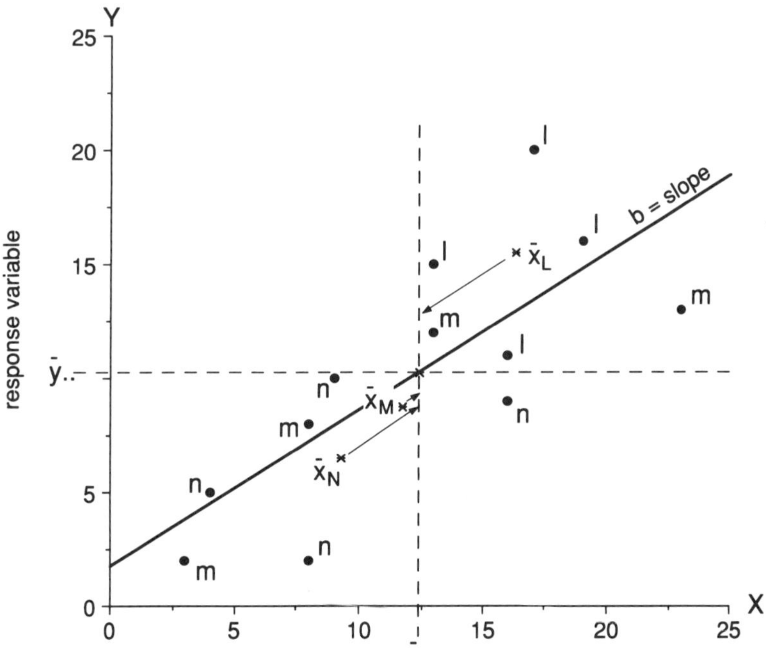

Figure 14.1: A GRAPHIC PORTRAYAL OF ANCOVA

mean of the **X** values). As the slope of the regression of **Y** on **X** approaches 0 (the horizontal line), the less is the effect of the adjustment. When the concomitant variable has no relation to **Y**, there is no difference between the unadjusted and adjusted means.

The spatial analyst almost always will be confronted with an unbalanced ANCOVA. Not surprisingly, all the assumptions affiliated with the unbalanced ANOVA pertain to the unbalanced ANCOVA. To retain the overall mean interpretation of the regression intercept parameter β_0, though, any interval/ratio variable included in an ANCOVA must be written as a difference between its individual areal unit values and its grand mean value (the arithmetic average calculated over *all* n areal units). In other words, a single interval/ratio variable is denoted by $(X_{ij1} - \bar{X}_{..1})$, where j denotes group membership, i denotes the observation number for a group, and 1 denotes being the first interval/ratio variable included in the regression equation. Accordingly $\bar{X}_{..1}$ is the overall or grand mean of variable X. The notion of interaction was introduced in Chapter 13 for multiple classification ANOVAs; this notion is extended in ANCOVA to capture any interaction between the nominal variables and the interval/ratio variables. Since the interaction between nominal and interval/ratio terms is considered, practical applications of ANCOVA have tended to be limited to the inclusion of at most four interval/ratio variables.

14.1. The ANCOVA model

Consider the simplest ANOVA, namely a one-way design. Introducing a single interval/ratio variable, called the **concomitant variable**, into this particular ANOVA produces the ANCOVA model

$$Y_{ij} = \mu_{..} + \mu_{.j}(C_{i1} - C_{i2}) + \beta_1(X_{ij1} - \bar{X}_{..1}) + \gamma(C_{i1} - C_{i2})\bullet(X_{ij1} - \bar{X}_{..1}) + \xi_{ij}$$

where $\mu_{..}$ is the mean of the dependent variable Y, $\mu_{.j}$ is the difference of regional means parameter, β_1 is the general slope of the regression line in the population, and γ is the interaction parameter. The regression t test statistics allow suitable inferences regarding these various parameters to be drawn.

ANCOVA is a conditional ANOVA in which the analysis of variance is performed on regression residuals when variable Y is regressed on the concomitant and interaction variables. Tabulated results from this analysis sometimes are presented using a table similar in format to that used for ANOVA results. Entries into this table are computed according to the extra sum of squares principle (see Chapter 13); recalling that this is a conditional computation, such sums of squares can be denoted by $SS(\beta_2|\beta_1)$, where the estimation of the vector of parameters β_2 is conditioned on the estimation of the vector of parameters β_1. The interaction row included in this table refers to interaction between the concomitant and indicator variables, rather than groups of classification variables; hence, significant interaction here does not preclude simplified interpretation (rather, it refers to a difference of slopes test). Making appropriate modifications to the ANOVA table presented in Section 11.3.1 produces the following table:

SINGLE COVARIATE ANCOVA

Source of variation	Sum of squares	Degrees of freedom	Mean squares
Regions	$SS(\mu.\|\mu..,\beta_1,\Delta)$	$k-1$	$SS(\mu.\|\mu..,\beta_1,\Delta)/(k-1)$
Concomitant variable	$SS(\beta_1\|\mu..,\mu.,\Delta)$	p	$SS(\beta_1\|\mu..,\mu.,\Delta)/p$
Interaction	$SS(\Delta\|\mu..,\mu.,\beta_1)$	$p(k-1)$	$SS(\Delta\|\mu..,\mu.,\beta_1)/[p(k-1)]$
Error	e^te	$n-k(p+1)$	$e^te/[n-k(p+1)]$
Corrected total	$Y^tY-(1^tY)^2/n$	$n-1$	

Either F ratios or their square roots, which are t statistics, may be computed from the mean square terms of this table, again using the mean square error value as the denominator.

EXAMPLE PR-14.1. ANCOVA of DCD for the San Juan and Mayaguez regions, using DFRM as the concomitant variable; examples PR-11.5 and PR-12.8 revisited.

The estimated model is

$$\begin{aligned}\text{DCD} &= \mu.. + \mu.(I_M - I_{SJ}) + \beta_1(\text{DFRM} - \overline{\text{DFRM}}) + \gamma(I_M - I_{SJ})(\text{DFRM} - \overline{\text{DFRM}}) \\ &\quad + \xi \\ &= (\mu.. - \beta_1 \bullet \overline{\text{DFRM}}) + (\mu. - \gamma \bullet \overline{\text{DFRM}})(I_M - I_{SJ}) + \beta_1 \bullet \text{DFRM} \\ &\quad + \gamma(I_M - I_{SJ})\text{DFRM} + \xi.\end{aligned}$$

From MINITAB-PC (see Appendix 14-A), the extra sum of squares for each of the three terms in this problem (see Example PR-13.6, for instance) produces the following table:

ANCOVA FOR DCD USING DFRM AS THE COVARIATE

Source of variation	Sum of squares	Degrees of freedom	Mean squares
Regions	1110.1	1	1110.1/1
Concomitant variable	971.8	1	971.8/1
Interaction	529.7	1	529.7/1
Error	3090.5	31	3090.5/31 = 99.7
Corrected total	6243.0	34	

Again note that the individual sums of squares fail to add to the corrected total (1110.1 + 971.8 + 529.7 + 3090.5 = 5702.1, which is not equal to 6243.0), indicating that the covariate and indicator variables contain redundant information and hence are not independent. Thus, whereas in ANOVA using either a one-way or a balanced design guarantees additivity, additivity for ANCOVA requires far more binding restrictions.

EXAMPLE CA-14.1. ANCOVA table for EXP$ based upon PPC, using SALE as the concomitant variable; Example CA-13.1 revisited.

The same aggregation of PPC is used in this example as was used in Examples CA-3.2 and CA-13.2. From MINITAB-PC, the extra sum of squares for each of the three terms in this problem (see Example PR-13.6, for instance) produces the following table:

ANCOVA FOR EXP$ USING SALE AS THE COVARIATE

Source of variation	Sum of squares	Degrees of freedom	Mean squares	F ratio
Regions	50673409	2	25336704.5	0.217
Concomitant variable	7561781459	1	7561781459.0	64.746
Interaction	1055300763	2	527650382.5	4.518
Error	5489215681	47	116791823.0	
Corrected total	15219331065	52		

Note that the F ratios are computed here with the error mean squares in the denominator (a fixed effects model).

14.2. Interpreting the ANCOVA table

Since ANCOVA is best executed as a regression problem, especially for spatial data series, its interpretation is presented here as that for regression analysis. Accordingly, the focus is shifted to inference about parameter estimates for the multiple regression mode.

Returning to the single concomitant variable and two-regions situation, the ANOVA model is appropriate when $\mu_{.j} \neq 0$, $\beta_1 = \gamma = 0$. This case is illustrated by Figure 14.2(a). If $\mu_{.j} = 0$, $\beta_1 \neq 0$, and $\gamma = 0$, then the regression equation is the same in both regions (a homogeneous process is operating over the geographic landscape). This case is illustrated by Figure 14.2(b). If $\mu_{.j} = 0$, then the regression equation intercept is the same in both regions. If $\mu_{.j} = 0$, $\beta_1 = 0$ or $\beta_1 \neq 0$ (β_1 can assume any value), and $\gamma \neq 0$, then a separate regression line exists for each region, intersecting at a common intercept. This case is illustrated by Figure 14.2(c). If $\mu_{.j} \neq 0$, $\beta_1 \neq 0$, and $\gamma = 0$, then each region has a different but parallel regression line (the regression equation slope is the same in both regions). This case is illustrated by Figure 14.2(d). If $\mu_{.j} \neq 0$, $\beta_1 = 0$ or $\beta_1 \neq 0$ (β_1 can assume any value), and $\gamma \neq 0$, then ANCOVA is equivalent to fitting two separate regression equations, one for each of the two regions. This case is illustrated by Figure 14.2(e). If $\mu_{.j} = \beta_1 = \gamma = 0$, then no spatial differentiation occurs (a homogeneous process is operating over the geographic landscape). As mentioned in the introduction to this chapter, the two primary advantages to retaining a single equation in this last case are that (1) simultaneous testing can be performed, and (2) a single common error variance estimate is obtained based upon more degrees of freedom.

EXAMPLE PR-14.2. Example PR-14.1 extended.

The estimated regression equation, using MINITAB-PC (see Appendix 14-A), together with test statistic information, for Example PR-14.1 are as follows (DFRM = 8.8829):

$$\widehat{DCD} = 24.5466 - 11.5473(I_{SJ} - I_M) + 1.0095 \bullet DFRM + 0.7453(I_{SJ} - I_M)DFRM$$

Parameter	Estimate	Standard error	t	Type I error probability
$\mu_{..}$	24.5466	3.460	7.09	0.000
$\mu_{.}$	-11.5473	3.460	-3.34	0.002
β_1	1.0095	0.3233	3.12	0.004
γ	0.7453	0.3233	2.31	0.028

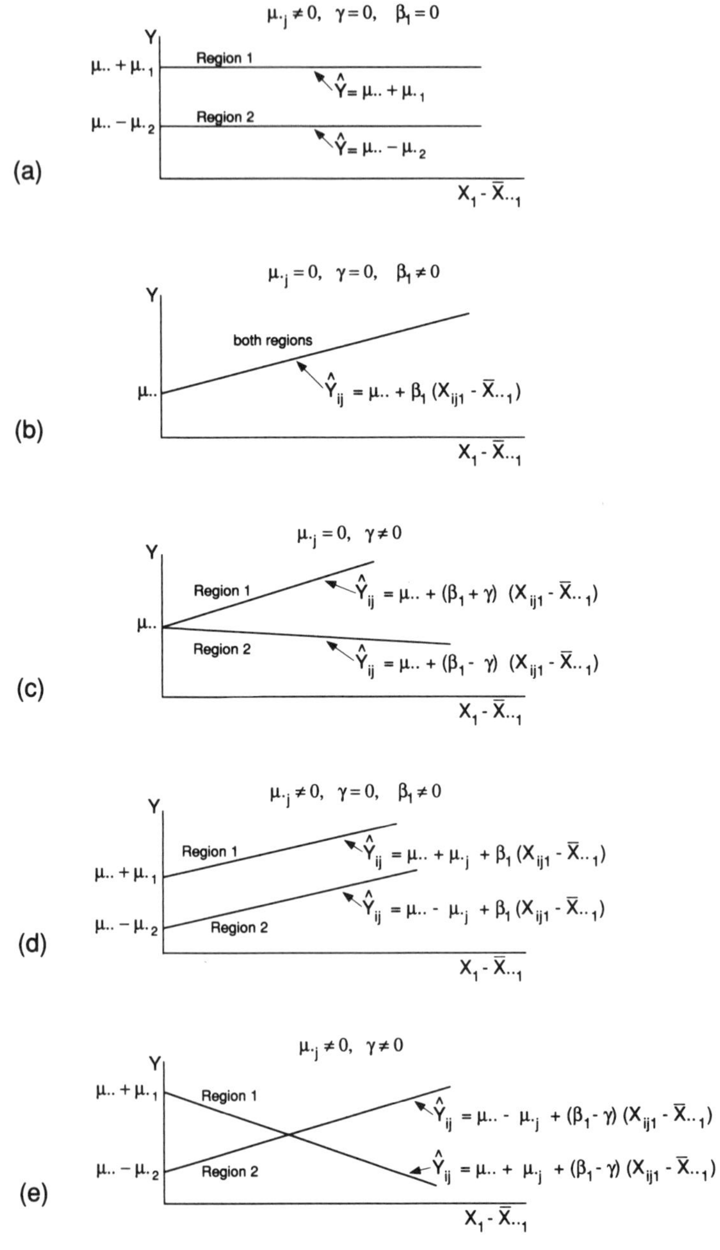

Figure 14.2: (a) SLOPES ZERO, INTERCEPTS DIFFER. (b) SLOPES IDENTICAL, INTERCEPTS DIFFER. (c) SLOPES DIFFER, INTERCEPTS IDENTICAL. (d) SLOPES AND INTERCEPTS IDENTICAL. (e) SLOPES AND INTERCEPTS DIFFER.

The inferences to be drawn here, using a 5% level of significance (the Type I error probability), are that (1) there is interaction between DFRM and the regional classification in the population (implying that the slope of the regression line is not the same in both regions), (2) as DFRM increases DCD tends to increase, in the population, and (3) the adjusted regional means of DCD are different in the population (implying that the intercept of the regression line changes from one region to the other). Consequently, in the population two separate regression lines would be expected.

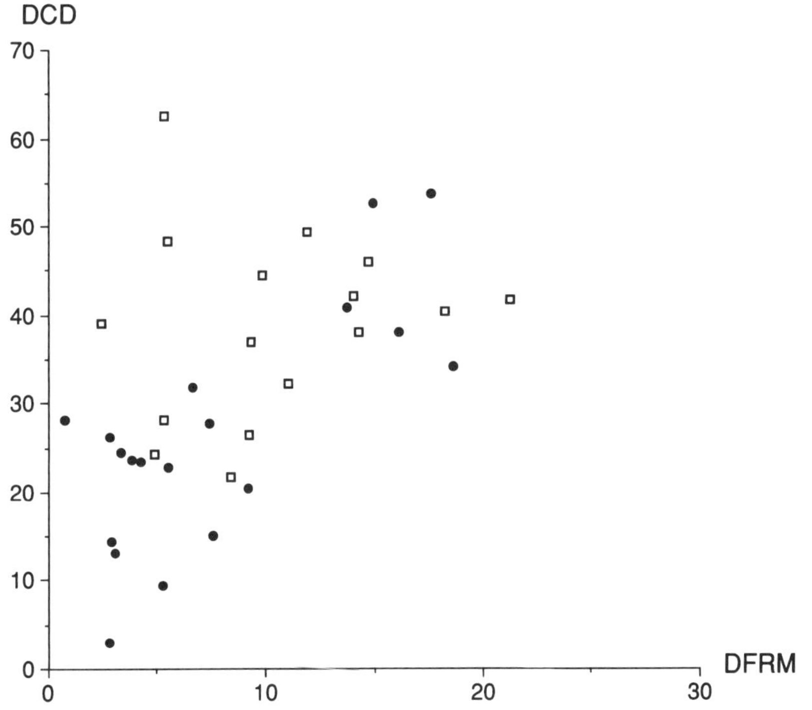

Figure 14.3: SCATTER DIAGRAM SHOWING THE MIXTURE OF MAYAGUEZ (BOXES) AND SAN JUAN (DOTS) MUNICIPIOS

The scatter diagram of these data appears in Figure 14.3 and helps to illustrate this notion. There is not a well-mixed set of dots (denoting the San Juan region) and boxes (denoting the Mayaguez region) in this scatter diagram. The dots tend to concentrate in the lower left-hand corner, spreading out toward the upper right-hand corner, whereas the boxes tend to concentrate across the top of this plot. Visual inspection suggests that the slope of the regression line for the San Juan region would be considerably steeper than would be the slope for the Mayaguez region, and that its intercept would be much lower than that for Mayaguez.

Of course, the soundness of inferences for ANCOVA is predicated upon how well the model assumptions are satisfied. In other words, a battery of diagnostics needs to be performed here, too.

EXAMPLE PR-14.2 continued.

As noted in Example PR-11.5, these two regional data sets appear to come from parent populations that are normally distributed and have the same variance; properties of the ANOVA errors continue to support these assumptions. But a problem with spatial autocorrelation in the error population for the San Juan region is uncovered in Example PR-12.8. In addition, the plot of e_i versus $\hat{y}_i$ appearing at the top of Figure 14.4, values that were obtained in PR-12.8, suggests that possibly several outliers are present. The introduction of variable DFRM into this analysis as a covariate does not compromise the normality or variance equality assumptions. Further, it dramatically reduces the Moran Coefficient value for the geographic distribution of regression residuals in the San Juan region (MC = 0.0895) without making much change in MC for the Mayaguez region (0.0146) (see Example PR-12.8). And, the plot of e_i versus $\hat{y}_i$ from the ANCOVA model (see the bottom of Figure 14.4) suggests that perhaps there is only a single outlier, which is positioned in the upper right-hand corner.

Moreover, Example PR-14.2 indicates that, rather than attempting to concoct some complex transformation in order to remove outliers and correct for the presence of spatial autocorrelation in the San Juan region, adjusting for the covariation between DCD and DFRM allows the assumptions of the linear regression model to be satisfied. Unfortunately the researcher continues to have work to do here; if a single outlier is present in this last analysis, then the inferential basis of ANCOVA still is somewhat compromised.

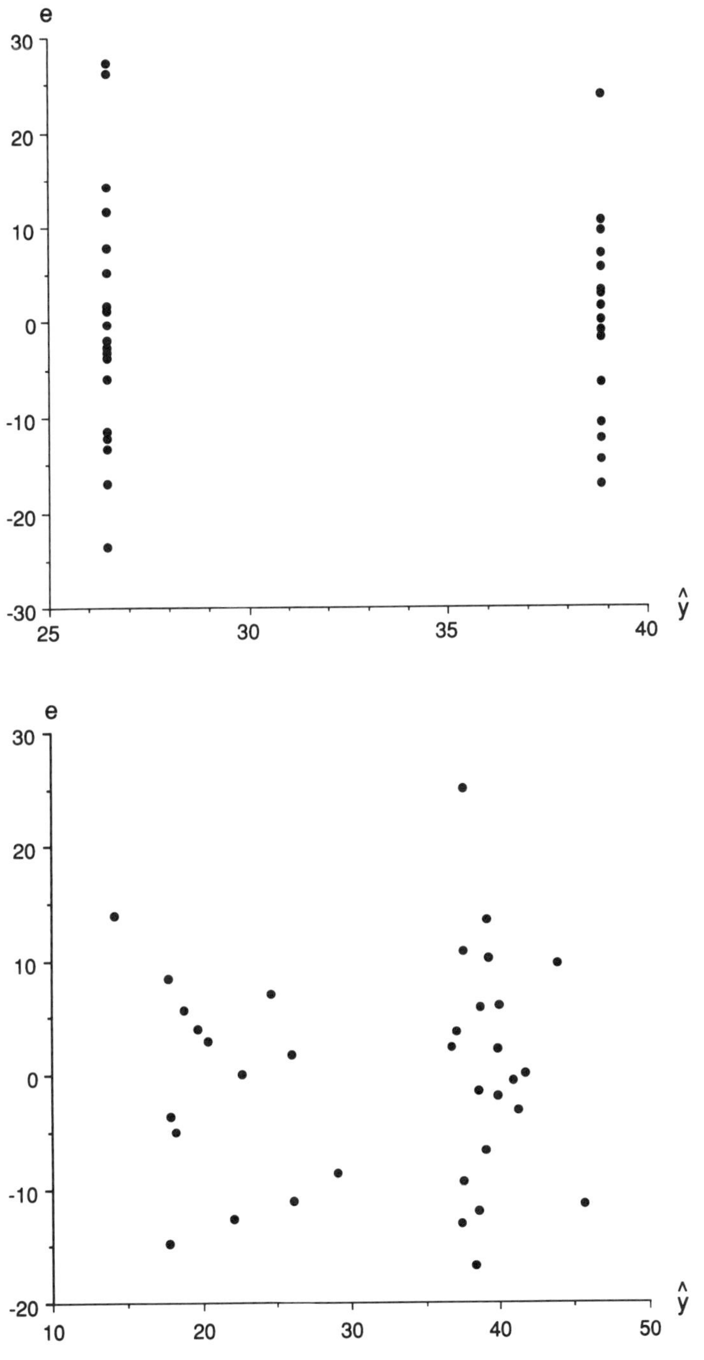

Figure 14.4: SCATTER DIAGRAM FOR EVALUATING CONSTANT VARIANCE; (TOP) FROM ANOVA, (BOTTOM) FROM ANCOVA

14.3. Summary

Of course, covariance analysis also can be employed in two- and three-way classification problems. But now the number of parameters to be estimated becomes quite problematic. For example, for the two-way ANCOVA there would be the intercept (1), the individual differences of means for the R regions and C groups in the two classification variables [(R - 1) + (C - 1)], the ANOVA interaction terms [(R - 1)(C - 1)], the covariates (p), and the ANCOVA interaction terms {p[(R-1) + (C - 1) + (R - 1)(C - 1)]}, for a grand total of [RC(p + 1) - 2] parameters.

14.4. Problems

1. Verify the calculations in Examples PR-14.1 and PR-14.2.

2. Complete an ANCOVA table for DCD, using DFRM as a concomitant variable, for the five agricultural administrative regions. Interpret the tabulated results.

14.5. Additional Readings

In addition to the texts listed in Chapter 1,

Silk, J. *Analysis of Covariance and Comparison of Regression Lines*. CATMOG 20.

Wildt, A. and O. Ahtola. *Analysis of Covariance*. SAGE 12.

Long, J. *Covariance Structure Models*. SAGE 34.

APPENDIX 14-A

COMPUTER CODE FOR EXAMPLE PR-14.1

MINITAB-PC code

The master macro file is as follows:

```
NOECHO
READ 'DATA.141' C1-C3
GLM C3 = C1 C2 C1*C2;
  COVARIATES C2;
  COEF C5.
PRINT C5
END
```

where the contents of the file called upon is as follows:

DATA.141

```
 3 21.2 41.7
 3  5.3 28.1
 3  9.3 37.0
 1 16.1 38.0
 1  5.2  9.4
 3  5.5 48.3
 1  5.5 22.7
 1  3.0 13.2
 1 14.9 52.6
 1 13.7 40.7
 1  2.7 26.2
 1  3.3 24.5
 3  2.4 39.0
 1  7.5 15.0
 3  9.8 44.5
 3 11.0 32.3
 3  5.3 62.5
 3 18.2 40.3
 1  3.8 23.7
 1  4.2 23.3
 3 14.0 42.0
 3  9.2 26.4
 3 14.3 37.9
```

```
1 18.6 34.2
1 17.6 53.6
3  8.4 21.6
1  2.8 14.3
3  4.9 24.3
3 11.9 49.4
1  2.7  2.9
3 14.7 45.9
1  7.4 27.7
1  0.7 28.2
1  9.2 20.4
1  6.6 31.7
```

SAS code

The SAS file, which contains the data, is as follows:

```
OPTIONS LINESIZE=80;
TITLE 'ANCOVA SAS CODE FOR EXAMPLE PR-14.1';
DATA GRIFFITH;
INPUT AAR DFRM DCD;
IF AAR=1 THEN IND=1;ELSE IND=-1;
LIST;
CARDS;
  3  21.2  41.7
  3   5.3  28.1
  3   9.3  37.0
  1  16.1  38.0
  1   5.2   9.4
  3   5.5  48.3
  1   5.5  22.7
  1   3.0  13.2
  1  14.9  52.6
  1  13.7  40.7
  1   2.7  26.2
  1   3.3  24.5
  3   2.4  39.0
  1   7.5  15.0
  3   9.8  44.5
  3  11.0  32.3
  3   5.3  62.5
  3  18.2  40.3
  1   3.8  23.7
  1   4.2  23.3
  3  14.0  42.0
  3   9.2  26.4
```

```
  3  14.3  37.9
  1  18.6  34.2
  1  17.6  53.6
  3   8.4  21.6
  1   2.8  14.3
  3   4.9  24.3
  3  11.9  49.4
  1   2.7   2.9
  3  14.7  45.9
  1   7.4  27.7
  1   0.7  28.2
  1   9.2  20.4
  1   6.6  31.7
;
PROC PRINT; RUN;
PROC GLM;
   MODEL DCD = IND DFRM IND*DFRM;
RUN;
ENDSAS;
```

CHAPTER 15

AN INTRODUCTION TO TREND SURFACE MODELS

A special case of multiple regression analysis most appropriate for geographic research, and which is closely related to spatial autocorrelation work, is trend surface modelling. Its overall purpose is to discern pattern in the general three-dimensional surface of a smoothed areal frequency distribution (see Chapter 5). In its simplest form, some variable Y is regressed on a pair of variables U and V, where their observed ordered pair (u_i, v_i) denotes the Cartesian coordinate of a centroid for areal unit i, or

$$Y = \beta_0 + \beta_1 U + \beta_2 V + \xi.$$

Note that the intercept term here, namely, β_0, sometimes is denoted as α, as was the case in Chapter 12 for the simple linear regression model.

For example, physical centroids, such as those presented in Chapter 5 (see Figure 5.2), can be computed and serve as two independent variables in a multiple regression equation. To retain the original meaning of the intercept term, and also to aid in minimizing the presence of multicollinearity, the regression equation needs to include $(U - \bar{U})$ and $(V - \bar{V})$ rather than U and V; this yields

$$Y = \beta_0 + \beta_1(U - \bar{U}) + \beta_2(V - \bar{V}) + \xi.$$

In fact, if the coordinate measurement scales involve sizeable magnitudes, a researcher should radically rescale the coordinate system by converting U and V to their respective z-score counterparts.

Trend surface modelling requires two additions to the regression analysis assumptions outlined in Chapter 12, namely

(h) the spatial distribution of variable Y is continuous over the geographic landscape; and,

(i) when the areal units do not conform to a regular grid pattern of cells, the centroids furnish a reasonably dispersed coverage of the region under study.

A trend surface regression equation also frequently requires modification of assumption (b), such that

(b) measurement errors in the centroid coordinates U and V are random and conform to normal distributions, whereas any measurement error in variable Y is random and independent of that found either in U or V.

The new assumption (b) could be made stronger by requiring that axes U and V be statistically independent, but the weaker statement that is made here is sufficient. Satisfying this modified assumption is a taxing feature of trend surface analysis. Digitization procedures probably do not introduce random measurement error, whereas selection of an inappropriate centroid measure probably will result in systematic measurement errors in U and V. The error in Y, however, is likely to be independent of any errors in U and V.

15.1. Trend surface models and spatial autocorrelation

The presence of nonzero slopes for U and V in the population regression equation implies that the mean response of Y is not constant over the geographic landscape. In other words, a nonstationary process is operating over this surface. Spatial autocorrelation assumes that the mean is constant across all areal units; recall that calculating MC and GR involves subtracting a single, constant mean (see Chapter 5). When trend surface terms are present in the population, but absent in sample computations, then specification error is introduced into an analysis. This specification error can result in the detection of spatial autocorrelation when none exists; for example, if values tend to increase in a systematic fashion moving from the southeast to the northwest of a region, then similar values will tend to cluster on the map.

A spatial analyst may undertake trend surface modelling with several different objectives in mind. A trend surface can be used to describe a spatial series, indicating a tendency for a geographic distribution to show some alignment. The trend surface is a plane that is parallel to the map surface when $\beta_1 = \beta_2 = 0$; the mean response of Y is constant over the geographic landscape (that is, $Y = \beta_0$). Deviations of these two parameters from 0 cause this plane to tilt. This description posits that the variation in variable Y is a function of the absolute locations of the areal units involved, rather than a function of relative location, as conceptualized with spatial autocorrelation.

Placing this analysis into a sampling context, a spatial analyst can pursue the drawing of inferences and testing of hypotheses about trend surfaces. Since the assumption often invoked in geography is that of a stationary process, then the most commonly tested null hypothesis is H_0: $\beta_j = 0$. A spatial analyst also might be interested in decomposing a spatial data series into regional and local components. This treatment labels the residuals as local fluctuations, and $\hat{Y}$ as the more general, unobscured pattern of spatial variation.

EXAMPLE PR-15.1. A linear trend surface adjustment to the measure of spatial autocorrelation for DCD in the San Juan Agricultural Administrative Region.

Considering the raw DCD data, MC = 0.3967. Following Chapter 5, the computed geometric centroids of the areal units for this region are

Areal unit	U	$U - \bar{U}$	V	$V - \bar{V}$
Barranquitas	7268.39	-1290.77	4654.44	-878.931
Bayamon/Catano	8088.97	-470.19	5685.80	152.429
Carolina	9336.64	777.48	5847.67	314.299
Ceiba	11089.80	2530.64	5126.88	-406.491
Comerio	7792.26	-766.90	4810.38	-722.991
Corozal	7139.82	-1419.34	5307.57	-225.801
Dorado	7397.24	-1161.92	6153.44	620.069
Fajardo	11069.90	2510.74	5538.63	5.259
Guaynabo	8402.48	-156.68	5603.48	70.109
Loiza/Canovanas	9733.79	1174.63	5806.60	273.229
Luquillo	10742.60	2183.44	5692.43	159.060
Naranjito	7588.13	-971.03	5211.26	-322.111
Orocovis	6520.93	-2038.23	4712.00	-821.371
Rio Grande	10211.90	1652.74	5698.94	165.569
San Juan/ Rio Piedras	8706.43	147.27	5892.52	359.149
Toa Alta	7607.43	-951.73	5691.55	158.179
Toa Baja	7787.66	-771.50	6144.99	611.620
Trujillo Alto	9091.54	532.38	5588.45	55.080
Vega Alta	7048.16	-1511.00	5967.01	433.639

The trend surface regression equation for these data, calculated with MINITAB-PC (see Appendix 15-A), is

$$\hat{DCD} = 26.4368 - 0.00440 \bullet U - 0.01519 \bullet V$$

which has a coefficient of determination (R^2) equal to 0.575, as well as the following inferential results:

Variable	Coefficient	Standard error	t	Type I error probability
Intercept	26.4368	2.123	12.45	0.000
U coordinate	- 0.004396	0.001543	- 2.85	0.012
V coordinate	- 0.015190	0.004932	- 3.08	0.007

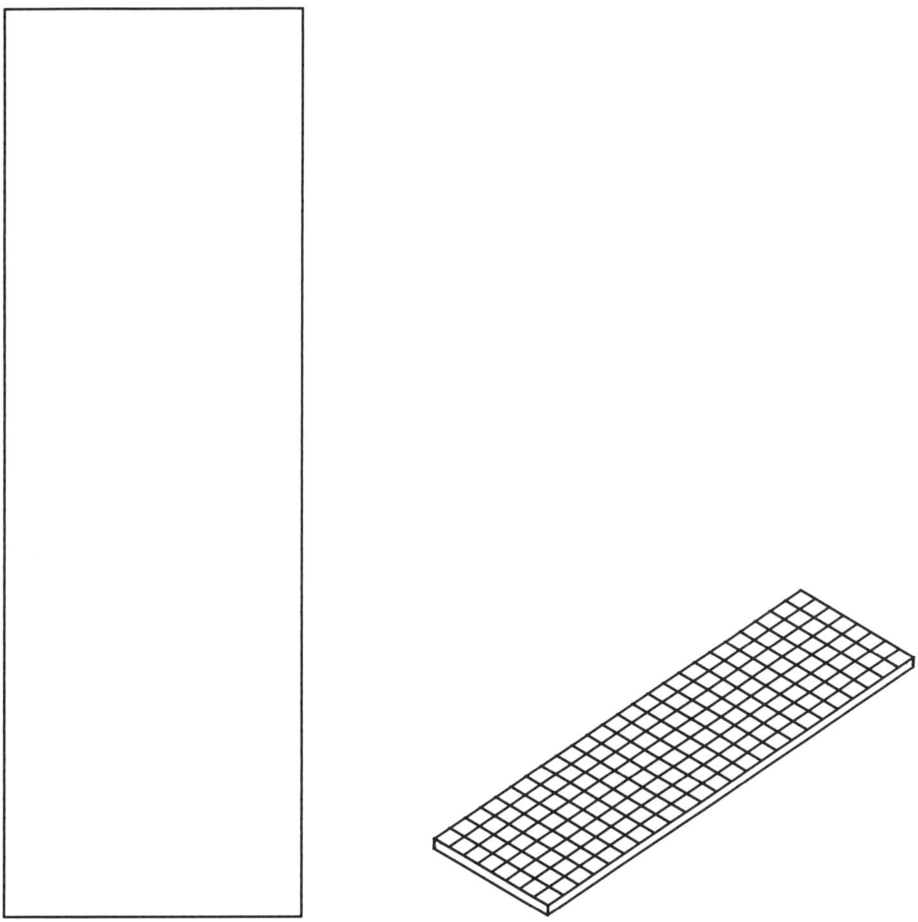

Figure 15.1: CONSTANT TREND SURFACE CONTOUR MAPS BASED UPON THE GEOGRAPHIC DISTRIBUTION OF DCF IN THE SAN JUAN AGRICULTURAL ADMINISTRATION REGION

These test statistics result in rejecting the null hypotheses of H_0: $\beta_0 = 0$ and H_0: $\beta_j = 0$, j= 1, 2. The inference suggested here is that there is an inclined plane across the San Juan Region for DCD, in the population, that slopes downward from west to east and south to north. Diagnostics for the residuals from this regression indicate a failure to reject the null hypothesis that the errors are from a parent population whose frequency distribution conforms to a normal distribution, and the spatial autocorrelation for the residuals produces an MC value of -0.0438.

An important consequence revealed in Example PR-15.1 is that the strong positive spatial autocorrelation originally detected in DCD is nothing more than an artifact of specification error (the mean response for DCD across areal units is not constant).

15.2. More complex spatial variation

The pattern of spatial variation over a region may be far more complex than that characterized by a simple inclined plane. If the pattern approximates a single dome or a single valley with one trough, then a second-order trend surface should be considered. This higher-order surface requires that the three terms U^2, UV, and V^2 be augmented to the multiple regression equation for a linear trend surface (i. e., containing U and V), converting the trend surface equation into

$$Y = \beta_0 + \beta_1(U - \bar{U}) + \beta_2(V - \bar{V}) + \beta_3(U - \bar{U})^2 + \beta_4(U - \bar{U})(V - \bar{V}) + \beta_5(V - \bar{V})^2 + \xi.$$

If the pattern of variation approximates a hill and a valley, each with a single inflection point, then a third-order trend surface should be considered. This higher-order surface requires that the four terms U^3, U^2V, UV^2, and V^3 be augmented to the multiple regression equation for a quadratic trend surface (i. e., containing U, V, U^2, UV, and V^2), converting the trend surface equation into

$$Y = \beta_0 + \beta_1(U - \bar{U}) + \beta_2(V - \bar{V}) + \beta_3(U - \bar{U})^2 + \beta_4(U - \bar{U})(V - \bar{V}) + \beta_5(V - \bar{V})^2$$
$$+ \beta_6(U - \bar{U})^3 + \beta_7(U - \bar{U})^2(V - \bar{V}) + \beta_8(U - \bar{U})(V - \bar{V})^2 + \beta_9(V - \bar{V})^3 + \xi.$$

Contour maps (based upon interpolations) exemplifying these three types of trend surfaces, together with a constant surface (a plane parallel to the UV surface), appear in Figures 15.1-15.4. A spatial analyst could consider increasingly higher-order trend surfaces in an attempt to represent an increasingly undulated landscape, but few go beyond a third-order or cubic surface. This constraint largely is due to the presence of multicollinearity in the polynomial power-series expansion of the coordinates. As was mentioned earlier, this multicollinearity is minimized by

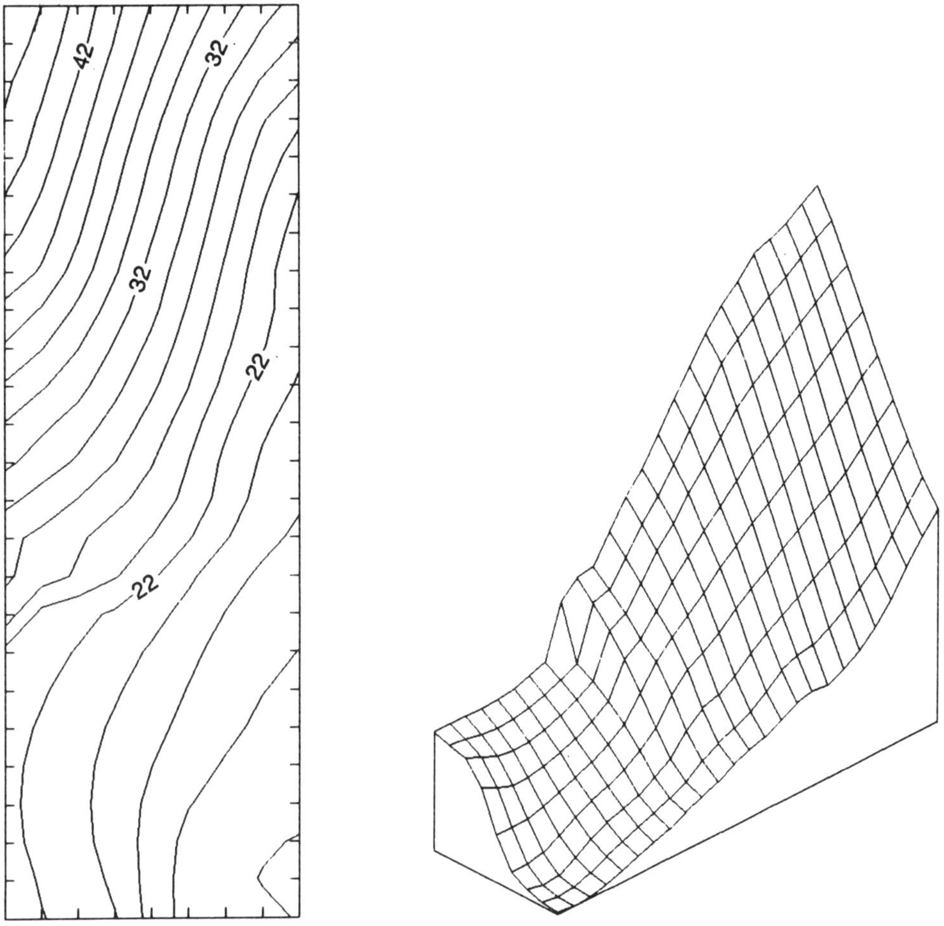

Figure 15.2: LINEAR TREND SURFACE CONTOUR MAPS BASED UPON THE GEOGRAPHIC DISTRIBUTION OF DCF IN THE SAN JUAN AGRICULTURAL ADMINISTRATION REGION

subtracting the mean of each coordinate scale. Latent nonzero correlation between, for instance, U, U^2, and U^3, should be apparent (for example, from Example PR-15.1, $r_{U^2,U} = 0.532$ and $r_{V^2,V} = -0.549$).

Here, again, the extra sum of squares principle provides the necessary test statistic information for drawing an inference regarding the appropriate order of trend surface. The sum of squares for all terms particular to a given order surface need to be collected and then an F ratio for their aggregate influence computed. This procedure means that for the linear trend surface there will be two terms contributing to the marginal sum of squares, and hence 2 degrees of freedom. For the quadratic

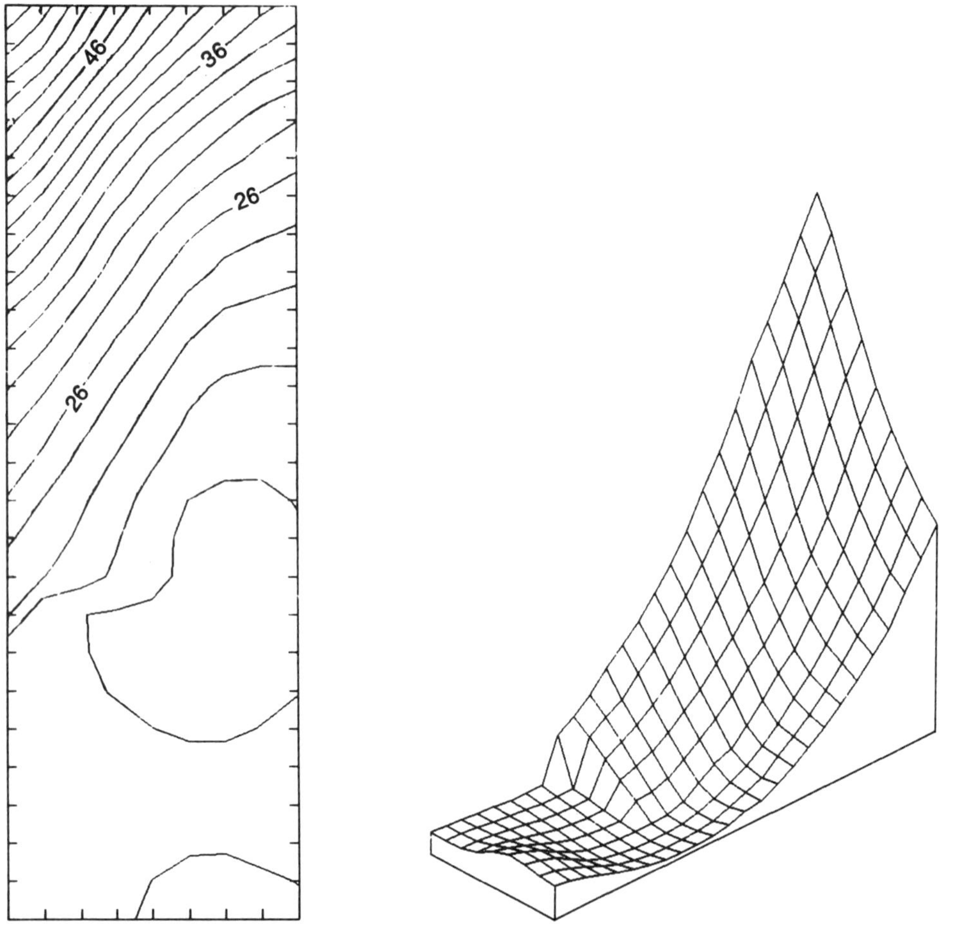

Figure 15.3: QUADRATIC TREND SURFACE CONTOUR MAPS BASED UPON THE GEOGRAPHIC DISTRIBUTION OF DCF IN THE SAN JUAN AGRICULTURAL ADMINISTRATION REGION

trend surface there will be three terms contributing to the marginal sum of squares, and hence 3 degrees of freedom. Finally, for the cubic trend surface there will be four terms contributing to the sum of squares, and hence 4 degrees of freedom. Testing is done sequentially, with order being important (linear, followed by linear and quadratic, followed by linear and quadratic and cubic). Since each of these sums is conditioned upon the lower-order trend surface sums already being accounted for, the sum of squares remains additive. Once a collection of terms is found to be insignificant, then in practice all higher-order trend surfaces usually are dismissed.

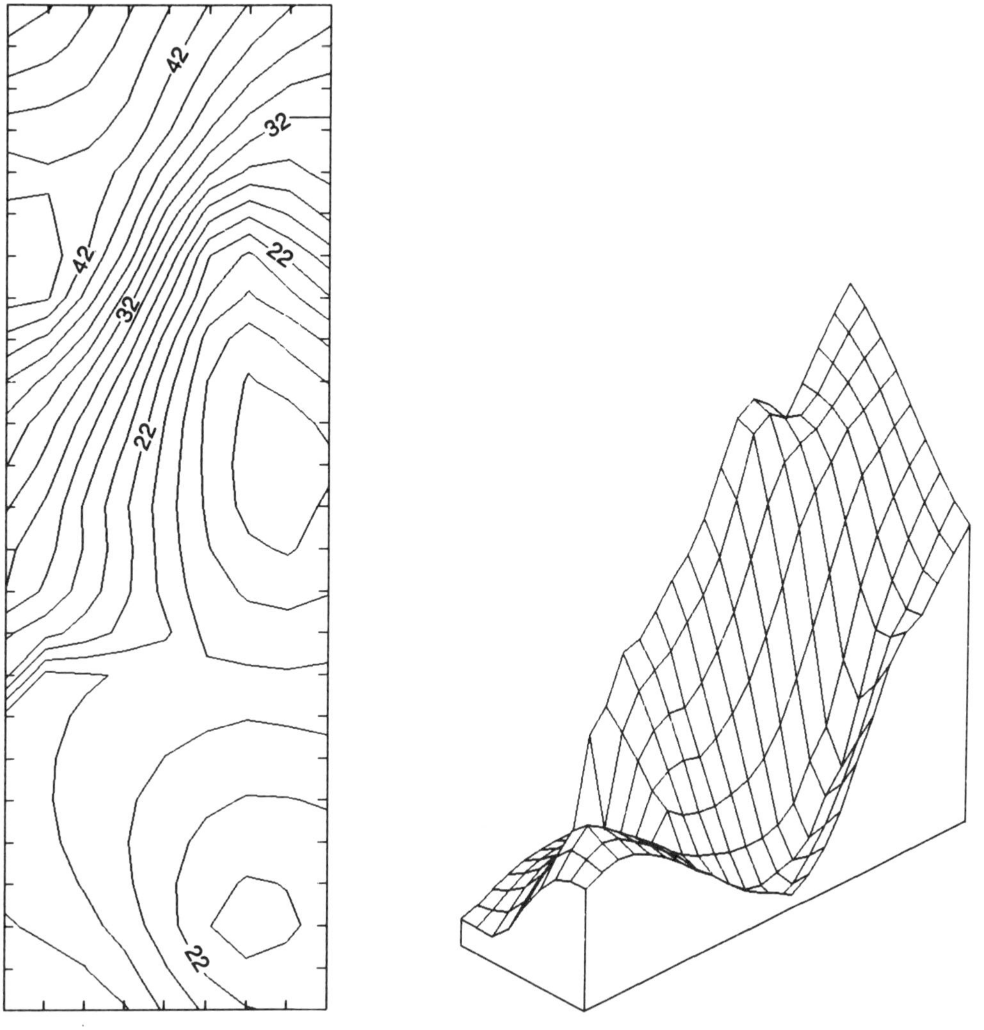

Figure 15.4: CUBIC TREND SURFACE CONTOUR MAPS BASED UPON THE GEOGRAPHIC DISTRIBUTION OF DCF IN THE SAN JUAN AGRICULTURAL ADMINISTRATION REGION

EXAMPLE PR-15.2. Example PR-15.1 extended.

Higher-order trend surface models are considered for Example PR-15.1. The denominator of the linear F ratio assumes that all other variation is attributable to random error; hence, the regression parameters for the quadratic and cubic terms would be zero, yielding (336.85 + 562.06 + 471.52)/(3 + 4 + 9) = 85.65. Using a 5% level of significance (see Table 11.1-A, found in the appendices), the null hypothesis that all the regression parameters for the linear terms equal zero is rejected. Introducing a quadratic surface (which is equivalent to hypothesizing that not all the regression parameters for these terms are zero) changes the denominator of the accompanying F ratio to (562.06 + 471.52)/(9 + 4) = 79.51. At a 5% level of significance (see Table 11.1-A), the null hypothesis that all the regression parameters for the quadratic terms equal zero is not rejected. Therefore, the conclusion is that a linear trend surface model is most appropriate. The results of this exploration have been tabulated as follows, based upon MINITAB-PC output (see Appendix 15-B):

EXTRA SUM OF SQUARES RESULTS

Trend surface terms	Degrees of freedom	Sum of squares	Mean squares	F ratio
Linear	2	1850.83	925.42	925.42/85.65 > 3.6337
Quadratic	3	336.85	112.28	112.28/79.51 < 3.4106
Cubic	4	562.06	140.52	140.52/52.39 < 3.6331
Error	9	471.52	52.39	
Corrected total	18	3221.26		

Since the quadratic terms are not significant in Example PR-15.2, a spatial analyst would not proceed to explore higher-order trend surfaces. This notion is exemplified by the test for cubic terms, for which the researcher also would fail to reject the null hypothesis. Notice that the cubic surface terms are not significant, either.

15.3. Prominent complications in the interpretation of trend surface results

Attempts to attach substantive interpretations to trend surface regression output encounter a number of serious drawbacks. Foremost is the problem that the U- and V-axes are correlated; accurate trend surface mapping requires an uncorrelated coordinate system. Areal units arranged on a regular grid preserve this property, at least for linear and quadratic terms. But irregular grid arrangements of areal unit centroids, especially where clustering occurs [violation of assumption (i)], fail to exhibit this property. This correlation means that (see Chapter 4) information pertaining to the variation in Y along the U axis is at least partly summarized by the information for this same variation along the V axis (redundant information). As mentioned earlier, the accompanying multicollinearity can only be minimized, rather than eliminated, by recentering an axis (subtracting its mean). Trend surface regression coefficients for this latter case should be interpreted with caution. In terms of Examples PR-15.1 and PR-15.2, $r_{U,V}$ = 0.186, meaning that interpretation is somewhat problematic.

A second pronounced restriction pertains to interpolation and extrapolation issues. On the one hand, interpolation makes little sense if assumption (h) is violated, and the geographic distribution of Y is such that only certain locations take on a positive value for the phenomenon under study. Extrapolation, on the other hand, refers to predictions about Y for areal units lying outside the study region. Such predictions should be viewed with extreme suspicion. As emphasized in Chapter 12, except for a handful of well-behaved cases, extrapolations to values for U and V that are positioned outside the range of those for the sample region should not be made. This is a general rule of thumb for virtually any regression analysis. This notion is reinforced by recalling that the regression equation best fits at the mean. The mean of the U and V coordinate axes will tend to be near the geometric center of the study region; a uniform distribution of (U_i, V_i) values will be closest to this geometric mean. Moving away from the mean point in a regression, the confidence intervals increase in width. Similarly, moving away from the center of a study region, the confidence intervals for $\hat{Y}$ of a trend surface equation increase, generating the widest confidence intervals for areal units located along the boundary of the study region.

A third problem concerns degrees of freedom. Each time a higher-order model is entertained, an increasing number of degrees of freedom are lost; a linear surface (r = 1) costs 2 degrees of freedom, a quadratic surface (r = 2) costs 5 degrees of freedom, a cubic surface (r = 3) costs 9 degrees of freedom, and in general an rth order surface costs [(r + 1)(r + 2)/2 - 1] degrees of freedom. Because so many variables are being introduced into a regression equation, without too much effort all that might remain is a model lacking any degrees of freedom!

Many other problems plague trend surface modelling. The interested reader should consult a more specialized text for fuller and more comprehensive discussions of these matters. Only the most troublesome problems have been touched upon here.

15.4. Summary

Trend surface techniques are resurfacing again, increasing in importance as microcomputers and sophisticated display hardware become more readily available. Knowledge about the shortcomings described in this chapter, as well as other more subtle weaknesses not reviewed in this chapter, should not preclude the use of these techniques; rather, this knowledge should provide a sound foundation for appropriately applying trend surface models to geo-referenced data. Unfortunately, all too often the *black box* of the computer, as well as the software it houses, provides attractive graphical output that is accepted uncritically, in part because the neatly printed output looks so authoritarian and contains generic, appealing titles. The contemporary statistician Searle (1989, p. 189) has noted that

> [t]oday's computing packages are octopus-like in their wide range of output values associated with their prime topic, and they produce them all so easily, speedily, and cheaply (cost per unit of arithmetic is minute) that we are continually being surrounded by computed statistics that we've not requested and may not need. The temptation to ask, and to answer "yes" to, "Shall I use this?" is therefore ever more pressing and more pervasive.

In his words, this chapter has focused on *what* to compute and *when* to compute, not just *how* to compute (p. 190). The final technique presented in Chapter 16, cluster analysis, is plagued with similar limitations that are often not presented with computer-generated results. The critical consumer of statistics (for example, one who understands the contents of this book) is well advised to beware of computer output that does not carefully describe the limitations of the underlying statistical results; Searle (p. 190) would like to see computer output peppered with the note "Get advice from a statistician on interpreting these results." This is true of trend surface models, and especially true of cluster analyses.

15.5. Problems

1. Verify the calculations in Examples PR-15.1 and PR-15.2.

2. Determine the appropriate trend surface model for DCD in the Mayaguez region (consult Figure 5.1 for centroid data).

15.6. References

Searle, S. (1989) "Statistical Computing Packages: Some Words of Caution." *The American Statistician*, 43: 189-90.

15.7. Additional Readings

Unwin, D. *An Introduction to Trend Surface Analysis*. CATMOG 5.

Wrigley, N. *Probability Surface Mapping. An Introduction with Examples and Fortran Programmes*. CATMOG 16.

Unwin, D. (1981) *Introductory Spatial Analysis*. New York: Methuen.

APPENDIX 15-A

COMPUTER CODE FOR EXAMPLE PR-15.1

MINITAB-PC code

The master macro file is as follows:

```
NOECHO
READ 'DATA.151' C2-C4
READ 'SANJUAN.CON' 19 19 M1
LET K1 = 19
LET C1=C2
EXEC 'MORAN.CAL'
CENTER C3 C4 C5 C6;
LOCATION.
REGRESS C2 2 C5 C6;
RESID C29;
COEF C30.
PRINT C30
LET C1=C29
EXEC 'MORAN.CAL'
END
```

where the contents of the files called upon are as follows:

SANJUAN.CON and MORAN.CAL are the same as in Appendix 13-A, and

DATA.151

```
38.0   7268.4  4654.44
 9.4   8089.0  5685.80
22.7   9336.6  5847.67
13.2  11089.8  5126.88
52.6   7792.3  4810.38
40.7   7139.8  5307.57
26.2   7397.2  6153.44
24.5  11069.9  5538.63
15.0   8402.5  5603.48
23.7   9733.8  5806.60
23.3  10742.6  5692.43
34.2   7588.1  5211.26
53.6   6520.9  4712.00
14.3  10211.9  5698.94
 2.9   8706.4  5892.52
```

```
27.7   7607.4  5691.55
28.2   7787.7  6144.99
20.4   9091.5  5588.45
31.7   7048.2  5967.01
```

SAS code

```
OPTIONS LINESIZE=80;
TITLE 'REGRESSION SAS CODE FOR EXAMPLE PR-15.1';
DATA STEP1;
INPUT DCD U V C1-C19;
LIST;
CARDS;
38.0  7268.39  4654.44   0 0 0 0 1 1 0 0 0 0 0 1 1 0 0 0 0 0 0
 9.4  8088.97  5685.80   0 0 0 0 1 0 0 0 1 0 0 1 0 0 0 1 1 0 0
22.7  9336.64  5847.67   0 0 0 0 0 0 0 0 0 1 0 0 0 0 1 0 0 1 0
13.2 11089.80  5126.88   0 0 0 0 0 0 0 1 0 0 1 0 0 1 0 0 0 0 0
52.6  7792.26  4810.38   1 1 0 0 0 0 0 0 0 0 0 1 0 0 0 0 0 0 0
40.7  7139.82  5307.57   1 0 0 0 0 0 0 0 0 0 0 1 1 0 0 1 0 0 1
26.2  7397.24  6153.44   0 0 0 0 0 0 0 0 0 0 0 0 0 0 0 1 1 0 1
24.5 11069.90  5538.63   0 0 0 1 0 0 0 0 0 0 1 0 0 0 0 0 0 0 0
15.0  8402.48  5603.48   0 1 0 0 0 0 0 0 0 0 0 0 0 0 1 0 0 0 0
23.7  9733.79  5806.60   0 0 1 0 0 0 0 0 0 0 0 0 0 1 0 0 0 0 0
23.3 10742.60  5692.43   0 0 0 1 0 0 0 1 0 0 0 0 0 1 0 0 0 0 0
34.2  7588.13  5211.26   1 1 0 0 1 1 0 0 0 0 0 0 0 0 0 1 0 0 0
53.6  6520.93  4712.00   1 0 0 0 0 1 0 0 0 0 0 0 0 0 0 0 0 0 0
14.3 10211.90  5698.94   0 0 0 1 0 0 0 0 0 1 1 0 0 0 0 0 0 0 0
 2.9  8706.43  5892.52   0 0 1 0 0 0 0 0 1 0 0 0 0 0 0 0 0 1 0
27.7  7607.43  5691.55   0 1 0 0 0 1 1 0 0 0 0 1 0 0 0 0 1 0 1
28.2  7787.66  6144.99   0 1 0 0 0 0 1 0 0 0 0 0 0 0 0 1 0 0 0
20.4  9091.54  5588.45   0 0 1 0 0 0 0 0 0 0 0 0 0 0 1 0 0 0 0
31.7  7048.16  5967.01   0 0 0 0 0 1 1 0 0 0 0 0 0 0 0 1 0 0 0
;
PROC PRINT; RUN;
PROC STANDARD OUT=TEMP1 MEAN=0;
   VAR DCD;
RUN;
PROC TRANSPOSE DATA=TEMP1 PREFIX=DCDT OUT=DCDTDATA;
   VAR DCD;
DATA STEP2;
SET TEMP1;
DCDSQ = DCD*DCD;
ARRAY CONN{19} C1-C19;
ARRAY DCDCONN{19} DCDC1-DCDC19;
   CSUM = 0;
   DO I=1 TO 19;
```

```
   CSUM = CSUM + CONN{I};
   DCDCONN{I} = DCD*CONN{I};
   END;
RUN;
DATA STEP3;
   SET STEP2;
   IF _N_=1 THEN SET DCDTDATA;
RUN;
DATA STEP4;
SET STEP3;
ARRAY DCDTMTX{19} DCDT1-DCDT19;
ARRAY DCDCONN{19} DCDC1-DCDC19;
   DCDTCDCD = 0;
   DO I=1 TO 19;
   DCDTCDCD = DCDTCDCD + DCDTMTX{I}*DCDCONN{I};
   END;
RUN;
PROC MEANS DATA=STEP4 MEAN;
   VAR CSUM DCDSQ DCDTCDCD;
   OUTPUT OUT=DCDSA MEAN=CSUM DCDSQ DCDTCDCD;
RUN;
DATA STEP5;
SET DCDSA;
MC = DCDTCDCD/(CSUM*DCDSQ);
PROC PRINT; RUN;
PROC STANDARD DATA=STEP1 MEAN=0;
   VAR U V;
RUN;
PROC REG;
   MODEL DCD = U V/P;
   OUTPUT OUT=SASREG
   PREDICTED = DCDHAT
   RESIDUAL = DCDRES;
RUN;
PROC PLOT DATA=SASREG;
   PLOT DCDRES*DCDHAT;
RUN;
PROC TRANSPOSE DATA=SASREG PREFIX=ET OUT=ETDATA;
   VAR DCDRES;
DATA STEP6;
SET SASREG;
ESQ = DCDRES*DCDRES;
ARRAY CONN{19} C1-C19;
```

```
ARRAY ECONN{19} EC1-EC19;
   CSUM = 0;
   DO I=1 TO 19;
   CSUM = CSUM + CONN{I};
   ECONN{I} = DCDRES*CONN{I};
   END;
RUN;
DATA STEP7;
   SET STEP6;
   IF _N_=1 THEN SET ETDATA;
RUN;
DATA STEP8;
SET STEP7;
ARRAY ETMTX{19} ET1-ET19;
ARRAY ECONN{19} EC1-EC19;
   ETCE = 0;
   DO I=1 TO 19;
   ETCE = ETCE + ETMTX{I}*ECONN{I};
   END;
RUN;
PROC MEANS DATA=STEP8 MEAN;
   VAR CSUM ESQ ETCE;
   OUTPUT OUT=REGSA MEAN=CSUM ESQ ETCE;
RUN;
DATA STEP9;
SET REGSA;
MC = ETCE/(CSUM*ESQ);
PROC PRINT; RUN;
ENDSAS;
```

APPENDIX 15-B

COMPUTER CODE FOR EXAMPLE PR-15.2

MINITAB-PC code

The master macro file is as follows:

```
NOECHO
READ 'DATA.151' C2-C4
CENTER C3 C4 C5 C6;
LOCATION.
LET C7 = C5**2
LET C8 = C5*C6
LET C9 = C6**2
LET C10 = C5**3
LET C11 = (C5**2)*C6
LET C12 = C5*(C6**2)
LET C13 = C6**3
REGRESS C2 9 C5-C13;
MSE K2.
LET K2 = 9*K2
REGRESS C2 5 C5-C9;
MSE K3.
LET K3 = 13*K3
REGRESS C2 2 C5 C6;
MSE K4.
LET K4 = 16*K4
LET K5 = K3-K2
LET K6 = K4-K3
CENTER C2 C1;
LOCATION.
LET K7 = SUM(C1**2)
LET K8=K7-K4
PRINT K8,K6,K5,K2,K7
END
```

where the contents of file 'DATA.151' are the same as in Appendix 15-A.

SAS code

```
OPTIONS LINESIZE=80;
TITLE 'TREND SURFACE MODEL SAS CODE FOR EXAMPLE PR-15.2';
DATA GRIFFITH;
INPUT DCD U V;
LIST;
CARDS;
  38.0    7268.4  4654.44
   9.4    8089.0  5685.80
  22.7    9336.6  5847.67
  13.2   11089.8  5126.88
  52.6    7792.3  4810.38
  40.7    7139.8  5307.57
  26.2    7397.2  6153.44
  24.5   11069.9  5538.63
  15.0    8402.5  5603.48
  23.7    9733.8  5806.60
  23.3   10742.6  5692.43
  34.2    7588.1  5211.26
  53.6    6520.9  4712.00
  14.3   10211.9  5698.94
   2.9    8706.4  5892.52
  27.7    7607.4  5691.55
  28.2    7787.7  6144.99
  20.4    9091.5  5588.45
  31.7    7048.2  5967.01
;
PROC STANDARD MEAN=0 OUT=NEWDATA1;
   VAR U V;
RUN;
PROC PRINT; RUN;
PROC REG DATA=NEWDATA1;
   MODEL DCD = U V;
RUN;
PROC RSREG DATA=NEWDATA1;
   MODEL DCD = U V;
RUN;
PROC GLM DATA=NEWDATA1;
   MODEL DCD = U V U*U U*V V*V U*U*U U*U*V U*V*V V*V*V;
RUN;
ENDSAS;
```

CHAPTER 16

CLUSTER ANALYSIS: AN INTRODUCTION TO OBJECTIVE GROUPING TECHNIQUES

The last technique to be presented in this text that bridges univariate and multivariate analyses deals with classification. Geography, like many other empirical sciences, has devoted considerable effort to taxonomic (classification) issues over the years. For example, choropleth mapping, city classification, and regionalization are concerned with grouping areal unit values into categories. The topic of this chapter is objective extraction of areal unit groupings from information contained in a single variable. Only classical grouping procedures are treated; techniques such as multidimensional scaling or pictorial representations (stars, Andrews plots, Chernoff faces) are not presented here. The coverage of this chapter is limited for two reasons. First, just a flavor of the topic is being sought. Second, classical clustering algorithms are closely akin to certain of the preceding statistical techniques, as well as location-allocation algorithms (and other geographical operations research techniques) that are used in spatial analysis.

16.1. The nature and scope of cluster analysis

Cluster analysis is a statistical technique that furnishes numerical methods for devising taxonomies. It is a procedure that searches data for some "natural" groupings of areal units; no assumptions are made concerning the number of groups or the group structure. Usually a spatial analyst needs to know enough about the substantive problem in question to distinguish "good" groupings from "bad" groupings. While this classification scheme tends to have some subjectivities affiliated with its selections, a particular selection must be defensible. The problem that cluster analysis is designed to solve may be stated as follows:

> given a sample of n areal units, each of which has a score on some interval/ratio variable, devise a scheme for grouping these areal units into K classes (i. e., clusters), where $1 < K << n$ and K almost always is unknown, such that similarities within clusters are maximized while similarities between clusters are minimized.

Both the number of clusters and group membership for each areal unit are unknown and must be simultaneously determined. K should be closer to 1 than to n, to capture generalities, with K being pulled toward n in order to reduce any within clusters dissimilarities; this problem is reminiscent of interval size determination for histogram construction. The package of solution methods formulated to solve the cluster analysis problem have remained purely descriptive. The indeterminacy of this problem continues to elude inferential statistical decision-making, so no statistical distribution assumptions are made concerning either the underlying variable or error, and to date virtually nothing is known about the corresponding sampling distributions.

Grouping is done on the basis of similarities, as posited in the problem statement, or dissimilarities, which alters the problem statement so that dissimilarities within clusters are minimized while dissimilarities between clusters are maximized. Many indices exist that measure these constructs (see Table 16.1); since they do not yield the same quantitative scale, their diversity illustrates that measurement error constantly is present. The history of a sequential grouping may be pictorially captured with a **dendogram**, or tree diagram; here results of cluster analysis are displayed graphically in the form of a two-dimensional diagram showing how groups successively are formed. The dendogram illustrates the mergers or divisions that have been made at successive levels; the branches in this tree represent clusters. The branches come together (merge) at nodes whose positions along a distance (or similarity) axis indicate the level at which the fusions occur. Next the dendogram is sliced to select the number of "natural" clusters. The general notion used to select an empirically implied but unknown number of clusters is to slice through a dendogram where a sizeable jump takes place in the quantitative grouping criterion; relatively large, sudden increases in this index value suggest that a slice should be made through the dendogram, with the number of clusters being determined by the groupings that have occurred below this cut. Presumably this point is where the dissimilarity of areal units within clusters is minimized, while the dissimilarity between clusters is maximized. Although numerical aspects impart some degree of objectivity to a solution, there still is a large subjective component in the assessment of results for any data set (choice of variable, of index, of algorithm, of dendogram slice location, and so on); many solutions are data dependent or sample specific. This entire procedure is illustrated in the examples included in this chapter.

TABLE 16.1

EXAMPLES OF SIMILARITY MEASURES FOR CLUSTERING BASED UPON THE PHI COEFFICIENT (SEE SECTION 4.3)

Measure	Description
(A + D)/(A + B + C + D)	Only 1-1 and 0-0 matches are important, and equally so, as a ratio of all matches and mismatches.
p(A + D)/[p(A + D) + B + C]	Only 1-1 and 0-0 matches are important, and equally so, as a ratio of all matches and mismatches, where matches are weighted more heavily than are mismatches; p > 1.
(A + D)/[A + D + p(B + C)]	Only 1-1 and 0-0 matches are important, and equally so, as a ratio of all matches and mismatches, where matches are weighted less heavily than are mismatches; p > 1.
D/(A + B + C + D)	Only 1-1 matches are important, as a ratio of all matches and mismatches.
D/(B + C + D)	Only 1-1 matches are important, as a ratio of 1-1 matches and all mismatches, while 0-0 matches are unimportant.
pD/(B + C + pD)	Only 1-1 matches are important, as a ratio of 1-1 matches and all mismatches, where mismatches are weighted less heavily, while 0-0 matches are unimportant; p > 1.
D/[p(B + C) + D]	Only 1-1 matches are important, as a ratio of 1-1 matches and all mismatches, where mismatches are weighted more heavily, while 0-0 matches are unimportant; p > 1.
D/(B + C)	Only 1-1 matches are important, as a ratio of all mismatches, while 0-0 matches are unimportant.

Note that none of the procedures discussed in this chapter includes a constraint that the regions joined to form a cluster share a boundary. In fact, some of the clusters formed in the following examples are far removed from each other. The constraint that the combined regions share a boundary, called a **contiguity constraint**, is obviously very important in certain spatial applications, for example, when combining a large number of areal units, say counties or census tracts, into a smaller number of larger areal units. This process has been referred to as an aggregation problem or the modifiable areal unit problem (see Openshaw in CATMOG 38). Many of the algorithms developed to solve this problem are context specific and not widely available (see Openshaw 1977 for an exception). Political redistricting algorithms exploit this constraint, too. Neither of these classes of algorithm will be reviewed here. For a survey of contiguity-constrained clustering procedures see Murthagh (1985).

16.2. Popular algorithms for solving the cluster analysis problem

Not only are there disagreements about which is the best global measure of similarity or dissimilarity for deciding upon clusters, but also no single optimization criterion has been agreed upon by methodologists. In addition, clustering can proceed in one of two completely opposite manners. One of these is known as the agglomerative hierarchical method, which has guided the foregoing discussions and which starts with individual areal units and successively merges similar groups (this is the method used by SAS). This method starts with n groups (initially each areal unit constitutes a separate group), sequentially combines these groups one at a time, and ends with one group. In contrast, the infrequently encountered divisive hierarchical method begins with a single group (comprised of all areal units), and successively divides groups into dissimilar subgroups until each areal unit is in its own separate group (this method ends with n groups). Computer algorithms for the agglomerative hierarchical method involve the following steps:

Step 1: Start with n groups or clusters and a matrix of dissimilarities.
Step 2: Search the dissimilarity matrix for the nearest pair of clusters (call this d_{uv} for groups U and V).
Step 3: Merge clusters U and V.
Step 4: Repeat steps 1-3 a total of (n - 1) times.

These steps are guaranteed to converge, since the total number of possible grouping paths is a combinatorial problem (see Chapter 6).

The first, and simplest algorithm, is **single linkage**, originally devised and later reinvented by a number of different scholars. Here groups are considered to be dissimilar in accordance with the minimum distance between all possible pairs of areal units, one from each of two clusters. Accordingly, this algorithm implements the following definition of distance between groups in step 3:

$$d_{(uv)w} = \text{MIN}\{d_{uw}, d_{vw}\}$$

where the quantities d_{uw} and d_{vw} are the distances between the nearest neighbor members of clusters U and W, and clusters V and W, respectively. Single linkage imposes no constraints on the shape of clusters, but sacrifices its ability to recover compact clusters in order to sharpen its ability to detect elongated and irregular clusters (often single linkage will produce chaining--the incremental addition of an areal unit to a single cluster in successive steps--in its dendogram). The technique cannot discern poorly separated clusters and tends to chop off the tails of distributions before separating main clusters.

EXAMPLE PR-16.1. The dendogram for the Arecibo Region, based upon DCD, using the single linkage clustering algorithm (from SAS), appears in Figure 16.1.

This solution strongly supports the existence of three groups, suggesting that Arecibo definitely should be isolated and placed in a group by itself. Besides the obvious change in index values here, this single linkage solution deviates from the four ensuing solutions in that the sequence of group construction changes; this finding again emphasizes that results produced by two different algorithms are not guaranteed to be the same.

EXAMPLE PR-16.1 (continued). Details for the single linkage algorithm.

Whereas the calculations necessary for many of the clustering algorithms are too complex to be presented easily in a textbook, the details for the single linkage algorithm are rather straightforward. The following is a matrix that presents the actual distances between the values for DCD in the Arecibo region (note that the algorithm uses standardized distances).

	B/Fl	Ca	Ci	H	L	Ma	Mo	Q	U	VB
A	6.9	14.3	9.4	27.8	21.9	9.2	25.3	11.3	21.2	7.1
B/Fl		7.4	2.5	20.9	15.0	2.3	18.4	4.4	14.3	0.2
Ca			4.9	13.5	7.6	5.1	11.0	3.0	6.9	7.2
Ci				18.4	12.5	0.2	15.9	1.9	11.8	2.3
H					5.9	18.6	2.5	16.5	6.6	20.7
L						12.7	3.4	10.6	0.7	14.8
Ma							16.1	2.1	12.0	2.1
Mo								14.0	4.1	18.2
Q									9.9	4.2
U										14.1

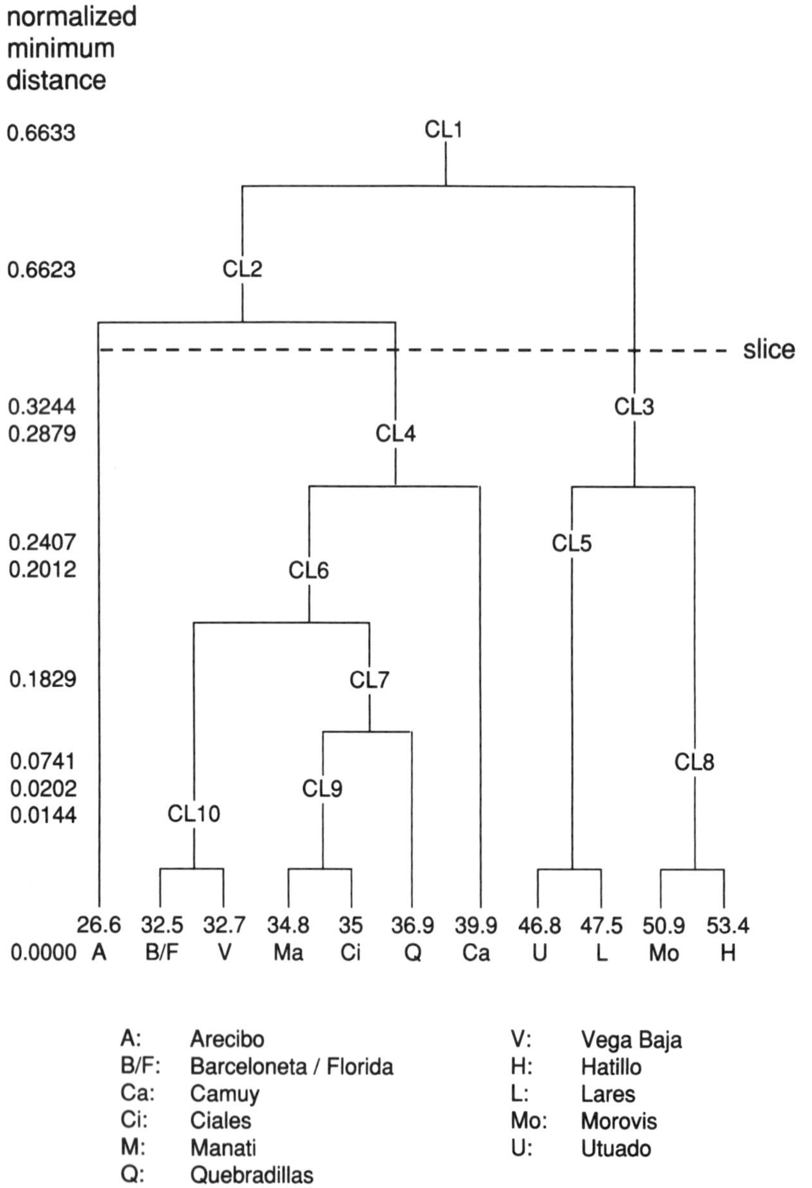

Figure 16.1: DENDOGRAM FROM SINGLE LINKAGE CLUSTERING ALGORITHM FOR THE ARECIBO REGION

Since the distance between municipios is reflexive, and the distance between a municipio and itself is zero, only half the matrix is filled in. The first row and the first column entries are identical, as are the second row and the second column entries, and so on. First the algorithm identifies that cell with the smallest distance and combines its row municipio with its column municipio to form a cluster. In this example, 0.2 is the smallest value, so either Ma and Ci, or B/Fl and VB, are combined to form a cluster; B/Fl and VB combine to form cluster 10 in the previous example. Now the distance matrix is modified by replacing the two rows/columns labelled B/Fl and VB by cluster 10, whose entries are the minimum distances to all other areal units, resulting in

	C10	Ca	Ci	H	L	Ma	Mo	Q	U
A	6.9	14.3	9.4	27.8	21.9	9.2	25.3	11.3	21.2
C10		7.2	2.3	20.7	14.8	2.1	18.2	4.2	14.1
Ca			4.9	13.5	7.6	5.1	11.0	3.0	6.9
Ci				18.4	12.5	0.2	15.9	1.9	11.8
H					5.9	18.6	2.5	16.5	6.6
L						12.7	3.4	10.6	0.7
Ma							16.1	2.1	12.0
Mo								14.0	4.1
Q									9.9

In the next pass of the algorithm the minimum distance is sought in this revised table of distances, and the process repeated; again that cell with the smallest distance is identified, and its row label unit is combined with its column label unit to form a cluster. In this example, 0.2 is the smallest value, so Ma and Ci are combined to form cluster 9. Now the distance matrix is modified by replacing the two rows/columns labelled Ma and Ci with cluster 9, whose entries are the minimum distances to all other areal units, resulting in

	C10	Ca	C9	H	L	Mo	Q	U
A	6.9	14.3	9.2	27.8	21.9	25.3	11.3	21.2
C10		7.2	2.1	20.7	14.8	18.2	4.2	14.1
Ca			4.9	13.5	7.6	11.0	3.0	6.9
C9				18.4	12.5	15.9	1.9	11.8
H					5.9	2.5	16.5	6.6
L						3.4	10.6	0.7
Mo							14.0	4.1
Q								9.9

The next cluster to be formed would combine municipios L and U, whose distance of 0.7 is now the smallest in the table. This process continues until every municipio has been joined, and a single cluster has been obtained (cluster 1 in the previous example).

Alternatively, the process of clustering with the single linkage algorithm involves locating each region along a number line according to the value of DCD (see Figure 16.1). The closest regions are joined first. The pair of elements (either an original region or an existing cluster) are then joined by a horizontal line. The value of a cluster is represented by a vertical line located halfway on the horizontal line joining the two cluster elements. The algorithm proceeds by sequentially joining the closest remaining elements. The algorithm continues until there is only one cluster. Example 16.1 is shown diagrammatically in Figure 16.1. *Note*: the numerical examples in this chapter, as in every other PR example in the book, are based on the data in Table 1.1A. However, the dendograms are constructed using the data in Appendix 16-A. Differences lie in the additional significant digit in the appendix data permitting better articulation of each dendogram. This situation illustrates nicely the idea of precision and the information content of data first introduced in Chapter 2.

The second algorithm is **average linkage**, originated by Sokal and Michener. Here groups are considered to be dissimilar in accordance with the average distance between pairs of areal units, one from each of two clusters. Accordingly, this algorithm implements the following definition of distance between groups in step 3:

$$d_{(uv)w} = \sum_{i=1}^{n_{uv}} \sum_{k=1}^{n_w} d_{ik}/[n_{uv}n_w]$$

where d_{ik} is the distance between object i in the cluster (UV) and object k in the cluster W, and n_{uv} and n_w are the number of items in clusters (UV) and W, respectively. This method tends to join clusters with small variances and has a slight predisposition to produce clusters with equal variances.

EXAMPLE PR-16.2. The dendogram for the Arecibo Region, based upon DCD, using the average linkage clustering algorithm (from SAS), appears in Figure 16.2.

In Example PR-16.2 either two or three groups may be identified, namely, {(Arecibo), (Barceloneta/Florida, Camuy, Ciales, Manati, Quebradillas, Vega Baja)} and (Hatillo, Lares, Morovis, Utuado). Although there is a sizeable jump in the normalized average distance in the last step, a relatively large jump occurs in the next to last step, too. Inspection of the original DCD values suggests that perhaps Arecibo is an outlier, which is why this clustering algorithm has isolated it. Analysts usually try to avoid single areal unit groups, and hence most would prefer the two-cluster solution. Note that although the cluster formations is the same with average linkage

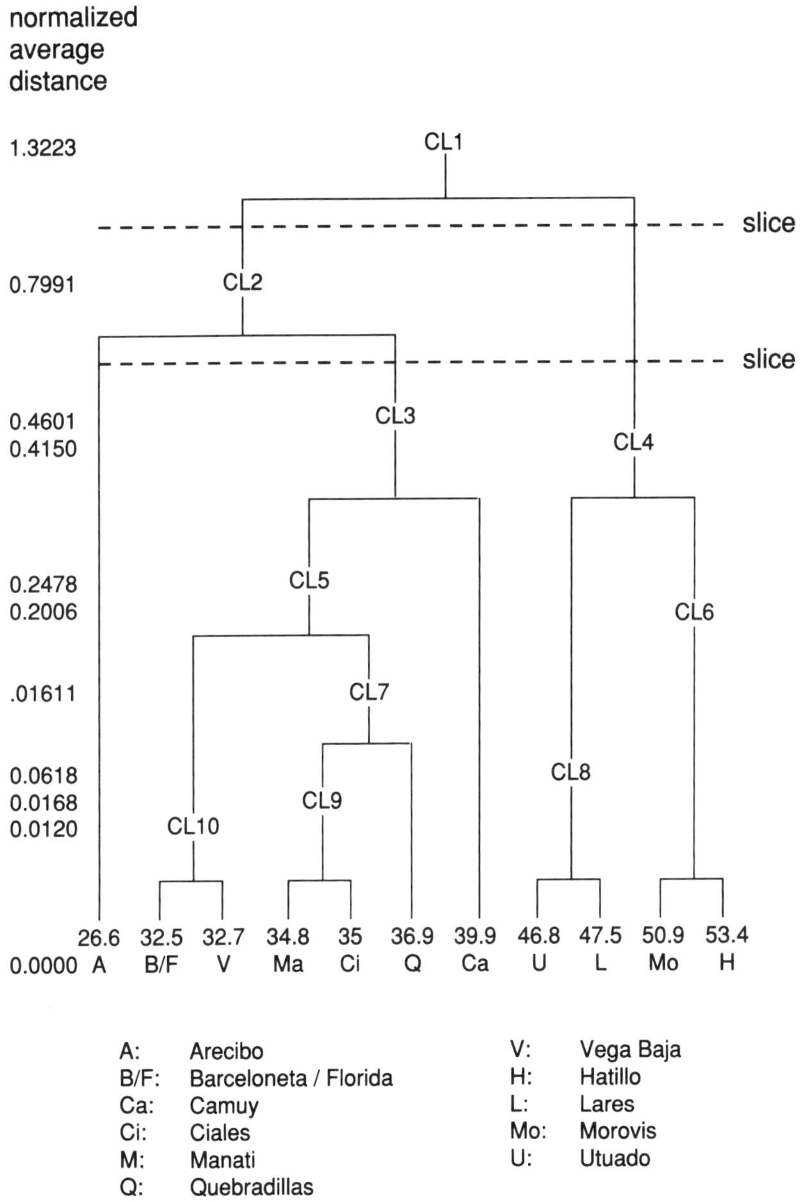

Figure 16.2: DENDOGRAM FROM AVERAGE LINKAGE CLUSTER ALGORITHM FOR THE ARECIBO REGION

as with single linkage, the order in which the clusters have been formed is different.

The third algorithm is **complete linkage**, devised first by Sorensen. Here groups are considered to be dissimilar in accordance with the maximum distance between all possible pairs of areal units, one from each of two clusters. Accordingly, this algorithm implements the following definition of distance between groups in step 3:

$$d_{(uv)w} = \text{MAX}\ \{d_{uw}, d_{vw}\}$$

where d_{uw} and d_{vw} are the distances between the most distant members of clusters U and W, and clusters V and W, respectively. This method has a strong predisposition toward producing clusters having equal ranges (see Chapter 4). Its solution can be severely distorted by the presence of moderate outliers.

EXAMPLE PR-16.3. The dendogram for the Arecibo region, based upon DCD, using the complete linkage clustering algorithm (from SAS) appears in Figure 16.3.

This solution more clearly articulates the separation of these 11 areal units into three groups; Arecibo is more markedly isolated. The other elements of the solution remain unchanged, though.

The fourth algorithm is the **centroid method**, also originated by Sokal and Michener. Here groups are considered to be dissimilar in accordance with the squared Euclidean distance between areal units and cluster means (centroids); each areal unit is assigned to that cluster having the nearest centroid. Accordingly, this algorithm implements the following sequence of steps:

Step 1: Randomly partition areal units into K initial clusters.
Step 2: Assign each areal unit to the cluster whose centroid (i. e., mean) is nearest to it, and recalculate the centroid for that cluster losing the areal unit, and that cluster gaining the areal unit.
Step 3: Repeat step 2 until no more reassignments take place.

The centroid solution tends to be more robust to the effects of outliers than solutions produced by most other clustering algorithms, but often is found to be inferior to the solution given by Ward's method (the following algorithm) or average linkage.

EXAMPLE PR-16.4. The dendogram for the Arecibo region, based upon DCD, using the centroid clustering algorithm (from SAS), appears in Figure 16.4.

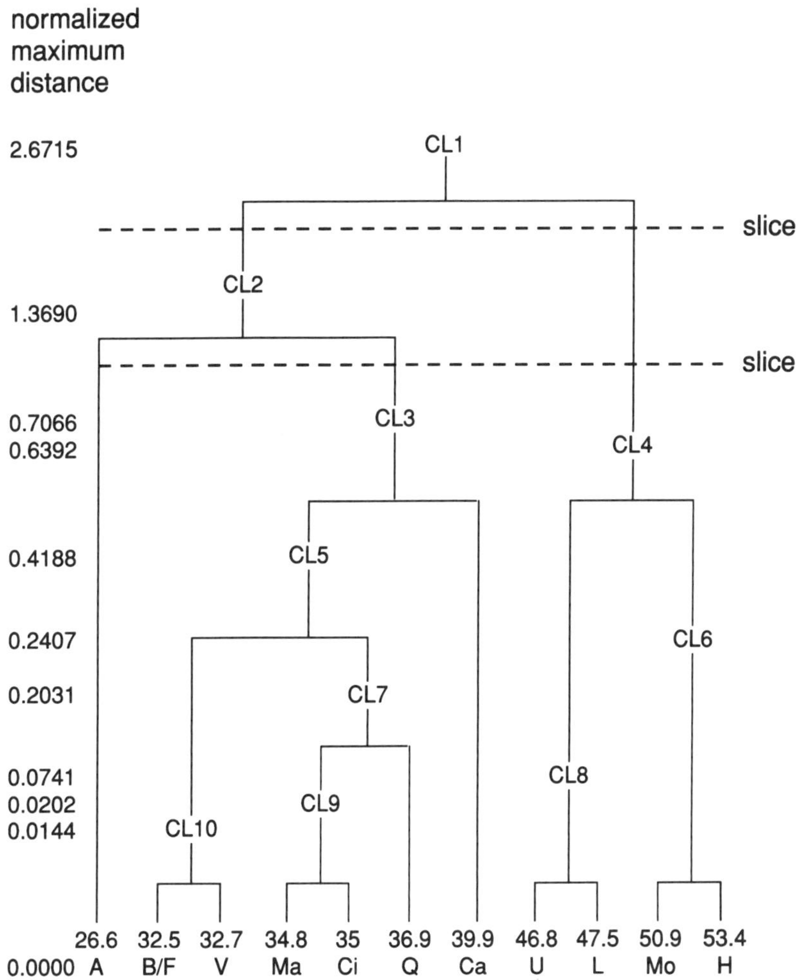

A:	Arecibo	V:	Vega Baja
B/F:	Barceloneta / Florida	H:	Hatillo
Ca:	Camuy	L:	Lares
Ci:	Ciales	Mo:	Morovis
M:	Manati	U:	Utuado
Q:	Quebradillas		

Figure 16.3: DENDOGRAM FROM COMPLETE LINKAGE CLUSTERING ALGORITHM FOR THE ARECIBO REGION

In comparison with the average linkage solution, the magnitude of values for the distance index here are slightly lower. However, the result is essentially the same as that for the average linkage; perhaps, then, Arecibo is not being isolated because it is an outlier.

The last, and perhaps the most frequently used, is **Ward's algorithm**, named after its designer. This algorithm is based upon the analysis of variance principle: in step 2 of Ward's algorithm, a search is made for those two groups that will add the smallest increment to the within-groups variation term. This procedure begins with n indicator variables, one for each areal unit, and sequentially combines these indicator variables as groups are merged (remember that these indicator variables represent groups that are mutually exclusive and collectively exhaustive). Accordingly, the multiple correlation coefficient is used to decide upon group identification. As the number of clusters decreases, R^2 tends to decrease until it reaches zero for a single cluster (the indicator variable is vector **1**). The major drawback of this method is that its groupings are not necessarily optimal beyond its first iterative step. Moreover, once a group has been formed, subsequent larger clusters must be built upon these smaller initial and intermediate groups. Once constructed, suboptimal groups are not dissolved at some later step. This method tends to join clusters with a small number of areal units, has a strong predisposition to produce clusters with roughly the same number of areal units, and tends to be acutely sensitive to outliers.

> EXAMPLE PR-16.5. The dendogram for the Arecibo region, based upon DCD, using Ward's clustering algorithm (from SAS), appears in Figure 16.5.

In Example PR-16.5 two groups are identified, namely, (Arecibo, Barceloneta-/Florida, Camuy, Ciales, Manati, Quebradillas, Vega Baja) and (Hatillo, Lares, Morovis, Utuado); relatively speaking, only the last change in R^2 is quite sizeable.

The SAS computer code for executing these algorithms appears in Appendix 16-A. Unfortunately, these five different algorithms failed to reproduce identical results for a given data set. In general, Milligan and Cooper (1985) have investigated four currently existing clustering algorithms coupled with 30 similarity indices and have concluded that for distinct, nonoverlapping population groups the algorithms perform well, whereas the scales for choosing a slice repeatedly have little validity. The spatial analyst, then, must survey the results, searching for commonalities, and choose the most sensible description available. Keep constantly in mind that the selected solution may well lack any counterpart in the real world.

16.3. Summary

Five clustering algorithms have been discussed that represent the principle, generic approaches to the classification problem. In practice, many applications that require customized clustering algorithms (modified objective functions and/or

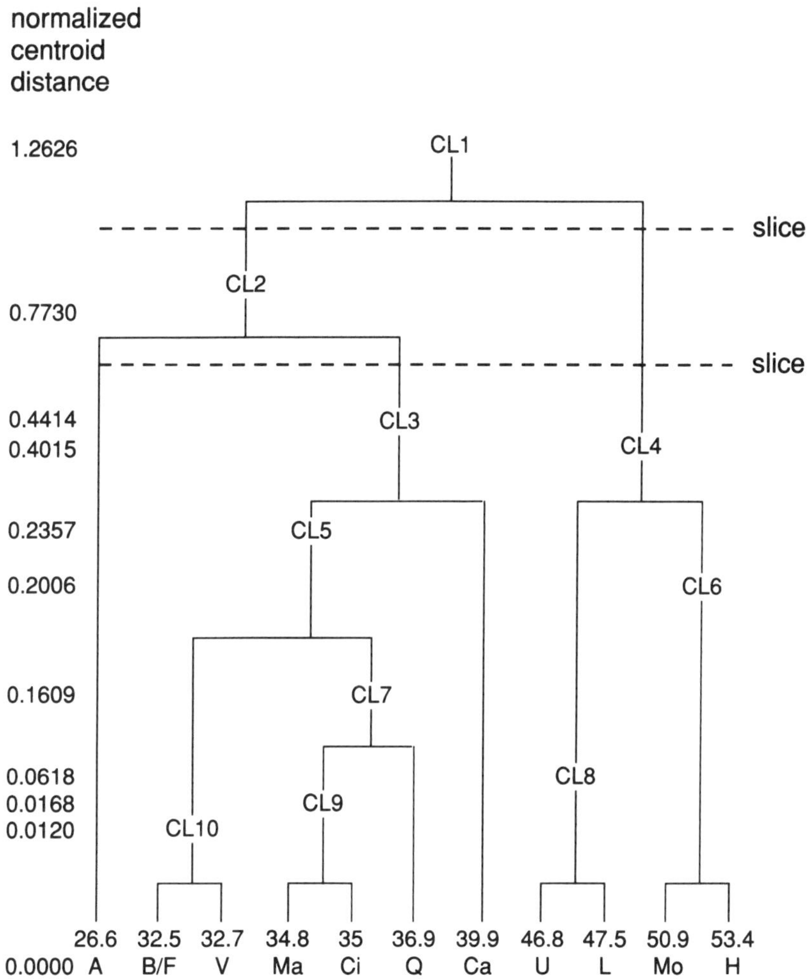

A:	Arecibo	V:	Vega Baja
B/F:	Barceloneta / Florida	H:	Hatillo
Ca:	Camuy	L:	Lares
Ci:	Ciales	Mo:	Morovis
M:	Manati	U:	Utuado
Q:	Quebradillas		

Figure 16.4: DENDOGRAM FROM CENTROID LINKAGE CLUSTERING ALGORITHM FOR THE ARECIBO REGION

measures to be optimized) have been developed; commercial statistical software packages such as SAS offer many of these others as options, too. The absence of a strong inferential foundation separates the material in this chapter from the last two chapters in this book. In fact, cluster analysis may be viewed as a descriptive technique that belongs in Part I. However, cluster analysis usually is multivariate in nature and also constitutes a noninferential introduction to some powerful ideas about exploratory data analysis. The inferential techniques that build upon these ideas form the core of companion volumes focussing on multivariate data analysis.

16.4. Problems

1. Verify the results in Examples PR-16.1 through PR-16.5.

2. Replicate the results in Examples PR-16.1 through PR-16.5 for the Ponce Agricultural Administrative Region.

3. Randomly sample two areal units from each of the five agricultural administration regions, and based upon the variable DCD, determine whether or not Ward's algorithm can uncover the underlying five groups.

16.5. References

S. Openshaw. *The Modifiable Areal Unit Problem.* CATMOG 38.

Milligan, G., and M. Cooper. (1985) "An Examination of Procedures for Determining the Number of Clusters in a Data Set." *Psychometrica*, 50: 159-79.

Murtagh, F. (1985) "A Survey of Algorithms for Contiguity-constrained Clustering and Related Problems." *The Computer Journal*, 28: 82-88.

Openshaw, S. (1977) "Algorithm 3: A Procedure to Generate Pseudorandom Aggregations of N Zones into M Zones, Where M is Less Than N." *Environment and Planning A*, 9: 169-84.

16.6. Additional Readings

M. Aldenderfir and R. Blashfield. *Cluster Analysis.* SAGE 44.

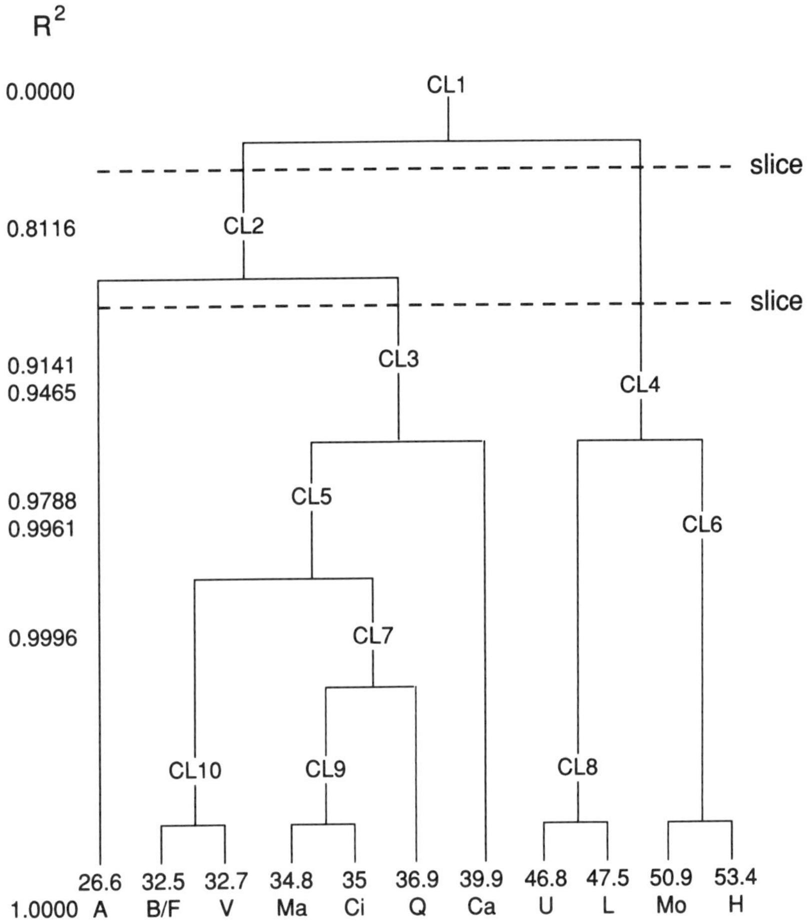

A:	Arecibo	V:	Vega Baja
B/F:	Barceloneta / Florida	H:	Hatillo
Ca:	Camuy	L:	Lares
Ci:	Ciales	Mo:	Morovis
M:	Manati	U:	Utuado
Q:	Quebradillas		

Figure 16.5: DENDOGRAM FROM WARD'S LINKAGE CLUSTERING ALGORITHM FOR THE ARECIBO REGION

APPENDIX 16-A

SAS CODE FOR EXECUTING SELECTED CLUSTER ANALYSIS ALGORITHMS

```
OPTION LINESIZE=80;
TITLE 'CLUSTER ANALYSIS EXAMPLE FOR PR-DCD, ARECIBO REGION';
DATA ARECIBO;
INPUT NUM DCD;
LIST;
CARDS;
   7  25.64
   9  32.52
  14  39.86
  18  34.97
  30  53.39
  38  47.52
  43  34.76
  48  50.89
  55  36.87
  68  46.75
  70  32.67
;
PROC PRINT; RUN;
PROC CLUSTER DATA=ARECIBO SIMPLE METHOD=WARD;
   VAR DCD; RUN;
PROC CLUSTER DATA=ARECIBO METHOD=AVERAGE;
   VAR DCD; RUN;
PROC CLUSTER DATA=ARECIBO METHOD=CENTROID;
   VAR DCD; RUN;
PROC CLUSTER DATA=ARECIBO METHOD=COMPLETE;
   VAR DCD; RUN;
PROC CLUSTER DATA=ARECIBO METHOD=SINGLE;
   VAR DCD; RUN;
RUN;
ENDSAS;
```

BIBLIOGRAPHY

1. REVIEW OF BASIC MATHEMATICS AND STATISTICS

Baggaley, A. (1969) *Mathematics for Introductory Statistics.* New York: Wiley.

Draper, N., and H. Smith. (1981) *Applied Regression Analysis*, 2nd ed. New York: Wiley.

Ferguson, G. (1971) *Statistical Analysis in Psychology and Education.* New York: McGraw-Hill.

Johnson, R. (1976) *Elementary Statistics.* North Scituate, Mass.: Duxbury Press.

Lindgren, B. (1968) *Statistical Theory*, 3rd ed. New York: Macmillan.

Lipschutz, S. (1965) *Schaum's Outline Of Theory and Problems of Probability.* New York: McGraw-Hill.

Montgomery, D., and E. Peck. (1982) *Introduction to Linear Regression Analysis.* New York: Wiley.

Siegel, S. (1956) *Nonparametric Statistics for the Behavioral Sciences.* New York: McGraw-Hill.

Summer, G. (1975) *Mathematics for Physical Geographers.* New York: Wiley.

Walker, H. (1964) *Mathematics Essential for Elementary Statistics*, rev. ed. New York: Holt, Rinehart & Winston.

Wilson, A., and M. Kirkby. (1980) *Mathematics for Geographers and Planners.* Oxford: Oxford University Press.

2. STATISTICAL TABLES

Arkin, H., and R. Colton. (1966) *Tables for Statisticians.* New York: Barnes & Noble (College Outline Series).

Beyer, W. (ed.). (1976) *Standard Mathematical Tables, 24th ed.* Cleveland: CRC Press.

Spiegel, M. (1968) *Mathematical Handbook of Formulas and Tables.* New York: McGraw-Hill (Schaum's Outline Series).

3a. STATISTICAL METHODS IN GEOGRAPHY AND RELATED FIELDS

Barber, G. (1988) *Elementary Statistics for Geographers.* New York: Guilford Press.

Bartlett, M. (1976) *The Statistical Analysis of Spatial Pattern.* London: Chapman and Hall.

Baxter, R. (1976) *Computer and Statistical Techniques for Planners.* London: Methuen.

Bennett, R. (1979) *Spatial Time Series.* London: Pion.

Berry, B., and A. Baker. (1968) "Geographic Sampling," in B. Berry and D. Marble (eds.) *Spatial Analysis: A Reader in Statistical Geography*, pp. 91-100. Englewood Cliffs, N.J.: Prentice Hall.

Berry, B., and D. Marble (eds.). (1968) *Spatial Analysis: A Reader in Statistical Geography.* Englewood Cliffs, N.J.: Prentice Hall.

Blalock, H. (1979) *Social Statistics*, 2nd rev. ed. New York: McGraw-Hill.

Bracken, I. (1981) *Urban Planning Methods: Research and Policy Analysis.* London: Methuen.

Budiansky, S. (1988) "The Numbers Racket: How Polls and Statistics Lie." *U. S. News and World Report*, 105 (No. 2; July 11): 44-47.

Chambers, J., W. Cleveland, B. Kleiner, and P. Tukey. (1983) *Graphical Methods for Data Analysis.* Belmont, Calif.: Wadsworth Publishing.

Clark, W., and P. Hosking. (1986) *Statistical Methods for Geographers.* New York: Wiley.

Cleveland, W. (1985) *The Elements of Graphing Data.* Monterey, Calif.: Wadsworth Publishing.

Cliff, A., and J. Ord. (1973) *Spatial Autocorrelation.* London: Pion.

Cliff, A., and J. Ord. (1981) *Spatial Processes, Models and Applications.* London: Pion.

Cliff, A., P. Haggett, J. Ord, K. Bassett, and R. Davies. (1975) *Elements of Spatial Structure: A Quantitative Approach.* Cambridge: Cambridge University Press.

Cole, J., and C. King. (1968) *Quantitative Geography: Techniques and Theories in Geography.* New York: Wiley.

Dickey, J., and J. Watts. (1978) *Analytic Techniques in Urban and Regional Planning.* New York: McGraw-Hill.

Duncan, O., R. Cuzzort, and B. Duncan. (1961) *Statistical Geography, Problems in Analyzing Areal Data.* Glencoe, Il.: The Free Press.

Ebdon, D. (1977; 2nd ed. 1985) *Statistics in Geography. A Practical Approach.* Oxford: Basil Blackwell.

Fitzgerald, B. (ed.). (1974) *Science in Geography.* Oxford: Oxford University Press.

Fitzgerald, B. *Developments in the Geographical Method*, vol. 1.

Daugherty, R. *Data Collection*, vol. 2.

David, P. *Data Description and Presentation*, vol. 3.

McCullagh, P. *Data Use and Interpretation*, vol. 4.

Getis, A., and B. Boots. (1978) *Models of Spatial Processes: An Approach to the Study of Point, Line and Area Patterns.* Cambridge: Cambridge University Press.

Golledge, R., and G. Rushton. (1972) *Multidimensional Scaling: Review and Geographical Applications.* Washington, D.C.: Association of American Geographers (Technical Paper No. 10).

Gould, P. (1970) "Is *Statistics Inferens* the Geographical Name for a Wild Goose Chase?" *Economic Geography*, 46: 439-48.

Greer-Wotten, B. (1972) *A Bibliography of Statistical Applications in Geography*. Washington, D.C.: Association of American Geographers (Technical Paper No. 11, Commission on College Geography).

Gregory, S. (1968) *Statistical Methods and the Geographer*, 3rd ed. New York: Longman.

Griffith, D. (1987) *Spatial Autocorrelation: A Primer*. Washington, D.C.: The Association of American Geographers.

Griffith, D. (1988) *Advanced Spatial Statistics*. Boston: Kluwer Academic Publishers.

Haggett, P., and C. Board. (1964) "Rotational and Parallel Transverses in the Rapid Integration of Geographic Areas." *Annals*, Association of American Geographers, 54: 406-410.

Haggett, P., A. Cliff, and A. Frey. (1977) *Locational Methods*, vol. 2 of 2nd ed. of *Location Analysis in Human Geography*. New York: Wiley, Halstead Press.

Hammond, R., and P. McCullagh. (1974) *Quantitative Techniques in Geography: An Introduction*. Oxford: Oxford University Press.

Harvey, M., and B. Holly. (1980) *Themes in Geographic Research*. London: Croom Helm.

Huggett, R. (1980) *Systems Analysis in Geography*. Oxford: Clarendon Press.

Johnston, R. (1978) *Multivariate Statistical Methods in Geography*. London: Longman.

King, L. (1969) *Statistical Analysis in Geography*. Englewood Cliffs, N.J.: Prentice-Hall.

Krueckeburg, D., and A. Silvers. (1973) *Urban Planning Analysis: Methods and Models*. New York: Wiley.

Krumbein, W., and F. Graybill. (1965) *Introduction to Statistical Methods in Geology*. New York: McGraw-Hill.

Lewis, P. (1977) *Maps and Statistics*. London: Methuen.

Mark, D., and M. Church. (1977) "On the Misuse of Regression in Earth Science." *Journal of the International Association of Mathematical Geology*, 9: 63-75.

Mark, D., and T. Peucker. (1978) "Regression Analysis and Geographic Models." *Canadian Geographer*, 22: 51-64.

Mather, P. (1976) *Computational Methods of Multivariate Analysis in Physical Geography*. New York: Wiley.

Matthews, J. (1981) *Quantitative and Statistical Approaches to Geography*. Oxford: Pergamon.

Milligan, G., and M. Cooper. (1985) "An Examination of Procedures for Determining the Number of Clusters in a Data Set." *Psychometrica*, 50: 159-79.

Murtagh, F. (1985) "A Survey of Algorithms for Contiguity-constrained Clustering and Related Problems." *The Computer Journal*, 28: 82-88.

Neft, D. (1966) *Statistical Analysis for Areal Distributions*. Philadelphia: Regional Science Research Institute (Monograph Series No. 2).

Norcliffe, G. (1977) *Inferential Statistics for Geographers*. New York: Halstead Press.

Odland, J. (1988) *Spatial Autocorrelation*. Beverly Hills, Calif.: Sage (Vol. 9 in the Scientific Geography Series).

Openshaw, S. (1977) "Algorithm 3: A Procedure to Generate Pseudorandom Aggregations of N Zones into M Zones, Where M is Less Than N." *Environment and Planning A*, 9: 169-84.

Ord, J. (1972) *Families of Frequency Distributions*. London: Griffin Limited.

Paelinck, J., and L. Klaassen. (1978) *Spatial Econometrics*. Westmead: Saxon House.

Rayner, J. (1971) *An Introduction to Spectral Analysis*. London: Pion.

Ripley, B. (1981) *Spatial Statistics*. New York: Wiley.

Rogers, A. (1974) *Statistical Analysis of Spatial Dispersion*. London: Pion.

Searle, S. (1989) "Statistical Computing Packages: Some Words of Caution." *The American Statistician*, 43: 189-90.

Shaw, G., and D. Wheeler. (1985) *Statistical Techniques in Geographical Analysis*. New York: Wiley.

Silk, J. (1979) *Statistical Concepts in Geography*. Winchester, Mass.: Allen & Unwin.

Smith, D. (1975) *Patterns in Human Geography: An Introduction to Numerical Methods*. New York: Crane-Russak.

Summerfield, M. (1983) "Population, Samples, and Statistical Inference in Geography." *The Professional Geographer*, 35: 143-49.

Taylor, P. (1977) *Quantitative Methods in Geography. An Introduction to Spatial Analysis*. Boston: Houghton-Mifflin.

Theakstone, W., and C. Harrison. (1969) *The Analysis of Geographical Data*. London: Heineman Educational Books.

Thomas, R., and R. Huggett. (1980) *Modelling in Geography: A Mathematical Approach*. London: Harper and Row.

Torgerson, W. (1958) *Theory and Methods of Scaling*. New York: Wiley.

Tufte, E. (1983) *The Visual Display of Quantitative Information*. Cheshire, Conn.: Graphics Press.

Upton, J., and B. Fingelton. (1985, 1989) *Spatial Data Analysis by Example*, vols. 1 & 2. New York: Wiley.

Unwin, D. (1981) *Introductory Spatial Analysis*. New York: Methuen.

Williams, K. (1975) *Statistics & Urban Planning*. New York: Halstead Press.

Wilson, A. (1981) *Geography and the Environment: Systems Analytic Models*. New York: Wiley.

Wilson, A., and M. Kirby. (1975) *Mathematics for Geographers and Planners*. Oxford: Clarendon Press.

Wrigley, N. (1979) *Statistical Applications in the Spatial Sciences.* London: Pion.

Wrigley, N. (1985) *Categorical Data Analysis for Geographers and Environmental Scientists.* New York: Longman.

Yeates, M. (1968) *An Introduction to Quantitative Methods in Economic Geography.* New York: McGraw-Hill.

Yeates, M. (1974) *An Introduction to Quantitative Analysis in Human Geography.* New York: McGraw-Hill.

3b. THE CATMOG SERIES

The CATMOG Series (Concepts and Techniques in Modern Geography) has around 50 paperback monographs each less than 50 pages, each introducing different methods. They are published by Geo Books, Regency House, 34 Duke Street, Norwich, England NR3 3AP. Listed here are the ones most relevant to an introductory statistics course.

No.	Author	Title
2	P. Taylor	*Distance Decay Models in Spatial Interaction*
5	D. Unwin	*An Introduction to Trend Surface Analysis*
6	R. Johnston	*Classification in Geography*
9	N. Davidson	*Causal Inferences from Dichotomous Variables*
10	N. Wrigley	*Introduction to the Use of Logit Models in Geography*
12	R. Thomas	*An Introduction to Quadrat Analysis*
13	N. Thrift	*An Introduction to Time-Geography*
14	K. Tinkler	*An Introduction to Graph Theoretical Models in Geography*
15	R. Ferguson	*Linear Regression in Geography*
16	N. Wrigley	*Probability Surface Mapping. An Introduction with Examples and Fortran Programmes*
17	C. Dixon and B. Leach	*Sampling Methods for Geographical Research*
18	C. Dixon and B. Leach	*Questionnaires and Interviews for Geographical Research*
19	V. Gardiner and G. Gardner	*Analysis of Frequency Distributions*

20	J. Silk	*Analysis of Covariance and Comparison of Regression Lines*
23	K. Richards	*Stochastic Processes in One-Dimensional Series: An Introduction*
25	G. Gaile and J. Burt	*Directional Statistics*
26	D. Rich	*Potential Models in Human Geography*
27	D. Pringle	*Causal Modelling: The Simon-Blalock Approach*
28	R. Bennett	*Statistical Forecasting*
30	J. Silk	*The Analysis of Variance*
31	R. Thomas	*Information Statistics in Geography*
32	A. Kellerman	*Centrographic Measures in Geography*
36	G. Aplin	*Order-neighbor Analysis*
38	S. Openshaw	*The Modifiable Areal Unit Problem*
41	R. Kirby	*Choice in Field Surveying*
46	S. Fotheringham and D. Knudsen	*Goodness-of-Fit Statistics*
47	M. Goodchild	*Spatial Autocorrelation*
48	K. Tinkler	*Introductory Matrix Algebra*

3c. THE SAGE SERIES

Listed next is a selection of short (100 pages) monographs on particular topics that you may wish to consult. They are part of the Sage University Paper series on Quantitative Applications in the Social Sciences.

No.	Author	Title
1	G. Iversen and N. Norporth	*Analysis of Variance*
3	H. Asher	*Causal Modelling*
4	R. Henkel	*Tests of Significance*
7	H. Reynolds	*Analysis of Nominal Data*
8	D. Hildebrand, J. Laing, and H. Rosenthal	*Analysis of Ordinal Data*
9	C. Ostrom, Jr.	*Time Series Analysis: Regression Techniques*
10	L. Langbein and A. Lichtman	*Ecological Inference*
12	A. Wildt and O. Ahtola	*Analysis of Covariance*
16	F. Hartwig and B. Dearing	*Explanatory Data Analysis*
22	M. Lewis-Beck	*Applied Regression*
23	P. Spector	*Research Designs*
29	C. Achen	*Interpreting and Using Regression*
34	J. Long	*Covariance Structure Models*

38	K. Namboodiri	*Matrix Algebra*
44	M. Aldenderfir and R. Blashfield	*Cluster Analysis*
45	J. Aldrich and F. Nelson	*Linear Probability, Logit and Probit Models*
50	W. Berry and S. Feldman	*Multiple Regression in Practice*
57	L. Schroeder, D. Sjoquist, and P. Stephan	*Understanding Regression Analysis*
65	P. Arabie, J. Carroll, and W. Desarbo	*Three-Way Scaling and Clustering*

4. OTHER WORKS ON METHODS AND MODELS FOR SPATIAL ANALYSIS

Bennett, R. (1979) *Spatial Time Series.* London: Pion.

Chorley, R. (Chairman) (1987) *Handling Geographic Information.* London: Her Majesty's Stationery Office, Report to the Select Committee on GIS.

Chorley, R. (ed.). (1972) *Spatial Analysis in Geomorphology.* Chicago: Methuen.

Chorley, R., and P. Haggett (eds.). (1966) *Models in Geography.* London: Methuen.

Cole, J., and C. King. (1968) *Quantitative Geography: Techniques and Theories in Geography.* New York: Wiley.

Dawson, J., and D. Unwin. (1976) *Computing for Geographers.* London: David & Charles.

Foot, D. (1981) *Operational Urban Models: An Introduction.* London: Methuen.

French, H., and J. Racine (eds.). (1971) *Quantitative and Qualitative Geography: La Necessite d'Un Dialogue.* Ottawa: University of Ottawa Press (Occasional Paper No. 1, Department of Geography, University of Ottawa).

French, H. (1971) "Quantitative Methods and Non-Parametric Statistics," in H. French and J. Racine, (eds.), *Quantitative and Qualitative Geography: La Necessite d'Un Dialogue*, pp. 119-128. Ottawa: University of Ottawa Press (Occasional Paper No. 1, Department of Geography, University of Ottawa).

Getis, A., and B. Boots. (1978) *Models of Spatial Processes.* Cambridge: Cambridge University Press.

Haggett, P., and R. Chorley. (1967) "Models, Paradigms and the New Geography," in R. Chorley and P. Haggett (eds.), *Models in Geography*, pp. 19-24. London: Methuen.

Haggett, P., A. Cliff, and A. Frey. (1977) *Locational Models*. London: Edward Arnold.

Haring, L., and J. Lounsbury. (1975) *Introduction to Scientific Geographic Research*. Dubuque, Iowa: Wm. C. Brown.

Huggett, R. (1980) *Systems Analysis in Geography*. Oxford: Clarendon Press.

Keyes, D., U. Basoglu, E. Kuhlmey, and M. Rhyner. (1976) "Comparison of Several Sampling Designs for Geographic Data." *Geographical Analysis*, 8(3): 295-304.

MacDougall, E. (1976) *Computer Programming for Spatial Problems*. London: Edward Arnold.

Masser, I. (1972) *Analytical Models for Urban and Regional Planning*. New York: Halstead Press.

Mather, P. (1976) *Computers in Geography: A Practical Approach*. Oxford: Blackwell.

Oppenheim, N. (1980) *Applied Models in Urban and Regional Analysis*. Englewood Cliffs, N.J.: Prentice Hall.

Thomas, R., and R. Huggett. (1980) *Modelling in Geography: A Mathematical Approach*. London: Harper and Row.

Wilson, A. (1974) *Urban and Regional Models in Geography and Planning*. New York: Wiley.

Wilson, A., and M. Kirby. (1980) *Mathematics for Geographers and Planners*, 2nd ed. Oxford: Clarendon Press.

5. BACKGROUND REFERENCE MATERIALS

Brower, B. (1988) "The Pernicious Power of the Polls." *Money*, 17 (March): 144-163.

Fuerbringer, J. (1989) "Accuracy in SHort Supply in Flood of U.S. Statistics." *The New York Times*, 139 (October 30): D1 and D4.

Gleick, J. (1988) "The Quest for True Randomness Finally Appears Successful." *The New York Times*, 138 (April 19): C1 and C8.

Hite, S. (1976) *The Hite Report: A Nationwide Study on Female Sexuality.* New York: Macmillan.

Huff, D. (1954) *How to Lie With Statistics.* New York: W. W. Norton.

Knight, N. (1988) "No Two Alike?" *Bulletin of the American Meteoroligical Society*, 69: 496.

Kolata, G. (1990a) "In Shuffling Cards, 7 is Winning Number." *The New York Times*, 140 (January 9): C1 and C12.

Kolata, G. (1990b) "1-in-a-trillion Coincidence, You Say? Not Really, Experts Find." *The New York Times*, 140 (February 27): C1 and C2.

Walsh, D. (1985) "Socrates in the Classroom." *American Educator*, 9 (Summer): 20-25.

APPENDICES

Appendices presented in this part of the book consist of four tables of data, which are first mentioned in Chapter 1, and eight statistical tables, which initially are mentioned in Chapters 7, 9, 10, and 11. These tables are designated with the extension "-A," which denotes their location in the appendices. Data tables have been labelled 1.1-A, 1.2-A, 1.3-A, and 1.4-A, since they are first encountered in Chapter 1. Each statistical table also is indexed according to that chapter in which it is first encountered. We believe that this indexing system allows a student to more easily refer introductory discussions of particular tables.

The tables of data presented in these appendices are

The statistical tables presented in these appendices are

TABLE 1.1-A: DATA SET PR

PUERTO RICO AGRICULTURAL DATA

No.	Name	ADM	LND	R-U	DFRM	DCD	DMLK	DSGR	DCF	DTB	DBN	DFAM
	SAN JUAN REGION (1)											
10	Barranquitas	1	1	0	16.1	38.0	2.0	0.0	3279.7	32.2	5.4	4.3
11	Bayamon/Catano	1	0	1	5.2	9.4	2.2	0.0	758.2	0.5	0.9	0.9
15	Carolina	1	0	1	5.5	22.7	43.0	4.0	11.0	0.0	0.2	1.1
17	Ceiba	1	0	1	3.0	13.2	20.8	2.1	23.7	0.0	0.2	0.7
21	Comerio	1	1	0	14.9	52.6	1.7	0.0	708.0	153.7	3.3	2.4
22	Corozal	1	1	0	13.7	40.7	28.6	0.0	2816.2	0.3	5.2	2.4
23	Dorado	1	0	1	2.7	26.2	66.6	7.6	8.5	0.5	0.2	2.5
24	Fajardo	1	0	1	3.3	24.5	12.0	10.0	15.6	0.0	0.3	0.5
28	Guaynabo	1	0	1	7.5	15.0	9.6	0.0	193.7	0.0	1.2	1.0
41	Loiza/Canovanas	1	0	1	3.8	23.7	19.2	4.2	720.3	0.0	0.8	2.5
42	Luquillo	1	0	0	4.2	23.3	45.1	1.8	22.7	0.0	0.4	1.1
50	Naranjito	1	1	0	18.6	34.2	3.1	0.0	1591.7	7.5	6.2	1.7
51	Orocovis	1	1	0	17.6	53.6	10.8	2.1	13265.8	19.7	19.3	4.4
57	Rio Grande	1	0	1	2.8	14.3	6.9	0.8	14.0	0.0	0.4	0.1
61	San Juan/Rio Piedras	1	0	1	2.7	2.9	0.0	0.0	31.1	0.0	0.2	0.1
65	Toa Alta	1	1	1	7.4	27.7	94.0	0.0	16.0	0.0	0.8	1.5
66	Toa Baja	1	0	1	0.7	28.2	91.1	8.3	0.0	0.0	0.0	1.4
67	Trujillo Alto	1	1	1	9.2	20.4	27.4	0.0	7.2	0.0	0.3	0.7
69	Vega Alta	1	0	1	6.6	31.7	49.1	2.7	59.0	0.0	0.3	1.0
	ARECIBO REGION (2)											
7	Arecibo	2	0	1	4.5	25.6	46.7	6.1	678.7	1.5	0.3	1.9
9	Barceloneta/Florida	2	0	0	5.6	32.5	32.3	10.9	268.6	0.3	0.6	1.5
14	Camuy	2	0	0	12.1	39.9	68.1	3.6	413.1	1.7	0.4	1.1
18	Ciales	2	1	0	8.5	35.0	4.9	0.1	12070.0	10.0	9.4	2.7
30	Hatillo	2	0	1	8.9	53.4	368.6	1.0	45.1	0.0	0.9	4.9
38	Lares	2	1	0	17.9	47.5	0.0	0.2	48874.9	0.0	25.6	7.2
43	Manati	2	0	1	3.0	34.8	89.2	3.1	80.5	0.0	0.1	1.5
48	Morovis	2	1	0	12.9	50.9	50.4	0.0	3158.9	23.7	3.4	4.5
55	Quebradillas	2	0	0	10.7	36.9	40.9	9.8	610.4	11.0	0.4	0.5
68	Utuado	2	1	0	18.3	46.8	2.7	0.0	26677.2	10.0	9.0	2.8
70	Vega Baja	2	0	1	4.8	32.7	39.1	9.1	227.9	0.3	0.3	1.1
	MAYAGUEZ REGION (3)											
2	Aguada	3	0	0	21.2	41.7	0.1	26.1	10286.9	0.0	2.2	1.3
3	Aguadilla	3	0	1	5.3	28.1	5.7	14.3	199.6	0.8	0.2	2.3
6	Anasco	3	0	0	9.3	37.0	0.0	14.7	11811.4	0.0	2.1	2.1
12	Cabo Rojo	3	0	0	5.5	48.3	22.5	14.8	7.0	0.0	0.2	3.6
25	Guanica	3	0	1	2.4	39.0	4.4	9.9	147.9	0.0	0.0	1.0
31	Hormigueros	3	1	1	9.8	44.5	0.1	27.7	4130.0	0.0	1.3	2.0
33	Isabela	3	0	0	11.0	32.3	39.6	5.8	683.6	8.3	0.5	2.0
37	Lajas	3	0	0	5.3	62.5	6.1	13.1	31.5	0.1	0.2	2.0
39	Las Marias	3	1	0	18.2	40.3	0.0	0.4	34270.9	3.2	9.3	2.5
44	Maricao	3	1	0	14.0	42.0	0.0	0.0	48140.1	0.0	14.3	4.8
46	Mayaguez	3	0	1	9.2	26.4	3.8	7.6	9763.0	0.3	2.9	1.6
47	Moca	3	1	0	14.3	37.9	2.1	15.3	6482.3	0.1	1.2	2.6
56	Rincon	3	0	0	8.4	21.6	0.0	7.5	2391.8	0.0	0.4	0.9
58	Sabana Grande	3	1	0	4.9	24.3	3.5	4.9	4898.7	0.0	0.1	1.7
60	San German	3	1	0	11.9	49.4	0.5	14.1	10993.9	5.8	1.7	4.3
63	San Sebastian	3	1	0	14.7	45.9	6.9	17.4	11689.8	1.8	1.8	2.7

DATA SET PR, Continued

No.	Name	ADM	LND	R-U	DFRM	DCD	DMLK	DSGR	DCF	DTB	DBN	DFAM
	PONCE REGION (4)											
1	Adjuntas	4	1	0	15.6	41.1	2.1	0.0	35710.4	0.0	10.8	3.2
8	Arroyo	4	0	0	9.3	41.4	0.0	13.4	2510.8	0.0	1.3	0.8
20	Coamo	4	1	0	5.3	43.6	10.9	0.0	2253.0	1.0	2.7	1.4
26	Guayama	4	0	1	3.7	36.8	5.3	9.7	1228.8	0.6	1.2	0.8
27	Guayanilla	4	0	0	6.5	62.9	0.0	7.6	9031.0	0.4	2.9	1.9
34	Jayuya	4	1	0	14.1	57.1	5.0	0.1	44081.6	0.1	32.4	5.2
35	Juana Diaz	4	0	0	2.0	49.5	8.6	12.9	3289.3	0.0	1.9	1.1
52	Patillas	4	0	0	8.3	26.7	0.0	1.5	485.9	1.5	1.2	2.0
53	Penuelas	4	0	0	5.5	44.6	5.3	8.8	5978.4	0.0	3.2	4.0
54	Ponce	4	0	1	4.0	37.5	4.6	8.2	7788.3	0.0	2.8	1.3
59	Salinas	4	0	0	1.3	51.7	7.8	15.2	117.5	0.4	0.0	0.6
64	Santa Isabel	4	0	0	0.7	48.6	1.7	30.0	0.0	0.0	0.1	0.8
71	Villalba	4	1	0	10.9	39.4	0.0	0.8	8650.7	0.0	2.6	2.8
73	Yauco	4	0	0	9.9	39.1	2.3	2.6	15325.7	0.3	7.4	2.9
	CAGUAS REGION (5)											
4	Aguas Buenas	5	1	0	15.4	30.4	10.7	0.0	4505.6	19.4	4.5	1.8
5	Aibonito	5	1	0	8.3	37.3	16.6	0.0	2437.4	49.0	1.1	1.3
13	Caguas	5	1	1	10.0	36.5	82.8	0.0	304.7	9.6	1.7	4.7
16	Cayey	5	1	1	11.1	43.1	18.0	0.0	3660.0	6.7	10.0	1.4
19	Cidra	5	1	0	11.5	36.2	23.7	0.2	493.9	63.0	1.6	2.5
29	Gurabo	5	1	1	6.9	39.5	75.2	6.1	42.0	3.6	0.4	2.3
32	Humacao	5	0	1	5.0	18.5	15.0	10.3	13.0	0.2	0.3	3.6
36	Juncos	5	1	1	7.8	39.8	134.4	0.2	87.4	0.8	0.1	5.7
40	Las Piedras	5	1	0	15.2	36.8	60.0	0.9	494.4	3.0	1.8	3.0
45	Maunabo	5	0	0	5.5	37.5	0.0	16.5	14.5	0.2	2.1	1.5
49	Naguabo	5	0	0	4.3	27.2	68.2	4.0	346.9	0.0	0.4	0.9
62	San Lorenzo	5	1	1	14.8	35.4	32.5	1.5	225.9	8.2	1.3	4.9
72	Yabucoa	5	0	0	15.3	43.0	15.2	9.7	402.3	1.5	2.0	2.5

Note: No. refers to the alphabetical order of the municipio names.

Sources: 1974 *U. S. Census of Agriculture*, and 1974/75 Commonwealth of Puerto Rico *Facts and Figures on Puerto Rico's Agriculture.*

TABLE 1.2-A: DATA SET CA

CANADIAN AUTOMOTIVE PARTS INDUSTRY DATA

No.	Firm name	PC1	PC2	FAC	EXP	MKT	PAR	SALE	EXP$	LWK	HWK	PPC	SPC
	EASTERN ONTARIO (K)												
80	GenWire	K	9A3	0	0	0	1	25000	0	100	249	7	7
192	Townsend	K	7G2	0	0	0	1	5000	1000	100	249	7	7
	CENTRAL ONTARIO (L)												
29	Bundy	L	6T1	1	0	0	1	25000	5000	250	499	0	1
41	Chalmers	L	6V2	1	1	0	0	1000	500	20	49	1	1
46	Chrys2	L	1S2	1	0	0	1	50000+	50000+	1000+	1000+	3	7
91	Hayes1	L	2R7	1	0	0	1	50000+	50000+	1000+	1000+	1	4
108	Jet	L	1S1	1	0	0	0	5000	1000	50	99	7	7
136	NWClay	L	6L5	1	1	0	0	5000	5000	100	249	3	7
140	PCDrop	L	3K5	0	0	0	0	5000	3000	100	249	7	7
164	Russo	L	4L1	1	1	0	0	500	1	10	19	6	7
182	Super7	L	3Y2	1	1	0	0	100	0	20	49	0	3
188	TheSKWel	L	4K2	1	0	0	0	3000	100	20	49	1	0
195	TRW1	L	8G3	1	0	0	1	50000+	25000	500	999	2	2
202	Waltec	L	4R4	1	0	0	0	5000	100	100	249	3	3
	METROPOLITAN TORONTO (M)												
4	Simpson	M	1H2	0	0	0	0	50000+	50000+	1000+	1000+	3	0
7	ABC2	M	9W4	1	0	0	0	50000+	50000+	1000+	1000+	3	3
26	Borden	M	1E4	0	0	0	1	50000+	1000	250	499	7	5
33	CanDecal	M	1P3	1	0	0	0	5000	25	50	99	7	7
59	Depco	M	8W3	1	0	0	0	50000+	25000	100	249	3	3
62	Donlee	M	9M1	0	0	0	0	25000	5000	500	999	3	3
77	GMA	M	1K2	0	0	0	0	5000	5000	100	249	7	1
83	Graphico	M	1H1	1	1	0	0	5000	5000	250	499	2	2
87	HRuhl	M	8Z2	1	0	0	0	5000	500	50	99	2	6
146	Philips	M	1B1	1	1	0	0	50000+	25	1000+	1000+	2	2
150	ProgMold	M	3J2	1	0	0	0	5000	500	100	249	3	3
160	Rock2	M	4A1	1	0	0	1	50000+	50000+	1000+	1000+	0	1
205	Wheel	M	8Z5	1	0	0	0	5000	100	50	99	1	6

DATA SET CA, Continued

No.	Firm name	PC1	PC2	FAC	EXP	MKT	PAR	SALE	EXP$	LWK	HWK	PPC	SPC
	SOUTHWESTERN ONTARIO (N)												
1	A&R	N	6N1	0	0	0	0	3000	1000	50	99	3	0
10	Ablo	N	3H4	1	0	0	0	5000	100	50	99	7	0
14	Apex	N	2G2	0	0	0	0	5000	5000	50	99	3	3
21	Bell	N	0G1	1	0	0	3	5000	25	100	249	7	7
36	CanGen	N	1R5	1	0	0	0	5000+	5000	250	499	7	7
51	Crown	N	3T5	1	0	0	0	1000	0	20	49	2	3
54	CustomTr	N	2J3	0	0	0	0	25000	25000	500	999	3	1
66	Elan	N	7M1	1	0	0	0	25000	5000	100	249	3	3
70	FAGBear	N	5A3	0	1	0	3	50000+	25000	500	999	7	7
73	Firest2	N	4S7	0	1	0	1	25000	1000	100	249	7	7
96	Hematite	N	1H7	0	0	0	0	3000	1000	50	99	7	4
99	Holmes	N	7T5	0	0	0	0	5000	5000	250	499	7	7
105	IntSint	N	1H1	0	0	0	0	5000	5000	50	99	0	0
112	Kendan	N	9A6	1	0	0	0	25000	25000	100	249	0	0
116	Libby	N	8N2	1	0	0	0	500	0	10	19	6	6
119	MacDon	N	8Y2	0	0	0	0	5000	100	50	99	5	5
123	Mastico	N	4G4	1	0	0	0	5000	3000	100	249	3	3
126	Mitten	N	1R5	0	0	0	0	5000	5000	50	99	3	3
130	NatAuto	N	8W1	0	0	0	0	50000+	50000+	500	999	3	0
133	Newmac	N	4S7	0	0	0	0	5000	1000	50	99	3	1
155	Relmech	N	3B2	1	0	0	0	1000	500	20	49	0	2
170	Spun	N	7G3	1	0	0	1	5000	5000	100	249	4	4
173	StnInduc	N	9C1	1	0	0	0	3000	500	50	99	0	7
176	StndTube	N	4S5	1	0	0	0	50000+	0	500	999	1	6
179	Steele's	N	1H1	0	0	0	0	3000	100	50	99	1	1
185	Teledyne	N	4S7	1	0	0	1	50000+	50000+	100	249	6	6

Source: Automotive Directorate's 1986 *Canadian Auto Parts and Vehicle Manufacturers Directory.*

TABLE 1.3-A: DATA SET OR

ONTARIO DRAINAGE BASINS RUNOFF DATA

No.	Drainage basin	AREA	Q	RUN	LAT	LONG	SNOW	PRECIP	TEMP
1	Asheweig	3240	132.2	215	53.5	89.3	76.0	22.0	25.0
2	Ausable	865	50.6	308	43.1	81.7	88.0	34.0	46.0
3	Bear Brook	440	39.3	471	45.4	75.2	87.0	34.5	42.5
4	Big Otter	342	18.0	277	47.5	80.4	60.0	35.5	46.0
5	Black R. (south)	1520	111.5	387	44.7	79.4	92.0	36.0	40.0
6	Black R.	1980	144.6	385	48.7	86.2	110.0	31.0	33.0
7	Blackwater	650	38.3	311	48.9	88.4	96.0	28.2	33.5
8	Blanche	1780	134.1	397	47.9	79.9	90.0	32.0	33.5
9	Brightsand	1170	31.1	140	49.6	90.6	92.0	28.5	33.0
10	East Oakville	199	9.5	252	43.5	79.8	48.0	30.0	46.0
11	Fawn R.	4350	80.3	97	47.4	79.6	88.0	31.8	36.5
12	Ganaraska	262	13.7	275	44.0	78.3	68.0	32.0	44.0
13	Goulais	1160	103.8	471	46.9	84.0	132.0	37.0	37.5
14	Humber	303	10.4	181	43.8	79.6	60.0	29.5	46.0
15	Indian	443	31.4	374	45.8	77.1	80.0	28.0	41.0
16	Kaministiquia	6480	190.5	155	48.5	89.6	88.0	28.5	34.0
17	Kwataboahegan	4250	220.0	273	51.2	80.8	112.0	30.5	30.0
18	Middle Maitland	648	51.5	419	43.8	81.3	108.0	35.0	45.0
19	Nith	1030	56.1	287	43.2	80.4	80.0	37.0	44.0
20	Nottawasaga	1180	48.7	218	44.2	79.8	72.0	31.0	44.5
21	Pineimuta	4900	251.9	271	52.3	88.8	88.0	25.0	28.0
22	Raisan R.	404	34.9	455	45.2	74.6	91.0	36.0	43.5
23	Saugeen	3960	278.0	370	44.5	81.3	120.0	36.0	44.0
24	Serpent	1350	105.1	410	46.2	82.5	85.0	32.5	40.0
25	Skootamata	712	46.0	341	44.5	77.3	80.0	34.0	42.0
26	Sturgeon	4450	142.4	169	50.2	91.6	90.0	26.5	33.5
27	Sydenham	730	33.6	243	42.8	81.8	64.0	35.0	47.0
28	Turtle	4870	144.8	157	48.8	92.7	70.0	26.5	36.0

Note: RUN = Q•t /AREA, where t is a 2-month time interval, Q is in cubic meters per second, and AREA is in square kilometers. Units are standardized to millimeters.

Sources: Drainage area and spring discharge data (22-year average computed for 1960-1982) are from the annual Environment Canada *Water Survey of Canadian Surface Water Data*. Latitude and longitude have been derived from topographic maps. Snowfall, temperature, and precipitation data are from Environment Canada *Atmospheric Environment Services Climate Summaries*.

TABLE 1.4-A: DATA SET SG

SENTINEL GLACIER DATA

DATE	ELA	BN	PRECIP	TEMP	DATE	ELA	BN	PRECIP	TEMP
1966	1820	-0.15	1136.8	12.4	1976	1670	1.47	1407.8	11.5
1967	1875	-0.18	1161.6	14.2	1977	1995	-1.33	626.1	12.2
1968	1720	0.38	1608.7	12.7	1978	1830	0.36	796.1	12.5
1969	1800	0.12	1178.3	13.9	1979	2060	-1.74	685.7	13.0
1970	2080	-1.30	657.3	13.3	1980	1830	0.27	953.4	12.3
1971	1795	0.57	1007.3	12.9	1981	1865	0.17	1043.8	12.8
1972	1825	0.26	1308.0	12.9	1982	1775	0.86	1103.9	13.1
1973	1750	0.78	884.5	12.6	1983	1690	1.20	1300.0	12.5
1974	1600	2.11	1397.4	12.9	1984	1780	0.85	971.1	11.7
1975	1790	0.88	994.5	13.2					

Source: Derived from published data from various government agencies. Field measurements are from the Sentinel Glacier, British Columbia, Canada (north of Vancouver, B.C.) and ground station Alta Lake, British Columbia, Canada (located just to the north of Sentinel Glacier).

TABLE 7.1-A

RANDOM NUMBERS FROM A UNIFORM DISTRIBUTION GENERATED WITH MINITAB-PC

9739451670	2274128233	0769716929	7467521573	3306541171	9826453360
4107193561	1379774306	3146248181	3937672270	8999280873	1454962338
1476303063	4598659026	1269049983	8076019039	7934305999	1215782865
5091828952	5081987131	1368186792	0123110197	4867851444	5367121910
3839846679	3259411270	6200299682	8747723452	0800317599	1638212558
6905713865	8116102899	7046929396	7517193777	3343709657	7435842594
7592726918	1218817054	9982279423	6510115736	3399444459	8152911952
9344209017	3050628741	3416629029	7538803467	6957201165	1422518247
5848476422	5672141214	8227801319	0971427825	9745634535	4071787084
5402065078	0886648812	2736066346	5550505204	5192920142	7785791263

2527610794	3442980320	2785091769	2540062925	3761865576	3086623965
1528367381	5957019134	7015151127	4840855001	0923181478	9464567379
2915416320	4903220501	2477758106	6247787446	6237330356	6536332626
7316449073	2474158487	2370104680	7789481222	7290456437	9224037671
6023212206	0077872617	6653752004	7431135209	6253079808	0871629416
3917749966	6113119736	6774277423	6252114265	3717953927	6074920894
2429429744	3263520709	7129638941	9469781495	5627220919	8189664796
8345333210	7134198997	7404046702	4165631610	3074886430	6798867375
3444234246	7162003069	0379642925	6749695825	5405031595	4998736489

4862378834	9782495051	3014821375	0810381873	5083325000	4229225440
6876831366	6072808788	0150839922	2854568194	5311311488	4650108343
3802858074	9499885990	6679549368	2213420302	6586409679	3297706513
3299261257	3168925329	2095980017	8994004810	0838548635	8790495354
8263392000	6133623639	7380115429	1811365601	3751441107	5870307310
6506076440	3518971459	0753313153	8171593726	5239202955	0891182928
3375340859	9256925815	8572335850	9997965690	9736767992	0305611619
4652782520	0193374852	7292497814	7871719754	4611007668	9797517911
3038461875	0663941424	0430072593	1036815680	7599916944	5305059909
7707336002	4864711936	0812909994	9369260623	5145430234	9682037768
0510353003	0215127296	4065268854	0673006552	5086574215	0164516182

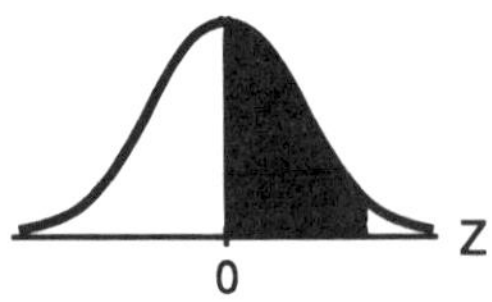

TABLE 7.2-A

STANDARD NORMAL CURVE CUMULATIVE PROBABILITIES (FROM z = 0)

z	0.0000	0.0100	0.0200	0.0300	0.0400	0.0500	0.0600	0.0700	0.0800	0.0900
0.00	0.0000	0.0040	0.0080	0.0120	0.0160	0.0199	0.0239	0.0279	0.0319	0.0359
0.10	0.0398	0.0438	0.0478	0.0517	0.0557	0.0596	0.0636	0.0675	0.0714	0.0753
0.20	0.0793	0.0832	0.0871	0.0910	0.0948	0.0987	0.1026	0.1064	0.1103	0.1141
0.30	0.1179	0.1217	0.1255	0.1293	0.1331	0.1368	0.1406	0.1443	0.1480	0.1517
0.40	0.1554	0.1591	0.1628	0.1664	0.1700	0.1736	0.1772	0.1808	0.1844	0.1879
0.50	0.1915	0.1950	0.1985	0.2019	0.2054	0.2088	0.2123	0.2157	0.2190	0.2224
0.60	0.2257	0.2291	0.2324	0.2357	0.2389	0.2422	0.2454	0.2486	0.2517	0.2549
0.70	0.2580	0.2611	0.2642	0.2673	0.2704	0.2734	0.2764	0.2794	0.2823	0.2852
0.80	0.2881	0.2910	0.2939	0.2967	0.2995	0.3023	0.3051	0.3078	0.3106	0.3133
0.90	0.3159	0.3186	0.3212	0.3238	0.3264	0.3289	0.3315	0.3340	0.3365	0.3389
1.00	0.3413	0.3438	0.3461	0.3485	0.3508	0.3531	0.3554	0.3577	0.3599	0.3621
1.10	0.3643	0.3665	0.3686	0.3708	0.3729	0.3749	0.3770	0.3790	0.3810	0.3830
1.20	0.3849	0.3869	0.3888	0.3907	0.3925	0.3944	0.3962	0.3980	0.3997	0.4015
1.30	0.4032	0.4049	0.4066	0.4082	0.4099	0.4115	0.4131	0.4147	0.4162	0.4177
1.40	0.4192	0.4207	0.4222	0.4236	0.4251	0.4265	0.4279	0.4292	0.4306	0.4319
1.50	0.4332	0.4345	0.4357	0.4370	0.4382	0.4394	0.4406	0.4418	0.4429	0.4441
1.60	0.4452	0.4463	0.4474	0.4484	0.4495	0.4505	0.4515	0.4525	0.4535	0.4545
1.70	0.4554	0.4564	0.4573	0.4582	0.4591	0.4599	0.4608	0.4616	0.4625	0.4633
1.80	0.4641	0.4649	0.4656	0.4664	0.4671	0.4678	0.4686	0.4693	0.4699	0.4706
1.90	0.4713	0.4719	0.4726	0.4732	0.4738	0.4744	0.4750	0.4756	0.4761	0.4767
2.00	0.4772	0.4778	0.4783	0.4788	0.4793	0.4798	0.4803	0.4808	0.4812	0.4817
2.10	0.4821	0.4826	0.4830	0.4834	0.4838	0.4842	0.4846	0.4850	0.4854	0.4857
2.20	0.4861	0.4864	0.4868	0.4871	0.4875	0.4878	0.4881	0.4884	0.4887	0.4890
2.30	0.4893	0.4896	0.4898	0.4901	0.4904	0.4906	0.4909	0.4911	0.4913	0.4916
2.40	0.4918	0.4920	0.4922	0.4925	0.4927	0.4929	0.4931	0.4932	0.4934	0.4936
2.50	0.4938	0.4940	0.4941	0.4943	0.4945	0.4946	0.4948	0.4949	0.4951	0.4952
2.60	0.4953	0.4955	0.4956	0.4957	0.4959	0.4960	0.4961	0.4962	0.4963	0.4964
2.70	0.4965	0.4966	0.4967	0.4968	0.4969	0.4970	0.4971	0.4972	0.4973	0.4974
2.80	0.4974	0.4975	0.4976	0.4977	0.4977	0.4978	0.4979	0.4979	0.4980	0.4981
2.90	0.4981	0.4982	0.4982	0.4983	0.4984	0.4984	0.4985	0.4985	0.4986	0.4986
3.00	0.4987	0.4987	0.4987	0.4988	0.4988	0.4989	0.4989	0.4989	0.4990	0.4990
3.10	0.4990	0.4991	0.4991	0.4991	0.4992	0.4992	0.4992	0.4992	0.4993	0.4993
3.20	0.4993	0.4993	0.4994	0.4994	0.4994	0.4994	0.4994	0.4995	0.4995	0.4995
3.30	0.4995	0.4995	0.4995	0.4996	0.4996	0.4996	0.4996	0.4996	0.4996	0.4997
3.40	0.4997	0.4997	0.4997	0.4997	0.4997	0.4997	0.4997	0.4997	0.4997	0.4998
3.50	0.4998	0.4998	0.4998	0.4998	0.4998	0.4998	0.4998	0.4998	0.4998	0.4998
3.60	0.4998	0.4998	0.4999	0.4999	0.4999	0.4999	0.4999	0.4999	0.4999	0.4999
3.70	0.4999	0.4999	0.4999	0.4999	0.4999	0.4999	0.4999	0.4999	0.4999	0.4999
3.80	0.4999	0.4999	0.4999	0.4999	0.4999	0.4999	0.4999	0.4999	0.4999	0.4999
3.90	0.5000	0.5000	0.5000	0.5000	0.5000	0.5000	0.5000	0.5000	0.5000	0.5000

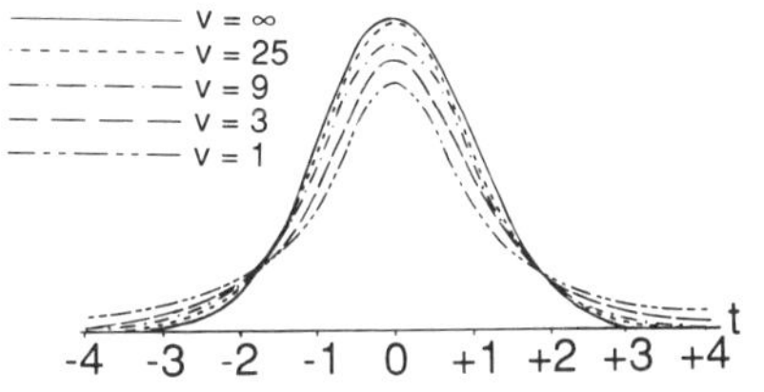

TABLE 9.1-A

t DISTRIBUTION FOR SELECTED DEGREES OF FREEDOM AND SELECTED CUMULATIVE PROBABILITY LEVELS (ONE-TAIL TEST)

Degrees of	Cumulative probabilities $P(T \le t)$				
freedom	0.9000	0.9500	0.9750	0.9900	0.9950
1	3.0777	6.3138	12.7062	31.8206	63.6570
2	1.8856	2.9200	4.3027	6.9646	9.9248
3	1.6378	2.3534	3.1825	4.5407	5.8410
4	1.5332	2.1318	2.7764	3.7470	4.6041
5	1.4759	2.0151	2.5706	3.3649	4.0321
6	1.4398	1.9432	2.4469	3.1427	3.7075
7	1.4149	1.8946	2.3646	2.9980	3.4995
8	1.3968	1.8595	2.3060	2.8965	3.3554
9	1.3830	1.8331	2.2622	2.8215	3.2498
10	1.3722	1.8124	2.2281	2.7638	3.1693
11	1.3634	1.7959	2.2010	2.7181	3.1058
12	1.3562	1.7823	2.1788	2.6810	3.0545
13	1.3502	1.7709	2.1604	2.6503	3.0123
14	1.3450	1.7613	2.1448	2.6245	2.9768
15	1.3406	1.7531	2.1315	2.6025	2.9467
16	1.3368	1.7459	2.1199	2.5835	2.9208
17	1.3334	1.7396	2.1098	2.5669	2.8982
18	1.3304	1.7341	2.1009	2.5524	2.8784
19	1.3278	1.7291	2.0930	2.5395	2.8610
20	1.3253	1.7247	2.0860	2.5280	2.8453
21	1.3232	1.7207	2.0796	2.5176	2.8314
22	1.3213	1.7171	2.0739	2.5083	2.8187
23	1.3194	1.7139	2.0687	2.4999	2.8074
24	1.3178	1.7109	2.0639	2.4921	2.7969
25	1.3164	1.7081	2.0595	2.4851	2.7874
26	1.3150	1.7056	2.0556	2.4786	2.7787
27	1.3137	1.7033	2.0519	2.4727	2.7707
28	1.3125	1.7011	2.0484	2.4671	2.7633
29	1.3114	1.6991	2.0452	2.4620	2.7564
30	1.3104	1.6972	2.0423	2.4573	2.7500
40	1.3031	1.6839	2.0211	2.4232	2.7045
50	1.2987	1.6759	2.0085	2.4033	2.6778
60	1.2958	1.6707	2.0003	2.3902	2.6604
70	1.2938	1.6669	1.9944	2.3808	2.6480
80	1.2922	1.6641	1.9901	2.3739	2.6387
90	1.2910	1.6620	1.9867	2.3685	2.6316
100	1.2901	1.6602	1.9840	2.3642	2.6259
∞	1.2816	1.6449	1.9600	2.3263	2.5758

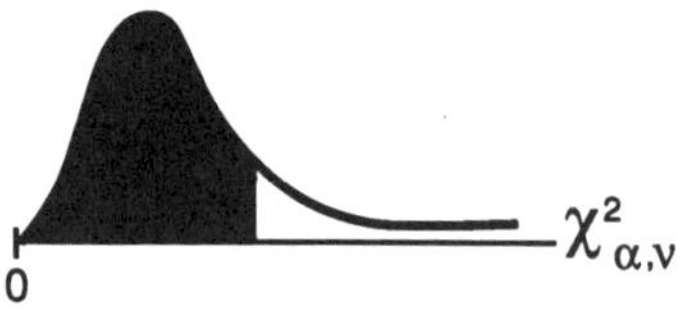

TABLE 10.1-A

CHI SQUARE DISTRIBUTION FOR SELECTED DEGREES OF FREEDOM AND SELECTED CUMULATIVE PROBABILITY LEVELS

df	0.005	0.010	0.025	0.050	0.100	0.900	0.950	0.975	0.990	0.995
1	0.0000	0.0002	0.0010	0.0039	0.0158	2.7055	3.8415	5.0239	6.6349	7.8795
2	0.0100	0.0201	0.0506	0.1026	0.2107	4.6052	5.9915	7.3778	9.2103	10.5966
3	0.0717	0.1148	0.2158	0.3518	0.5844	6.2514	7.8147	9.3484	11.3449	12.8382
4	0.2070	0.2971	0.4844	0.7107	1.0636	7.7794	9.4877	11.1433	13.2767	14.8603
5	0.4117	0.5543	0.8312	1.1455	1.6103	9.2364	11.0705	12.8325	15.0863	16.7496
6	0.6757	0.8721	1.2373	1.6354	2.2041	10.6446	12.5916	14.4494	16.8119	18.5476
7	0.9893	1.2390	1.6899	2.1674	2.8331	12.0170	14.0671	16.0128	18.4753	20.2778
8	1.3444	1.6465	2.1797	2.7326	3.4895	13.3616	15.5073	17.5346	20.0902	21.9550
9	1.7349	2.0879	2.7004	3.3251	4.1682	14.6837	16.9190	19.0228	21.6660	23.5893
10	2.1559	2.5582	3.2470	3.9403	4.8652	15.9872	18.3070	20.4832	23.2093	25.1882
11	2.6032	3.0535	3.8157	4.5748	5.5778	17.2750	19.6751	21.9201	24.7250	26.7569
12	3.0738	3.5706	4.4038	5.2260	6.3038	18.5493	21.0261	23.3367	26.2170	28.2996
13	3.5650	4.1069	5.0088	5.8919	7.0415	19.8119	22.3620	24.7356	27.6882	29.8194
14	4.0747	4.6604	5.6287	6.5706	7.7895	21.0641	23.6848	26.1190	29.1413	31.3194
15	4.6009	5.2293	6.2621	7.2609	8.5468	22.3071	24.9958	27.4884	30.5779	32.8013
16	5.1422	5.8122	6.9077	7.9616	9.3122	23.5418	26.2963	28.8454	32.0001	34.2674
17	5.6972	6.4078	7.5642	8.6718	10.0852	24.7690	27.5871	30.1910	33.4085	35.7182
18	6.2648	7.0149	8.2307	9.3905	10.8649	25.9894	28.8693	31.5264	34.8053	37.1564
19	6.8440	7.6327	8.9065	10.1170	11.6509	27.2036	30.1435	32.8523	36.1907	38.5820
20	7.4338	8.2604	9.5908	10.8508	12.4426	28.4120	31.4104	34.1696	37.5662	39.9968
21	8.0337	8.8972	10.2829	11.5913	13.2396	29.6151	32.6706	35.4789	38.9322	41.4011
22	8.6427	9.5425	10.9823	12.3380	14.0415	30.8133	33.9245	36.7808	40.2895	42.7960
23	9.2604	10.1957	11.6886	13.0905	14.8480	32.0069	35.1724	38.0755	41.6382	44.1808
24	9.8862	10.8564	12.4012	13.8484	15.6587	33.1962	36.4150	39.3641	42.9798	45.5586
25	10.5197	11.5240	13.1197	14.6114	16.4734	34.3816	37.6525	40.6466	44.3144	46.9285
26	11.1602	12.1981	13.8439	15.3792	17.2919	35.5632	38.8852	41.9233	45.6419	48.2903
27	11.8076	12.8785	14.5734	16.1514	18.1139	36.7412	40.1133	43.1946	46.9631	49.6452
28	12.4613	13.5647	15.3079	16.9279	18.9392	37.9159	41.3372	44.4608	48.2783	50.9936
29	13.1211	14.2565	16.0471	17.7084	19.7677	39.0875	42.5570	45.7224	49.5881	52.3360
30	13.7867	14.9535	16.7908	18.4927	20.5992	40.2560	43.7730	46.9793	50.8922	53.6720
31	14.4578	15.6555	17.5387	19.2806	21.4336	41.4218	44.9854	48.2320	52.1915	55.0030
32	15.1340	16.3622	18.2908	20.0719	22.2706	42.5847	46.1942	49.4804	53.4856	56.3278
33	15.8153	17.0735	19.0467	20.8665	23.1102	43.7451	47.3998	50.7250	54.7753	57.6479
34	16.5013	17.7891	19.8063	21.6643	23.9523	44.9031	48.6023	51.9659	56.0608	58.9636
35	17.1918	18.5089	20.5694	22.4650	24.7967	46.0588	49.8019	53.2035	57.3424	60.2755
36	17.8867	19.2327	21.3359	23.2686	25.6433	47.2122	50.9985	54.4373	58.6193	61.5813
37	18.5858	19.9602	22.1056	24.0749	26.4921	48.3635	52.1925	55.6683	59.8932	62.8847
38	19.2889	20.6914	22.8785	24.8839	27.3430	49.5127	53.3837	56.8958	61.1628	64.1827
39	19.9959	21.4262	23.6543	25.6954	28.1958	50.6598	54.5724	58.1203	62.4287	65.4767
40	20.7065	22.1643	24.4330	26.5093	29.0505	51.8051	55.7586	59.3420	63.6914	66.7673
41	21.4208	22.9056	25.2145	27.3256	29.9071	52.9485	56.9424	60.5606	64.9501	68.0528
42	22.1385	23.6501	25.9987	28.1441	30.7654	54.0902	58.1241	61.7769	66.2065	69.3365
43	22.8595	24.3976	26.7854	28.9647	31.6255	55.2301	59.3034	62.9902	67.4589	70.6151
44	23.5837	25.1480	27.5746	29.7875	32.4871	56.3685	60.4808	64.2014	68.7093	71.8921

CHI SQUARE DISTRIBUTION, Continued

df	0.005	0.010	0.025	0.050	0.100	0.900	0.950	0.975	0.990	0.995
45	24.3110	25.9013	28.3662	30.6123	33.3504	57.5053	61.6563	65.4102	69.9570	73.1663
46	25.0413	26.6572	29.1600	31.4390	34.2152	58.6405	62.8295	66.6162	71.2007	74.4353
47	25.7746	27.4158	29.9562	32.2676	35.0814	59.7742	64.0010	67.8204	72.4428	75.7031
48	26.5106	28.1770	30.7545	33.0981	35.9491	60.9065	65.1706	69.0223	73.6820	76.9675
49	27.2493	28.9406	31.5549	33.9303	36.8182	62.0375	66.3386	70.2224	74.9195	78.2307
50	27.9907	29.7067	32.3574	34.7643	37.6886	63.1671	67.5048	71.4201	76.1537	79.4896
51	28.7347	30.4750	33.1618	35.5999	38.5604	64.2954	68.6693	72.6160	77.3860	80.7468
52	29.4812	31.2457	33.9681	36.4371	39.4334	65.4224	69.8321	73.8098	78.6156	82.0005
53	30.2300	32.0185	34.7763	37.2759	40.3076	66.5482	70.9934	75.0018	79.8432	83.2522
54	30.9813	32.7935	35.5863	38.1162	41.1830	67.6729	72.1534	76.1924	81.0697	84.5036
55	31.7348	33.5705	36.3981	38.9580	42.0596	68.7962	73.3115	77.3805	82.2923	85.7493
56	32.4905	34.3495	37.2116	39.8013	42.9373	69.9187	74.4686	78.5677	83.5146	86.9960
57	33.2484	35.1305	38.0267	40.6459	43.8162	71.0397	75.6238	79.7523	84.7330	88.2368
58	34.0084	35.9134	38.8435	41.4919	44.6960	72.1596	76.7774	80.9349	85.9486	89.4738
59	34.7704	36.6983	39.6619	42.3393	45.5770	73.2791	77.9309	82.1181	87.1674	90.7185
60	35.5345	37.4849	40.4818	43.1880	46.4589	74.3972	79.0823	83.2984	88.3810	91.9547
61	36.3005	38.2732	41.3031	44.0379	47.3418	75.5139	80.2318	84.4758	89.5898	93.1833
62	37.0684	39.0633	42.1260	44.8890	48.2257	76.6300	81.3807	85.6531	90.8001	94.4159
63	37.8382	39.8551	42.9503	45.7414	49.1106	77.7456	82.5291	86.8302	92.0115	95.6520
64	38.6098	40.6486	43.7760	46.5949	49.9963	78.8597	83.6753	88.0042	93.2172	96.8787
65	39.3831	41.4436	44.6030	47.4496	50.8829	79.9728	84.8203	89.1764	94.4204	98.1019
66	40.1582	42.2402	45.4314	48.3054	51.7705	81.0856	85.9650	90.3492	95.6263	99.3315
67	40.9350	43.0384	46.2610	49.1623	52.6589	82.1974	87.1085	91.5202	96.8298	100.5579
68	41.7135	43.8380	47.0920	50.0202	53.5481	83.3078	88.2501	92.6884	98.0280	101.7751
69	42.4935	44.6392	47.9242	50.8792	54.4381	84.4180	89.3914	93.8568	99.2284	102.9978
70	43.2751	45.4417	48.7575	51.7393	55.3289	85.5269	90.5310	95.0227	100.4242	104.2130
71	44.0583	46.2456	49.5921	52.6003	56.2205	86.6346	91.6688	96.1860	101.6153	105.4204
72	44.8430	47.0510	50.4279	53.4623	57.1129	87.7426	92.8074	97.3515	102.8128	106.6410
73	45.6292	47.8577	51.2648	54.3253	58.0060	88.8492	93.9441	98.5140	104.0046	107.8518
74	46.4169	48.6656	52.1028	55.1891	58.8999	89.9550	95.0796	99.6749	105.1942	109.0595
75	47.2060	49.4750	52.9419	56.0540	59.7945	91.0614	96.2165	100.8391	106.3924	110.2845
76	47.9965	50.2856	53.7821	56.9198	60.6898	92.1661	97.3509	101.9992	107.5823	111.4950
77	48.7883	51.0974	54.6233	57.7864	61.5858	93.2692	98.4826	103.1548	108.7632	112.6892
78	49.5815	51.9104	55.4655	58.6539	62.4824	94.3725	99.6151	104.3126	109.9503	113.8963
79	50.3761	52.7246	56.3089	59.5222	63.3798	95.4757	100.7478	105.4712	111.1404	115.1097
80	51.1719	53.5401	57.1532	60.3914	64.2778	96.5782	101.8794	106.6284	112.3284	116.3203
81	51.9690	54.3566	57.9984	61.2614	65.1764	97.6790	103.0085	107.7816	113.5083	117.5165
82	52.7673	55.1743	58.8446	62.1322	66.0757	98.7798	104.1378	108.9355	114.6908	118.7184
83	53.5668	55.9930	59.6917	63.0038	66.9755	99.8793	105.2651	110.0864	115.8674	119.9100
84	54.3676	56.8130	60.5398	63.8763	67.8761	100.9799	106.3946	111.2418	117.0556	121.1244
85	55.1696	57.6339	61.3888	64.7494	68.7771	102.0789	107.5217	112.3933	118.2355	122.3242
86	55.9727	58.4559	62.2386	65.6233	69.6788	103.1772	108.6478	113.5435	119.4137	123.5213
87	56.7769	59.2789	63.0893	66.4978	70.5810	104.2744	109.7722	114.6908	120.5862	124.7085
88	57.5823	60.1030	63.9409	67.3732	71.4838	105.3722	110.8979	115.8413	121.7668	125.9119
89	58.3887	60.9280	64.7933	68.2493	72.3871	106.4687	112.0216	116.9883	122.9405	127.1031
90	59.1963	61.7540	65.6465	69.1259	73.2910	107.5642	113.1438	118.1331	124.1099	128.2867
91	60.0049	62.5810	66.5006	70.0034	74.1955	108.6601	114.2671	119.2804	125.2860	129.4841
92	60.8145	63.4090	67.3555	70.8816	75.1005	109.7555	115.3895	120.4266	126.4605	130.6790
93	61.6252	64.2378	68.2112	71.7603	76.0058	110.8490	116.5089	121.5675	127.6238	131.8534
94	62.4369	65.0676	69.0676	72.6397	76.9118	111.9430	117.6295	122.7111	128.7941	133.0417
95	63.2496	65.8983	69.9248	73.5198	77.8184	113.0373	118.7509	123.8567	129.9698	134.2410
96	64.0633	66.7299	70.7828	74.4005	78.7254	114.1304	119.8704	124.9991	131.1389	135.4288
97	64.8780	67.5624	71.6415	75.2818	79.6329	115.2231	120.9894	126.1411	132.3080	136.6170
98	65.6935	68.3957	72.5009	76.1638	80.5408	116.3150	122.1071	127.2809	133.4731	137.7983
99	66.5101	69.2299	73.3611	77.0463	81.4492	117.4066	123.2246	128.4209	134.6392	138.9821
100	67.3275	70.0648	74.2219	77.9294	82.3581	118.4973	124.3407	129.5586	135.8008	140.1584

TABLE 10.2-A

KOLMOGOROV-SMIRNOV DISTRIBUTION
FOR SELECTED SAMPLE SIZES
AND SELECTED CUMULATIVE PROBABILITY LEVELS
(ONE SAMPLE; TWO-TAIL TEST)

Sample Size n	0.80	0.85	0.90	0.95	0.99
1	0.900	0.925	0.950	0.975	0.995
2	0.684	0.726	0.776	0.842	0.929
3	0.565	0.597	0.642	0.708	0.828
4	0.494	0.525	0.564	0.624	0.733
5	0.446	0.474	0.510	0.565	0.669
6	0.410	0.436	0.470	0.521	0.618
7	0.381	0.405	0.438	0.486	0.577
8	0.358	0.381	0.411	0.457	0.543
9	0.339	0.360	0.388	0.432	0.514
10	0.322	0.342	0.368	0.410	0.490
11	0.307	0.326	0.352	0.391	0.468
12	0.295	0.313	0.338	0.375	0.450
13	0.284	0.302	0.325	0.361	0.433
14	0.274	0.292	0.314	0.349	0.418
15	0.266	0.283	0.304	0.338	0.404
16	0.258	0.274	0.295	0.328	0.392
17	0.250	0.266	0.286	0.318	0.381
18	0.244	0.259	0.278	0.309	0.371
19	0.237	0.252	0.272	0.301	0.363
20	0.231	0.246	0.264	0.294	0.356
25	0.21	0.22	0.24	0.27	0.32
30	0.19	0.20	0.22	0.24	0.29
35	0.18	0.19	0.21	0.23	0.27
Over 35	$1.07/\sqrt{n}$	$1.14/\sqrt{n}$	$1.22/\sqrt{n}$	$1.36/\sqrt{n}$	$1.63/\sqrt{n}$

Source: Adapted from F. Massey, 1951, "The Kolmogorov-Smirnov Test for Goodness of Fit," *Journal of the American Statistical Association*, 46: 70; with the permission of the author and the publisher.

TABLE 10.3-A

RUN STATISTIC CRITICAL VALUES FOR SMALL SAMPLE SIZES AND A CRITICAL REGION OF 5% (TWO-TAIL TEST)

The smaller of n_1 and n_2	The larger of n_1 and n_2 5	6	7	8	9	10	11	12	13	14	15	16	17	18	19	20
2								2	2	2	2	2	2	2	2	2
								6	6	6	6	6	6	6	6	6
3		2	2	2	2	2	2	2	2	2	3	3	3	3	3	3
		8	8	8	8	8	8	8	8	8	8	8	8	8	8	8
4	2	2	2	3	3	3	3	3	3	3	3	4	4	4	4	4
	9	9	10	10	10	10	10	10	10	10	10	10	10	10	10	10
5	2	3	3	3	3	3	4	4	4	4	4	4	4	5	5	5
	10	10	11	11	12	12	12	12	12	12	12	12	12	12	12	12
6		3	3	3	4	4	4	4	5	5	5	5	5	5	6	6
		11	12	12	13	13	13	13	14	14	14	14	14	14	14	14
7			3	4	4	5	5	5	5	5	6	6	6	6	6	6
			13	13	14	14	14	14	15	15	15	16	16	16	16	16
8				4	5	5	5	6	6	6	6	6	7	7	7	7
				14	14	15	15	16	16	16	16	17	17	17	17	17
9					5	5	6	6	6	7	7	7	7	8	8	8
					15	16	16	16	17	17	18	18	18	18	18	18
10						6	6	7	7	7	7	8	8	8	8	9
						16	17	17	18	18	18	19	19	19	20	20
11							7	7	7	8	8	8	9	9	9	9
							17	18	19	19	19	20	20	20	21	21
12								7	8	8	8	9	9	9	10	10
								19	19	20	20	21	21	21	22	22
13									8	9	9	9	10	10	10	10
									20	20	21	21	22	22	23	23

RUN STATISTIC CRITICAL VALUES, Continued

The smaller of n_1 and n_2	The larger of n_1 and n_2															
	5	6	7	8	9	10	11	12	13	14	15	16	17	18	19	20
14										9	9	10	10	10	11	11
										21	22	22	23	23	23	24
15											10	10	11	11	11	12
											22	23	23	24	24	25
16												11	11	11	12	12
												23	24	25	25	25
17													11	12	12	13
													25	25	26	26
18														12	13	13
														26	26	27
19															13	13
															27	27
20																14
																28

Source: Adapted from C. Eisenhart and F. Swed, 1943, "Tables for Testing Randomness of Grouping in a Sequence of Alternatives", *The Annals of Statistics*, 14: 66-87; with the permission of the Institute of Mathematical Statistics.

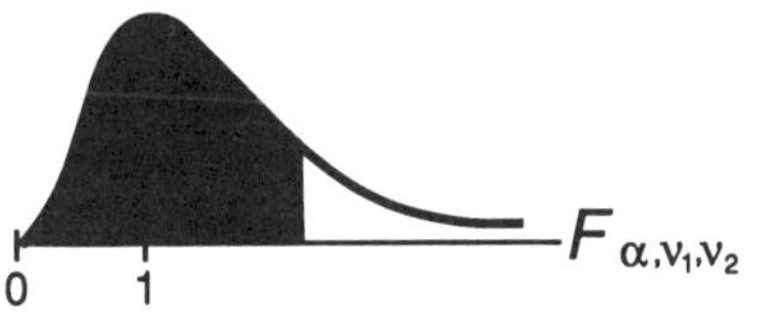

TABLE 11.1-A

F DISTRIBUTION FOR SELECTED DEGREES OF FREEDOM AND SELECTED CUMULATIVE PROBABILITY LEVELS (ONE-TAIL TEST)

df	1 − α	1	2	3	4	5	6	7	8
1	0.900	39.8634	49.5000	53.5932	55.8329	57.2401	58.2044	58.9059	59.4389
	0.950	161.4475	199.4999	215.7073	224.5832	230.1617	233.9859	236.7681	238.8823
	0.975	647.7870	799.4977	864.1609	899.5812	921.8446	937.1086	948.2143	956.6523
	0.990	4052.1350	4999.4487	5403.2935	5624.5195	5763.5850	5858.9238	5928.2920	5980.9927
	0.995	16210.3525	19999.0605	21614.2441	22499.0957	23055.2617	23436.6035	23714.0508	23924.8438
2	0.900	8.5263	9.0000	9.1618	9.2434	9.2926	9.3255	9.3491	9.3668
	0.950	18.5128	19.0000	19.1643	19.2468	19.2964	19.3295	19.3532	19.3710
	0.975	38.5063	38.9999	39.1654	39.2484	39.2982	39.3314	39.3551	39.3730
	0.990	98.5020	98.9995	99.1657	99.2488	99.2988	99.3321	99.3558	99.3737
	0.995	198.4991	198.9978	199.1642	199.2476	199.2974	199.3307	199.3545	199.3724
3	0.900	5.5383	5.4624	5.3910	5.3426	5.3090	5.2848	5.2657	5.2520
	0.950	10.1280	9.5521	9.2761	9.1169	9.0129	8.9419	8.8856	8.8450
	0.975	17.4434	16.0441	15.4395	15.1029	14.8840	14.7374	14.6274	14.5421
	0.990	34.1161	30.8164	29.4606	28.7088	28.2387	27.9075	27.6723	27.5048
	0.995	55.5515	49.7989	47.4555	46.1732	45.4171	44.8536	44.4028	44.1064
4	0.900	4.5448	4.3246	4.1910	4.1074	4.0507	4.0099	3.9792	3.9549
	0.950	7.7086	6.9443	6.5913	6.3881	6.2559	6.1627	6.0939	6.0412
	0.975	12.2178	10.6491	9.9789	9.6037	9.3636	9.1969	9.0742	8.9801
	0.990	21.1976	17.9999	16.6927	15.9760	15.5202	15.2064	14.9773	14.8014
	0.995	31.3326	26.2841	24.2579	23.1518	22.4552	21.9769	21.6207	21.3526
5	0.900	4.0604	3.7797	3.6196	3.5201	3.4529	3.4044	3.3678	3.3392
	0.950	6.6079	5.7861	5.4097	5.1920	5.0505	4.9504	4.8761	4.8186
	0.975	10.0070	8.4336	7.7637	7.3882	7.1467	6.9781	6.8536	6.7572
	0.990	16.2581	13.2739	12.0596	11.3910	10.9668	10.6727	10.4566	10.2904
	0.995	22.7847	18.3137	16.5302	15.5558	14.9397	14.5140	14.2009	13.9593
6	0.900	3.7759	3.4633	3.2887	3.1808	3.1074	3.0546	3.0144	2.9831
	0.950	5.9874	5.1433	4.7572	4.5337	4.3872	4.2837	4.2065	4.1468
	0.975	8.8131	7.2599	6.5986	6.2274	5.9876	5.8199	5.6956	5.6000
	0.990	13.7450	10.9247	9.7791	9.1488	8.7454	8.4657	8.2597	8.1022
	0.995	18.6349	14.5440	12.9157	12.0284	11.4629	11.0726	10.7863	10.5662
7	0.900	3.5894	3.2574	3.0740	2.9605	2.8833	2.8273	2.7850	2.7515
	0.950	5.5914	4.7374	4.3468	4.1204	3.9716	3.8660	3.7870	3.7256
	0.975	8.0727	6.5415	5.8900	5.5228	5.2852	5.1187	4.9952	4.8995
	0.990	12.2464	9.5466	8.4511	7.8469	7.4601	7.1918	6.9927	6.8403
	0.995	16.2355	12.4039	10.8828	10.0510	9.5220	9.1551	8.8855	8.6784
8	0.900	3.4579	3.1131	2.9239	2.8064	2.7264	2.6683	2.6242	2.5893
	0.950	5.3177	4.4590	4.0662	3.8377	3.6874	3.5806	3.5004	3.4382
	0.975	7.5709	6.0595	5.4158	5.0528	4.8174	4.6519	4.5285	4.4330
	0.990	11.2586	8.6491	7.5909	7.0062	6.6315	6.3706	6.1775	6.0291
	0.995	14.6881	11.0424	9.5961	8.8052	8.3020	7.9522	7.6942	7.4963

F DISTRIBUTION, Continued

df	1 - α	1	2	3	4	5	6	7	8
9	0.900	3.3603	3.0065	2.8129	2.6926	2.6106	2.5508	2.5053	2.4695
	0.950	5.1174	4.2565	3.8625	3.6331	3.4817	3.3737	3.2928	3.2295
	0.975	7.2093	5.7147	5.0781	4.7181	4.4843	4.3197	4.1970	4.1019
	0.990	10.5614	8.0215	6.9920	6.4221	6.0567	5.8015	5.6127	5.4670
	0.995	13.6136	10.1067	8.7172	7.9556	7.4714	7.1341	6.8850	6.6930
10	0.900	3.2850	2.9245	2.7277	2.6054	2.5217	2.4606	2.4140	2.3771
	0.950	4.9646	4.1028	3.7082	3.4781	3.3259	3.2173	3.1354	3.0717
	0.975	6.9367	5.4564	4.8257	4.4682	4.2360	4.0720	3.9497	3.8548
	0.990	10.0443	7.5594	6.5521	5.9942	5.6362	5.3859	5.2003	5.0566
	0.995	12.8264	9.4270	8.0810	7.3429	6.8721	6.5443	6.3028	6.1159
11	0.900	3.2252	2.8595	2.6602	2.5362	2.4512	2.3890	2.3416	2.3040
	0.950	4.8443	3.9823	3.5875	3.3567	3.2039	3.0946	3.0123	2.9481
	0.975	6.7241	5.2559	4.6299	4.2750	4.0441	3.8806	3.7587	3.6638
	0.990	9.6460	7.2057	6.2167	5.6683	5.3162	5.0692	4.8859	4.7445
	0.995	12.2263	8.9122	7.6004	6.8807	6.4219	6.1018	5.8646	5.6824
12	0.900	3.1765	2.8068	2.6055	2.4801	2.3941	2.3310	2.2828	2.2446
	0.950	4.7472	3.8853	3.4904	3.2591	3.1058	2.9962	2.9134	2.8485
	0.975	6.5538	5.0959	4.4742	4.1213	3.8911	3.7283	3.6066	3.5118
	0.990	9.3302	6.9266	5.9525	5.4119	5.0644	4.8205	4.6397	4.4995
	0.995	11.7542	8.5096	7.2255	6.5212	6.0711	5.7569	5.5244	5.3450
13	0.900	3.1362	2.7632	2.5602	2.4337	2.3468	2.2830	2.2341	2.1954
	0.950	4.6672	3.8056	3.4106	3.1791	3.0255	2.9152	2.8320	2.7669
	0.975	6.4143	4.9653	4.3472	3.9960	3.7666	3.6042	3.4828	3.3879
	0.990	9.0738	6.7010	5.7395	5.2053	4.8615	4.6204	4.4411	4.3021
	0.995	11.3735	8.1865	6.9256	6.2334	5.7911	5.4821	5.2531	5.0760
14	0.900	3.1022	2.7265	2.5223	2.3947	2.3070	2.2425	2.1932	2.1539
	0.950	4.6001	3.7389	3.3440	3.1123	2.9582	2.8478	2.7643	2.6986
	0.975	6.2979	4.8567	4.2416	3.8920	3.6635	3.5013	3.3799	3.2854
	0.990	8.8616	6.5149	5.5638	5.0354	4.6949	4.4559	4.2778	4.1400
	0.995	11.0602	7.9216	6.6805	5.9986	5.5624	5.2573	5.0312	4.8566
15	0.900	3.0732	2.6952	2.4897	2.3614	2.2730	2.2081	2.1581	2.1186
	0.950	4.5431	3.6823	3.2874	3.0556	2.9012	2.7905	2.7066	2.6407
	0.975	6.1995	4.7650	4.1527	3.8043	3.5765	3.4147	3.2933	3.1988
	0.990	8.6831	6.3589	5.4169	4.8933	4.5556	4.3183	4.1415	4.0046
	0.995	10.7980	7.7007	6.4760	5.8028	5.3722	5.0706	4.8472	4.6742
16	0.900	3.0481	2.6682	2.4619	2.3327	2.2438	2.1784	2.1281	2.0880
	0.950	4.4940	3.6337	3.2388	3.0069	2.8525	2.7413	2.6572	2.5910
	0.975	6.1151	4.6867	4.0768	3.7293	3.5022	3.3407	3.2194	3.1249
	0.990	8.5309	6.2262	5.2923	4.7725	4.4375	4.2017	4.0259	3.8897
	0.995	10.5754	7.5138	6.3034	5.6378	5.2117	4.9134	4.6921	4.5205
17	0.900	3.0262	2.6446	2.4374	2.3077	2.2182	2.1524	2.1017	2.0613
	0.950	4.4513	3.5915	3.1967	2.9647	2.8100	2.6987	2.6142	2.5479
	0.975	6.0420	4.6189	4.0111	3.6649	3.4379	3.2766	3.1557	3.0611
	0.990	8.3997	6.1121	5.1850	4.6690	4.3359	4.1016	3.9267	3.7909
	0.995	10.3841	7.3536	6.1557	5.4967	5.0744	4.7791	4.5593	4.3892
18	0.900	3.0070	2.6239	2.4160	2.2857	2.1959	2.1296	2.0785	2.0379
	0.950	4.4139	3.5546	3.1600	2.9277	2.7728	2.6613	2.5767	2.5101
	0.975	5.9780	4.5597	3.9540	3.6083	3.3819	3.2209	3.0999	3.0053
	0.990	8.2854	6.0129	5.0920	4.5791	4.2480	4.0146	3.8407	3.7055
	0.995	10.2181	7.2148	6.0277	5.3747	4.9561	4.6628	4.4449	4.2759
19	0.900	2.9899	2.6056	2.3970	2.2663	2.1759	2.1094	2.0580	2.0171
	0.950	4.3807	3.5219	3.1273	2.8951	2.7401	2.6283	2.5435	2.4768
	0.975	5.9216	4.5075	3.9033	3.5586	3.3328	3.1718	3.0509	2.9562
	0.990	8.1849	5.9259	5.0104	4.5003	4.1708	3.9385	3.7652	3.6305
	0.995	10.0725	7.0934	5.9160	5.2681	4.8527	4.5613	4.3447	4.1771

F DISTRIBUTION, Continued

df	1 - α	1	2	3	4	5	6	7	8
20	0.900	2.9747	2.5893	2.3800	2.2490	2.1582	2.0913	2.0396	1.9986
	0.950	4.3512	3.4928	3.0985	2.8661	2.7109	2.5989	2.5140	2.4471
	0.975	5.8715	4.4613	3.8588	3.5146	3.2890	3.1283	3.0074	2.9127
	0.990	8.0959	5.8489	4.9382	4.4306	4.1027	3.8715	3.6987	3.5644
	0.995	9.9439	6.9864	5.8176	5.1743	4.7616	4.4721	4.2569	4.0898
25	0.900	2.9177	2.5283	2.3170	2.1843	2.0921	2.0241	1.9713	1.9292
	0.950	4.2417	3.3852	2.9912	2.7586	2.6030	2.4903	2.4047	2.3370
	0.975	5.6864	4.2909	3.6943	3.3530	3.1287	2.9685	2.8479	2.7530
	0.990	7.7698	5.5680	4.6756	4.1773	3.8549	3.6271	3.4568	3.3239
	0.995	9.4753	6.5982	5.4616	4.8351	4.4328	4.1500	3.9394	3.7759
50	0.900	2.8086	2.4119	2.1967	2.0608	1.9660	1.8955	1.8404	1.7963
	0.950	4.0343	3.1826	2.7900	2.5572	2.4004	2.2864	2.1992	2.1300
	0.975	5.3403	3.9749	3.3902	3.0544	2.8326	2.6735	2.5531	2.4579
	0.990	7.1706	5.0566	4.1993	3.7195	3.4077	3.1864	3.0202	2.8901
	0.995	8.6257	5.9016	4.8259	4.2316	3.8486	3.5786	3.3764	3.2189
100	0.900	2.7564	2.3564	2.1394	2.0019	1.9057	1.8339	1.7778	1.7324
	0.950	3.9361	3.0873	2.6955	2.4626	2.3053	2.1906	2.1025	2.0323
	0.975	5.1786	3.8284	3.2496	2.9166	2.6961	2.5374	2.4168	2.3215
	0.990	6.8953	4.8239	3.9837	3.5127	3.2059	2.9877	2.8233	2.6943
	0.995	8.2406	5.5892	4.5424	3.9634	3.5895	3.3252	3.1271	2.9722
∞	0.900	2.7055	2.3026	2.0838	1.9449	1.8473	1.7741	1.7167	1.6702
	0.950	3.8414	2.9957	2.6049	2.3719	2.2141	2.0986	2.0096	1.9384
	0.975	5.0238	3.6889	3.1162	2.7858	2.5665	2.4082	2.2875	2.1918
	0.990	6.6348	4.6052	3.7816	3.3191	3.0172	2.8020	2.6393	2.5112
	0.995	7.8797	5.2983	4.2794	3.7152	3.3499	3.0912	2.8968	2.7444

F DISTRIBUTION, Continued

df	1 - α	9	10	11	12	13	14	15	16
1	0.900	59.8576	60.1950	60.4725	60.7051	60.9028	61.0724	61.2204	61.3497
	0.950	240.5433	241.8820	242.9829	243.9056	244.6901	245.3629	245.9504	246.4634
	0.975	963.2825	968.6258	973.0213	976.7036	979.8357	982.5214	984.8649	986.9140
	0.990	6022.4160	6055.7900	6083.2393	6106.2515	6125.8130	6142.5864	6157.2251	6170.0220
	0.995	24090.5059	24223.9824	24333.7832	24425.7969	24504.0391	24571.1289	24629.6953	24680.8789
2	0.900	9.3805	9.3916	9.4006	9.4081	9.4145	9.4200	9.4247	9.4288
	0.950	19.3848	19.3959	19.4049	19.4125	19.4189	19.4244	19.4291	19.4333
	0.975	39.3869	39.3979	39.4070	39.4146	39.4209	39.4264	39.4312	39.4353
	0.990	99.3876	99.3987	99.4076	99.4154	99.4216	99.4272	99.4320	99.4360
	0.995	199.3864	199.3974	199.4063	199.4142	199.4203	199.4258	199.4309	199.4347
3	0.900	5.2402	5.2297	5.2226	5.2159	5.2104	5.2047	5.1995	5.1961
	0.950	8.8127	8.7875	8.7611	8.7448	8.7261	8.7164	8.7013	8.6943
	0.975	14.4720	14.4217	14.3716	14.3330	14.3043	14.2709	14.2591	14.2266
	0.990	27.3365	27.2169	27.1195	27.0339	26.9946	26.9228	26.8530	26.8488
	0.995	43.8433	43.7331	43.4811	43.3825	43.2906	43.1966	43.1296	43.0041
4	0.900	3.9359	3.9197	3.9068	3.8958	3.8859	3.8776	3.8702	3.8640
	0.950	5.9989	5.9650	5.9351	5.9109	5.8902	5.8727	5.8578	5.8441
	0.975	8.9047	8.8451	8.7934	8.7519	8.7167	8.6841	8.6566	8.6326
	0.990	14.6591	14.5498	14.4522	14.3724	14.3099	14.2464	14.1945	14.1535
	0.995	21.1392	20.9733	20.8291	20.7021	20.5994	20.5078	20.4276	20.3718
5	0.900	3.3163	3.2972	3.2815	3.2682	3.2566	3.2468	3.2382	3.2304
	0.950	4.7724	4.7349	4.7042	4.6774	4.6551	4.6361	4.6186	4.6041
	0.975	6.6806	6.6196	6.5670	6.5254	6.4882	6.4558	6.4268	6.4034
	0.990	10.1577	10.0516	9.9630	9.8897	9.8265	9.7686	9.7221	9.6802
	0.995	13.7702	13.6211	13.4884	13.3850	13.2958	13.2153	13.1473	13.0863
6	0.900	2.9578	2.9369	2.9196	2.9046	2.8920	2.8810	2.8713	2.8627
	0.950	4.0992	4.0598	4.0272	4.0002	3.9765	3.9562	3.9383	3.9226
	0.975	5.5231	5.4611	5.4099	5.3666	5.3291	5.2970	5.2686	5.2443
	0.990	7.9755	7.8736	7.7901	7.7177	7.6575	7.6043	7.5594	7.5184
	0.995	10.3901	10.2488	10.1332	10.0346	9.9489	9.8766	9.8141	9.7573
7	0.900	2.7246	2.7026	2.6839	2.6682	2.6544	2.6428	2.6321	2.6230
	0.950	3.6765	3.6366	3.6030	3.5745	3.5504	3.5293	3.5106	3.4942
	0.975	4.8233	4.7610	4.7094	4.6661	4.6286	4.5957	4.5681	4.5424
	0.990	6.7187	6.6195	6.5386	6.4694	6.4102	6.3588	6.3138	6.2749
	0.995	8.5144	8.3797	8.2701	8.1758	8.0970	8.0278	7.9667	7.9156
8	0.900	2.5613	2.5380	2.5185	2.5020	2.4877	2.4751	2.4642	2.4546
	0.950	3.3881	3.3470	3.3128	3.2838	3.2589	3.2373	3.2183	3.2016
	0.975	4.3571	4.2954	4.2435	4.1995	4.1622	4.1294	4.1014	4.0762
	0.990	5.9106	5.8144	5.7341	5.6667	5.6088	5.5592	5.5154	5.4768
	0.995	7.3385	7.2111	7.1049	7.0152	6.9377	6.8715	6.8135	6.7632
9	0.900	2.4403	2.4163	2.3962	2.3788	2.3639	2.3509	2.3396	2.3296
	0.950	3.1789	3.1373	3.1024	3.0728	3.0476	3.0256	3.0060	2.9891
	0.975	4.0260	3.9639	3.9121	3.8680	3.8306	3.7978	3.7693	3.7443
	0.990	5.3513	5.2565	5.1779	5.1115	5.0549	5.0050	4.9621	4.9236
	0.995	6.5411	6.4175	6.3146	6.2274	6.1525	6.0888	6.0326	5.9824
10	0.900	2.3473	2.3226	2.3017	2.2841	2.2687	2.2553	2.2434	2.2330
	0.950	3.0203	2.9783	2.9430	2.9131	2.8871	2.8647	2.8450	2.8276
	0.975	3.7791	3.7167	3.6649	3.6211	3.5832	3.5504	3.5217	3.4963
	0.990	4.9427	4.8491	4.7712	4.7058	4.6494	4.6007	4.5579	4.5204
	0.995	5.9679	5.8470	5.7458	5.6617	5.5883	5.5260	5.4710	5.4221
11	0.900	2.2735	2.2483	2.2269	2.2087	2.1930	2.1792	2.1672	2.1563
	0.950	2.8963	2.8537	2.8179	2.7876	2.7614	2.7386	2.7186	2.7008
	0.975	3.5878	3.5257	3.4736	3.4295	3.3916	3.3587	3.3301	3.3045
	0.990	4.6315	4.5393	4.4625	4.3976	4.3415	4.2932	4.2508	4.2134
	0.995	5.5369	5.4183	5.3195	5.2363	5.1647	5.1029	5.0490	5.0010

F DISTRIBUTION, Continued

df	1 - α	9	10	11	12	13	14	15	16
12	0.900	2.2135	2.1878	2.1660	2.1474	2.1314	2.1173	2.1049	2.0939
	0.950	2.7963	2.7534	2.7173	2.6867	2.6602	2.6372	2.6168	2.5988
	0.975	3.4357	3.3735	3.3216	3.2774	3.2394	3.2062	3.1772	3.1514
	0.990	4.3874	4.2959	4.2197	4.1552	4.0999	4.0516	4.0098	3.9723
	0.995	5.2020	5.0857	4.9882	4.9060	4.8359	4.7746	4.7214	4.6743
13	0.900	2.1638	2.1376	2.1155	2.0966	2.0802	2.0659	2.0532	2.0419
	0.950	2.7143	2.6710	2.6346	2.6036	2.5770	2.5535	2.5331	2.5150
	0.975	3.3121	3.2498	3.1976	3.1532	3.1152	3.0818	3.0527	3.0270
	0.990	4.1911	4.1002	4.0245	3.9603	3.9052	3.8573	3.8152	3.7784
	0.995	4.9351	4.8198	4.7240	4.6430	4.5735	4.5129	4.4602	4.4133
14	0.900	2.1219	2.0955	2.0730	2.0537	2.0371	2.0224	2.0096	1.9980
	0.950	2.6458	2.6021	2.5655	2.5342	2.5072	2.4836	2.4630	2.4446
	0.975	3.2092	3.1469	3.0946	3.0502	3.0118	2.9785	2.9494	2.9235
	0.990	4.0297	3.9394	3.8641	3.8003	3.7451	3.6974	3.6558	3.6185
	0.995	4.7174	4.6032	4.5086	4.4279	4.3590	4.2995	4.2470	4.2004
15	0.900	2.0862	2.0594	2.0366	2.0171	2.0001	1.9853	1.9722	1.9605
	0.950	2.5876	2.5437	2.5068	2.4752	2.4481	2.4243	2.4035	2.3849
	0.975	3.1226	3.0603	3.0078	2.9632	2.9249	2.8915	2.8620	2.8361
	0.990	3.8949	3.8050	3.7298	3.6661	3.6115	3.5638	3.5221	3.4853
	0.995	4.5364	4.4235	4.3295	4.2496	4.1814	4.1220	4.0699	4.0235
16	0.900	2.0553	2.0281	2.0051	1.9854	1.9683	1.9532	1.9399	1.9281
	0.950	2.5376	2.4935	2.4563	2.4246	2.3972	2.3732	2.3523	2.3336
	0.975	3.0487	2.9862	2.9338	2.8891	2.8505	2.8170	2.7875	2.7614
	0.990	3.7804	3.6909	3.6161	3.5528	3.4982	3.4506	3.4088	3.3721
	0.995	4.3839	4.2719	4.1785	4.0993	4.0313	3.9723	3.9203	3.8746
17	0.900	2.0283	2.0009	1.9777	1.9578	1.9403	1.9251	1.9117	1.8997
	0.950	2.4942	2.4499	2.4125	2.3807	2.3531	2.3290	2.3078	2.2888
	0.975	2.9848	2.9222	2.8696	2.8250	2.7863	2.7527	2.7230	2.6968
	0.990	3.6824	3.5931	3.5186	3.4553	3.4008	3.3533	3.3117	3.2747
	0.995	4.2535	4.1424	4.0495	3.9710	3.9032	3.8446	3.7928	3.7472
18	0.900	2.0047	1.9770	1.9535	1.9333	1.9158	1.9005	1.8868	1.8747
	0.950	2.4563	2.4118	2.3742	2.3420	2.3142	2.2900	2.2687	2.2495
	0.975	2.9291	2.8664	2.8137	2.7688	2.7302	2.6964	2.6667	2.6404
	0.990	3.5972	3.5081	3.4338	3.3705	3.3162	3.2688	3.2272	3.1905
	0.995	4.1411	4.0305	3.9382	3.8599	3.7925	3.7341	3.6827	3.6373
19	0.900	1.9836	1.9557	1.9320	1.9117	1.8940	1.8785	1.8647	1.8524
	0.950	2.4226	2.3779	2.3402	2.3079	2.2801	2.2557	2.2341	2.2149
	0.975	2.8801	2.8172	2.7645	2.7195	2.6807	2.6470	2.6170	2.5906
	0.990	3.5225	3.4337	3.3595	3.2964	3.2421	3.1950	3.1532	3.1166
	0.995	4.0427	3.9329	3.8409	3.7631	3.6960	3.6378	3.5866	3.5413
20	0.900	1.9648	1.9367	1.9129	1.8923	1.8745	1.8588	1.8449	1.8325
	0.950	2.3929	2.3478	2.3099	2.2776	2.2494	2.2249	2.2033	2.1840
	0.975	2.8366	2.7736	2.7208	2.6759	2.6369	2.6030	2.5731	2.5465
	0.990	3.4566	3.3682	3.2942	3.2311	3.1768	3.1296	3.0880	3.0513
	0.995	3.9565	3.8471	3.7555	3.6780	3.6110	3.5531	3.5020	3.4568
25	0.900	1.8946	1.8657	1.8412	1.8200	1.8015	1.7852	1.7708	1.7579
	0.950	2.2820	2.2364	2.1979	2.1649	2.1362	2.1111	2.0888	2.0690
	0.975	2.6767	2.6135	2.5603	2.5149	2.4757	2.4412	2.4110	2.3839
	0.990	3.2171	3.1294	3.0558	2.9931	2.9389	2.8918	2.8501	2.8133
	0.995	3.6446	3.5370	3.4470	3.3705	3.3044	3.2468	3.1964	3.1515
50	0.900	1.7598	1.7292	1.7028	1.6801	1.6602	1.6426	1.6269	1.6127
	0.950	2.0734	2.0261	1.9860	1.9516	1.9214	1.8950	1.8713	1.8503
	0.975	2.3809	2.3168	2.2626	2.2162	2.1759	2.1404	2.1090	2.0810
	0.990	2.7849	2.6981	2.6250	2.5625	2.5084	2.4609	2.4190	2.3816
	0.995	3.0921	2.9874	2.8997	2.8247	2.7598	2.7031	2.6531	2.6086

F DISTRIBUTION, Continued

df	1 - α	9	10	11	12	13	14	15	16
100	0.900	1.6949	1.6632	1.6361	1.6124	1.5916	1.5731	1.5566	1.5417
	0.950	1.9748	1.9267	1.8857	1.8503	1.8194	1.7919	1.7676	1.7457
	0.975	2.2439	2.1793	2.1245	2.0774	2.0363	2.0001	1.9679	1.9391
	0.990	2.5898	2.5033	2.4301	2.3675	2.3132	2.2654	2.2230	2.1852
	0.995	2.8472	2.7439	2.6569	2.5825	2.5179	2.4614	2.4113	2.3665
∞	0.900	1.6315	1.5987	1.5705	1.5458	1.5240	1.5046	1.4871	1.4714
	0.950	1.8799	1.8307	1.7887	1.7522	1.7202	1.6918	1.6664	1.6435
	0.975	2.1136	2.0483	1.9927	1.9447	1.9027	1.8656	1.8326	1.8028
	0.990	2.4073	2.3209	2.2477	2.1848	2.1298	2.0815	2.0385	2.0000
	0.995	2.6210	2.5188	2.4324	2.3583	2.2938	2.2371	2.1868	2.1417

F DISTRIBUTION, Continued

df	1 − α	17	18	19	20	25	50	100	∞
1	0.900	61.4641	61.5664	61.6576	61.7402	62.0542	62.6876	63.0069	63.3273
	0.950	246.9171	247.3231	247.6849	248.0125	249.2591	251.7720	253.0393	254.3181
	0.975	988.7254	990.3468	991.7906	993.0981	998.0759	1008.1072	1013.1629	1018.2869
	0.990	6181.3423	6191.4609	6200.4868	6208.6514	6239.7300	6302.4102	6334.0132	6366.2332
	0.995	24726.1289	24766.6289	24802.6797	24835.3516	24959.7070	25210.3594	25336.6504	25474.3160
2	0.900	9.4325	9.4358	9.4387	9.4413	9.4513	9.4712	9.4812	9.4912
	0.950	19.4370	19.4402	19.4431	19.4458	19.4558	19.4756	19.4858	19.4959
	0.975	39.4390	39.4422	39.4451	39.4478	39.4580	39.4776	39.4879	39.4981
	0.990	99.4398	99.4429	99.4457	99.4486	99.4589	99.4781	99.4889	99.4985
	0.995	199.4386	199.4413	199.4440	199.4472	199.4581	199.4762	199.4877	199.4987
3	0.900	5.1939	5.1889	5.1857	5.1854	5.1741	5.1546	5.1444	5.1337
	0.950	8.6823	8.6740	8.6675	8.6625	8.6327	8.5810	8.5541	8.5265
	0.975	14.2170	14.1958	14.1842	14.1696	14.1141	14.0098	13.9565	13.9021
	0.990	26.8134	26.7696	26.6846	26.6724	26.5840	26.3540	26.2406	26.1257
	0.995	42.8968	42.9208	42.8683	42.6840	42.5050	42.2126	42.0220	41.8289
4	0.900	3.8588	3.8532	3.8480	3.8448	3.8277	3.7952	3.7781	3.7607
	0.950	5.8321	5.8220	5.8104	5.8028	5.7692	5.6995	5.6640	5.6281
	0.975	8.6091	8.5908	8.5746	8.5586	8.5023	8.3807	8.3194	8.2574
	0.990	14.1158	14.0768	14.0500	14.0238	13.9054	13.6895	13.5769	13.4631
	0.995	20.3113	20.2466	20.2144	20.1686	19.9856	19.6671	19.4967	19.3267
5	0.900	3.2236	3.2174	3.2115	3.2069	3.1873	3.1471	3.1263	3.1050
	0.950	4.5900	4.5787	4.5684	4.5582	4.5208	4.4444	4.4051	4.3650
	0.975	6.3813	6.3625	6.3434	6.3286	6.2679	6.1436	6.0800	6.0153
	0.990	9.6442	9.6079	9.5778	9.5520	9.4483	9.2378	9.1299	9.0205
	0.995	13.0314	12.9859	12.9372	12.9046	12.7582	12.4534	12.2995	12.1439
6	0.900	2.8549	2.8481	2.8422	2.8364	2.8150	2.7697	2.7463	2.7222
	0.950	3.9080	3.8958	3.8842	3.8738	3.8353	3.7531	3.7118	3.6689
	0.975	5.2213	5.2020	5.1847	5.1682	5.1071	4.9817	4.9154	4.8492
	0.990	7.4839	7.4505	7.4223	7.3965	7.2947	7.0917	6.9867	6.8801
	0.995	9.7103	9.6622	9.6268	9.5873	9.4481	9.1704	9.0257	8.8797
7	0.900	2.6147	2.6076	2.6009	2.5947	2.5714	2.5229	2.4971	2.4708
	0.950	3.4797	3.4671	3.4553	3.4448	3.4036	3.3190	3.2749	3.2298
	0.975	4.5202	4.5006	4.4834	4.4663	4.4041	4.2764	4.2101	4.1424
	0.990	6.2406	6.2095	6.1811	6.1562	6.0574	5.8587	5.7546	5.6496
	0.995	7.8672	7.8258	7.7872	7.7537	7.6244	7.3529	7.2165	7.0762
8	0.900	2.4459	2.4381	2.4309	2.4247	2.3999	2.3483	2.3208	2.2926
	0.950	3.1867	3.1732	3.1614	3.1504	3.1081	3.0209	2.9747	2.9276
	0.975	4.0538	4.0335	4.0157	3.9995	3.9369	3.8066	3.7393	3.6702
	0.990	5.4426	5.4120	5.3840	5.3595	5.2627	5.0659	4.9633	4.8589
	0.995	6.7175	6.6780	6.6408	6.6083	6.4812	6.2201	6.0875	5.9507
9	0.900	2.3204	2.3123	2.3049	2.2984	2.2724	2.2179	2.1892	2.1592
	0.950	2.9736	2.9600	2.9476	2.9365	2.8931	2.8031	2.7556	2.7067
	0.975	3.7217	3.7013	3.6834	3.6669	3.6035	3.4716	3.4034	3.3329
	0.990	4.8905	4.8599	4.8324	4.8083	4.7132	4.5170	4.4150	4.3106
	0.995	5.9382	5.8994	5.8636	5.8321	5.7089	5.4548	5.3223	5.1878
10	0.900	2.2237	2.2152	2.2076	2.2008	2.1740	2.1172	2.0869	2.0554
	0.950	2.8120	2.7981	2.7853	2.7740	2.7299	2.6374	2.5884	2.5379
	0.975	3.4738	3.4536	3.4349	3.4184	3.3548	3.2218	3.1517	3.0798
	0.990	4.4872	4.4570	4.4299	4.4058	4.3110	4.1151	4.0137	3.9091
	0.995	5.3789	5.3401	5.3056	5.2736	5.1533	4.9031	4.7721	4.6387
11	0.900	2.1467	2.1380	2.1301	2.1231	2.0954	2.0365	2.0050	1.9721
	0.950	2.6850	2.6710	2.6580	2.6465	2.6013	2.5068	2.4568	2.4045
	0.975	3.2818	3.2612	3.2430	3.2261	3.1616	3.0269	2.9557	2.8828
	0.990	4.1803	4.1502	4.1235	4.0988	4.0048	3.8102	3.7073	3.6025
	0.995	4.9588	4.9202	4.8862	4.8552	4.7356	4.4879	4.3579	4.2256

F DISTRIBUTION, Continued

df	1 - α	17	18	19	20	25	50	100	∞
12	0.900	2.0840	2.0751	2.0669	2.0597	2.0313	1.9705	1.9379	1.9036
	0.950	2.5828	2.5684	2.5554	2.5436	2.4976	2.4010	2.3496	2.2962
	0.975	3.1286	3.1081	3.0895	3.0727	3.0077	2.8712	2.7997	2.7250
	0.990	3.9390	3.9095	3.8829	3.8585	3.7650	3.5688	3.4669	3.3609
	0.995	4.6320	4.5944	4.5608	4.5297	4.4111	4.1652	4.0377	3.9041
13	0.900	2.0319	2.0227	2.0145	2.0070	1.9778	1.9152	1.8815	1.8462
	0.950	2.4988	2.4842	2.4710	2.4588	2.4124	2.3137	2.2615	2.2064
	0.975	3.0041	2.9832	2.9645	2.9477	2.8823	2.7444	2.6711	2.5955
	0.990	3.7451	3.7156	3.6890	3.6644	3.5708	3.3751	3.2717	3.1654
	0.995	4.3714	4.3343	4.3006	4.2704	4.1525	3.9076	3.7790	3.6466
14	0.900	1.9878	1.9785	1.9701	1.9624	1.9326	1.8686	1.8338	1.7973
	0.950	2.4282	2.4135	2.4000	2.3879	2.3406	2.2404	2.1872	2.1307
	0.975	2.9003	2.8794	2.8606	2.8437	2.7777	2.6385	2.5645	2.4873
	0.990	3.5857	3.5561	3.5296	3.5051	3.4117	3.2156	3.1118	3.0041
	0.995	4.1592	4.1222	4.0890	4.0586	3.9417	3.6975	3.5693	3.4360
15	0.900	1.9501	1.9407	1.9321	1.9243	1.8938	1.8284	1.7930	1.7551
	0.950	2.3682	2.3533	2.3399	2.3275	2.2798	2.1779	2.1232	2.0659
	0.975	2.8129	2.7920	2.7730	2.7560	2.6893	2.5488	2.4739	2.3954
	0.990	3.4523	3.4229	3.3960	3.3719	3.2781	3.0813	2.9772	2.8685
	0.995	3.9825	3.9458	3.9126	3.8824	3.7664	3.5225	3.3944	3.2603
16	0.900	1.9174	1.9078	1.8991	1.8913	1.8603	1.7935	1.7571	1.7182
	0.950	2.3168	2.3016	2.2880	2.2756	2.2272	2.1241	2.0684	2.0096
	0.975	2.7381	2.7170	2.6980	2.6808	2.6138	2.4719	2.3959	2.3163
	0.990	3.3391	3.3098	3.2830	3.2587	3.1649	2.9673	2.8623	2.7529
	0.995	3.8338	3.7971	3.7640	3.7340	3.6180	3.3744	3.2461	3.1116
17	0.900	1.8889	1.8792	1.8703	1.8624	1.8309	1.7628	1.7256	1.6856
	0.950	2.2720	2.2567	2.2429	2.2305	2.1816	2.0769	2.0203	1.9604
	0.975	2.6732	2.6521	2.6331	2.6159	2.5485	2.4053	2.3283	2.2474
	0.990	3.2419	3.2125	3.1857	3.1618	3.0676	2.8695	2.7638	2.6531
	0.995	3.7065	3.6703	3.6373	3.6073	3.4917	3.2484	3.1189	2.9840
18	0.900	1.8637	1.8539	1.8450	1.8368	1.8049	1.7356	1.6977	1.6567
	0.950	2.2326	2.2172	2.2034	2.1906	2.1412	2.0353	1.9780	1.9168
	0.975	2.6168	2.5956	2.5766	2.5591	2.4911	2.3467	2.2692	2.1869
	0.990	3.1576	3.1281	3.1015	3.0771	2.9830	2.7843	2.6780	2.5660
	0.995	3.5966	3.5605	3.5278	3.4979	3.3822	3.1386	3.0095	2.8733
19	0.900	1.8414	1.8314	1.8225	1.8142	1.7818	1.7114	1.6725	1.6308
	0.950	2.1978	2.1822	2.1682	2.1556	2.1057	1.9985	1.9402	1.8780
	0.975	2.5670	2.5456	2.5264	2.5089	2.4407	2.2954	2.2169	2.1333
	0.990	3.0836	3.0542	3.0274	3.0033	2.9089	2.7092	2.6022	2.4893
	0.995	3.5008	3.4645	3.4319	3.4021	3.2866	3.0430	2.9133	2.7762
20	0.900	1.8214	1.8114	1.8022	1.7938	1.7611	1.6895	1.6501	1.6074
	0.950	2.1667	2.1511	2.1371	2.1241	2.0740	1.9656	1.9064	1.8432
	0.975	2.5228	2.5013	2.4821	2.4645	2.3960	2.2494	2.1700	2.0853
	0.990	3.0183	2.9886	2.9620	2.9376	2.8433	2.6431	2.5355	2.4212
	0.995	3.4163	3.3801	3.3474	3.3177	3.2027	2.9585	2.8281	2.6904
25	0.900	1.7463	1.7359	1.7263	1.7175	1.6831	1.6072	1.5645	1.5176
	0.950	2.0514	2.0353	2.0208	2.0075	1.9555	1.8422	1.7794	1.7110
	0.975	2.3599	2.3381	2.3185	2.3004	2.2304	2.0786	1.9955	1.9055
	0.990	2.7802	2.7506	2.7238	2.6994	2.6041	2.4000	2.2888	2.1694
	0.995	3.1115	3.0755	3.0428	3.0132	2.8981	2.6521	2.5192	2.3766
50	0.900	1.6000	1.5884	1.5779	1.5681	1.5294	1.4410	1.3885	1.3266
	0.950	1.8313	1.8141	1.7984	1.7841	1.7274	1.5994	1.5249	1.4382
	0.975	2.0557	2.0329	2.0122	1.9933	1.9185	1.7519	1.6559	1.5453
	0.990	2.3481	2.3179	2.2904	2.2652	2.1666	1.9490	1.8247	1.6832
	0.995	2.5686	2.5327	2.5000	2.4702	2.3533	2.0967	1.9513	1.7865

F DISTRIBUTION, Continued

df	1 - α	17	18	19	20	25	50	100	∞
100	0.900	1.5283	1.5160	1.5047	1.4943	1.4528	1.3548	1.2934	1.2142
	0.950	1.7259	1.7078	1.6915	1.6764	1.6163	1.4772	1.3917	1.2832
	0.975	1.9131	1.8896	1.8682	1.8486	1.7705	1.5917	1.4832	1.3474
	0.990	2.1511	2.1204	2.0924	2.0667	1.9652	1.7353	1.5977	1.4276
	0.995	2.3264	2.2901	2.2572	2.2271	2.1079	1.8401	1.6809	1.4858
∞	0.900	1.4570	1.4439	1.4318	1.4206	1.3753	1.2633	1.1849	1.0000
	0.950	1.6228	1.6039	1.5865	1.5705	1.5061	1.3501	1.2434	1.0000
	0.975	1.7759	1.7515	1.7291	1.7085	1.6259	1.4284	1.2956	1.0000
	0.990	1.9652	1.9336	1.9048	1.8783	1.7726	1.5231	1.3582	1.0000
	0.995	2.1011	2.0643	2.0306	1.9999	1.8771	1.5898	1.4019	1.0000

Source: Except for those values associated with an infinite number of degrees of freedom, all values in this table have been generated using the IMSL10 subroutine DFIN, which computes the inverse of the F distribution function using double precision. Those values for an infinite number of degrees of freedom were obtained using the IMSL10 subroutine DCHIIN, which computes the inverse of the chi squared distribution function using double precision.

TABLE 11.2-A

MANN-WHITNEY U STATISTIC CRITICAL VALUES FOR SMALL SAMPLE SIZES AND SELECTED CRITICAL REGIONS

n_2 \ n_1	2	3	4	5	6	7	8	9	10	11	12	13	14	15	16	17	18	19	20
α = 0.025 for a one-tail test; α = 0.05 for a two-tail test																			
2							0	0	0	0	1	1	1	1	1	2	2	2	2
3				0	1	1	2	2	3	3	4	4	5	5	6	6	7	7	8
4			0	1	2	3	4	4	5	6	7	8	9	10	11	11	12	13	13
5		0	1	2	3	5	6	7	8	9	11	12	13	14	15	17	18	19	20
6		1	2	3	5	6	8	10	11	13	14	16	17	19	21	22	24	25	27
7		1	3	5	6	8	10	12	14	16	18	20	22	24	26	28	30	32	34
8	0	2	4	6	8	10	13	15	17	19	22	24	26	29	31	34	36	38	41
9	0	2	4	7	10	12	15	17	20	23	26	28	31	34	37	39	42	45	48
10	0	3	5	8	11	14	17	20	23	26	29	33	36	39	42	45	48	52	55
11	0	3	6	9	13	16	19	23	26	30	33	37	40	44	47	51	55	58	62
12	1	4	7	11	14	18	22	26	29	33	37	41	45	49	53	57	61	65	69
13	1	4	8	12	16	20	24	28	33	37	41	45	50	54	59	63	67	72	76
14	1	5	9	13	17	22	26	31	36	40	45	50	55	59	64	67	74	78	83
15	1	5	10	14	19	24	29	34	39	44	49	54	59	64	70	75	80	85	90
16	1	6	11	15	21	26	31	37	42	47	53	59	64	70	75	81	86	92	98
17	2	6	11	17	22	28	34	39	45	51	57	63	67	75	81	87	93	99	105
18	2	7	12	18	24	30	36	42	48	55	61	67	74	80	86	93	99	106	112
19	2	7	13	19	25	32	38	45	52	58	65	72	78	85	92	99	106	113	119
20	2	8	13	20	27	34	41	48	55	62	69	76	83	90	98	105	112	119	127
α = 0.05 for a one-tail test; α = 0.10 for a two-tail test																			
2				0	0	0	1	1	1	1	2	2	2	3	3	3	4	4	4
3		0	0	1	2	2	3	3	4	5	5	6	7	7	8	9	9	10	11
4		0	1	2	3	4	5	6	7	8	9	10	11	12	14	15	16	17	18
5	0	1	2	4	5	6	8	9	11	12	13	15	16	18	19	20	22	23	25
6	0	2	3	5	7	8	10	12	14	16	17	19	21	23	25	26	28	30	32
7	0	2	4	6	8	11	13	15	17	19	21	24	26	28	30	33	35	37	39
8	1	3	5	8	10	13	15	18	20	23	26	28	31	33	36	39	41	44	47
9	1	3	6	9	12	15	18	21	24	27	30	33	36	39	42	45	48	51	54
10	1	4	7	11	14	17	20	24	27	31	34	37	41	44	48	51	55	58	62
11	1	5	8	12	16	19	23	27	31	34	38	42	46	50	54	57	61	65	69
12	2	5	9	13	17	21	26	30	34	38	42	47	51	55	60	64	68	72	77
13	2	6	10	15	19	24	28	33	37	42	47	51	56	61	65	70	75	80	84
14	2	7	11	16	21	26	31	36	41	46	51	56	61	66	71	77	82	87	92
15	3	7	12	18	23	28	33	39	44	50	55	61	66	72	77	83	88	94	100
16	3	8	14	19	25	30	36	42	48	54	60	65	71	77	83	89	95	101	107
17	3	9	15	20	26	33	39	45	51	57	64	70	77	83	89	96	102	109	115
18	4	9	16	22	28	35	41	48	55	61	68	75	82	88	95	102	109	116	123
19	4	10	17	23	30	37	44	51	58	65	72	80	87	94	101	109	116	123	130
20	4	11	18	25	32	39	47	54	62	69	77	84	92	100	107	115	123	130	138

Source: Adapted from H. Mann and D. Whitney, 1947, "On a Test of Whether One or Two Random Variables is Stochastically Larger Than the Other," *Annals of Statistics*, 18: 52-54; with the permission of the Institute of Mathematical Statistics.

INDEX